Lecture Notes in Computer Science 11884

More information about this series at http://www.springer.com/series/7407

Robert Mark Hierons · Mohamed Mosbah (Eds.)

Theoretical Aspects of Computing – ICTAC 2019

16th International Colloquium
Hammamet, Tunisia, October 31 – November 4, 2019
Proceedings

Springer

Editors
Robert Mark Hierons
University of Sheffield
Sheffield, UK

Mohamed Mosbah
University of Bordeaux
Talence, France

ISSN 0302-9743 ISSN 1611-3349 (electronic)
Lecture Notes in Computer Science
ISBN 978-3-030-32504-6 ISBN 978-3-030-32505-3 (eBook)
https://doi.org/10.1007/978-3-030-32505-3

LNCS Sublibrary: SL1 – Theoretical Computer Science and General Issues

This Springer imprint is published by the registered company Springer Nature Switzerland AG
The registered company address is: Gewerbestrasse 11, 6330 Cham, Switzerland

Preface

This volume contains the proceedings of the 16th International Colloquium on Theoretical Aspects of Computing (ICTAC 2019), which was held in Hammamet, Tunisia, during October 31 – November 4, 2019. Established in 2004 by the International Institute for Software Technology of the United Nations University (UNU-IIST), the ICTAC conference series aims at bringing together researchers and practitioners from academia, industry, and government to present research and exchange ideas and experience addressing challenges in both theoretical aspects of computing and the exploitation of theory through methods and tools for system development. ICTAC also specifically aims to promote research cooperation between developing and industrial countries.

We received a total of 138 submissions, of which 20 were accepted for publication (19 regular papers and one tool paper). The acceptance rate was therefore approximately 14.5%. Papers were assigned to three reviewers, with the reviewing phase being followed by discussions. Reviewing was single-blind. The program also included three keynote presentations from Patrick Cousot (Courant Institute of Mathematical Sciences, New York University, USA), Thomas A. Henzinger (Institute of Science and Technology, Austria), and Dominique Mery (University of Lorraine, France). There was also a collocated summer school and two workshops.

We are grateful for the support provided by the many people who contributed to ICTAC 2019. We received invaluable advice from the Steering Committee at a number of important times, particularly from the Steering Committee chair (Ana Cavalcanti), and the incoming chair (Martin Leucker). We also received advice and support from the general chair (Mohamed Jmaiel). We would like to thank the members of the Program Committee for their timely and high-quality reviews as well as their contributions to the discussions. We also received invaluable support from Saoussen Cheikhrouhou who helped us to publicize the event.

Naturally, the conference could not have taken place without presentations. We would like to thank our three keynotes (Patrick Cousot, Thomas A. Henzinger, and Dominique Mery) for accepting our invitations – it was a pleasure and honor to have such a distinguished set of keynotes. We would also like to thank the authors for submitting and then revising a set of high-quality papers. Finally, we appreciate the support and advice provided by Alfred Hofmann and Anna Kramer from Springer.

November 2019

Robert M. Hierons
Mohamed Mosbah

Organization

General Chairs

Mohamed Jmaiel	University of Sfax, Tunisia
Walid Gaaloul	University of Paris-Saclay, France

Program Committee Chairs

Robert M. Hierons	University of Sheffield, UK
Mohamed Mosbah	LaBRI, Bordeaux INP, France

Steering Committee

Ana Cavalcanti (Chair)	University of York, UK
Martin Leucker	Universität zu Lübeck, Germany
Zhiming Liu	Southwest University, China
Tobias Nipkow	Technische Universität München, Germany
Augusto Sampaio	Universidade Federal de Pernambuco, Brazil
Natarajan Shankar	SRI International, USA

Program Committee

Éric Badouel	IRISA, France
Kamel Barkaoui	CNAM, France
Frédéric Blanqui	Inria, France
Eduardo Bonelli	Universidad Nacional de Quilmes, Argentina
Ana Cavalcanti	University of York, UK
Uli Fahrenberg	LIX, France
Adrian Francalanza	University of Malta, Malta
Ahmed Hadj Kacem	University of Sfax, Tunisia
Edward H. Haeusler	Pontifícia Universidade Católica do Rio de Janeiro, Brazil
Ross Horne	University of Luxembourg, Luxembourg
David Janin	University of Bordeaux, France
Jan Kretinsky	Technische Universität München, Germany
Martin Leucker	Universität zu Lübeck, Germany
Radu Mardare	Aalborg Universitet, Denmark
Dominique Méry	LORIA, France
Mohammad Mousavi	University of Leicester, UK
Tobias Nipkow	Technische Universität München, Germany
Maciej Piróg	Wrocław University, Poland
Sanjiva Prasad	IIT Delhi, India

Riadh Robbana	INSAT, Tunisia
Augusto Sampaio	Universidade Federal de Pernambuco, Brazil
Georg Struth	University of Sheffield, UK
Cong Tian	Xidian University, China
Tarmo Uustalu	Reykjavik University, IS/Tallinn University of Technology, Iceland

Additional Reviewers

Accattoli Beniamino
Ahmadi Sharar
Ait Sadoune Idir
Albano Michele
Araujo Hugo
Attiogbe Christian
Bacci Giorgio
Bacci Giovanni
Barnett Will
Baxter James
Ben Hafaiedh Imen
Benes Nikola
Berrima Mouhebeddine
Bodeveix Jean-Paul
Bollig Benedikt
Caillaud Benoît
Capobianco Silvio
Diego Damasceno Carlos
Denise Alain
Doherty Simon
Eisentraut Julia
Fonseca Paulo
Foughali Mohammed
Freitas Leo
Gazda Maciej
Guerrini Stefano
Hélouët Loïc
Immler Fabian
Kharraz Karam
Kleijn Jetty
Lepigre Rodolphe
Lima Lucas
Lopes Bruno
Maarand Hendrik
Markey Nicolas
Markin Grigory
Meggendorfer Tobias
Merz Stephan
Micallef Mark
Naik Vinayak
Oliveira Bruno C. D. S.
Ouaknine Joel
Pedro Andre
Penelle Vincent
Pinchinat Sophie
Puerto Aubel Adrian
Rosa Nelson
Saivasan Prakash
Sambale Maurice
Sampaio Gabriela
Santos Jefferson
Scheffel Torben
Schnoebelen Philippe
Sfaxi Lilia
Sickert Salomon
Smith Stephen
Smith Zach
Sokolova Ana
Stümpel Annette
van den Brand Mark
Vella Mark
Vighio Saleem
Vojdani Vesal
Weininger Maximilian
Wu Xi
Yurkov Semen

Contents

Verification and Concurrency

Privacy and Security

Equations, Types, and Programming Languages

Keynote Papers

Calculational Design of a Regular Model Checker by Abstract Interpretation*

Patrick Cousot

CS, CIMS, NYU, New York, NY, USA
pcousot@cims.nyu.edu

Abstract. Security monitors have been used to check for safety program properties at runtime, that is for any given execution trace. Such security monitors check a safety temporal property specified by a finite automaton or, equivalently, a regular expression. Checking this safety temporal specification for all possible execution traces, that is the program semantics, is a static analysis problem, more precisely a model checking problem, since model checking specializes in temporal properties. We show that the model checker can be formally designed by calculus, by abstract interpretation of a formal trace semantics of the programming language. The result is a structural sound and complete model checker, which proceeds by induction on the program syntax (as opposed to the more classical approach using computation steps formalized by a transition system). By Rice theorem, further hypotheses or abstractions are needed to get a realistic model checking algorithm.

Keywords: Abstract interpretation · Calculational design · Model checking.

1 Introduction

Model checking [9,33] consists in proving that a model of a given program/computer system satisfies a temporal specification[1]. Traditionally, the model of the given program/computer system is a transition system and its semantics is the set of traces generated by the transition system. The temporal specification is usually one of the many variants of temporal logics such as the Linear Time Temporal logic (LTL), the Computation Tree Logic (CTL), or the combination CTL* of the two. The semantics of the temporal specification is a set of traces. The problem is therefore to check that the set of traces of the semantics of the given program/computer system is included in the set of traces of the semantics of the temporal specification. This is a Galois connection-based abstraction and so a model checking algorithm can be designed by calculus. To show that we consider a non-conventional temporal specification using regular expressions [25] and a structural fixpoint prefix-closed trace semantics which diffrs from the

* Supported by NSF Grant CCF-1617717.

[1] We define model checking as the verification of temporal properties and do not reduce it to the reachability analysis (as done *e.g.* in [10, Ch. 15, 16, 17, *etc.*]) since reachability analysis predates model checking [14] including for the use of transition systems [12].

R. M. Hierons and M. Mosbah (Eds.): ICTAC 2019, LNCS 11884, pp. 3–21, 2019.
https://doi.org/10.1007/978-3-030-32505-3_1

traditional small-step operational semantics specified by a transition system. There are properties of traces that are not expressible in temporal logic but are easily expressible using regular expressions [36].

2 Syntax and Trace Semantics of the Programming Language

Syntax Programs are a subset of C with the following context-free syntax.

$\mathsf{x}, \mathsf{y}, \ldots \in \mathbb{X}$	variable ($\mathbb{X}$ not empty)
$\mathsf{A} \in \mathbb{A} ::= \mathtt{1} \mid \mathsf{x} \mid \mathsf{A}_1 - \mathsf{A}_2$	arithmetic expression
$\mathsf{B} \in \mathbb{B} ::= \mathsf{A}_1 < \mathsf{A}_2 \mid \mathsf{B}_1\ \mathsf{nand}\ \mathsf{B}_2$	boolean expression
$\mathsf{E} \in \mathbb{E} ::= \mathsf{A} \mid \mathsf{B}$	expression
$\mathsf{S} \in \mathbb{S} ::=$	statement
`x = A ;`	assignment
`| ;`	skip
`| if (B) S	if (B) S else S`	conditionals
`| while (B) S	break ;`	iteration and break
`| { Sl }`	compound statement
$\mathsf{Sl} \in \mathbb{Sl} ::= \mathsf{Sl}\ \ \mathsf{S} \mid \epsilon$	statement list
$\mathsf{P} \in \mathbb{P} ::= \mathsf{Sl}$	program

A break exits the closest enclosing loop, if none this is a syntactic error. If P is a program then `int main () { P }` is a valid C program. We call "[program] component" $\mathsf{S} \in \mathbb{P}c \triangleq \mathbb{S} \cup \mathbb{Sl} \cup \mathbb{P}$ either a statement, a statement list, or a program. We let $\lhd$ be the syntactic relation between immediate syntactic components. For example, if $\mathsf{S} = \mathtt{if\ (B)}\ \mathsf{S}_t\ \mathtt{else}\ \mathsf{S}_f$ then $\mathsf{B} \lhd \mathsf{S}$, $\mathsf{S}_t \lhd \mathsf{S}$, and $\mathsf{S}_f \lhd \mathsf{S}$.

Program labels Labels are not part of the language, but useful to discuss program points reached during execution. For each program component S, we define

- $\mathsf{at}[\![\mathsf{S}]\!]$ the program point at which execution of S starts;
- $\mathsf{aft}[\![\mathsf{S}]\!]$ the program exit point after S, at which execution of S is supposed to normally terminate, if ever;
- $\mathsf{esc}[\![\mathsf{S}]\!]$ a boolean indicating whether or not the program component S contains a **break ;** statement escaping out of that component S;
- $\mathsf{brk\text{-}to}[\![\mathsf{S}]\!]$ the program point at which execution of the program component S goes to when a **break ;** statement escapes out of that component S;
- $\mathsf{brks\text{-}of}[\![\mathsf{S}]\!]$ the set of labels of all **break ;** statements that can escape out of S;
- $\mathsf{in}[\![\mathsf{S}]\!]$ the set of program points inside S (including $\mathsf{at}[\![\mathsf{S}]\!]$ but excluding $\mathsf{aft}[\![\mathsf{S}]\!]$ and $\mathsf{brk\text{-}to}[\![\mathsf{S}]\!]$);
- $\mathsf{labs}[\![\mathsf{S}]\!]$ the potentially reachable program points while executing S either at, in, or after the statement, or resulting from a break.

Prefix trace semantics Prefix traces are non-empty finite sequences $\pi \in \mathbb{S}^+$ of states where states $\langle \ell, \rho \rangle \in \mathbb{S} \triangleq (\mathbb{L} \times \mathbb{E}\mathrm{v})$ are pairs of a program label $\ell \in \mathbb{S}$ designating the next action to be executed in the program and an environment $\rho \in \mathbb{E}\mathrm{v} \triangleq \mathbb{X} \to \mathbb{V}$ assigning values $\rho(\mathsf{x}) \in \mathbb{V}$ to variables $\mathsf{x} \in \mathbb{X}$. A trace π can be finite $\pi \in \mathbb{S}^+$ or infinite $\pi \in \mathbb{S}^\infty$ (recording a non-terminating computation) so $\mathbb{S}^{+\infty} \triangleq \mathbb{S}^+ \cup \mathbb{S}^\infty$. Trace concatenation $\mathbin{\bar{\cdot}}$ is defined as follows

$\pi_1\sigma_1 \mathbin{\bar{\cdot}} \sigma_2\pi_2$ undefined if $\sigma_1 \neq \sigma_2$ $\quad \pi_1 \mathbin{\bar{\cdot}} \sigma_2\pi_2 \triangleq \pi_1$ if $\pi_1 \in \mathbb{S}^\infty$ is infinite

$\pi_1\sigma_1 \mathbin{\bar{\cdot}} \sigma_1\pi_2 \triangleq \pi_1\sigma_1\pi_2$ if $\pi_1 \in \mathbb{T}^+$ is finite

In pattern matching, we sometimes need the empty trace $\ni$. For example if $\sigma\pi\sigma' = \sigma$ then $\pi = \ni$ and so $\sigma = \sigma'$.

Formal definition of the prefix trace semantics The prefix trace semantics $\mathcal{S}^*[\![\mathsf{S}]\!]$ is given structurally (by induction on the syntax) using fixpoints for the iteration.

- *The prefix traces of an assignment statement* $\mathsf{S} ::= \ell\ \mathsf{x = A}\ ;$ (where $\mathsf{at}[\![\mathsf{S}]\!] = \ell$) either stops in an initial state $\langle \ell, \rho \rangle$ or is this initial state $\langle \ell, \rho \rangle$ followed by the next state $\langle \mathsf{aft}[\![\mathsf{S}]\!], \rho[\mathsf{x} \leftarrow \mathcal{A}[\![\mathsf{A}]\!]\rho] \rangle$ recording the assignment of the value $\mathcal{A}[\![\mathsf{A}]\!]\rho$ of the arithmetic expression to variable x when reaching the label $\mathsf{aft}[\![\mathsf{S}]\!]$ after the assignment.

$$\mathcal{S}^*[\![\mathsf{S}]\!] = \{\langle \ell, \rho \rangle \mid \rho \in \mathbb{E}\mathrm{v}\} \cup \{\langle \ell, \rho \rangle \langle \mathsf{aft}[\![\mathsf{S}]\!], \rho[\mathsf{x} \leftarrow \mathcal{A}[\![\mathsf{A}]\!]\rho] \rangle \mid \rho \in \mathbb{E}\mathrm{v}\} \tag{1}$$

The value of an arithmetic expression A in environment $\rho \in \mathbb{E}\mathrm{v} \triangleq \mathbb{X} \to \mathbb{V}$ is $\mathcal{A}[\![\mathsf{A}]\!]\rho \in \mathbb{V}$:

$$\mathcal{A}[\![\mathsf{1}]\!]\rho \triangleq 1 \qquad \mathcal{A}[\![\mathsf{x}]\!]\rho \triangleq \rho(\mathsf{x}) \qquad \mathcal{A}[\![\mathsf{A}_1 - \mathsf{A}_2]\!]\rho \triangleq \mathcal{A}[\![\mathsf{A}_1]\!]\rho - \mathcal{A}[\![\mathsf{A}_2]\!]\rho \tag{2}$$

- *The prefix trace semantics of a break statement* $\mathsf{S} ::= \ell\ \mathbf{break}\ ;$ either stops at ℓ or goes on to the break label $\mathsf{brk\text{-}to}[\![\mathsf{S}]\!]$ (which is defined as the exit label of the closest enclosing iteration).

$$\mathcal{S}^*[\![\mathsf{S}]\!] \triangleq \{\langle \ell, \rho \rangle \mid \rho \in \mathbb{E}\mathrm{v}\} \cup \{\langle \ell, \rho \rangle \langle \mathsf{brk\text{-}to}[\![\mathsf{S}]\!], \rho \rangle \mid \rho \in \mathbb{E}\mathrm{v}\} \tag{3}$$

- *The prefix trace semantics of a conditional statement* $\mathsf{S} ::= \mathbf{if}\ \ell\ (\mathsf{B})\ \mathsf{S}_t$ is
  - either the trace $\langle \ell, \rho \rangle$ when the observation of the execution stops on entry of the program component for initial environment ρ;
  - or, when the value of the boolean expression B for ρ is false ff, the initial state $\langle \ell, \rho \rangle$ followed by the state $\langle \mathsf{aft}[\![\mathsf{S}]\!], \rho \rangle$ at the label $\mathsf{aft}[\![\mathsf{S}]\!]$ after the conditional statement;
  - or finally, when the value of the boolean expression B for ρ is true tt, the initial state$\langle \ell, \rho \rangle$ followed by a prefix trace of S_t starting $\mathsf{at}[\![\mathsf{S}_t]\!]$ in environment ρ (and possibly ending $\mathsf{aft}[\![\mathsf{S}_t]\!] = \mathsf{aft}[\![\mathsf{S}]\!]$).

$$\begin{aligned}\widehat{\mathcal{S}}^*[\![\mathsf{S}]\!] \triangleq\ & \{\langle \ell, \rho \rangle \mid \rho \in \mathbb{E}\mathrm{v}\} \cup \{\langle \ell, \rho \rangle \langle \mathsf{aft}[\![\mathsf{S}]\!], \rho \rangle \mid \mathcal{B}[\![\mathsf{B}]\!]\rho = \mathrm{ff}\} \\ & \cup \{\langle \ell, \rho \rangle \langle \mathsf{at}[\![\mathsf{S}_t]\!], \rho \rangle \pi \mid \mathcal{B}[\![\mathsf{B}]\!]\rho = \mathrm{tt} \wedge \langle \mathsf{at}[\![\mathsf{S}_t]\!], \rho \rangle \pi \in \widehat{\mathcal{S}}^*[\![\mathsf{S}_t]\!]\}\end{aligned} \tag{4}$$

Observe that definition (4) includes the case of a conditional within an iteration and containing a break statement in the true branch S_t. Since $\mathsf{brk\text{-}to}[\![\mathsf{S}]\!] =$

$\mathsf{brk\text{-}to}[\![\mathsf{S}_t]\!]$, from $\langle \mathsf{at}[\![\mathsf{S}_t]\!], \rho\rangle\pi\langle \mathsf{brk\text{-}to}[\![\mathsf{S}_t]\!], \rho'\rangle \in \boldsymbol{\mathcal{S}}^*[\![\mathsf{S}_t]\!]$ and $\boldsymbol{\mathcal{B}}[\![\mathsf{B}]\!]\rho = \mathrm{tt}$, we infer that $\langle \mathsf{at}[\![\mathsf{S}]\!], \rho\rangle\langle \mathsf{at}[\![\mathsf{S}_t]\!], \rho\rangle\pi\langle \mathsf{brk\text{-}to}[\![\mathsf{S}]\!], \rho'\rangle \in \boldsymbol{\mathcal{S}}^*[\![\mathsf{S}]\!]$.

- The prefix traces of the *prefix trace semantics of a non-empty statement list* $\mathsf{Sl} ::= \mathsf{Sl}'\ \mathsf{S}$ are the prefix traces of Sl' or the finite maximal traces of Sl' followed by a prefix trace of S.

$$\widehat{\boldsymbol{\mathcal{S}}}^*[\![\mathsf{Sl}]\!] \triangleq \widehat{\boldsymbol{\mathcal{S}}}^*[\![\mathsf{Sl}']\!] \cup \widehat{\boldsymbol{\mathcal{S}}}^*[\![\mathsf{Sl}']\!] \mathbin{\widehat{\cdot}} \boldsymbol{\mathcal{S}}^*[\![\mathsf{S}]\!] \tag{5}$$
$$\boldsymbol{\mathcal{S}} \mathbin{\widehat{\cdot}} \boldsymbol{\mathcal{S}}' \triangleq \{\pi \mathbin{\widehat{\cdot}} \pi' \mid \pi \in \boldsymbol{\mathcal{S}} \wedge \pi' \in \boldsymbol{\mathcal{S}}' \wedge \pi \mathbin{\widehat{\cdot}} \pi' \text{ is well-defined}\}$$

Notice that if $\pi \in \widehat{\boldsymbol{\mathcal{S}}}^*[\![\mathsf{Sl}']\!]$, $\pi' \in \boldsymbol{\mathcal{S}}^*[\![\mathsf{S}]\!]$, and $\pi \mathbin{\widehat{\cdot}} \pi' \in \widehat{\boldsymbol{\mathcal{S}}}^*[\![\mathsf{Sl}]\!]$ then the last state of π must be the first state of π' and this state is $\mathsf{at}[\![\mathsf{S}]\!] = \mathsf{aft}[\![\mathsf{Sl}']\!]$ and so the trace π must be a maximal terminating execution of Sl' *i.e.* S is executed only if Sl' terminates.

- *The prefix finite trace semantic definition* $\boldsymbol{\mathcal{S}}^*[\![\mathsf{S}]\!]$ (6) *of an iteration statement of the form* $\mathsf{S} ::= \mathbf{while}\ \ell\ (\mathsf{B})\ \mathsf{S}_b$ where $\ell = \mathsf{at}[\![\mathsf{S}]\!]$ is the $\subseteq$-least solution $\mathsf{lfp}^{\subseteq}\, \boldsymbol{\mathcal{F}}^*[\![\mathsf{S}]\!]$ to the equation $X = \boldsymbol{\mathcal{F}}^*[\![\mathsf{S}]\!](X)$. Since $\boldsymbol{\mathcal{F}}^*[\![\mathsf{S}]\!] \in \wp(\mathbb{S}^+) \to \wp(\mathbb{S}^+)$ is $\subseteq$- monotone (if $X \subseteq X'$ then $\boldsymbol{\mathcal{F}}^*[\![\mathsf{S}]\!](X) \subseteq \boldsymbol{\mathcal{F}}^*[\![\mathsf{S}]\!](X')$ and $\langle \wp(\mathbb{S}^+), \subseteq, \emptyset, \mathbb{S}^+, \cup, \cap\rangle$ is a complete lattice, $\mathsf{lfp}^{\subseteq}\, \boldsymbol{\mathcal{F}}^*[\![\mathsf{S}]\!]$ exists by Tarski's fixpoint theorem and can be defined as the limit of iterates [15]. In definition (6) of the transformer $\boldsymbol{\mathcal{F}}^*[\![\mathsf{S}]\!]$, case (6.a) corresponds to a loop execution observation stopping on entry, (6.b) corresponds to an observation of a loop exiting after 0 or more iterations, and (6.c) corresponds to a loop execution observation that stops anywhere in the body S_b after 0 or more iterations. This last case covers the case of an iteration terminated by a break statement (to $\mathsf{aft}[\![\mathsf{S}]\!]$ after the iteration statement).

$$\boldsymbol{\mathcal{S}}^*[\![\mathbf{while}\ \ell\ (\mathsf{B})\ \mathsf{S}_b]\!] = \mathsf{lfp}^{\subseteq}\, \boldsymbol{\mathcal{F}}^*[\![\mathbf{while}\ \ell\ (\mathsf{B})\ \mathsf{S}_b]\!] \tag{6}$$

$$\begin{aligned}
\boldsymbol{\mathcal{F}}^*_{\mathbb{S}}[\![\mathbf{while}\ \ell\ (\mathsf{B})\ \mathsf{S}_b]\!]\, X \triangleq{}& \{\langle \ell, \rho\rangle \mid \rho \in \mathbb{E}\mathrm{v}\} && \text{(a)}\\
&\cup \{\pi_2\langle \ell', \rho\rangle\langle \mathsf{aft}[\![\mathsf{S}]\!], \rho\rangle \mid \pi_2\langle \ell', \rho\rangle \in X \wedge \boldsymbol{\mathcal{B}}[\![\mathsf{B}]\!]\,\rho = \mathrm{ff} \wedge \ell' = \ell\}^2 && \text{(b)}\\
&\cup \{\pi_2\langle \ell', \rho\rangle\langle \mathsf{at}[\![\mathsf{S}_b]\!], \rho\rangle \cdot \pi_3 \mid \pi_2\langle \ell', \rho\rangle \in X \wedge \boldsymbol{\mathcal{B}}[\![\mathsf{B}]\!]\,\rho = \mathrm{tt} \wedge && \text{(c)}\\
&\qquad \langle \mathsf{at}[\![\mathsf{S}_b]\!], \rho\rangle \cdot \pi_3 \in \boldsymbol{\mathcal{S}}^*[\![\mathsf{S}_b]\!] \wedge \ell' = \ell\}
\end{aligned}$$

- The other cases are similar.

Semantic properties As usual in abstract interpretation [16], we represent properties of entities in a universe $\mathbb{U}$ by a subset of this universe. So a property of elements of $\mathbb{U}$ belongs to $\wp(\mathbb{U})$. For example "to be a natural" is the property $\mathbb{N} \triangleq \{n \in \mathbb{Z} \mid n \geqslant 0\}$ of the integers $\mathbb{Z}$. The property "n is a natural" is "$n \in \mathbb{N}$". By program component (safety) property, we understand a property of their prefix trace semantics $\boldsymbol{\mathcal{S}}^*[\![\mathsf{S}]\!] \in \wp(\mathbb{S}^+)$. So program properties belong to $\wp(\wp(\mathbb{S}^+))$. The *collecting semantics* is the strongest program property, that is the singleton $\{\boldsymbol{\mathcal{S}}^*[\![\mathsf{S}]\!]\}$.

[2] A definition of the form $d(\vec{x}) \triangleq \{f(\vec{x}') \mid P(\vec{x}', \vec{x})\}$ has the variables $\vec{x}'$ in $P(\vec{x}', \vec{x})$ bound to those of $f(\vec{x}')$ whereas $\vec{x}$ is free in $P(\vec{x}', \vec{x})$ since it appears neither in $f(\vec{x}')$ nor (by assumption) under quantifiers in $P(\vec{x}', \vec{x})$. The $\vec{x}$ of $P(\vec{x}', \vec{x})$ is therefore bound to the $\vec{x}$ of $d(\vec{x})$.

3 Specifying computations by regular expressions

Stephen Cole Kleene introduced regular expressions and finite automata to specify execution traces (called events) of automata (called nerve nets) [25]. Kleene proved in [25] that regular expressions and (non-deterministic) finite automata can describe exactly the same classes of languages (see [34, Ch. 1, Sect. 4]). He noted that not all computable execution traces of nerve nets can be (exactly) represented by a regular expression. The situation is the same for programs for which regular expressions (or equivalently finite automata) can specify a superset of the prefix state trace semantics $\mathcal{S}^*[\![\mathsf{S}]\!]$ of program components $\mathsf{S} \in \mathbb{P}c$.

Example 1 (Security monitors). An example is Fred Schneider's security monitors [35,29] using a finite automata specification to state requirements of hardware or software systems. They have been used to check for safety program properties at runtime, that is for any given execution trace in the semantics $\mathcal{S}^*[\![\mathsf{S}]\!]$. The safety property specified by the finite automaton or, equivalently, an regular expression is temporal *i.e.* links events occurring at different times in the computation (such as a file must be opened before being accessed and must eventually be closed). □

Syntax of regular expressions Classical regular expressions denote sets of strings using constants (empty string ε, literal characters a, b, *etc.*) and operator symbols (concatenation $\bullet$, alternation $\mid$, repetition zero or more times * or one or more times $^+$). We replace the literal characters by invariant specifications $\mathsf{L} : \mathsf{B}$ stating that boolean expression B should be true whenever control reaches any program point in the set L of program labels. The boolean expression B may depend on program variables $\mathsf{x}, \mathsf{y}, \ldots \in \mathbb{X}$ and their initial values denoted $\underline{\mathsf{x}}, \underline{\mathsf{y}}, \ldots \in \underline{\mathbb{X}}$ where $\underline{\mathbb{X}} \triangleq \{\underline{\mathsf{x}} \mid \mathsf{x} \in \mathbb{X}\}$.

$$
\begin{array}{rlr}
\mathsf{L} & \in \wp(\mathbb{L}) & \text{sets of program labels} \\
\mathsf{x}, \mathsf{y}, \ldots & \in \mathbb{X} & \text{program variables} \\
\underline{\mathsf{x}}, \underline{\mathsf{y}}, \ldots & \in \underline{\mathbb{X}} & \text{initial values of variables} \\
\mathsf{B} & \in \mathbb{B} & \text{boolean expressions such that } \mathsf{vars}[\![\mathsf{B}]\!] \subseteq \mathbb{X} \cup \underline{\mathbb{X}}
\end{array}
$$

$$
\begin{array}{llr}
\mathsf{R} \in \mathbb{R} & & \text{regular expressions} \quad (7) \\
\mathsf{R} ::= \varepsilon & & \text{empty} \\
\quad \mid \mathsf{L} : \mathsf{B} & & \text{invariant } \mathsf{B} \text{ at } \mathsf{L} \\
\quad \mid \mathsf{R}_1\mathsf{R}_2 & (\text{or } \mathsf{R}_1 \bullet \mathsf{R}_2) & \text{concatenation} \\
\quad \mid \mathsf{R}_1 \mid \mathsf{R}_2 & & \text{alternative} \\
\quad \mid \mathsf{R}_1^* \mid \mathsf{R}_1^+ & & \text{zero/one or more occurrences of } \mathsf{R} \\
\quad \mid (\mathsf{R}_1) & & \text{grouping}
\end{array}
$$

We use abbreviations to designate sets of labels such as $? : \mathsf{B} \triangleq \mathbb{L} : \mathsf{B}$ so that B is invariant, $\ell : \mathsf{B} \triangleq \{\ell\} : \mathsf{B}$ so that B is invariant at program label ℓ, $\neg\ell : \mathsf{B} \triangleq \mathbb{L} \setminus \{\ell\} : \mathsf{B}$ when B holds everywhere but at program point ℓ, *etc.*

Example 2. $(\texttt{?} : \texttt{tt})^*$ holds for any program. $(\texttt{?} : \texttt{x >= 0})^*$ states that the value of x is always positive or zero during program execution. $(\texttt{?} : \texttt{x >=} \underline{\texttt{x}})^*$ states that the value of x is always greater than or equal to its initial value $\underline{\texttt{x}}$ during execution. $(\texttt{?} : \texttt{x >= 0})^* \bullet \ell : \texttt{x == 0} \bullet (\texttt{?} : \texttt{x < 0})^*$ states that the value of x should be positive or zero and if program point ℓ is ever reached then x should be 0, and if computations go on after program point ℓ then x should be negative afterwards. □

Example 3. Continuing Ex. 1 for security monitors, the basic regular expressions are names a of program actions. We can understand such an action a as designating the set L of labels of all its occurrences in the program. If necessary, the boolean expression B can be used to specify the parameters of the action. □

There are many regular expressions denoting the language {∋} containing only the empty sequence ∋ (such that ε, εε, ε*, *etc.*), as shown by the following grammar.

$$\mathbb{R}_\varepsilon \ni \mathsf{R} ::= \varepsilon \mid \mathsf{R}_1\mathsf{R}_2 \mid \mathsf{R}_1 \,|\, \mathsf{R}_2 \mid \mathsf{R}_1^* \mid \mathsf{R}_1^+ \mid (\mathsf{R}_1) \qquad \text{empty regular expressions}$$

For specification we use only non-empty regular expressions $\mathsf{R} \in \mathbb{R}^+$ since traces cannot be empty.

$$\mathbb{R}^+ \ni \mathsf{R} ::= \mathsf{L} : \mathsf{B} \mid \varepsilon\mathsf{R}_2 \mid \mathsf{R}_1\varepsilon \mid \mathsf{R}_1\mathsf{R}_2 \mid \mathsf{R}_1 \,|\, \mathsf{R}_2 \mid \mathsf{R}_1^+ \mid (\mathsf{R}_1) \qquad \text{non-empty r.e.}$$

We also have to consider regular expressions $\mathsf{R} \in \mathbb{R}^{\nmid}$ containing no alternative |.

$$\mathbb{R}^{\nmid} \ni \mathsf{R} ::= \varepsilon \mid \mathsf{L} : \mathsf{B} \mid \mathsf{R}_1\mathsf{R}_2 \mid \mathsf{R}_1^* \mid \mathsf{R}_1^+ \mid (\mathsf{R}_1) \qquad |\text{-free regular expressions}$$

Relational semantics of regular expressions The semantics (2) of expressions is changed as follows ($\underline{\varrho}(\mathsf{x})$ denotes the initial values $\underline{\mathsf{x}}$ of variables x and $\rho(\mathsf{x})$ their current value, ↑ is the alternative denial logical operation)

$$\begin{array}{ll}
\mathcal{A}[\![1]\!]\underline{\varrho}, \rho \triangleq 1 & \mathcal{A}[\![\mathsf{A}_1 - \mathsf{A}_2]\!]\underline{\varrho}, \rho \triangleq \mathcal{A}[\![\mathsf{A}_1]\!]\underline{\varrho}, \rho - \mathcal{A}[\![\mathsf{A}_2]\!]\underline{\varrho}, \rho \\
\mathcal{A}[\![\underline{\mathsf{x}}]\!]\underline{\varrho}, \rho \triangleq \underline{\varrho}(\mathsf{x}) & \mathcal{B}[\![\mathsf{A}_1 < \mathsf{A}_2]\!]\underline{\varrho}, \rho \triangleq \mathcal{A}[\![\mathsf{A}_1]\!]\underline{\varrho}, \rho < \mathcal{A}[\![\mathsf{A}_2]\!]\underline{\varrho}, \rho \\
\mathcal{A}[\![\mathsf{x}]\!]\underline{\varrho}, \rho \triangleq \rho(\mathsf{x}) & \mathcal{B}[\![\mathsf{B}_1 \text{ nand } \mathsf{B}_2]\!]\underline{\varrho}, \rho \triangleq \mathcal{B}[\![\mathsf{B}_1]\!]\underline{\varrho}, \rho \uparrow \mathcal{B}[\![\mathsf{B}_2]\!]\underline{\varrho}, \rho
\end{array} \tag{8}$$

We represent a non-empty finite sequence $\sigma_1 \dots \sigma_n \in \mathbb{S}^+ \triangleq \bigcup_{n \in \mathbb{N}\setminus\{0\}} [1, n] \to \mathbb{S}$ of states $\sigma_i \in \mathbb{S} \triangleq (\mathbb{L} \times \mathbb{E}\mathbf{v})$ by a map $\sigma \in [1, n] \to \mathbb{S}$ (which is the empty sequence $\sigma = \ni$ when $n = 0$).

The relational semantics $\mathcal{S}^{\mathsf{r}}[\![\mathsf{R}]\!] \in \wp(\mathbb{E}\mathbf{v} \times \mathbb{S}^*)$ of regular expressions R relates an arbitrary initial environment $\underline{\varrho} \in \mathbb{E}\mathbf{v}$ to a trace $\pi \in \mathbb{S}^*$ by defining how the states of the trace π are related to that initial environment $\underline{\varrho}$.

$$\begin{array}{ll}
\mathcal{S}^{\mathsf{r}}[\![\varepsilon]\!] \triangleq \{\langle\underline{\varrho}, \ni\rangle \mid \underline{\varrho} \in \mathbb{E}\mathbf{v}\} & \mathcal{S}^{\mathsf{r}}[\![\mathsf{R}]\!]^1 \triangleq \mathcal{S}^{\mathsf{r}}[\![\mathsf{R}]\!] \\
\mathcal{S}^{\mathsf{r}}[\![\mathsf{L} : \mathsf{B}]\!] \triangleq \{\langle\underline{\varrho}, \langle\ell, \rho\rangle\rangle \mid \ell \in \mathsf{L} \wedge \mathcal{B}[\![\mathsf{B}]\!]\underline{\varrho}, \rho\} & \mathcal{S}^{\mathsf{r}}[\![\mathsf{R}]\!]^{n+1} \triangleq \mathcal{S}^{\mathsf{r}}[\![\mathsf{R}]\!]^n \circledcirc \mathcal{S}^{\mathsf{r}}[\![\mathsf{R}]\!] \\
\mathcal{S}^{\mathsf{r}}[\![\mathsf{R}_1\mathsf{R}_2]\!] \triangleq \mathcal{S}^{\mathsf{r}}[\![\mathsf{R}_1]\!] \circledcirc \mathcal{S}^{\mathsf{r}}[\![\mathsf{R}_2]\!] & \mathcal{S}^{\mathsf{r}}[\![\mathsf{R}^*]\!] \triangleq \bigcup_{n \in \mathbb{N}} \mathcal{S}^{\mathsf{r}}[\![\mathsf{R}]\!]^n \\
\mathcal{S} \circledcirc \mathcal{S}' \triangleq \{\langle\underline{\varrho}, \pi \cdot \pi'\rangle \mid \langle\underline{\varrho}, \pi\rangle \in \mathcal{S} \wedge \langle\underline{\varrho}, \pi'\rangle \in \mathcal{S}'\} & \mathcal{S}^{\mathsf{r}}[\![\mathsf{R}^+]\!] \triangleq \bigcup_{n \in \mathbb{N}\setminus\{0\}} \mathcal{S}^{\mathsf{r}}[\![\mathsf{R}]\!]^n \\
\mathcal{S}^{\mathsf{r}}[\![\mathsf{R}_1 \,|\, \mathsf{R}_2]\!] \triangleq \mathcal{S}^{\mathsf{r}}[\![\mathsf{R}_1]\!] \cup \mathcal{S}^{\mathsf{r}}[\![\mathsf{R}_2]\!] & \mathcal{S}^{\mathsf{r}}[\![(\mathsf{R})]\!] \triangleq \mathcal{S}^{\mathsf{r}}[\![\mathsf{R}]\!]
\end{array} \tag{9}$$

Example 4. The semantics of the regular expression $\mathsf{R} \triangleq \ell : \mathsf{x} = \underline{\mathsf{x}} \bullet \ell' : \mathsf{x} = \underline{\mathsf{x}} + 1$ is $\mathcal{S}^r[\![\mathsf{R}]\!] = \{\langle \underline{\varrho}, \langle \ell, \rho\rangle\langle \ell', \rho'\rangle\rangle \mid \rho(\mathsf{x}) = \underline{\varrho}(\mathsf{x}) \wedge \rho'(\mathsf{x}) = \underline{\varrho}(\mathsf{x}) + 1\}$. □

4 Definition of regular model checking

Let the prefix closure $\mathsf{prefix}(\Pi)$ of a set $\Pi \in \wp(\mathbb{E}\mathbb{v} \times \mathbb{S}^+)$ of stateful traces be

$$\mathsf{prefix}(\Pi) \triangleq \{\langle \underline{\varrho}, \pi\rangle \mid \pi \in \mathbb{S}^+ \wedge \exists \pi' \in \mathbb{S}^* \,.\, \langle \underline{\varrho}, \pi \cdot \pi'\rangle \in \Pi\} \qquad \text{prefix closure.}$$

The following Def. 1 defines the model checking problem $\mathsf{P}, \underline{\varrho} \models \mathsf{R}$ as checking that the semantics of the given program $\mathsf{P} \in \mathbb{P}$ meets the regular specification $\mathsf{R} \in \mathbb{R}^+$ for the initial environment $\underline{\varrho}$[3].

Definition 1 (Model checking).

$$\mathsf{P}, \underline{\varrho} \models \mathsf{R} \triangleq (\{\underline{\varrho}\} \times \mathcal{S}^*[\![\mathsf{P}]\!]) \quad\subseteq\quad \mathsf{prefix}(\mathcal{S}^r[\![\mathsf{R} \bullet (? : \mathsf{tt})^*]\!])$$

The prefix closure prefix allows the regular specification R to specify traces satisfying a prefix of the specification only, as in $\ell\ \mathsf{x = x + 1}\ ;\ell' \models \ell : \mathsf{x} = \underline{\mathsf{x}} \bullet \ell' : \mathsf{x} = \underline{\mathsf{x}} + 1 \bullet \ell'' : \mathsf{x} = \underline{\mathsf{x}} + 3$. The extension of the specification by $(? : \mathsf{tt})^*$ allows for the regular specification R to specify only a prefix of the traces, as in $\ell\ \mathsf{x = x + 1}\ ;\ell'\ \mathsf{x = x + 2}\ ;\ell'' \models \ell : \mathsf{x} = \underline{\mathsf{x}} \bullet \ell' : \mathsf{x} = \underline{\mathsf{x}} + 1$. Model checking is a boolean abstraction $\langle \wp(\mathbb{S}^+), \subseteq\rangle \underset{\alpha_{\underline{\varrho},\mathsf{R}}}{\overset{\gamma_{\underline{\varrho},\mathsf{R}}}{\leftrightarrows}} \langle \mathbb{B}, \Leftarrow\rangle$ where $\alpha_{\underline{\varrho},\mathsf{R}}(\Pi) \triangleq (\{\underline{\varrho}\} \times \Pi) \subseteq \mathsf{prefix}(\mathcal{S}^r[\![\mathsf{R} \bullet (? : \mathsf{tt})^*]\!])$.

5 Properties of regular expressions

Equivalence of regular expressions We say that regular expressions are equivalent when they have the same semantics *i.e.* $\mathsf{R}_1 \simeq \mathsf{R}_2 \triangleq (\mathcal{S}^r[\![\mathsf{R}_1]\!] = \mathcal{S}^r[\![\mathsf{R}_2]\!])$.

Disjunctive normal form dnf of regular expressions As noticed by Kleene [25, p. 14], regular expressions can be put in the equivalent disjunctive normal form of Hilbert—Ackermann. A regular expression is in disjunctive normal form if it is of the form $(\mathsf{R}_1 \mid \ldots \mid \mathsf{R}_n)$ for some $n \geqslant 1$, in which none of the R_i, for $1 \leqslant i \leqslant n$, contains an occurrence of $\mid$. Any regular expression R has a disjunctive normal form $\mathsf{dnf}(\mathsf{R})$ defined as follows.

$$\mathsf{dnf}(\varepsilon) \triangleq \varepsilon \qquad\qquad \mathsf{dnf}(\mathsf{L} : \mathsf{B}) \triangleq \mathsf{L} : \mathsf{B}$$

$$\mathsf{dnf}(\mathsf{R}_1 \mid \mathsf{R}_2) \triangleq \mathsf{dnf}(\mathsf{R}_1) \mid \mathsf{dnf}(\mathsf{R}_2) \qquad\qquad \mathsf{dnf}(\mathsf{R}^+) \triangleq \mathsf{dnf}(\mathsf{R}\mathsf{R}^*)$$

$$\mathsf{dnf}(\mathsf{R}^*) \triangleq \mathsf{let}\ \mathsf{R}^1 \mid \ldots \mid \mathsf{R}^n = \mathsf{dnf}(\mathsf{R})\ \mathsf{in}\ ((\mathsf{R}^1)^* \ldots (\mathsf{R}^n)^*)^* \qquad \mathsf{dnf}((\mathsf{R})) \triangleq (\mathsf{dnf}(\mathsf{R}))$$

$$\mathsf{dnf}(\mathsf{R}_1\mathsf{R}_2) \triangleq \mathsf{let}\ \mathsf{R}_1^1 \mid \ldots \mid \mathsf{R}_1^{n_1} = \mathsf{dnf}(\mathsf{R}_1)\ \mathsf{and}\ \mathsf{R}_2^1 \mid \ldots \mid \mathsf{R}_2^{n_2} = \mathsf{dnf}(\mathsf{R}_2)\ \mathsf{in}\ \Big|_{i=1}^{n_1} \Big|_{j=1}^{n_2} \mathsf{R}_1^i\mathsf{R}_2^j$$

3 We understand "regular model checking" as checking temporal specifications given by a regular expression. This is different from [1] model checking transition systems which states are regular word or tree languages.

The Lem. 1 below shows that normalization leaves the semantics unchanged. It uses the fact that $(\mathsf{R}_1 \mid \mathsf{R}_2)^* \mathrel{\approx} (\mathsf{R}_1{}^*\mathsf{R}_2{}^*)^*$ where the R_1 and R_2 do not contain any $\mid$ [23, Sect. 3.4.6, p. 118]. It shows that normalization in (10) can be further simplified by $\varepsilon\mathsf{R} \mathrel{\approx} \mathsf{R}\varepsilon \mathrel{\approx} \mathsf{R}$ and $(\varepsilon)^* \mathrel{\approx} \varepsilon$ which have equivalent semantics.

Lemma 1. $\mathsf{dnf}(\mathsf{R}) \mathrel{\approx} \mathsf{R}$.

first and next of regular expressions Janusz Brzozowski [7] introduced the notion of derivation for regular expressions (extended with arbitrary Boolean operations). The derivative of a regular expression R with respect to a symbol a, typically denoted as $D_a(\mathsf{R})$ or $a^{-1}\mathsf{R}$, is a regular expression given by a simple recursive definition on the syntactic structure of R. The crucial property of these derivatives is that a string of the form $a\sigma$ (starting with the symbol a) matches an expression R iff the suffix σ matches the derivative $D_a(\mathsf{R})$ [7,32,2].

Following this idea, assume that a non-empty regular expression $\mathsf{R} \in \mathbb{R}^+$ has been decomposed into disjunctive normal form $(\mathsf{R}_1 \mid \ldots \mid \mathsf{R}_n)$ for some $n \geqslant 1$, in which none of the R_i, for $i \in [1, n]$, contains an occurrence of $\mid$. We can further decompose each $\mathsf{R}_i \in \mathbb{R}^+ \cap \mathbb{R}^{\nmid}$ into $\langle \mathsf{L} : \mathsf{B},\ \mathsf{R}_i' \rangle = \mathsf{fstnxt}(\mathsf{R}_i)$ such that

- $\mathsf{L} : \mathsf{B}$ recognizes the first state of sequences of states recognized by R_i;
- the regular expression R_i' recognizes sequences of states after the first state of sequences of states recognized by R_i.

We define fstnxt for non-empty $\mid$-free regular expressions $\mathsf{R} \in \mathbb{R}^+ \cap \mathbb{R}^{\nmid}$ by structural induction, as follows.

$$\mathsf{fstnxt}(\mathsf{L} : \mathsf{B}) \triangleq \langle \mathsf{L} : \mathsf{B},\ \varepsilon \rangle \tag{10}$$

$$\mathsf{fstnxt}(\mathsf{R}_1\mathsf{R}_2) \triangleq \mathsf{fstnxt}(\mathsf{R}_2) \qquad \text{if } \mathsf{R}_1 \in \mathbb{R}_\varepsilon$$

$$\mathsf{fstnxt}(\mathsf{R}_1\mathsf{R}_2) \triangleq \mathsf{let}\ \langle \mathsf{R}_1^f,\ \mathsf{R}_1^n \rangle = \mathsf{fstnxt}(\mathsf{R}_1)\ \mathsf{in}\ [\![\mathsf{R}_1^n \in \mathbb{R}_\varepsilon \ ?\ \langle \mathsf{R}_1^f,\ \mathsf{R}_2 \rangle \ :\ \langle \mathsf{R}_1^f,\ \mathsf{R}_1^n \bullet \mathsf{R}_2 \rangle]\!] \qquad \text{if } \mathsf{R}_1 \notin \mathbb{R}_\varepsilon$$

$$\mathsf{fstnxt}(\mathsf{R}^+) \triangleq \mathsf{let}\ \langle \mathsf{R}^f,\ \mathsf{R}^n \rangle = \mathsf{fstnxt}(\mathsf{R})\ \mathsf{in}\ [\![\mathsf{R}^n \in \mathbb{R}_\varepsilon \ ?\ \langle \mathsf{R}^f,\ \mathsf{R}^* \rangle \ :\ \langle \mathsf{R}^f,\ \mathsf{R}^n \bullet \mathsf{R}^* \rangle]\!]$$

$$\mathsf{fstnxt}((\mathsf{R})) \triangleq \mathsf{fstnxt}(\mathsf{R})$$

The following Lem. 2 shows the equivalence of an alternative-free regular expression and its first and next decomposition.

Lemma 2. *Let* $\mathsf{R} \in \mathbb{R}^+ \cap \mathbb{R}^{\nmid}$ *be a non-empty* $\mid$*-free regular expression and* $\langle \mathsf{L} : \mathsf{B},\ \mathsf{R}' \rangle = \mathsf{fstnxt}(\mathsf{R})$. *Then* $\mathsf{R}' \in \mathbb{R}^{\nmid}$ *is* $\mid$*-free and* $\mathsf{R} \mathrel{\approx} \mathsf{L} : \mathsf{B} \bullet \mathsf{R}'$.

6 The model checking abstraction

The model checking abstraction in Section 4 is impractical for structural model checking since *e.g.* when checking that a trace concatenation $\pi_1 \mathbin{\hat{}} \pi_2$ of a statement list $\mathsf{Sl} ::= \mathsf{Sl}'\ \mathsf{S}$ for a specification R where π_1 is a trace of Sl' and π_2 is a trace of S, we first check that π_1 satisfies R and then we must check π_2 for a continuation R_2 of R which should be derived from π_1 and R. This is not provided by the boolean abstraction $\alpha_{\underline{\varrho},\mathsf{R}}$ which needs to be refined as shown below.

Example 5. Assume we want to check ℓ_1 `x = x + 1 ;`ℓ_2 `x = x + 2 ;`ℓ_3 for the regular specification $? : \mathsf{x} = \underline{\mathsf{x}} \bullet ? : \mathsf{x} = \underline{\mathsf{x}} + 1 \bullet ? : \mathsf{x} = \underline{\mathsf{x}} + 3$ by first checking the first statement and then the second. Knowing the boolean information that ℓ_1 `x = x + 1 ;`ℓ_2 model checks for $? : \mathsf{x} = \underline{\mathsf{x}} \bullet ? : \mathsf{x} = \underline{\mathsf{x}} + 1$ is not enough. We must also know what to check the continuation ℓ_2 `x = x + 2 ;`ℓ_3 for. (This is $? : \mathsf{x} = \underline{\mathsf{x}} + 1 \bullet ? : \mathsf{x} = \underline{\mathsf{x}} + 3$ that is if x is equal to the initial value plus 1 at ℓ_2, it is equal to this initial value plus 3 at ℓ_3.) □

The model-checking $\mathcal{M}^t\langle\underline{\varrho}, \mathsf{R}\rangle\pi$ of a trace π with initial environment $\underline{\varrho}$ for a $\mid$-free specification $\mathsf{R} \in \mathbb{R}^{\nmid}$ is a pair $\langle b, \mathsf{R}'\rangle$ where the boolean b states whether the specification R holds for the trace π and R' specifies the possible continuations of π according to R, ε if none.

Example 6. For $\mathsf{Sl} = \ell_1$ `x = x + 1 ;`ℓ_2 `x = x + 2 ;`ℓ_3, we have $\mathcal{S}^*[\![\mathsf{Sl}]\!] = \{\langle\ell_1, \rho\rangle\langle\ell_2, \rho[\mathsf{x} \leftarrow \rho(\mathsf{x})+1]\rangle\langle\ell_3, \rho[\mathsf{x} \leftarrow \rho(\mathsf{x})+3]\rangle \mid \rho \in \mathbb{E}\mathrm{v}\}$ and $\mathcal{M}^t\langle\rho, ? : \mathsf{x} = \underline{\mathsf{x}} \bullet ? : \mathsf{x} = \underline{\mathsf{x}}+1 \bullet ? : \mathsf{x} = \underline{\mathsf{x}} + 3\rangle(\langle\ell_1, \rho\rangle\langle\ell_2, \rho[\mathsf{x} \leftarrow \rho(\mathsf{x}) + 1]\rangle\langle\ell_3, \rho[\mathsf{x} \leftarrow \rho(\mathsf{x}) + 3]\rangle) = \langle\mathsf{tt}, \varepsilon\rangle$ (we have ignored the initial empty statement list in Sl to simplify the specification). □

The fact that $\mathcal{M}^t\langle\underline{\varrho}, \mathsf{R}\rangle\pi$ returns a pair $\langle b, \mathsf{R}'\rangle$ where R' is to be satisfied by continuations of π allows us to perform program model checking by structural induction on the program in Section **8**. The formal definition is the following.

Definition 2 (Regular model checking).

– *Trace model checking ($\underline{\varrho} \in \mathbb{E}\mathrm{v}$ is an initial environment and $\mathsf{R} \in \mathbb{R}^+ \cap \mathbb{R}^{\nmid}$ is a non-empty and $\mid$-free regular expression):*

$$\mathcal{M}^t\langle\underline{\varrho}, \varepsilon\rangle\pi \triangleq \langle\mathsf{tt}, \varepsilon\rangle \tag{11}$$
$$\mathcal{M}^t\langle\underline{\varrho}, \mathsf{R}\rangle\ni \triangleq \langle\mathsf{tt}, \mathsf{R}\rangle$$
$$\mathcal{M}^t\langle\underline{\varrho}, \mathsf{R}\rangle\pi \triangleq \mathsf{let}\ \langle\ell_1, \rho_1\rangle\pi' = \pi\ \mathsf{and}\ \langle\mathsf{L} : \mathsf{B}, \mathsf{R}'\rangle = \mathsf{fstnxt}(\mathsf{R})\ \mathsf{in} \qquad \pi \neq \ni$$
$$(\!|\ \langle\underline{\varrho}, \langle\ell_1, \rho_1\rangle\rangle \in \mathcal{S}^r[\![\mathsf{L} : \mathsf{B}]\!]\ ⨾\ \mathcal{M}^t\langle\underline{\varrho}, \mathsf{R}'\rangle\pi'\ ⨾\ \langle\mathsf{ff}, \mathsf{R}'\rangle\ |\!)$$

– *Set of traces model checking (for an $\mid$-free regular expression $\mathsf{R} \in \mathbb{R}^{\nmid}$):*

$$\mathcal{M}^{\nmid}\langle\underline{\varrho}, \mathsf{R}\rangle\Pi \triangleq \{\langle\pi, \mathsf{R}'\rangle \mid \pi \in \Pi \wedge \langle\mathsf{tt}, \mathsf{R}'\rangle = \mathcal{M}^t\langle\underline{\varrho}, \mathsf{R}\rangle\pi\} \tag{12}$$

– *Program component $\mathsf{S} \in \mathbb{P}\mathrm{c}$ model checking (for an $\mid$-free regular expression $\mathsf{R} \in \mathbb{R}^{\nmid}$):*

$$\mathcal{M}^{\nmid}[\![\mathsf{S}]\!]\langle\underline{\varrho}, \mathsf{R}\rangle \triangleq \mathcal{M}^{\nmid}\langle\underline{\varrho}, \mathsf{R}\rangle(\mathcal{S}^*[\![\mathsf{S}]\!]) \tag{13}$$

– *Set of traces model checking (for regular expression $\mathsf{R} \in \mathbb{R}$):*

$$\mathcal{M}\langle\underline{\varrho}, \mathsf{R}\rangle\Pi \triangleq \mathsf{let}\ (\mathsf{R}_1 \mid \ldots \mid \mathsf{R}_n) = \mathsf{dnf}(\mathsf{R})\ \mathsf{in} \tag{14}$$
$$\bigcup_{i=1}^{n} \{\pi \mid \exists \mathsf{R}' \in \mathbb{R}\ .\ \langle\pi, \mathsf{R}'\rangle \in \mathcal{M}^{\nmid}\langle\underline{\varrho}, \mathsf{R}_i\rangle\Pi\}$$

– *Model checking of a program component* $\mathsf{S} \in \mathbb{P}\mathbb{c}$ *(for regular expression* $\mathsf{R} \in \mathbb{R}$*):*

$$\boldsymbol{\mathcal{M}}[\![\mathsf{S}]\!]\langle \underline{\varrho}, \mathsf{R}\rangle \triangleq \boldsymbol{\mathcal{M}}\langle \underline{\varrho}, \mathsf{R}\rangle(\boldsymbol{\mathcal{S}}^*[\![\mathsf{S}]\!]) \tag{15}$$ □

The model checking $\boldsymbol{\mathcal{M}}^t\langle \underline{\varrho}, \mathsf{R}\rangle\pi$ of a stateful trace π in (11) returns a pair $\langle b, \mathsf{R}'\rangle$ specifying whether π satisfies the specification R (when $b = \mathrm{tt}$) or not (when $b = \mathrm{ff}$). So if $\boldsymbol{\mathcal{M}}^t\langle \rho, \mathsf{R}\rangle(\pi) = \langle \mathrm{ff}, \mathsf{R}'\rangle$ in (12) then the trace π is a counter example to the specification R. R' specifies what a continuation π' of π would have to satisfy for $\pi \mathbin{\frown} \pi'$ to satisfy R (nothing specific when $\mathsf{R}' = \varepsilon$).

Notice that $\boldsymbol{\mathcal{M}}^t\langle \underline{\varrho}, \mathsf{R}\rangle\pi$ checks whether the given trace π satisfies the regular specification R for initial environment $\underline{\varrho}$. Because only one trace is involved, this check can be done at runtime using a monitoring of the program execution. This is the case Fred Schneider's security monitors [35] in Ex. 1 (using an equivalent specification by finite automata).

The set of traces model checking $\boldsymbol{\mathcal{M}}^{\nmid}\langle \underline{\varrho}, \mathsf{R}\rangle\Pi$ returns the subset of traces of Π satisfying the specification R for the initial environment $\underline{\varrho}$. Since all program executions $\boldsymbol{\mathcal{S}}^*[\![\mathsf{P}]\!]$ are involved, the model checking $\boldsymbol{\mathcal{M}}^{\nmid}[\![\mathsf{P}]\!]\langle \underline{\varrho}, \mathsf{R}\rangle$ of a program P becomes, by Rice theorem, undecidable.

The regular specification R is relational in that it may relate the initial and current states (or else may only assert a property of the current states when R never refer to the initial environment $\underline{\varrho}$). If $\pi\langle \ell, \rho\rangle\pi' \in \boldsymbol{\mathcal{S}}^*[\![\mathsf{S}]\!]$ is an execution trace satisfying the specification R then R in (15) determines a relationship between the initial environment $\underline{\varrho}$ and the current environment ρ. For example $\mathsf{R} = \langle\{\mathsf{at}[\![\mathsf{S}]\!]\}, \mathsf{B}\rangle \bullet \mathsf{R}'$ with $\boldsymbol{\mathcal{B}}[\![\mathsf{B}]\!]\underline{\varrho}, \rho = \forall \mathsf{x} \in \mathcal{X} \,.\, \underline{\varrho}(\underline{\mathsf{x}}) = \rho(\mathsf{x})$ expresses that the initial values of variables x are denoted $\underline{\mathsf{x}}$. $\boldsymbol{\mathcal{B}}[\![\mathsf{B}]\!]\underline{\varrho}, \rho = \mathrm{tt}$ would state that there is no constraint on the initial value of variables. The difference with the invariant specifications of is that the order of computations is preserved. R can specify in which order program points may be reached, which is impossible with invariants [4].

The model checking abstraction (12) which, given an initial environment $\underline{\varrho} \in \mathbb{E}\mathrm{v}$ and an $\mid$-free regular specification $\mathsf{R} \in \mathbb{R}^{\nmid}$, returns the set of traces satisfying this specification is the lower adjoint of the Galois connection

$$\langle \wp(\mathbb{S}^+), \subseteq\rangle \xrightleftharpoons[\boldsymbol{\mathcal{M}}^{\nmid}\langle \underline{\varrho}, \mathsf{R}\rangle]{\gamma_{\boldsymbol{\mathcal{M}}^{\nmid}\langle \underline{\varrho}, \mathsf{R}\rangle}} \langle \wp(\mathbb{S}^+), \subseteq\rangle \quad \text{for } \mathsf{R} \in \mathbb{R}^{\nmid} \text{ in (12)} \tag{16}$$

[4] By introduction of an auxiliary variable C incremented at each program step one can simulate a trace with invariants. But then the reasoning is not on the original program P but on a transformed program $\overline{\mathsf{P}}$. Invariants in $\overline{\mathsf{P}}$ holding for a given value of c of C also hold at the position c of the traces in P. This king of indirect reasoning is usually heavy and painful to maintain when programs are modified since values of counters are no longer the same. The use of temporal specifications has the advantage of avoiding the reasoning on explicit positions in the trace.

If $\langle C, \leqslant\rangle$ is a poset, $\langle \mathcal{A}, \sqsubseteq, \sqcup, \sqcap\rangle$ is a complete lattice, $\forall i \in [1,n]\,.\,\langle C, \leqslant\rangle \xrightleftharpoons[\alpha_i]{\gamma_i} \langle \mathcal{A}, \sqsubseteq\rangle$ then $\langle C, \leqslant\rangle \xrightleftharpoons[\alpha]{\gamma} \langle \mathcal{A}, \sqsubseteq\rangle$ where $\alpha \triangleq \dot{\bigsqcup_{i=1}^{n}} \alpha_i$ and $\gamma = \dot{\bigsqcap_{i=1}^{n}} \gamma_i$, pointwise. This implies that

$$\langle \wp(\mathbb{S}^+), \subseteq\rangle \xrightleftharpoons[\mathcal{M}\langle \underline{\varrho}, \mathsf{R}\rangle]{\gamma_{\mathcal{M}\langle \underline{\varrho}, \mathsf{R}\rangle}} \langle \wp(\mathbb{S}^+), \subseteq\rangle \quad \text{for } \mathsf{R} \in \mathbb{R} \text{ in (14)} \tag{17}$$

To follow the tradition that model checking returns a boolean answer this abstraction can be composed with the boolean abstraction

$$\langle \wp(\mathbb{S}^+), \subseteq\rangle \xrightleftharpoons[\alpha_{\mathcal{M}\langle \underline{\varrho}, \mathsf{R}\rangle}]{\gamma_{\mathcal{M}\langle \underline{\varrho}, \mathsf{R}\rangle}} \langle \mathbb{B}, \Leftarrow\rangle \tag{18}$$

where $\alpha_{\mathcal{M}\langle \underline{\varrho}, \mathsf{R}\rangle}(X) \triangleq (\{\underline{\varrho}\} \times X) \subseteq \mathcal{M}\langle \underline{\varrho}, \mathsf{R}\rangle(X)$.

7 Soundness and completeness of the model checking abstraction

The following Th. 1 shows that the Def. 1 of model checking a program semantics for a regular specification is a sound and complete abstraction of this semantics.

Theorem 1 (Model checking soundness ($\Leftarrow$) and completeness ($\Rightarrow$)).

$$\mathsf{P}, \underline{\varrho} \models \mathsf{R} \Leftrightarrow \alpha_{\mathcal{M}\langle \underline{\varrho}, \mathsf{R}\rangle}(\mathcal{S}^*[\![\mathsf{P}]\!])$$

At this point we know, by (15) and Th. 1 that a model checker $\mathcal{M}[\![\mathsf{S}]\!]\langle \underline{\varrho}, \mathsf{R}\rangle$ is a sound and complete abstraction $\mathcal{M}\langle \underline{\varrho}, \mathsf{R}\rangle(\mathcal{S}^*[\![\mathsf{S}]\!])$ of the program component semantics $\mathcal{S}^*[\![\mathsf{S}]\!]$ which provides a counter example in case of failure. This allows us to derive a structural model checker $\widehat{\mathcal{M}}[\![\mathsf{P}]\!]\langle \underline{\varrho}, \mathsf{R}\rangle$ in Section **8** by calculational design.

8 Structural model checking

By Def. 1 of the model checking of $\mathsf{S}, \underline{\varrho} \models \mathsf{R}$ of a program $\mathsf{P} \in \mathbb{P}$ for a regular specification $\mathsf{R} \in \mathbb{R}^+$ and initial environment $\underline{\varrho}$, Th. 1 shows that a model checker $\mathcal{M}[\![\mathsf{P}]\!]\langle \underline{\varrho}, \mathsf{R}\rangle$ is a sound and complete abstraction $\mathcal{M}\langle \underline{\varrho}, \mathsf{R}\rangle(\mathcal{S}^*[\![\mathsf{P}]\!])$ of the program semantics $\mathcal{S}^*[\![\mathsf{P}]\!]$. This abstraction does not provide a model checking algorithm specification.

The standard model checking algorithms [10] use a transition system (or a Kripke structure variation [26]) for hardware and software modeling and proceeding by induction on computation steps.

In contrast, we proceed by structural induction on programs, which will be shown in Th. 2 to be logically equivalent (but maybe more efficient since fixpoints are computed locally). The structural model checking $\widehat{\mathcal{M}}[\![\mathsf{P}]\!]\langle \underline{\varrho}, \mathsf{R}\rangle$ of the program P proceeds by structural induction on the program structure:

$$\begin{cases}\widehat{\mathcal{M}}[\![\mathtt{S}]\!]\langle\underline{\varrho},\ \mathtt{R}\rangle \triangleq \widehat{\mathcal{F}}[\![\mathtt{S}]\!](\prod_{\mathtt{S}'\lhd\mathtt{S}} \widehat{\mathcal{M}}[\![\mathtt{S}']\!])\langle\underline{\varrho},\ \mathtt{R}\rangle\\ \mathtt{S}\in\mathbb{P}\mathtt{c}\end{cases}$$

where the transformer $\widehat{\mathcal{F}}$ uses the results of model checking of the immediate components $\mathtt{S}' \lhd \mathtt{S}$ and involves a fixpoint computation for iteration statements.

The following Th. 2 shows that the algorithm specification is correct, that is $\widehat{\mathcal{M}}[\![\mathtt{S}]\!] = \mathcal{M}[\![\mathtt{S}]\!]$ for all program components $\mathtt{S}$. So together with Th. 1, the structural model checking is proved sound and complete.

Theorem 2. *$\forall \mathtt{S} \in \mathbb{P}\mathtt{c}, \mathtt{R} \in \mathbb{R}, \underline{\varrho} \in \mathbb{E}\mathtt{v}\ .\ \widehat{\mathcal{M}}^{\dagger}[\![\mathtt{S}]\!]\langle\underline{\varrho},\ \mathtt{R}\rangle = \mathcal{M}^{\dagger}[\![\mathtt{S}]\!]\langle\underline{\varrho},\ \mathtt{R}\rangle$ and $\widehat{\mathcal{M}}[\![\mathtt{S}]\!]\langle\underline{\varrho},\ \mathtt{R}\rangle = \mathcal{M}[\![\mathtt{S}]\!]\langle\underline{\varrho},\ \mathtt{R}\rangle$.*

The proof of Th. 2 is by calculational design and proceeds by structural induction on the program component $\mathtt{S}$. Assuming $\mathcal{M}[\![\mathtt{S}']\!] = \widehat{\mathcal{M}}[\![\mathtt{S}']\!]$ for all immediate components $\mathtt{S}' \lhd \mathtt{S}$ of statement, the proof for each program component $\mathtt{S}$ has the following form.

$$\begin{aligned}
&\mathcal{M}[\![\mathtt{S}]\!]\langle\underline{\varrho},\ \mathtt{R}\rangle\\
\triangleq\ &\mathcal{M}\langle\underline{\varrho},\ \mathtt{R}\rangle(\mathcal{S}^*[\![\mathtt{S}]\!]) && \text{\{(15)\}}\\
=\ &\mathcal{M}\langle\underline{\varrho},\ \mathtt{R}\rangle(\mathcal{F}[\![\mathtt{S}]\!](\prod_{\mathtt{S}'\lhd\mathtt{S}} \mathcal{S}^*[\![\mathtt{S}']\!])\langle\underline{\varrho},\ \mathtt{R}\rangle)\\
&\quad\text{\{by structural definition } \mathcal{S}^*[\![\mathtt{S}]\!] = \mathcal{F}[\![\mathtt{S}]\!](\textstyle\prod_{\mathtt{S}'\lhd\mathtt{S}} \mathcal{S}^*[\![\mathtt{S}']\!]) \text{ of the stateful}\\
&\quad\text{prefix trace semantics in Section } \mathbf{2}\text{\}}\\
=\ &\dots \quad\text{\{calculus to expand definitions, rewrite and simplify formulæ by}\\
&\quad\text{algebraic laws\}}\\
=\ &\widehat{\mathcal{F}}[\![\mathtt{S}]\!](\prod_{\mathtt{S}'\lhd\mathtt{S}} \mathcal{M}[\![\mathtt{S}']\!])\langle\underline{\varrho},\ \mathtt{R}\rangle\\
&\quad\text{\{by calculational design to commute the model checking abstraction}\\
&\quad\text{on the result to the model checking of the arguments of } \mathcal{S}^*[\![\mathtt{S}]\!]\text{\}}\\
=\ &\widehat{\mathcal{F}}[\![\mathtt{S}]\!](\prod_{\mathtt{S}'\lhd\mathtt{S}} \widehat{\mathcal{M}}[\![\mathtt{S}']\!])\langle\underline{\varrho},\ \mathtt{R}\rangle && \text{\{ind. hyp.\}}\\
\triangleq\ &\widehat{\mathcal{M}}[\![\mathtt{S}]\!]\langle\underline{\varrho},\ \mathtt{R}\rangle && \text{\{by defining } \widehat{\mathcal{M}}[\![\mathtt{S}]\!] \triangleq \widehat{\mathcal{F}}[\![\mathtt{S}]\!](\textstyle\prod_{\mathtt{S}'\lhd\mathtt{S}} \widehat{\mathcal{M}}[\![\mathtt{S}']\!])\text{\}}
\end{aligned}$$

For iteration statements, $\mathcal{F}[\![\mathtt{S}]\!](\prod_{\mathtt{S}'\lhd\mathtt{S}} \mathcal{S}^*[\![\mathtt{S}']\!])\langle\underline{\varrho},\ \mathtt{R}\rangle$ is a fixpoint, and this proof involves the fixpoint transfer theorem [16, Th. 7.1.0.4 (3)] based on the commutation of the concrete and abstract transformer with the abstraction. The calculational design of the structural model checking $\widehat{\mathcal{M}}[\![\mathtt{S}]\!]$ is shown below.

Definition 3 (Structural model checking).

– *Model checking a program* $\mathtt{P} ::= \mathtt{Sl}\ \ell$ *for a temporal specification* $\mathtt{R} \in \mathbb{R}$ *with alternatives.*

$$\widehat{\mathcal{M}}[\![\mathtt{P}]\!]\langle\underline{\varrho},\ \mathtt{R}\rangle \triangleq \mathsf{let}\,(\mathtt{R}_1 \mid \dots \mid \mathtt{R}_n) = \mathsf{dnf}(\mathtt{R})\ \mathsf{in} \tag{19}$$

$$\bigcup_{i=1}^{n} \{\pi \mid \exists \mathsf{R}' \in \mathbb{R} \,.\, \langle \pi,\, \mathsf{R}' \rangle \in \widehat{\mathcal{M}}^{\nmid}[\![\mathsf{Sl}]\!]\langle \underline{\varrho},\, \mathsf{R}_i \rangle\}$$

Proof. — In case (19) of a program $\mathsf{P} ::= \mathsf{Sl}\ \ell$, the calculational design is as follows.

$$\begin{aligned}
&\mathcal{M}[\![\mathsf{P}]\!]\langle \underline{\varrho},\, \mathsf{R} \rangle\\
\triangleq\ &\mathcal{M}\langle \underline{\varrho},\, \mathsf{R} \rangle(\mathcal{S}^*[\![\mathsf{P}]\!]) && (15)\\
=\ &\mathsf{let}\,(\mathsf{R}_1 \mid \ldots \mid \mathsf{R}_n) = \mathsf{dnf}(\mathsf{R})\ \mathsf{in} \bigcup_{i=1}^{n} \{\pi \mid \exists \mathsf{R}' \in \mathbb{R} \,.\, \langle \pi,\, \mathsf{R}' \rangle \in \mathcal{M}^{\nmid}\langle \underline{\varrho},\, \mathsf{R}_i \rangle(\mathcal{S}^*[\![\mathsf{P}]\!])\}\\
& && (14)\\
=\ &\mathsf{let}\,(\mathsf{R}_1 \mid \ldots \mid \mathsf{R}_n) = \mathsf{dnf}(\mathsf{R})\ \mathsf{in} \bigcup_{i=1}^{n} \{\pi \mid \exists \mathsf{R}' \in \mathbb{R} \,.\, \langle \pi,\, \mathsf{R}' \rangle \in \mathcal{M}^{\nmid}\langle \underline{\varrho},\, \mathsf{R}_i \rangle(\mathcal{S}^*[\![\mathsf{Sl}]\!])\}\\
& && \text{def. of } \mathcal{S}^*[\![\mathsf{P}]\!] \triangleq \mathcal{S}^*[\![\mathsf{Sl}]\!]\\
=\ &\mathsf{let}\,(\mathsf{R}_1 \mid \ldots \mid \mathsf{R}_n) = \mathsf{dnf}(\mathsf{R})\ \mathsf{in} \bigcup_{i=1}^{n} \{\pi \mid \exists \mathsf{R}' \in \mathbb{R} \,.\, \langle \pi,\, \mathsf{R}' \rangle \in \widehat{\mathcal{M}}^{\nmid}\langle \underline{\varrho},\, \mathsf{R}_i \rangle(\mathcal{S}^*[\![\mathsf{Sl}]\!])\}\\
& && \text{ind. hyp.}\\
=\ &\mathsf{let}\,(\mathsf{R}_1 \mid \ldots \mid \mathsf{R}_n) = \mathsf{dnf}(\mathsf{R})\ \mathsf{in} \bigcup_{i=1}^{n} \{\pi \mid \exists \mathsf{R}' \in \mathbb{R} \,.\, \langle \pi,\, \mathsf{R}' \rangle \in \widehat{\mathcal{M}}^{\nmid}[\![\mathsf{Sl}]\!]\langle \underline{\varrho},\, \mathsf{R}_i \rangle\} && (13)\\
=\ &\widehat{\mathcal{M}}[\![\mathsf{P}]\!]\langle \underline{\varrho},\, \mathsf{R} \rangle && (19)\ \square
\end{aligned}$$

Definition 3 (Structural model checking, contn'd)

– Model checking an empty temporal specification ε.

$$\widehat{\mathcal{M}}^{\nmid}[\![\mathsf{S}]\!]\langle \underline{\varrho},\, \varepsilon \rangle \triangleq \{\langle \pi,\, \varepsilon \rangle \mid \pi \in \mathcal{S}^*[\![\mathsf{S}]\!]\} \tag{20}$$

– It is assumed below that $\mathsf{R} \in \mathbb{R}^{\nmid} \cap \mathbb{R}^{+}$ is a non-empty, alternative $\mid$-free regular expression.

– Model checking a statement list $\mathsf{Sl} ::= \mathsf{Sl}'\ \mathsf{S}$

$$\begin{aligned}
\widehat{\mathcal{M}}^{\nmid}[\![\mathsf{Sl}]\!]\langle \underline{\varrho},\, \mathsf{R} \rangle \triangleq\ &\widehat{\mathcal{M}}^{\nmid}[\![\mathsf{Sl}']\!]\langle \underline{\varrho},\, \mathsf{R} \rangle\\
&\cup \{\langle \pi \cdot \langle \mathsf{at}[\![\mathsf{S}]\!],\, \rho \rangle \cdot \pi',\, \mathsf{R}'' \rangle \mid \langle \pi \cdot \langle \mathsf{at}[\![\mathsf{S}]\!],\, \rho \rangle,\, \mathsf{R}' \rangle \in \widehat{\mathcal{M}}^{\nmid}[\![\mathsf{Sl}']\!]\langle \underline{\varrho},\, \mathsf{R} \rangle \wedge\\
&\qquad \langle \langle \mathsf{at}[\![\mathsf{S}]\!],\, \rho \rangle \cdot \pi',\, \mathsf{R}'' \rangle \in \widehat{\mathcal{M}}^{\nmid}[\![\mathsf{S}]\!]\langle \underline{\varrho},\, \mathsf{R}' \rangle\}
\end{aligned} \tag{21}$$

– Model checking an empty statement list $\mathsf{Sl} ::= \epsilon$

$$\begin{aligned}
\widehat{\mathcal{M}}^{\nmid}[\![\mathsf{Sl}]\!]\langle \underline{\varrho},\, \mathsf{R} \rangle \triangleq\ &\mathsf{let}\ \langle \mathsf{L} : \mathsf{B},\, \mathsf{R}' \rangle = \mathsf{fstnxt}(\mathsf{R})\ \mathsf{in}\\
&\{\langle \langle \mathsf{at}[\![\mathsf{Sl}]\!],\, \rho \rangle,\, \mathsf{R}' \rangle \mid \langle \underline{\varrho},\, \langle \mathsf{at}[\![\mathsf{Sl}]\!],\, \rho \rangle \rangle \in \mathcal{S}^{r}[\![\mathsf{L} : \mathsf{B}]\!]\}
\end{aligned} \tag{22}$$

(In practice the empty statement list ϵ needs not be specified so we could eliminate that need by ignoring ϵ in the specification R and defining $\widehat{\mathcal{M}}^{\nmid}[\![\mathsf{Sl}]\!]\langle \underline{\varrho},\, \mathsf{R} \rangle \triangleq \{\langle \langle \mathsf{at}[\![\mathsf{Sl}]\!],\, \rho \rangle,\, \mathsf{R} \rangle \mid \rho \in \mathbb{E}\mathbb{v}\}$.)

– Model checking an assignment statement $\mathsf{S} ::= \ell\ \mathsf{x} = \mathsf{A}\ ;$

$$\widehat{\mathcal{M}}^{\dagger}[\![\mathsf{S}]\!]\langle \underline{\varrho},\ \mathsf{R}\rangle \triangleq \text{let}\ \langle \mathsf{L} : \mathsf{B},\ \mathsf{R}'\rangle = \text{fstnxt}(\mathsf{R})\ \text{in} \tag{23}$$

$$\{\langle\langle \text{at}[\![\mathsf{S}]\!],\ \rho\rangle,\ \mathsf{R}'\rangle \mid \langle \underline{\varrho},\ \langle \text{at}[\![\mathsf{S}]\!],\ \rho\rangle\rangle \in \mathcal{S}^{r}[\![\mathsf{L} : \mathsf{B}]\!]\} \quad \text{(a)}$$

$$\cup \{\langle\langle \text{at}[\![\mathsf{S}]\!],\ \rho\rangle\langle \text{aft}[\![\mathsf{S}]\!],\ \rho[\mathsf{x} \leftarrow \mathcal{A}[\![\mathsf{A}]\!]\rho]\rangle,\ \varepsilon\rangle \mid \mathsf{R}' \in \mathbb{R}_{\varepsilon} \wedge \langle \underline{\varrho},\ \langle \text{at}[\![\mathsf{S}]\!],\ \rho\rangle\rangle \in \mathcal{S}^{r}[\![\mathsf{L} : \mathsf{B}]\!]\} \quad \text{(b)}$$

$$\cup \{\langle\langle \text{at}[\![\mathsf{S}]\!],\ \rho\rangle\langle \text{aft}[\![\mathsf{S}]\!],\ \rho[\mathsf{x} \leftarrow \mathcal{A}[\![\mathsf{A}]\!]\rho]\rangle,\ \mathsf{R}''\rangle \mid \mathsf{R}' \notin \mathbb{R}_{\varepsilon} \wedge \langle \underline{\varrho},\ \langle \text{at}[\![\mathsf{S}]\!],\ \rho\rangle\rangle \in \mathcal{S}^{r}[\![\mathsf{L} : \mathsf{B}]\!] \wedge \langle \mathsf{L}' : \mathsf{B}',\ \mathsf{R}''\rangle = \text{fstnxt}(\mathsf{R}') \wedge \langle \underline{\varrho},\ \langle \text{aft}[\![\mathsf{S}]\!],\ \rho[\mathsf{x} \leftarrow \mathcal{A}[\![\mathsf{A}]\!]\rho]\rangle\rangle \in \mathcal{S}^{r}[\![\mathsf{L}' : \mathsf{B}']\!]\} \quad \text{(c)}$$

For the assignment $\mathsf{S} ::= \ell\ \mathsf{x} = \mathsf{A}\ ;$ in (23), case (a) checks the prefixes that stops at ℓ whereas (b) and (c) checks the maximal traces stopping after the assignment. In each trace checked for the specification R, the states are checked successively and the continuation specification is returned together with the checked trace, unless the check fails. Checking the assignment $\mathsf{S} ::= \ell\ \mathsf{x} = \mathsf{A}\ ;$ in (23) for $\langle \mathsf{L} : \mathsf{B},\ \mathsf{R}'\rangle = \text{fstnxt}(\mathsf{R})$ consists in first checking $\mathsf{L} : \mathsf{B}$ at ℓ and then checking on R' after the statement. In case (b), R' is empty so trivially satisfied. Otherwise, in case (c), $\langle \mathsf{L}' : \mathsf{B}',\ \mathsf{R}''\rangle = \text{fstnxt}(\mathsf{R}')$ so $\mathsf{L}' : \mathsf{B}'$ is checked after the statement while R'' is the continuation specification.

Proof. — In case (23) of an assignment statement $\mathsf{S} ::= \ell\ \mathsf{x} = \mathsf{A}\ ;$, the calculational design is as follows.

$$\mathcal{M}^{\dagger}[\![\mathsf{S}]\!]\ \langle \underline{\varrho},\ \mathsf{R}\rangle$$

$$= \{\langle \pi,\ \mathsf{R}'\rangle \mid \pi \in \mathcal{S}^{*}[\![\mathsf{Sl}]\!] \wedge \langle \text{tt},\ \mathsf{R}'\rangle = \mathcal{M}^{t}\langle \underline{\varrho},\ \mathsf{R}\rangle\pi\} \qquad \text{(13) and (12)}$$

$$= \{\langle \pi, \mathsf{R}'\rangle \mid \pi \in \{\langle \ell, \rho\rangle \mid \rho \in \mathbb{E}\mathbb{v}\} \cup \{\langle \ell, \rho\rangle\langle \text{aft}[\![\mathsf{S}]\!], \rho[\mathsf{x} \leftarrow v]\rangle \mid \rho \in \mathbb{E}\mathbb{v} \wedge v = \mathcal{A}[\![\mathsf{A}]\!]\rho \wedge \langle \text{tt},\ \mathsf{R}'\rangle = \mathcal{M}^{t}\langle \underline{\varrho},\ \mathsf{R}\rangle\pi\} \qquad \text{(1)}$$

$$= \{\langle\langle \ell,\ \rho\rangle,\ \mathsf{R}'\rangle \mid \rho \in \mathbb{E}\mathbb{v} \wedge \langle \text{tt},\ \mathsf{R}'\rangle = \mathcal{M}^{t}\langle \underline{\varrho},\ \mathsf{R}\rangle\langle \ell,\ \rho\rangle\} \cup \{\langle\langle \ell,\ \rho\rangle\langle \text{aft}[\![\mathsf{S}]\!],\ \rho[\mathsf{x} \leftarrow v]\rangle,\ \mathsf{R}'\rangle \mid \rho \in \mathbb{E}\mathbb{v} \wedge v = \mathcal{A}[\![\mathsf{A}]\!]\rho\langle \text{tt},\ \mathsf{R}'\rangle = \mathcal{M}^{t}\langle \underline{\varrho},\ \mathsf{R}\rangle\langle \ell,\ \rho\rangle\langle \text{aft}[\![\mathsf{S}]\!],\ \rho[\mathsf{x} \leftarrow v]\rangle\} \qquad \text{def. } \cup \text{ and } \in$$

$$= \{\langle\langle \ell,\ \rho\rangle,\ \mathsf{R}'\rangle \mid \langle \text{tt},\ \mathsf{R}'\rangle = \text{let}\ \langle \mathsf{L} : \mathsf{B},\ \mathsf{R}''\rangle = \text{fstnxt}(\mathsf{R})\ \text{in}\ (\!|\ \langle \underline{\varrho},\ \langle \ell,\ \rho\rangle\rangle \in \mathcal{S}^{r}[\![\mathsf{L} : \mathsf{B}]\!]\ ?\ \langle \text{tt},\ \mathsf{R}''\rangle : \langle \text{ff},\ \mathsf{R}'\rangle\ |\!)\} \cup \{\langle\langle \ell,\ \rho\rangle\langle \text{aft}[\![\mathsf{S}]\!],\ \rho[\mathsf{x} \leftarrow v]\rangle,\ \mathsf{R}'\rangle \mid v = \mathcal{A}[\![\mathsf{A}]\!]\rho \wedge \langle \text{tt},\ \mathsf{R}'\rangle = \text{let}\ \langle \mathsf{L} : \mathsf{B},\ \mathsf{R}''\rangle = \text{fstnxt}(\mathsf{R})\ \text{in}\ (\!|\ \langle \underline{\varrho},\ \langle \ell,\ \rho\rangle\rangle \in \mathcal{S}^{r}[\![\mathsf{L} : \mathsf{B}]\!]\ ?\ \mathcal{M}^{t}\langle \underline{\varrho},\ \mathsf{R}''\rangle\langle \text{aft}[\![\mathsf{S}]\!],\ \rho[\mathsf{x} \leftarrow v]\rangle : \langle \text{ff},\ \mathsf{R}''\rangle\ |\!)\} \qquad \text{(11)}$$

$$= \{\langle\langle \ell,\ \rho\rangle,\ \mathsf{R}'\rangle \mid \langle \mathsf{L} : \mathsf{B},\ \mathsf{R}'\rangle = \text{fstnxt}(\mathsf{R}) \wedge \langle \underline{\varrho},\ \langle \ell,\ \rho\rangle\rangle \in \mathcal{S}^{r}[\![\mathsf{L} : \mathsf{B}]\!]\} \cup \{\langle\langle \ell, \rho\rangle\langle \text{aft}[\![\mathsf{S}]\!], \rho[\mathsf{x} \leftarrow v]\rangle, \mathsf{R}'\rangle \mid v = \mathcal{A}[\![\mathsf{A}]\!]\rho \wedge \exists \mathsf{R}'' \in \mathbb{R}\ .\ \langle \mathsf{L} : \mathsf{B},\ \mathsf{R}''\rangle = \text{fstnxt}(\mathsf{R}) \wedge \langle \underline{\varrho},\ \langle \ell,\ \rho\rangle\rangle \in \mathcal{S}^{r}[\![\mathsf{L} : \mathsf{B}]\!] \wedge (\!|\ \mathsf{R}'' \in \mathbb{R}_{\varepsilon}\ ?\ \text{tt} : \mathcal{M}^{t}\langle \underline{\varrho},\ \mathsf{R}''\rangle\langle \text{aft}[\![\mathsf{S}]\!],\ \rho[\mathsf{x} \leftarrow v]\rangle = \langle \text{tt},\ \mathsf{R}'\rangle\ |\!)\} \qquad \text{def. = and } \mathcal{M}^{t}\langle \underline{\varrho},\ \varepsilon\rangle\pi \triangleq \langle \text{tt},\ \varepsilon\rangle \text{ by (11)}$$

$= \{\langle\langle \ell, \rho\rangle, \mathsf{R}'\rangle \mid \langle \mathsf{L} : \mathsf{B}, \mathsf{R}'\rangle = \mathsf{fstnxt}(\mathsf{R}) \wedge \langle \underline{\varrho}, \langle \ell, \rho\rangle\rangle \in \mathcal{S}^{\mathsf{r}}[\![\mathsf{L} : \mathsf{B}]\!]\} \cup$
$\{\langle\langle \ell, \rho\rangle\langle \mathsf{aft}[\![\mathsf{S}]\!], \rho[\mathsf{x} \leftarrow v]\rangle, \mathsf{R}'\rangle \mid v = \mathcal{A}[\![\mathsf{A}]\!]\rho \wedge \exists \mathsf{R}'' \in \mathbb{R} \,.\, \langle \mathsf{L} : \mathsf{B}, \mathsf{R}''\rangle = \mathsf{fstnxt}(\mathsf{R}) \wedge \langle \underline{\varrho}, \langle \ell, \rho\rangle\rangle \in \mathcal{S}^{\mathsf{r}}[\![\mathsf{L} : \mathsf{B}]\!] \wedge (\!| \mathsf{R}'' \in \mathbb{R}_\varepsilon \,?\, \mathsf{tt} \,\colon \mathsf{let}\ \langle \mathsf{L}' : \mathsf{B}', \mathsf{R}'''\rangle = \mathsf{fstnxt}(\mathsf{R}'')\ \mathsf{in}\ \langle \underline{\varrho}, \langle \mathsf{aft}[\![\mathsf{S}]\!], \rho[\mathsf{x} \leftarrow v]\rangle\rangle \in \mathcal{S}^{\mathsf{r}}[\![\mathsf{L}' : \mathsf{B}']\!] \,|\!)\}$ $\quad\wr(11)\int$

$= \mathsf{let}\ \langle \mathsf{L} : \mathsf{B}, \mathsf{R}'\rangle = \mathsf{fstnxt}(\mathsf{R})\ \mathsf{in}$
$\quad \{\langle\langle \ell, \rho\rangle, \mathsf{R}'\rangle \mid \langle \underline{\varrho}, \langle \ell, \rho\rangle\rangle \in \mathcal{S}^{\mathsf{r}}[\![\mathsf{L} : \mathsf{B}]\!]\}$
$\quad \cup \{\langle\langle \ell, \rho\rangle\langle \mathsf{aft}[\![\mathsf{S}]\!], \rho[\mathsf{x} \leftarrow v]\rangle, \varepsilon\rangle \mid v = \mathcal{A}[\![\mathsf{A}]\!]\rho \wedge \langle \underline{\varrho}, \langle \ell, \rho\rangle\rangle \in \mathcal{S}^{\mathsf{r}}[\![\mathsf{L} : \mathsf{B}]\!] \wedge \mathsf{R}' \in \mathbb{R}_\varepsilon\}$
$\quad \cup \{\langle\langle \ell, \rho\rangle\langle \mathsf{aft}[\![\mathsf{S}]\!], \rho[\mathsf{x} \leftarrow v]\rangle, \mathsf{R}''\rangle \mid v = \mathcal{A}[\![\mathsf{A}]\!]\rho \wedge \langle \underline{\varrho}, \langle \ell, \rho\rangle\rangle \in \mathcal{S}^{\mathsf{r}}[\![\mathsf{L} : \mathsf{B}]\!] \wedge \mathsf{R}' \notin \mathbb{R}_\varepsilon \wedge \mathsf{let}\ \langle \mathsf{L}' : \mathsf{B}', \mathsf{R}''\rangle = \mathsf{fstnxt}(\mathsf{R}')\ \mathsf{in}\ \langle \underline{\varrho}, \langle \mathsf{aft}[\![\mathsf{S}]\!], \rho[\mathsf{x} \leftarrow v]\rangle\rangle \in \mathcal{S}^{\mathsf{r}}[\![\mathsf{L}' : \mathsf{B}']\!]\}$
$\quad\wr$def. $\cup\int$

$= \widehat{\mathcal{M}}^{\dagger}[\![\mathsf{S}]\!]\,\mathsf{R}$ $\quad\wr(23)\int$ □

Definition 3 (Structural model checking, continued)

– Model checking a conditional statement $\mathsf{S} ::= \mathtt{if}\ ^{\ell}\ \mathtt{(B)}\ \mathsf{S}_t$

$$\begin{aligned}
\widehat{\mathcal{M}}^{\dagger}[\![\mathsf{S}]\!]\langle \underline{\varrho}, \mathsf{R}\rangle \triangleq\ & \mathsf{let}\ \langle \mathsf{L}' : \mathsf{B}', \mathsf{R}'\rangle = \mathsf{fstnxt}(\mathsf{R})\ \mathsf{in} \\
& \{\langle\langle \mathsf{at}[\![\mathsf{S}]\!], \rho\rangle, \mathsf{R}'\rangle \mid \langle \underline{\varrho}, \langle \mathsf{at}[\![\mathsf{S}]\!], \rho\rangle\rangle \in \mathcal{S}^{\mathsf{r}}[\![\mathsf{L}' : \mathsf{B}]\!]\} \\
& \cup \{\langle\langle \mathsf{at}[\![\mathsf{S}]\!], \rho\rangle\langle \mathsf{at}[\![\mathsf{S}_t]\!], \rho\rangle\pi, \mathsf{R}''\rangle \mid \mathcal{B}[\![\mathsf{B}]\!]\rho = \mathsf{tt} \wedge \\
& \quad \langle \underline{\varrho}, \langle \mathsf{at}[\![\mathsf{S}]\!], \rho\rangle\rangle \in \mathcal{S}^{\mathsf{r}}[\![\mathsf{L}' : \mathsf{B}']\!] \wedge \\
& \quad \langle\langle \mathsf{at}[\![\mathsf{S}_t]\!], \rho\rangle\pi, \mathsf{R}''\rangle \in \widehat{\mathcal{M}}^{\dagger}[\![\mathsf{S}_t]\!]\langle \underline{\varrho}, \mathsf{R}'\rangle\} \\
& \cup \{\langle\langle \mathsf{at}[\![\mathsf{S}]\!], \rho\rangle\langle \mathsf{aft}[\![\mathsf{S}]\!], \rho\rangle, \varepsilon\rangle \mid \mid \mathcal{B}[\![\mathsf{B}]\!]\rho = \mathsf{ff} \wedge \mathsf{R}' \in \mathbb{R}_\varepsilon \wedge \\
& \quad \langle \underline{\varrho}, \langle \mathsf{at}[\![\mathsf{S}]\!], \rho\rangle\rangle \in \mathcal{S}^{\mathsf{r}}[\![\mathsf{L}' : \mathsf{B}]\!]\} \\
& \cup \{\langle\langle \mathsf{at}[\![\mathsf{S}]\!], \rho\rangle\langle \mathsf{aft}[\![\mathsf{S}]\!], \rho\rangle, \mathsf{R}''\rangle \mid \mathcal{B}[\![\mathsf{B}]\!]\rho = \mathsf{ff} \wedge \mathsf{R}' \notin \mathbb{R}_\varepsilon \wedge \\
& \quad \langle \underline{\varrho}, \langle \mathsf{at}[\![\mathsf{S}]\!], \rho\rangle\rangle \in \mathcal{S}^{\mathsf{r}}[\![\mathsf{L}' : \mathsf{B}']\!] \wedge \langle \mathsf{L}'' : \mathsf{B}'', \mathsf{R}''\rangle = \mathsf{fstnxt}(\mathsf{R}') \wedge \\
& \quad \langle \underline{\varrho}, \langle \mathsf{aft}[\![\mathsf{S}]\!], \rho\rangle\rangle \in \mathcal{S}^{\mathsf{r}}[\![\mathsf{L}'' : \mathsf{B}'']\!]\}
\end{aligned} \tag{24}$$

– Model checking a break statement $\mathsf{S} ::= {}^{\ell}\mathtt{break}\ ;$

$$\begin{aligned}
\widehat{\mathcal{M}}^{\dagger}[\![\mathsf{S}]\!]\langle \underline{\varrho}, \mathsf{R}\rangle \triangleq\ & \mathsf{let}\ \langle \mathsf{L} : \mathsf{B}, \mathsf{R}'\rangle = \mathsf{fstnxt}(\mathsf{R})\ \mathsf{in} \\
& \{\langle\langle \mathsf{at}[\![\mathsf{S}]\!], \rho\rangle, \mathsf{R}'\rangle \mid \langle \underline{\varrho}, \langle \mathsf{at}[\![\mathsf{S}]\!], \rho\rangle\rangle \in \mathcal{S}^{\mathsf{r}}[\![\mathsf{L} : \mathsf{B}]\!]\} \\
& \cup \{\langle\langle \mathsf{at}[\![\mathsf{S}]\!], \rho\rangle\langle \mathsf{brk\text{-}to}[\![\mathsf{S}]\!], \rho\rangle, \varepsilon\rangle \mid \mathsf{R}' \in \mathbb{R}_\varepsilon \wedge \\
& \quad \langle \underline{\varrho}, \langle \mathsf{at}[\![\mathsf{S}]\!], \rho\rangle\rangle \in \mathcal{S}^{\mathsf{r}}[\![\mathsf{L} : \mathsf{B}]\!]\} \\
& \cup \{\langle\langle \mathsf{at}[\![\mathsf{S}]\!], \rho\rangle\langle \mathsf{brk\text{-}to}[\![\mathsf{S}]\!], \rho\rangle, \mathsf{R}''\rangle \mid \mathsf{R}' \notin \mathbb{R}_\varepsilon \wedge \\
& \quad \langle \underline{\varrho}, \langle \mathsf{at}[\![\mathsf{S}]\!], \rho\rangle\rangle \in \mathcal{S}^{\mathsf{r}}[\![\mathsf{L} : \mathsf{B}]\!] \wedge \langle \mathsf{L}' : \mathsf{B}', \mathsf{R}''\rangle = \mathsf{fstnxt}(\mathsf{R}') \wedge \\
& \quad \langle \underline{\varrho}, \langle \mathsf{brk\text{-}to}[\![\mathsf{S}]\!], \rho\rangle\rangle \in \mathcal{S}^{\mathsf{r}}[\![\mathsf{L}' : \mathsf{B}']\!]\}
\end{aligned} \tag{25}$$

– Model checking an iteration statement $\mathsf{S} ::= \mathtt{while}\ ^{\ell}\ \mathtt{(B)}\ \mathsf{S}_b$

$$\widehat{\mathcal{M}}^{\dagger}[\![\mathsf{S}]\!]\langle\underline{\varrho},\ \mathsf{R}\rangle \triangleq \mathsf{lfp}^{\subseteq}\,(\widehat{\mathcal{F}}^{\dagger}[\![\mathsf{S}]\!]\langle\underline{\varrho},\ \mathsf{R}\rangle) \tag{26}$$

$$\begin{aligned}
\widehat{\mathcal{F}}^{\dagger}[\![\mathsf{S}]\!]\langle\underline{\varrho},\ \mathsf{R}\rangle\, X \triangleq\ & \mathsf{let}\ \langle \mathsf{L}'\ :\ \mathsf{B}',\ \mathsf{R}'\rangle = \mathsf{fstnxt}(\mathsf{R})\ \mathsf{in} && (27)\\
& \{\langle\langle \mathsf{at}[\![\mathsf{S}]\!],\ \rho\rangle,\ \mathsf{R}'\rangle \mid \rho \in \mathbb{E}\mathbb{v} \wedge \langle\underline{\varrho},\ \langle \mathsf{at}[\![\mathsf{S}]\!],\ \rho\rangle\rangle \in \mathcal{S}^{\mathsf{r}}[\![\mathsf{L}'\ :\ \mathsf{B}']\!]\} && \text{(a)}\\
& \cup \{\langle \pi_2\langle \mathsf{at}[\![\mathsf{S}]\!],\ \rho\rangle\langle \mathsf{aft}[\![\mathsf{S}]\!],\ \rho\rangle,\ \varepsilon\rangle \mid \langle \pi_2\langle \mathsf{at}[\![\mathsf{S}]\!],\ \rho\rangle,\ \varepsilon\rangle \in X \wedge \\
& \qquad \mathcal{B}[\![\mathsf{B}]\!]\,\rho = \mathsf{ff}\}\\
& \cup \{\langle \pi_2\langle \mathsf{at}[\![\mathsf{S}]\!],\ \rho\rangle\langle \mathsf{aft}[\![\mathsf{S}]\!],\ \rho\rangle,\ \varepsilon\rangle \mid \langle \pi_2\langle \mathsf{at}[\![\mathsf{S}]\!],\ \rho\rangle,\ \mathsf{R}''\rangle \in X \wedge && \text{(b)}\\
& \qquad \mathcal{B}[\![\mathsf{B}]\!]\,\rho = \mathsf{ff} \wedge \mathsf{R}'' \notin \mathbb{R}_{\varepsilon} \wedge \langle \mathsf{L}'\ :\ \mathsf{B}',\ \mathsf{R}'\rangle = \mathsf{fstnxt}(\mathsf{R}'') \wedge \mathsf{R}' \in \mathbb{R}_{\varepsilon} \wedge \\
& \qquad\quad \langle\underline{\varrho},\ \langle \mathsf{at}[\![\mathsf{S}]\!],\ \rho\rangle\rangle \in \mathcal{S}^{\mathsf{r}}[\![\mathsf{L}'\ :\ \mathsf{B}']\!]\}\\
& \cup \{\langle \pi_2\langle \mathsf{at}[\![\mathsf{S}]\!],\ \rho\rangle\langle \mathsf{aft}[\![\mathsf{S}]\!],\ \rho\rangle,\ \mathsf{R}'\rangle \mid \langle \pi_2\langle \mathsf{at}[\![\mathsf{S}]\!],\ \rho\rangle,\ \mathsf{R}''\rangle \in X \wedge && \text{(c)}\\
& \qquad \mathcal{B}[\![\mathsf{B}]\!]\,\rho = \mathsf{ff} \wedge \mathsf{R}'' \notin \mathbb{R}_{\varepsilon} \wedge \langle \mathsf{L}'\ :\ \mathsf{B}',\ \mathsf{R}'''\rangle = \mathsf{fstnxt}(\mathsf{R}'') \wedge \mathsf{R}''' \notin \mathbb{R}_{\varepsilon} \wedge \\
& \qquad\quad \langle\underline{\varrho},\ \langle \mathsf{at}[\![\mathsf{S}]\!],\ \rho\rangle\rangle \in \mathcal{S}^{\mathsf{r}}[\![\mathsf{L}'\ :\ \mathsf{B}']\!] \wedge \langle \mathsf{L}''\ :\ \mathsf{B}'',\ \mathsf{R}'\rangle = \mathsf{fstnxt}(\mathsf{R}''') \wedge \\
& \qquad\qquad \langle\underline{\varrho},\ \langle \mathsf{aft}[\![\mathsf{S}]\!],\ \rho\rangle\rangle \in \mathcal{S}^{\mathsf{r}}[\![\mathsf{L}''\ :\ \mathsf{B}'']\!]\}\\
& \cup \{\langle \pi_2\langle \mathsf{at}[\![\mathsf{S}]\!],\ \rho\rangle\langle \mathsf{at}[\![\mathsf{S}_b]\!],\ \rho\rangle\pi_3,\ \varepsilon\rangle \mid \langle \pi_2\langle \mathsf{at}[\![\mathsf{S}]\!],\ \rho\rangle,\ \varepsilon\rangle \in X \wedge && \text{(d)}\\
& \qquad \mathcal{B}[\![\mathsf{B}]\!]\,\rho = \mathsf{tt} \wedge \langle \mathsf{at}[\![\mathsf{S}_b]\!],\ \rho\rangle\pi_3 \in \mathcal{S}^{*}[\![\mathsf{S}_b]\!]\}\\
& \cup \{\langle \pi_2\langle \mathsf{at}[\![\mathsf{S}]\!],\ \rho\rangle\langle \mathsf{at}[\![\mathsf{S}_b]\!],\ \rho\rangle\pi_3,\ \varepsilon\rangle \mid \langle \pi_2\langle \mathsf{at}[\![\mathsf{S}]\!],\ \rho\rangle,\ \mathsf{R}''\rangle \in X \wedge && \text{(e)}\\
& \qquad \mathcal{B}[\![\mathsf{B}]\!]\,\rho = \mathsf{tt} \wedge \mathsf{R}'' \notin \mathbb{R}_{\varepsilon} \wedge \langle \mathsf{L}\ :\ \mathsf{B},\ \varepsilon\rangle = \mathsf{fstnxt}(\mathsf{R}'') \wedge \\
& \qquad\quad \langle\underline{\varrho},\ \langle \mathsf{at}[\![\mathsf{S}]\!],\ \rho\rangle\rangle \in \mathcal{S}^{\mathsf{r}}[\![\mathsf{L}\ :\ \mathsf{B}]\!] \wedge \langle \mathsf{at}[\![\mathsf{S}_b]\!],\ \rho\rangle\pi_3 \in \mathcal{S}^{*}[\![\mathsf{S}_b]\!]\}\\
& \cup \{\langle \pi_2\langle \mathsf{at}[\![\mathsf{S}]\!],\ \rho\rangle\langle \mathsf{at}[\![\mathsf{S}_b]\!],\ \rho\rangle\pi_3,\ \mathsf{R}'\rangle \mid \langle \pi_2\langle \mathsf{at}[\![\mathsf{S}]\!],\ \rho\rangle,\ \mathsf{R}''\rangle \in X \wedge && \text{(f)}\\
& \qquad \mathcal{B}[\![\mathsf{B}]\!]\,\rho = \mathsf{tt} \wedge \mathsf{R}'' \notin \mathbb{R}_{\varepsilon} \wedge \langle \mathsf{L}\ :\ \mathsf{B},\ \mathsf{R}''''\rangle = \mathsf{fstnxt}(\mathsf{R}'') \wedge \\
& \qquad\quad \langle\underline{\varrho},\ \langle \mathsf{at}[\![\mathsf{S}]\!],\ \rho\rangle\rangle \in \mathcal{S}^{\mathsf{r}}[\![\mathsf{L}\ :\ \mathsf{B}]\!] \wedge \mathsf{R}'''' \notin \mathbb{R}_{\varepsilon} \wedge \\
& \qquad\qquad \langle \mathsf{L}'\ :\ \mathsf{B}',\ \mathsf{R}'''\rangle = \mathsf{fstnxt}(\mathsf{R}'''') \wedge \langle\underline{\varrho},\ \langle \mathsf{at}[\![\mathsf{S}_b]\!],\ \rho\rangle\rangle \in \mathcal{S}^{\mathsf{r}}[\![\mathsf{L}'\ :\ \mathsf{B}']\!] \wedge \\
& \qquad\qquad \langle\langle \mathsf{at}[\![\mathsf{S}_b]\!],\ \rho\rangle\pi_3,\ \mathsf{R}'\rangle \in \widehat{\mathcal{M}}^{\dagger}[\![\mathsf{S}_b]\!]\langle\underline{\varrho},\ \mathsf{R}'''\rangle\}
\end{aligned}$$

— The model checking of an iteration statement $\mathsf{S} ::= \mathbf{while}\,\ell\ (\mathsf{B})\ \mathsf{S}_b$ in (27) checks one more iteration (after checking the previous ones as recorded by X) while the fixpoint (26) repeats this check for all iterations. Case (a) checks the prefixes that stops at loop entry ℓ. (b) and (c) check the exit of an iteration when the iteration condition is false, (b) when the specification stops at loop entry ℓ before leaving and (c) when the specification goes further. (d), (e) and (f) check one more iteration when the iteration condition is true. In case (d), the continuation after the check of the iterates is empty so trivially satisfied by any continuation of the prefix trace. In case (e), the continuation after the check of the iterates just impose to verify $\mathsf{L}\ :\ \mathsf{B}$ on iteration entry and nothing afterwards. In case (f) the continuation after the check of the iterates requires to verify $\mathsf{L}\ :\ \mathsf{B}$ at the loop entry, $\mathsf{L}'\ :\ \mathsf{B}'$ at the body entry, and the rest R''' of the specification for the loop body (which returns the possibly empty continuation specification R'). The cases (b) to (f) are mutually exclusive.

9 Notes on implementations and expressivity

Of course further hypotheses and refinements would be necessary to get an effective algorithm as specified by the Def. 3 of structural model checking. A common hypothesis in model checking is that the set of states $\mathbb{S}$ is finite. Traces may still be infinite so the fixpoint computation (26) may not converge. However, infinite traces on finite states must involve an initial finite prefix followed by a finite cycle (often called a lasso). It follows that the infinite set of prefix traces can be finitely represented by a finite set of maximal finite traces and finite lassos. Regular expressions `L : B` can be attached to the states as determined by the analysis, and there are finitely many of them in the specification. These finiteness properties can be taken into account to ensure the convergence of the fixpoint computation in (26).

A symbolic representation of the states in finite/lasso stateful traces may be useful as in symbolic execution [24] or using BDDs [6] for boolean encodings of programs. By Kleene theorem [34, Theorem 2.1, p. 87], a convenient representation of regular expressions is by (deterministic) finite automata *e.g.* [28]. Symbolic automata-based algorithms can be used to implement a data structure for operations over sets of sequences [22].

Of course the hypothesis that the state space is finite and small enough to scale up and limit the combinatorial blow up of the finite state-space is unrealistic [11]. In practice, the set of states $\mathbb{S}$ is very large, so abstraction and a widening/dual narrowing are necessary. A typical trivial widening is bounded model checking (*e.g.* widen to all states after n fixpoint iterations) [5]. Those of [30] are more elaborated.

10 Conclusion

We have illustrated the idea that model checking is an abstract interpretation, as first introduced in [17]. This point of view also yields specification-preserving abstract model checking [18] as well as abstraction refinement algorithms [20].

Specifications by temporal logics are not commonly accepted by programmers. For example, in [31], the specifications had to be written by academics. Regular expressions or path expressions [8] or more expressive extensions might turn out to be more familiar. Moreover, for security monitors the false alarms of the static analysis can be checked at runtime [35,29].

Convergence of model checking requires expressivity restrictions on both the considered models of computation and the considered temporal logics. For some expressive models of computation and temporal logics, state finiteness is not enough to guarantee termination of model checking [17,21]. Finite enumeration is limited, even with symbolic encodings. Beyond finiteness, scalability is always a problem with model checking and the regular software model checking algorithm $\widehat{\mathcal{M}}$ is no exception, so abstraction and induction are ultimately required to reason on programs.

Most often, abstract model checking uses homomorphic/partitioning abstractions *e.g.* [4]. This is because the abstraction of a transition system on concrete states is a transition system on abstract states so model checkers are reusable in the abstract. However, excluding edgy abstractions as in [13], state-based finite abstraction is very restrictive [21] and do not guarantee scalability (*e.g.* SLAM [3]). Such restrictions on abstractions do not apply to structural model checking so that abstractions more powerful than partitioning can be considered.

As an alternative approach, a regular expression can be automatically extracted by static analysis of the program trace semantics that recognizes all feasible execution paths and usually more [19]. Then model-checking a regular specification becomes a regular language inclusion problem [27].

References

1. Abdulla, P.A., Jonsson, B., Nilsson, M., Saksena, M.: A survey of regular model checking. In: Gardner, P., Yoshida, N. (eds.) CONCUR 2004. LNCS, vol. 3170, pp. 35–48. Springer, Heidelberg (2004). https://doi.org/10.1007/978-3-540-28644-8_3
2. Alur, R., Mamouras, K., Ulus, D.: Derivatives of quantitative regular expressions. In: Aceto, L., Bacci, G., Bacci, G., Ingólfsdóttir, A., Legay, A., Mardare, R. (eds.) Models, Algorithms, Logics and Tools. LNCS, vol. 10460, pp. 75–95. Springer, Cham (2017). https://doi.org/10.1007/978-3-319-63121-9_4
3. Ball, T., Levin, V., Rajamani, S.K.: A decade of software model checking with SLAM. Commun. ACM **54**(7), 68–76 (2011)
4. Ball, T., Podelski, A., Rajamani, S.K.: Boolean and cartesian abstraction for model checking C programs. In: Margaria, T., Yi, W. (eds.) TACAS 2001. LNCS, vol. 2031, pp. 268–283. Springer, Heidelberg (2001). https://doi.org/10.1007/3-540-45319-9_19
5. Biere, A., Cimatti, A., Clarke, E.M., Strichman, O., Zhu, Y.: Bounded model checking. Adv. Comput. **58**, 117–148 (2003)
6. Bryant, R.E.: Graph-based algorithms for boolean function manipulation. IEEE Trans. Computers **35**(8), 677–691 (1986)
7. Brzozowski, J.A.: Derivatives of regular expressions. J. ACM **11**(4), 481–494 (1964)
8. Campbell, R.H., Habermann, A.N.: The specification of process synchronization by path expressions. In: Gelenbe, E., Kaiser, C. (eds.) OS 1974. LNCS, vol. 16, pp. 89–102. Springer, Heidelberg (1974). https://doi.org/10.1007/BFb0029355
9. Clarke, E.M., Emerson, E.A.: Design and synthesis of synchronization skeletons using branching time temporal logic. In: Kozen, D. (ed.) Logic of Programs 1981. LNCS, vol. 131, pp. 52–71. Springer, Heidelberg (1982). https://doi.org/10.1007/BFb0025774
10. Clarke, E.M., Henzinger, T.A., Veith, H., Bloem, R. (eds.): Handbook of Model Checking. Springer, Cham (2018). https://doi.org/10.1007/978-3-319-10575-8
11. Clarke, E.M., Klieber, W., Nováček, M., Zuliani, P.: Model checking and the state explosion problem. In: Meyer, B., Nordio, M. (eds.) LASER 2011. LNCS, vol. 7682, pp. 1–30. Springer, Heidelberg (2012). https://doi.org/10.1007/978-3-642-35746-6_1
12. Cousot, P.: Méthodes itératives de construction et d'approximation de points fixes d'opérateurs monotones sur un treillis, analyse sémantique de programmes (in French). Thèse d'Étatès sciences mathématiques, Université de Grenoble Alpes, Grenoble, France, 21 March 1978
13. Cousot, P.: Partial completeness of abstract fixpoint checking. In: Choueiry, B.Y., Walsh, T. (eds.) SARA 2000. LNCS (LNAI), vol. 1864, pp. 1–25. Springer, Heidelberg (2000). https://doi.org/10.1007/3-540-44914-0_1
14. Cousot, P., Cousot, R.: Abstract interpretation: a unified lattice model for static analysis of programs by construction or approximation of fixpoints. In: POPL, pp. 238–252. ACM (1977)
15. Cousot, P., Cousot, R.: Constructive versions of Tarski's fixed point theorems. Pac. J. Math. **81**(1), 43–57 (1979)

16. Cousot, P., Cousot, R.: Systematic design of program analysis frameworks. In: POPL, pp. 269–282. ACM Press (1979)
17. Cousot, P., Cousot, R.: Temporal abstract interpretation. In: POPL, pp. 12–25. ACM (2000)
18. Crafa, S., Ranzato, F., Tapparo, F.: Saving space in a time efficient simulation algorithm. Fundam. Inform. **108**(1–2), 23–42 (2011)
19. Cyphert, J., Breck, J., Kincaid, Z., Reps, T.: Refinement of path expressions for static analysis. PACMPL **3**(POPL), 45 (2019)
20. Giacobazzi, R., Quintarelli, E.: Incompleteness, counterexamples, and refinements in abstract model-checking. In: Cousot, P. (ed.) SAS 2001. LNCS, vol. 2126, pp. 356–373. Springer, Heidelberg (2001). https://doi.org/10.1007/3-540-47764-0_20
21. Giacobazzi, R., Ranzato, F.: Incompleteness of states w.r.t. traces in model checking. Inf. Comput. **204**(3), 376–407 (2006)
22. Heizmann, M., Hoenicke, J., Podelski, A.: Software model checking for people who love automata. In: Sharygina, N., Veith, H. (eds.) CAV 2013. LNCS, vol. 8044, pp. 36–52. Springer, Heidelberg (2013). https://doi.org/10.1007/978-3-642-39799-8_2
23. Hopcroft, J.E., Motwani, R., Ullman, J.D.: Introduction to Automata Theory, Languages, and Computation - International Edition, 2nd edn. Addison-Wesley, Boston (2003)
24. King, J.C.: Symbolic execution and program testing. Commun. ACM **19**(7), 385–394 (1976)
25. Kleene, S.C.: Representation of events in nerve nets and finite automata. In: Shannon, C.D., McCarthy, J. (eds.) Automata Studies, pp. 3–42. Princeton University Press, Princeton (1951)
26. Kripke, S.A.: Semantical considerations on modal logic. Acta Philosophica Fennica **16**, 83–94 (1963). Proceedings of a Colloquium on Modal and Many-Valued Logics, Helsinki, 23–26 August 1962
27. Kupferman, O.: Automata theory and model checking. In: Clarke, E., Henzinger, T., Veith, H., Bloem, R. (eds.) Handbook of Model Checking, pp. 107–151. Springer, Cham (2018). https://doi.org/10.1007/978-3-319-10575-8_4
28. Lichtenstein, O., Pnueli, A.: Checking that finite state concurrent programs satisfy their linear specification. In: POPL, pp. 97–107. ACM Press (1985)
29. Mallios, Y., Bauer, L., Kaynar, D., Ligatti, J.: Enforcing more with less: formalizing target-aware run-time monitors. In: Jøsang, A., Samarati, P., Petrocchi, M. (eds.) STM 2012. LNCS, vol. 7783, pp. 17–32. Springer, Heidelberg (2013). https://doi.org/10.1007/978-3-642-38004-4_2
30. Mauborgne, L.: Binary decision graphs. In: Cortesi, A., Filé, G. (eds.) SAS 1999. LNCS, vol. 1694, pp. 101–116. Springer, Heidelberg (1999). https://doi.org/10.1007/3-540-48294-6_7
31. Miller, S.P., Whalen, M.W., Cofer, D.D.: Software model checking takes off. Commun. ACM **53**(2), 58–64 (2010)
32. Owens, S., Reppy, J.H., Turon, A.: Regular-expression derivatives re-examined. J. Funct. Program. **19**(2), 173–190 (2009)
33. Queille, J.P., Sifakis, J.: Specification and verification of concurrent systems in CESAR. In: Dezani-Ciancaglini, M., Montanari, U. (eds.) Programming 1982. LNCS, vol. 137, pp. 337–351. Springer, Heidelberg (1982). https://doi.org/10.1007/3-540-11494-7_22
34. Sakarovitch, J.: Elements of Automata Theory. Cambridge University Press, Cambridge (2009)
35. Schneider, F.B.: Enforceable security policies. ACM Trans. Inf. Syst. Secur. **3**(1), 30–50 (2000)
36. Wolper, P.: Temporal logic can be more expressive. Inf. Control **56**(1/2), 72–99 (1983)

Verification by Construction of Distributed Algorithms

Dominique Méry(✉)

Université de Lorraine, LORIA UMR CNRS 7503,
Campus scientifique - BP 239, 54506 Vandœuvre-lès-Nancy, France
dominique.mery@loria.fr
http://members.loria.fr/Mery

Abstract. The verification of distributed algorithms is a challenge for formal techniques supported by tools, as model checkers and proof assistants. The difficulties, even for powerful tools, lie in the derivation of proofs of required properties, such as safety and eventuality, for distributed algorithms. Verification by construction can be achieved by using a formal framework in which models are constructed at different levels of abstraction; each level of abstraction is refined by the one below, and this refinement relationships is documented by an abstraction relation namely a gluing invariant. The highest levels of abstraction are used to express the required behavior in terms of the problem domain and the lowest level of abstraction corresponds to an implementation from which an efficient implementation can be derived automatically. In this paper, we describe a methodology based on the general concept of refinement and used for developing distributed algorithms satisfying a given list of safety and liveness properties. The modelling methodology is defined in the Event-B modelling language using the IDE Rodin.

Keywords: Correct-by-construction · Modelling · Refinement · Distributed algorithms · Verification · Proof assistant

1 Introduction

The verification of distributed algorithms is a challenge for formal techniques supported by tools, as model checkers and proof assistants. The difficulties, even for powerful tools, lie in the derivation of proofs of required properties, such as safety and eventuality, for distributed algorithms. Verification by construction can be achieved by using a formal framework in which models are constructed at different levels of abstraction; each level of abstraction is refined by the one below, and this refinement relationships is documented by an abstraction relation namely a gluing invariant. The highest levels of abstraction are used to express the required behavior in terms of the problem domain and the lowest level of

This work was supported by grant ANR-17-CE25-0005 (The DISCONT Project http://discont.loria.fr) from the Agence Nationale de la Recherche (ANR).

R. M. Hierons and M. Mosbah (Eds.): ICTAC 2019, LNCS 11884, pp. 22–38, 2019.
https://doi.org/10.1007/978-3-030-32505-3_2

abstraction corresponds to an implementation from which an efficient implementation can be derived automatically. In this paper, we describe a methodology based on the general concept of refinement and used for developing distributed algorithms satisfying a given list of safety and liveness properties. The modelling methodology is defined in the Event-B modelling language using the IDE Rodin. More precisely, we show how Event-B models can be developed for specific problems and how they can be simply reused by controlling the composition of state-based models through the refinement relationship. Following Polya [34], we have identified patterns [29] expressed as Event-B models which can be replayed for developing distributed algorithms from requirements. Consequently, we obtain a (re)development of *correct-by-construction* existing distributed algorithms and a framework for deriving new distributed algorithms (by integrating models) and for ensuring the correctness of resulting distributed algorithms by construction. We illustrate our methodology using classical problems as communication in a network, leader election protocol, self-stabilisation. Patterns are guiding the derivation of solutions from problems and we are introducing a new pattern in the case of the dynamic networks. We illustrate the methodology on algorithms dealing with dynamic topology as for instance the management of spanning trees [14] where the leader election process is still possible. The development in [18] has already addressed the case study of using patterns [21] in Event-B for distributed algorithms operating in a dynamic network and our current work is considering a simpler way to develop the protocols acting whil the network is modified and by niot using the pattern plugin. Our pattern is in fact a technique based on the observation of invariant and on the properties of the underlying structures. In our case, we have made a constant as a variable and we have played with the property of being a forest. Let us recall that the main idea is to guide the user to show how a pattern or a recipe can be used for deriving a new correct algorithmic solution from a previous development by reusing as much as possible previous proofs required by our refinement-based technique. Section 2 summarizes related works and modelling techniques integrating patterns. Section 3 shortly describes the Event-B modelling language and its IDE Rodin [2] as well as tools added to Rodin as plugins. In Sect. 4, we review the service as event paradigm which is expressing principles used for developing distributed algorithms and we summarize two patterns (the distributed pattern and the PCAM pattern) used for developing the previous problems. In Sect. 5, we add a new pattern, which is making a simple transformation of static properties to variables of the model. Finally, Sect. 6 summarizes the technique for developing the distributed algorithm and its proof of correctness. We discuss possible further developments and works on distributed algorithms using refinement.

2 Patterns as Methodological Supports

Patterns [20] have greatly improved the development of programs and software by identifying practices that could be replayed and reused in different software projects. Moreover, they help to communicate new and robust solutions for developing a software for instance; it is clear that design patterns are a set of recipes that are improving software production. When developing (formal) system models, we are waiting for adequate patterns for developing models and later for translating models into programs or even software. Abrial et al. [21] have already addressed the definition of patterns in the Event-B modelling language and have proposed a plugin which is implementing the instantiation of a pattern. Cansell et al. [10] propose a way to reuse and to instantiate patterns. Moreover, patterns intends to make the refinement-based development simpler and the tool BART [16] provides commands for automatic refinement using the AtelierB toolbox [15]. The BART process is rule-based so that the user can *drive* refinement. We aim to develop patterns which are following Pólya's approach in a smooth application of Event-B models corresponding to classes of problems to solve as for instance an iterative algorithm, a recursive algorithm [27], a distributed algorithm ... Moreover, no plugin is necessary for applying our patterns [29], which are organized with respect to paradigms identified in our refinement-based development. A paradigm [29] is a distinct set of patterns, including theories, research methods, postulates, and standards for what constitutes legitimate contributions to designing programs. A pattern [29] for modelling in Event-B is a set (project) of contexts and machines that have parameters as sets, constants, variables ... The notion of pattern has been introduced progressively in the Event-B process for improving the derivation of formal models and for facilitating the task of the person who is developing a model. In our work, students are the main target for testing and using these patterns. Our definition is very general but we do not want a very precise definition since the notion of pattern should be as simple as possible and should be helpful. We [29] have listed and documented a list of paradigms as the inductive paradigm, the call-as-event paradigm and the service-as-event paradigm; each paradigm gathers parametrized patterns which can be applied for developing a given algorithmic solution.

3 The Modelling Framework: Event-B for Step-Wise Development

This section describes the essential components of the modelling framework. In particular, we will use the Event-B modelling language [1] for modelling systems in a progressive way. Event-B has two main components: *context* and *machine*. A *context* is a formal static structure that is composed of several other clauses, such as *carrier sets*, *constants*, *axioms* and *theorems*. A *machine* is a formal structure composed of *variables, invariants, theorems, variants* and *events*; it expresses state-related properties. A machine and a context can be connected with the *sees* relationship.

Events play an important role for modelling the functional behaviour of a system and are observed. An event is a state transition that contains two main components: *guard* and *action*. A *guard* is a predicate based on the state variables that defines a necessary condition for enabling the event. An *action* is also a predicate that allows modifying the state variables when the given guard becomes true. A set of invariants defines required safety properties that must be satisfied by all the defined state variables. There are several proof obligations, such as invariant preservation, non-deterministic action feasibility, guard strengthening in refinements, simulation, variant, well-definiteness, that must be checked during the modelling and verification process.

Event-B allows us modelling a complex system gradually using *refinement*. The refinement enables us to introduce more detailed behaviour and the required safety properties by transforming an abstract model into a concrete version. At each refinement step, events can be refined by: (1) keeping the event as it is; (2) splitting an event into several events; or (3) refining by introducing another event to maintain state variables. Note that the refinement always preserves a relation between an abstract model and its corresponding concrete model. The newly generated proof obligations related to refinement ensures that the given abstract model is correctly refined by its concrete version. Note that the refined version of the model always reduces the degree of non-determinism by strengthening the guards and/or predicates. The modelling framework has a very good tool support (Rodin) for project management, model development, conducting proofs, model checking and animation, and automatic code generation. There are numerous publications and books available for an introduction to Event-B and related refinement strategies [1].

Since models may generate very *tough* proof obligations to automatically discharge, the development of proved models can be improved by the refinement process. The key idea is to combine models and elements of requirements using the refinement. The refinement [7,8] of a machine allows us to enrich a model in a *step-by-step* approach, and is the foundation of our *correct-by-construction* approach. Refinement provides a way to strengthen the invariant and to add details to a model. It is also used to transform an abstract model into a more concrete version by modifying the state description. This is done by extending the list of state variables, by refining each abstract event into a corresponding concrete version, and by adding new events. The next diagram illustrates the refinement-based relationship among events and models:

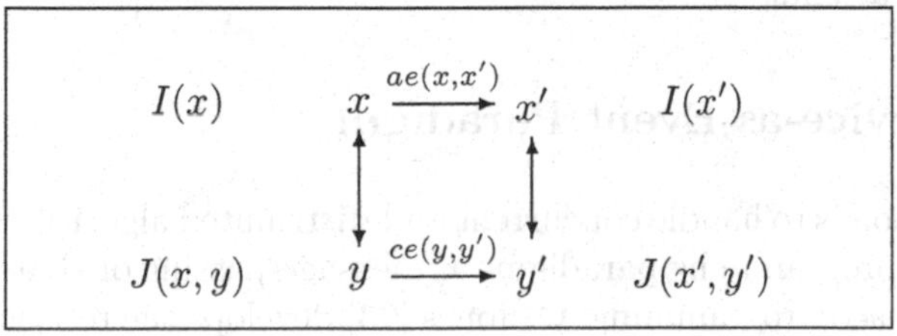

We suppose that an abstract model AM with variables x and invariant $I(x)$ is refined by a concrete model CM with variables y and gluing invariant $J(x, y)$. The abstract state variables, x, and the concrete ones, y, are linked together by means of the, so-called, *gluing invariant* $J(x, y)$. A number of proof obligations ensure that (1) each abstract event of AM is correctly refined by its corresponding concrete version of CM, (2) each new event of CM refines *skip*, which is intending to model *hidden* actions over variables appearing in the refinement model CM. More formally, if $BA(ae)(x, x')$ and $BA(ce)(y, y')$ are respectively the abstract and concrete before-after predicates of events, we say that ce in CM refines ae in AM or that ce simulates ae, if one proves the following statement corresponding to proof obligation: $I(x) \wedge J(x,y) \wedge BA(ce)(y,y') \Rightarrow \exists x' \cdot (BA(ae)(x,x') \wedge J(x',y'))$. To summarise, refinement guarantees that the set of traces of the abstract model AM contains (modulo stuttering) the traces of the concrete model CM.

The next diagram summarises links between contexts (CC extends AC); AC defines the set-theoretical logical and problem-based theory of level i called Th_i, which is extended by the set-theoretical logical and problem-based theory of level i called Th_{i+1}, which is defined by CC. Each machine (AM, CM) sees set-theoretical and logical objects defined from the problem statement and located in the CONTEXTS models (AC, CC). The abstract model AM of the level i is refined by CM; state variables of AM is x and satisfies the invariant $I(x)$. The refinement of AM by CM is checking the invariance of $J(x,y)$ and does need to prove the invariance of $I(x)$, since it is obtained freely from the checking of AM.

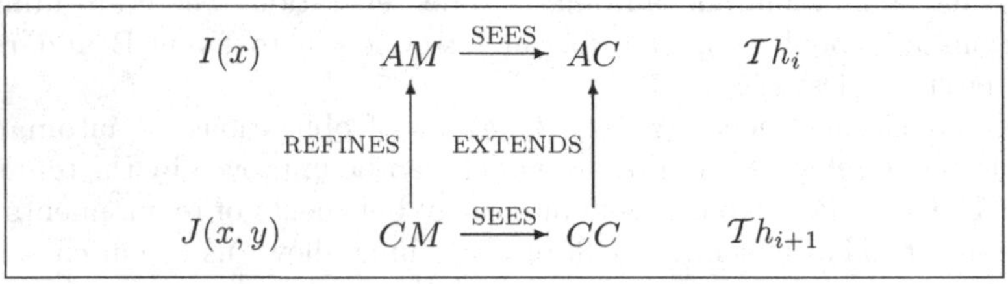

The management of proof obligations is a technical task supported by the Rodin tool [2], which provides an environment for developing correct-by-construction models for software-based systems according to the diagram. Moreover, the Rodin platform integrates ProB, a tool for animating EVENT-B models and for model-checking finite configurations of EVENT-B models at different steps of refinement. ProB is used for checking deadlock-freedom and for helping in the discovery of invariants.

4 The Service-as-Event Paradigm

The next question is to handle concurrent and distributed algorithms corresponding to different programming paradigms as message-passing or shared-memory or coordination-based programming. C. Jones [23] develops the rely/guarantee concept for handling (possible and probably wanted) interferences among sequential programs. Rely/Guarantee intends to make *implicit* [4,9] interferences as well as cooperation proofs in a proof system. In other methods as Owicki and Gries [33], the management of non-interference proofs among annotated processes leads to an important amount of extra proof obligations: checking interference freeness is explicitly expressed in the inferences rules. When considering an event as modelling a call of function or a call of a procedure, we implicitly express a computation and a sequence of states. We [31] propose a temporal extension of

Event-B to express liveness properties. The extension is a small bridge between Event-B and TLA/TLA$^+$ [24] with a refinement perspective. As C. Jones in rely/guarantee, we express implicit properties of the environment on the protocol under description by extending the call-as-event paradigm by a service-as-event paradigm. In [5,6], the service-as-event paradigm is explored on two different classes of distributed programs/algorithms/applications: the snapshot problem and the self-healing P2P by Marquezan et al. [26]. The self-healing problem is belonging to the larger class of self-$\star$ systems [17].

We [29] identify one event which *simulates* the execution of an algorithm either as an iterative version or as a recursive version and we separate the problem to solve into three problem domains: the domain for expressing pre/post specifications, the domain of Event-B models and the domain of programs/algorithms. The translation function generates effective algorithms producing the same traces of states. We are introducing patterns which are representatives of the service-as-event paradigm.

4.1 The PCAM Pattern

Coordination [13] is a paradigm that allows programmers to develop distributed systems; web services are using this paradigm for organising interactions among services and processes. In parallel programming, coordination plays also a central role and I. Foster [19] has proposed the PCAM methodology for designing concurrent programs from a problem statement: PCAM emphasizes a decomposition into four steps corresponding to analysis of the problem and leading to a machine-independant solution. Clearly, the goal of I. Foster is to make concurrent programming based on abstractions, which are progressively adding details leading to specific concurrent programming notation as, for instance MPI (http://www.open-mpi.org/). The PCAM methodology identifies four distinct stages corresponding to a Partition of identified tasks from the problem statement and which are concurrently executed. A problem is possibly an existing complex C or Fortran code for a computing process requiring processors and concurrent executions. Communication is introduced by an appropriate coordination among tasks and then two final steps, Agglomeration and Mapping complete the methodology steps. The PCAM methodology includes features related to the functional requirements in the two first stages and to the implementation in the two last stages. I. Foster has developed the PCAM methodology together with tools for supporting the implementation of programs on different architectures. The success of the design is mainly due to the coordination paradigm which allows us to freely organise the stages of the development.

The PCAM methodology includes features related to the functional requirements in the two first stages and to the implementation in the two last stages. The general approach is completely described in [28].

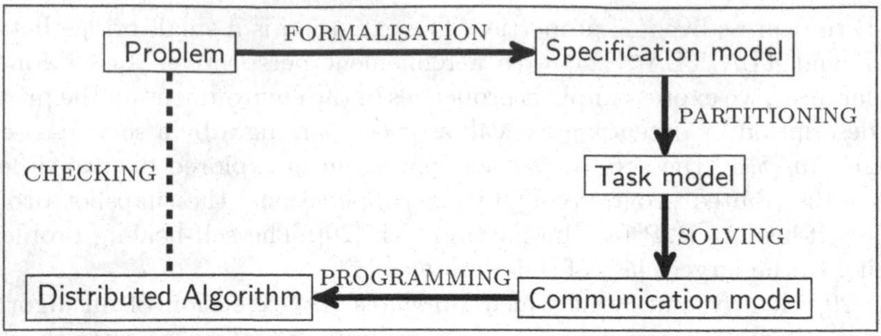

We consider the two first stages (Partitioning, Communication) for producing state-based models satisfying functional requirements and which will be a starting point for generating a concurrent program following the AM last suffix. We have described a general methodology for developing correct by construction concurrent algorithms and we have developed a solution specified by a unique event.

4.2 The Distributed Pattern

The main idea is to analyse a problem as a pre/post specification which is then refined by a machine corresponding to the simulation of a recursive function or procedure. The class of algorithms is the class of sequential algorithms and there is no concurrent or distributed interpretation of an event. However, an event can be observed in a complex environment. The environment may be active and should be expressed by a set of events which are simulating the environment. Since the systems under consideration are reactive, it means that we should be able to model a service that a system should ensure. For instance, a communication protocol is a service which allows to transfer a file of a process A into a file of a process B.

Figure 1 sketches the distributed pattern. The machine SERVICE is modelling services of the protocol; the machine PROCESS is refining each service considered as an event and makes the (computing) process explicit. The machine COMMUNICATION is defining the communications among the different agents of the possible network. Finally the machine LOCALALGO is localizing events of the protocol. The distributed pattern is used for expressing *phases* of the target distributed algorithm (for instance, requesting mutual exclusion) and to have a separate refinement of each phase. We sketch the service-as-event paradigm as follows. We consider one service. The target algorithm $\mathcal{A}$ is first described by a machine M0 with variables x satisfying the invariant $I(x)$.

The first step is to list the services e $S \mathrel{\widehat{=}} \{s_0, s_1, \ldots, s_m\}$ provided by the algorithm $\mathcal{A}$ and to state for each service s_i a liveness property $P_i \leadsto Q_i$. We characterise by $\Phi_0 \mathrel{\widehat{=}} \{P_0 \leadsto Q_0, P_1 \leadsto Q_1, \ldots, P_m \leadsto Q_m\}$. We add a list of safety properties defined by $\Sigma_0 = \{Safety_0, Safety_1, \ldots, Safety_n\}$. An event is defined for each liveness property and standing for the eventuality of e by a fairness assumption which is supposed on e. Liveness properties can be visualised by assertions diagrams helping to understand the relationship among phases.

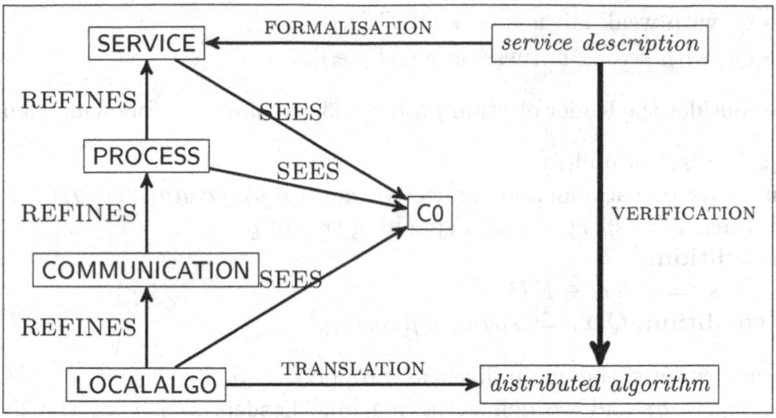

Fig. 1. The distributed pattern

The second step is its refinement M1 with variables y glued properties in by $J(x, y)$ using the Event-B refinement and using the REF refinement which is defined using the temporal proof rules for expanding liveness properties. $P \leadsto Q$ in Φ_0 is proved from a list of Φ_1 using temporal rules. For instance, $P \leadsto Q$ in Φ_0 is then refined by $P \leadsto R, R \leadsto Q$, if $P \leadsto R, R \leadsto Q \vdash P \leadsto Q$. If we consider C as the context and M as the machine, C, M satisfies $P \leadsto Q$ and C, M satisfies $\Box Safety$. We use a temporal semantics relating contexts, machines and properties [31]. The link called LIVE expresses the satisfaction relationship. The next diagram is summarising the relationship among models.

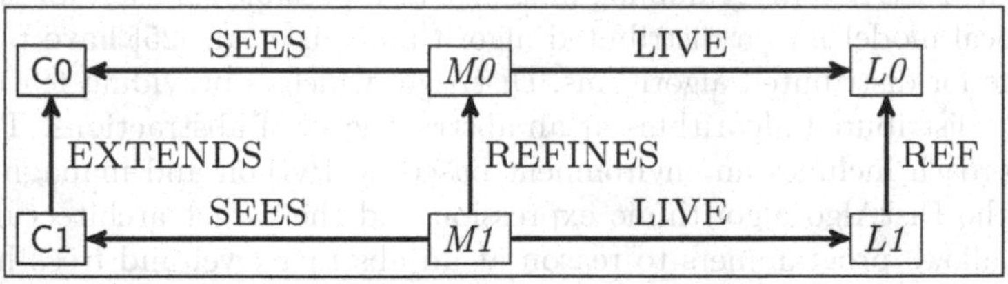

Liveness properties can be gathered in *assertions diagrams*. For instance, $P \xrightarrow{e} Q$ means that

- $\forall x, x' \cdot P(x) \wedge I(x) \wedge BA(e)(x, x') \Rightarrow Q(x')$
- $\forall x \cdot P(x) \wedge I(x) \Rightarrow (\exists x' \cdot BA(e)(x, x'))$
- $\forall f \neq e \cdot \forall x, x' \cdot P(x) \wedge I(x) \wedge BA(f)(x, x') \Rightarrow (P(x') \vee Q(x'))$

$P \xrightarrow{e} Q$ expresses implicitly that tyhe event e is under weak fairness. Each liveness property $P_i \leadsto Q_i$ in Φ_0 is modelled by an event:

$$\textbf{EVENT } \mathsf{e}_i \mathrel{\widehat{=}} \textbf{WHEN } P_i(x) \textbf{ THEN } x : |Q_i(x') \textbf{ END}$$

We can add some fairness assumption over the event:

- $P_i \xrightarrow{\mathsf{e_i}} Q_i$ with weak fairness on e ($WF_x(\mathsf{e_i})$),
- $P_i \stackrel{\mathsf{e_i}}{\Rightarrow} Q_i$, with strong fairness on e ($SF_x(\mathsf{e_i})$).

If we consider the leader election protocol [3], we have the following elements:

- **Sets:** ND (set of nodes).
- **Constants:** g is acyclic and connected ($acyclic(g) \wedge connected(g)$).
- **Variables:** $x = (sp, rt)$ (sp is a spanning tree of g).
- **Precondition:**
  $P(x) \mathrel{\widehat{=}} sp = \varnothing \wedge rt \in ND$
- **Postcondition:** $Q(x) \mathrel{\widehat{=}} spanning(sp, rt, g)$

We can express the main liveness property: $(sp = \varnothing \wedge rt \in ND) \leadsto spanning(sp, rt, g)$ and we define the machine $\mathsf{Leader_0}$ satisfying the liveness property:

```
EVENT election0 ≙
  BEGIN
    sp, rt : |spanning(sp′, rt′, g)
  END
```

$$\mathsf{C_0} \xleftarrow{\mathsf{SEES}} \mathsf{Leader_0} \xrightarrow{\mathsf{LIVE}} (WF_x(\mathsf{election_0}), \{P \leadsto Q\})$$

We have introduced the service specification which should be refined separately from events of the machine *M0*. The next refinement should first introduce details of a computing process and then introduce communications in a very abstract way. The last refinement intends to localise the events. The model LOCALALGO is in fact an expression of a distributed algorithm. A current work explores the DistAlgo programming language as a possible solution for translating the local model into a distributed algorithm. Liu et al. [25] have proposed a language for distributed algorithms, DistAlgo, which is providing features for expressing distributed algorithms at an abstract level of abstractions. The DistAlgo approach includes an environment based on Python and managing links between the DistAlgo algorithmic expression and the target architecture. The language allows programmers to reason at an abstract level and frees her/him from architecture-based details. According to experiments of authors with students, DistAlgo improves the development of distributed applications. From our point of view, it is an application of the coordination paradigm based on a given level of abstraction separating the concerns.

4.3 Applying the Distributed Pattern

The distributed pattern (Fig. 1) is applied for the famous *sliding window protocol.* The *service description* is expressing that a process P is sending a file IN to a process Q and the received file is stored in a variable OUT. The service is simply expressed by the liveness property $(at(P, s) \wedge IN \in 0..n \rightarrow D) \leadsto (at(Q, r) \wedge OUT = IN)$ and the event **EVENT** communication $\mathrel{\widehat{=}}$ **WHEN** $at(P, s) \wedge IN \in 0..n \rightarrow D$ **THEN** $OUT := IN$ **END** is defining the service. $at(P, s)$

means that P is at the sending statement called s and $at(Q, r)$ means that Q is at the receiving statement r. The context C0 and the machine SERVICE are defined in Fig. 1. The next step is to decompose the liveness property using one of the possible inference rules of the temporal framework as transitivity, induction, confluence of the *leadsto* operatior. In this configuration, we have to introduce the computation process which is simulating the protocol. Obviously, we use and induction rule to express that te file *IN* is sent item per item and we introduce sending and receiving events and the sliding events. In the new machine PROTOCOL, variables are OUT, i, chan, ack, got and satisfied the following invariant:

INVARIANTS
$inv1 : OUT \in \mathbb{N} \nrightarrow D$
$inv2 : i \in 0 \,..\, n+1$
$inv3 : 0 \,..\, i-1 \subseteq dom(OUT) \wedge dom(OUT) \subseteq 0 \,..\, n$
$inv7 : chan \in \mathbb{N} \nrightarrow D$
$inv8 : ack \subseteq \mathbb{N}$
$inv9 : got \subseteq i \,..\, i+l \cap 0 \,..\, n$
$inv10 : got \subseteq \mathbb{N}$
$inv12 : dom(chan) \subseteq 0 \,..\, i+l \cap 0 \,..\, n$
$inv13 : got \subseteq dom(OUT)$
$inv14 : ack \subseteq dom(OUT)$
$inv16 : 0 \,..\, i-1 \lhd OUT = 0 \,..\, i-1 \lhd IN$
$inv17 : chan \subseteq IN$
$inv18 : OUT \subseteq IN$
$inv19 : ack \subseteq 0 \,..\, i+l \cap 0 \,..\, n$

Name	Total	Auto	Inter
protocol-swp	124	101	23
C0	1	1	0
SERVICE	4	2	20
PROCESS	63	51	12
WINDOW	19	13	6
BUFFER	21	18	3
LOCAL	16	16	0

The variable *got* is simulating a window identified by the values between i and i+l in the variables chan, got and ack. The sliding window is in fact defined by the variable i which is sliding or incrementing, when the value OUT(i) is received or equivalently when *iinack*. The events are send, receive, receiveack, sliding together with events which are modelling possible loss of messages. The machine PROCESS is simulating the basic mechanism of the sliding window protocol and is expressing the environment of the protocol. The next refinement WINDOW is introducing an explicit *window* variable satisfying the invariant $w \in \mathbb{N} \nrightarrow D \wedge w \subseteq chan \wedge dom(w) \subseteq i..i+l$. The events are enriched by guards and actions over the variable *window*. The *window* variable is still an abstract view of the window which is contained in a buffer b. The buffer b is introduced in the refinement called BUFFER. The new variable b is preparing the localisation and introduced the explicit communications: $b \in 0..l \nrightarrow D \wedge \forall k \cdot k \in dom(b) \Rightarrow i+k \in dom(w) \wedge b(k) = w(i+k) \wedge \forall h \cdot h \in dom(w) \Rightarrow h-i \in dom(b) \wedge w(h) = b(h-i)$. The visible variables of the machine are OUT, i, chan, ack, got, w and b and in the next refinement, we obtain a local model called LOCAL with OUT, i, chan, ack, got and b: the window is not part of the implementation of the protocol. The events are localised by hiding the variable w and the final model can now be transformed into the Sliding Window Protocol. The proof obligations summary shows that proof obligations for the machine PROCESS correspond to the main effort of proof, when the induction is introduced. However, we have not checked the liveness properties using the temporal proof system namely TLA and it remains to be effectively supported by the toolbox for TLA/TLA$^+$. We use the temporal proof rules to as guidelines for decomposing liveness properties while we are refining events in Event-B. The technique has been already used for developing population protocols [31] and it was also a way to deal with questions related to dynamic topologies. In the newt section, we focus on the (re-)development of the leader election while the topology is modified with respect to given operations.

5 Modifying the Topology

5.1 Problem Statement

The paper [14] is addressing the problem of building and maintaining a forest of spanning trees in highly dynamic networks. The question is to allow modifications of the topology, while maintaining the property to relain a forest which is silply to express that there is no cycle introduced by actions over topology as adding a link or deleting a link: topological events can occur at any time and any rate. Moreover, we can not assume any stable periods. The algorithm operates at a coarse-grain level, using atomic pairwise interactions as population protocol or graph relabeling systems and it works forever. The algorithm is merging trees following some rules:

- Each tree in the forest has exactly one token (also called root) that performs a random walk inside the tree, switching parent- child relationships as it crosses edges.
- When two tokens are located on both sides of a same edge, their trees are merged upon this edge and one token disappears.
- Whenever an edge that belongs to a tree disappears, its child endpoint regenerates a new token instantly.

In [14], the two following properties are proved to be satisfied by the algorithm:

- Lemma 1: At any time, there is at least one token per tree.
- Lemma 2: At any time, there is at most one token per tree.

The problem is in fact generalizing the *leader election* protocol and illustrates a protocol in a dynamic network. The distributed pattern (Fig. 1) has been used and we are proposing a refinement-based development for the algorithm. The idea is to transform a constant of the model into a variable and we are keeping the time reference as an implicit feature. The development is based on a superposition of three operations (op1, op2, op3) defining the new protocol on a dynamic network managed by two specific operations (topop1, topo2). The final model should satisfy the property proved in [14] *at any time, there is one and only token per tree.*

5.2 Scenario of the Protocol

We give a short description of the protocol operations over a graph.

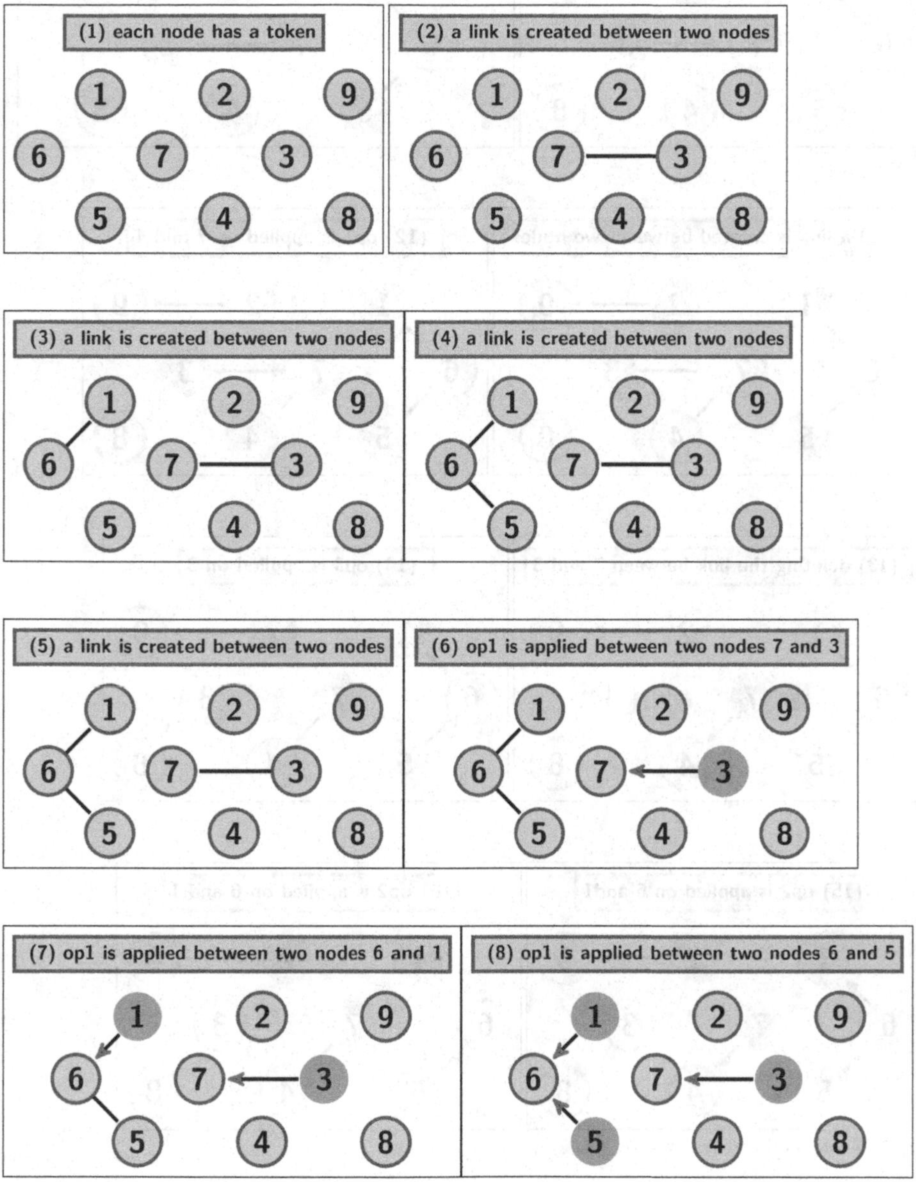

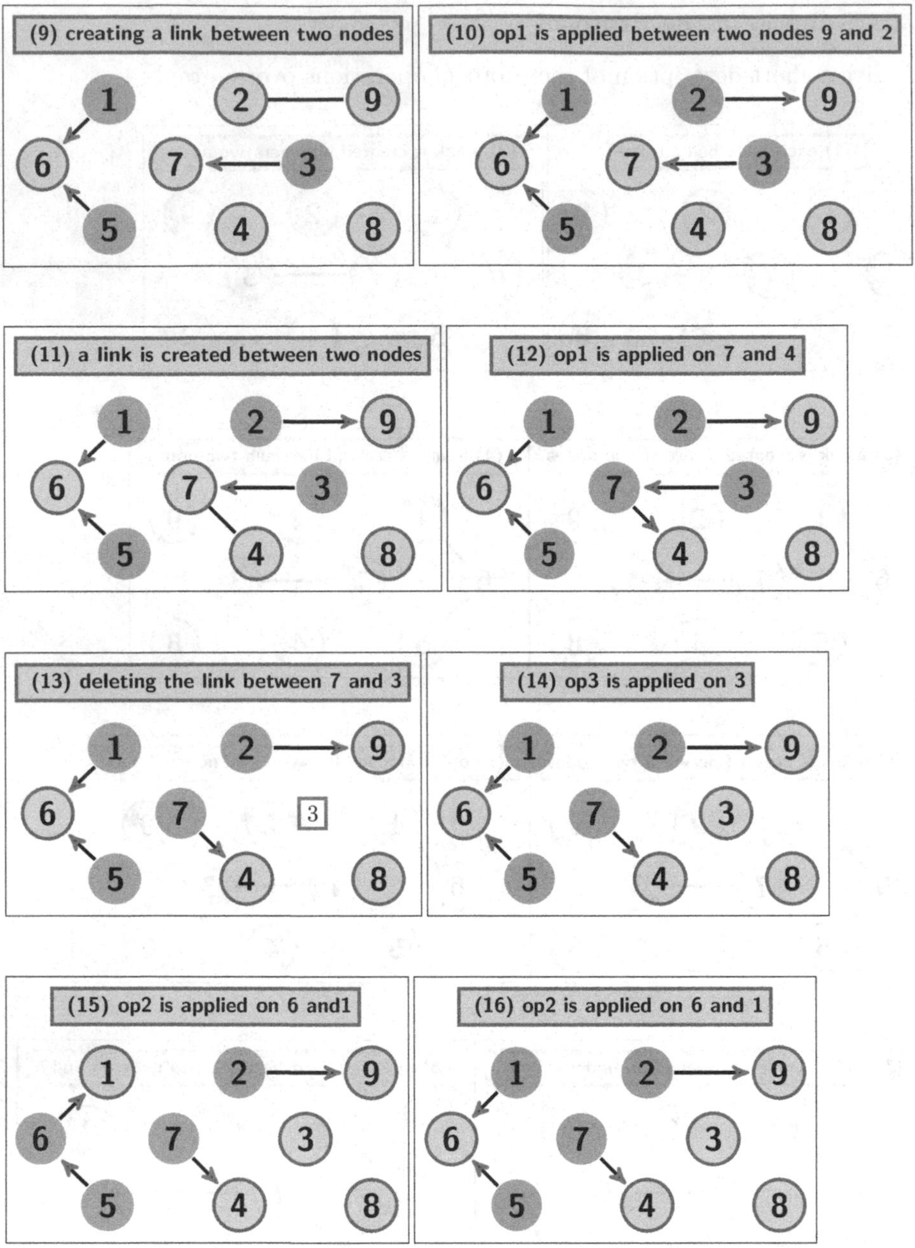

5.3 First Model Dynamic Network Management

The first machine is expressing the *environment* of the protocol to develop and we transform the constant g (see [3]) into a variable g which is an undirected graph; the variable t is modelling the forest. The event operation is modelling the abstract view of the protocol which is maintaining the invariant: the event operation is *keeping* the invariant. The two other operations are expressing the two possible operations over the graph either adding a link or deleting a link.

```
axm1 : g ∈ N ↔ N
axm2 : g = g⁻¹
axm3 : id ∩ g = ∅
inv1; : t ∈ N ⇸ N
inv3; : t ∩ t⁻¹ = ∅
inv11 : ∀S·(S ⊆ N ∧ S ⊆ t⁻¹[S] ⇒ S = ∅)
inv21 : t ⊆ g
inv22 : token ⊆ N
inv23 : dom(t) ⊆ N \ token
```

```
EVENT operation
  BEGIN
                              ( g′ ∈ N ↔ N∧
                              ( g′ = g′⁻¹∧
                              ( id ∩ g′ = ∅∧
                              ( t′ ∈ N ⇸ N∧
    act1 : g, t, token :|     ( t′ ∩ t′⁻¹ = ∅∧
                              ( ∀S·(S ⊆ N∧
                              ( S ⊆ t′⁻¹[S] ⇒ S = ∅)∧
                              ( t′ ⊆ g′ ∧ token′ ⊆ N∧
                              ( dom(t′) ⊆ N \ token′
  END
```

```
EVENT removing
  ANY
    x, y
  WHERE
    grd1 : x ↦ y ∈ g
  THEN
    act1 : g := g \ {x ↦ y, y ↦ x}
    act2 : t := t \ {x ↦ y, y ↦ x}
  END
```

```
EVENT adding
  ANY
    x, y
  WHERE
    grd1 : x ∈ N
    grd2 : y ∈ N
    grd3 : x ↦ y ∉ g
    grd4 : x ≠ y
  THEN
    act1 : g := g ∪ {x ↦ y, y ↦ x}
  END
```

5.4 Second Model Network Reconfiguration

The new model is adding three operations which are maintaining the invariant and the safety property $dom(t) \subseteq N \setminus token$. The safety property is expressing that there is at most one token per tree. However, we can not prove that there is at least one token per tree and it means that our modelling technique is not a programming technique which is what is assumed in [14]. More precisely, the operation op3 (see 14) is not *observed immediatly* after the deletion of an edge and the fact that a node is no more related to another node.

```
EVENT op1
REFINES operation
  ANY
    x, y
  WHERE
    grd1 : x ↦ y ∈ g
    grd2 : x ∉ dom(t)
    grd3 : y ∉ dom(t)
  THEN
    act1 : t := t ∪ {x ↦ y}
    act2 : token := token \ {x}
  END
```

```
EVENT op2
  REFINES operation
  ANY
    x, y
  WHERE
    grd1 : x ↦ y ∈ t
    grd2 : y ∉ dom(t)
  THEN
    act1 : t := (t \ {x ↦ y}) ∪ {y ↦ x}
    act2 : token := (token \ {y}) ∪ {x}
  END
```

```
EVENT op3
  REFINES operation
  ANY
    x
  WHERE
    grd1 : x ∉ token
    grd2 : x ∉ dom(t)
    grd3 : x ∉ dom(g)
  THEN
    act1 : token := token ∪ {x}
  END
```

Name	Total	Auto	Inter
dynamic-leader	34	29	5
C0	1	1	0
DM1	25	23	2
DM2	8	5	3

The summary of proof obligations shows that the proofs are mainly automatic. Thanks to the previous development [3] which is using properties over graph and which is describing the leader election protocol as a forest leading to a tree. Moreover, the new operations are preserving the property of the forest topology.

The next step is to refine the current model into an Event-B-like local model to get a set of rules in Visidia [36]. We have simply re-applied the properties over trees by generalizing the graph as a variable and the summary of proof obligations is very good since we have only 15% (5 out of 34) of manually discharged proof obligations.

6 Conclusion and Perspectives

The refinement-based modelling technique combines modelling and proving through discharging proof obligations and helps to discover invariants when developing distributed algorithms [3,11,12,22,29,32]. Moreover, the refinement-based methodology can integrate different computation models as the local model computation [18,30,35,36]. Our contribution aims to assist anyone who wants to obtain a completely checked Event-B project for a given problem with less toil. The toil is related to the use of the Rodin platform [2]: it is a real and useful proof companion but it requires a specific skill in proof development. Following the ideas of Pólya, we enrich the library of patterns [29] for providing guidelines for defining fully proved Event-B models, when considering problems to solve defined by explicit inductive definitions.

Our illustration is based on a protocol which is applying rules called operations which can modify the graph while the topology of the graph may evolve. We show that the operations are maintaining the property of the graph to be a forest which is expressed as an acyclic graph. Our solution is simplifying the solution that was proposed in [18]. Moreover, the use of Event-B tool supporting the patterns is not necessary. Proofs obligations are simple to verify and are mainly automatic but they are derived from properties over graphs which have been developed separatly.

Archives of Event-B projects are available at the following link http://eb2all.loria.fr and are used by students of the MsC programme at Université de Lorraine and Telecom Nancy. In a current project namely Atlas of Correct-by-Construction Distributed Algorithms, we aim to develop a complete library of correct-by-construction distributed algorithms for the main distributed algorithms that are fundamental as communication, security, computation, election ... Others links as http://rimel.loria.fr or http://visidia.labri.fr can be also used. Finally, the translation from Event-B models into a distributed algorithm should be improved and we plan to explore distributed programming languages with a high level of abstraction as for instance DistAlgo [25].

Acknowledgement. The author thanks the organizers and the chairs of the conference ICTAC for the invitation to give a keynote. He especially thanks Mohamed Mosbah, Yves Métivier, Pierre Castéran, Mohamed Tounsi and researchers who have worked in the ANR project RIMEL (http://rimel.loria.fr) and who have made distributed algorithms simpler.

References

1. Abrial, J.R.: Modeling in Event-B: System and Software Engineering. Cambridge University Press, Cambridge (2010)

2. Abrial, J.R., Butler, M., Hallerstede, S., Hoang, T.S., Mehta, F., Voisin, L.: Rodin: an open toolset for modelling and reasoning in Event-B. Int. J. Softw. Tools Technol. Transf. **12**(6), 447–466 (2010)
3. Abrial, J.R., Cansell, D., Méry, D.: A mechanically proved and incremental development of ieee 1394 tree identify protocol. Formal Asp. Comput. **14**(3), 215–227 (2003)
4. Ameur, Y.A., Méry, D.: Making explicit domain knowledge in formal system development. Sci. Comput. Program. **121**, 100–127 (2016). https://doi.org/10.1016/j.scico.2015.12.004
5. Andriamiarina, M.B., Méry, D., Singh, N.K.: Analysis of self-⋆ and P2P systems using refinement. In: Ameur, Y.A., Schewe, K. (eds.) ABZ 2014. LNCS, vol. 8477, pp. 117–123. Springer, USA (2014). https://doi.org/10.1007/978-3-662-43652-3_9
6. Andriamiarina, M.B., Méry, D., Singh, N.K.: Revisiting snapshot algorithms by refinement-based techniques. Comput. Sci. Inf. Syst. **11**(1), 251–270 (2014). https://doi.org/10.2298/CSIS130122007A
7. Back, R.: On correct refinement of programs. Int. J. Softw. Tools Technol. Transf. **23**(1), 49–68 (1979)
8. Back, R.: A calculus of refinements for program derivations. Acta Informatica **25**, 593–624 (1998)
9. Bjørner, D.: Domain analysis & description - the implicit and explicit semantics problem. In: Laleau, R., Méry, D., Nakajima, S., Troubitsyna, E. (eds.) Proceedings Joint Workshop on Handling IMPlicit and EXplicit Knowledge in Formal System Development (IMPEX) and Formal and Model-Driven Techniques for Developing Trustworthy Systems (FM&MDD), Xi'An, China, 16th November 2017. Electronic Proceedings in Theoretical Computer Science, vol. 271, pp. 1–23. Open Publishing Association (2018). https://doi.org/10.4204/EPTCS.271.1
10. Cansell, D., Gibson, J.P., Méry, D.: Formal verification of tamper-evident storage for e-voting. In: Fifth IEEE International Conference on Software Engineering and Formal Methods (SEFM 2007), London, England, UK, 10–14 September 2007, pp. 329–338. IEEE Computer Society (2007). https://doi.org/10.1109/SEFM.2007.21, https://doi.org/10.1109/SEFM.2007.21
11. Cansell, D., Méry, D.: Formal and incremental construction of distributed algorithms: on the distributed reference counting algorithm. Theor. Comput. Sci. **364**(3), 318–337 (2006). https://doi.org/10.1016/j.tcs.2006.08.015
12. Cansell, D., Méry, D.: Designing old and new distributed algorithms by replaying an incremental proof-based development. In: Abrial, J.-R., Glässer, U. (eds.) Rigorous Methods for Software Construction and Analysis. LNCS, vol. 5115, pp. 17–32. Springer, Heidelberg (2009). https://doi.org/10.1007/978-3-642-11447-2_2
13. Carriero, N., Gelernter, D.: A computational model of everything. Commun. ACM **44**(11), 77–81 (2001). https://doi.org/10.1145/384150.384165
14. Casteigts, A., Chaumette, S., Guinand, F., Pigné, Y.: Distributed maintenance of anytime available spanning trees in dynamic networks. CoRR abs/0904.3087 (2009). http://arxiv.org/abs/0904.3087
15. Clearsy System Engineering: Atelier B (2002–2019). http://www.atelierb.eu/
16. Clearsy System Engineering: BART (2010). http://tools.clearsy.com/tools/bart/
17. Dolev, S.: Self-Stabilization. MIT Press, Cambridge (2000)
18. Fakhfakh, F., Tounsi, M., Mosbah, M., Méry, D., Kacem, A.H.: Proving distributed coloring of forests in dynamic networks. Computación y Sistemas **21**(4) (2017). http://www.cys.cic.ipn.mx/ojs/index.php/CyS/article/view/2857
19. Foster, I.T.: Designing and Building Parallel Programs - Concepts and Tools for Parallel Software Engineering. Addison-Wesley, Boston (1995)

20. Gamma, E., Helm, R., Johnson, R., Vlissides, R., Gamma, P.: Design Patterns: Elements of Reusable Object-Oriented Software Design Patterns. Addison-Wesley Professional Computing, Reading (1994)
21. Hoang, T.S., Fürst, A., Abrial, J.: Event-B patterns and their tool support. Softw. Syst. Model. **12**(2), 229–244 (2013). https://doi.org/10.1007/s10270-010-0183-7
22. Hoang, T.S., Kuruma, H., Basin, D.A., Abrial, J.: Developing topology discovery in Event-B. Sci. Comput. Program. **74**(11–12), 879–899 (2009). https://doi.org/10.1016/j.scico.2009.07.006
23. Jones, C.B.: Tentative steps toward a development method for interfering programs. ACM Trans. Program. Lang. Syst. **5**(4), 596–619 (1983). https://doi.org/10.1145/69575.69577
24. Lamport, L.: The temporal logic of actions. ACM Trans. Program. Lang. Syst. **16**(3), 872–923 (1994)
25. Liu, Y.A., Stoller, S.D., Lin, B.: From clarity to efficiency for distributed algorithms. ACM Trans. Program. Lang. Syst. **39**(3), 12:1–12:41 (2017). https://doi.org/10.1145/2994595
26. Marquezan, C.C., Granville, L.Z.: Self-* and P2P for Network Management - Design Principles and Case Studies. Springer Briefs in Computer Science. Springer, London (2012). https://doi.org/10.1007/978-1-4471-4201-0
27. Méry, D.: Refinement-based guidelines for algorithmic systems. Int. J. Softw. Inform. **3**(2–3), 197–239 (2009). http://www.ijsi.org/ch/reader/view_abstract.aspx?file_no=197&flag=1
28. Méry, D.: Playing with state-based models for designing better algorithms. Future Gener. Comput. Syst. **68**, 445–455 (2017). https://doi.org/10.1016/j.future.2016.04.019
29. Méry, D.: Modelling by patterns for correct-by-construction process. In: Margaria, T., Steffen, B. (eds.) ISoLA 2018. LNCS, vol. 11244, pp. 399–423. Springer, Cham (2018). https://doi.org/10.1007/978-3-030-03418-4_24
30. Méry, D., Mosbah, M., Tounsi, M.: Refinement-based verification of local synchronization algorithms. In: Butler, M., Schulte, W. (eds.) FM 2011. LNCS, vol. 6664, pp. 338–352. Springer, Heidelberg (2011). https://doi.org/10.1007/978-3-642-21437-0_26
31. Méry, D., Poppleton, M.: Towards an integrated formal method for verification of liveness properties in distributed systems: with application to population protocols. Softw. Syst. Model. **16**(4), 1083–1115 (2017). https://doi.org/10.1007/s10270-015-0504-y
32. Méry, D., Singh, N.K.: Analysis of DSR protocol in Event-B. In: Défago, X., Petit, F., Villain, V. (eds.) SSS 2011. LNCS, vol. 6976, pp. 401–415. Springer, Heidelberg (2011). https://doi.org/10.1007/978-3-642-24550-3_30
33. Owicki, S., Gries, D.: An axiomatic proof technique for parallel programs I. Acta Informatica **6**, 319–340 (1976)
34. Polya, G.: How to Solve It, 2nd edn. Princeton University Press, Princeton (1957). ISBN 0-691-08097-6
35. Tounsi, M., Mosbah, M., Méry, D.: Proving distributed algorithms by combining refinement and local computations. ECEASST **35** (2010). https://doi.org/10.14279/tuj.eceasst.35.442
36. ViSiDiA (2006–2019). http://visidia.labri.fr

Models and Transition Systems

The Linear Time-Branching Time Spectrum of Equivalences for Stochastic Systems with Non-determinism

Arpit Sharma(✉)

Department of Electrical Engineering and Computer Science, Indian Institute of Science Education and Research Bhopal, Bhopal, India
arpit@iiserb.ac.in

Abstract. This paper studies linear time-branching time spectrum of equivalences for interactive Markov chains (IMCs). We define several variants of trace equivalence by performing button pushing experiments on stochastic trace machines. We establish the relation between these equivalences and also compare them with bisimulation for IMCs. Next, we define several variants of stutter trace equivalence for IMCs. We perform button pushing experiments with stutter insensitive stochastic trace machines to obtain these equivalences. We investigate the relationship among these stutter equivalences and also compare them with weak bisimulation for IMCs. Finally, we discuss the relation between several strong and weak equivalences defined in this paper.

Keywords: Markov · Equivalence · Trace · Stutter · Bisimulation · Stochastic

1 Introduction

Equivalence relations are widely used for comparing and relating the behavior of system models. For example, equivalences have been used to efficiently check if the implementation is an approximation of specification of the expected behavior. Additionally, equivalences are also used for reducing the size of models by combining equivalent states into a single state. The reduced state space obtained under an equivalence relation, called a quotient system, can then be used for formal verification provided it preserves a rich class of the properties of interest [4,5,22]. For non-deterministic and probabilistic models, one usually distinguishes between linear time and branching time equivalence relations [4–6,34]. Trace equivalence is one of the most widely used equivalence relations to compare the linear time behavior of models. For non-deterministic models, two states are trace equivalent if the sets of words starting from these states are the same [4,34]. Additionally, for Markov chains, similar words need to have the same probability [4,6]. Similarly, in the weak setting, stutter trace equivalence has been proposed where two words are said to be equivalent if they differ in at most the number of times a set of propositions may adjacently repeat [4]. In the branching time setting, bisimulation relations are used to compare the branching time behavior of system models [4,5,20,23,26]. Bisimulation can also be used to substantially reduce the state-space

R. M. Hierons and M. Mosbah (Eds.): ICTAC 2019, LNCS 11884, pp. 41–58, 2019.
https://doi.org/10.1007/978-3-030-32505-3_3

of models to be verified [22]. The condition to exhibit identical stepwise behavior is slightly relaxed in case of simulation relations [4,5]. Stuttering variants of bisimulation and simulation pre-orders have also been defined for non-deterministic and Markovian models [4,5].

In this paper we focus on studying the linear time-branching time spectrum of equivalence relations for interactive Markov chains (IMCs) [19,20]. IMCs extend labeled transition systems (LTSs) with stochastic aspects. IMCs thus support both reasoning about non-deterministic behaviors as in LTSs [4] and stochastic phenomena as in continuous-time Markov chains (CTMCs) [3]. IMCs are compositional and are widely used for performance and dependability analysis of complex distributed systems, e.g., shared memory mutual exclusion protocols [24]. They have been used as semantic model for amongst others dynamic fault trees [10,11], architectural description languages such as AADL [9,13], generalized stochastic Petri nets [21] and Statemate [8]. They are also used for modeling and analysis of GALS (Globally Asynchronous Locally Synchronous) hardware design [14]. For analysis, an IMC is *closed*[1] followed by application of model checking algorithms [18,25] to compute the probability of linear or branching real-time objectives, e.g., extremal time-bounded reachability probabilities [20,25] and expected time [18].

We define several variants of trace equivalence for closed IMC models using button pushing experiments with stochastic trace machines. Since schedulers are used to resolve non-deterministic choices in IMCs, for every class of IMC scheduler, we define a corresponding variant of trace equivalence. Roughly speaking, two IMCs $\mathcal{I}_1, \mathcal{I}_2$ are trace equivalent (w.r.t. scheduler class $\mathcal{C}$), denoted $\equiv_{\mathcal{C}}$, if for every scheduler $\mathcal{D}$ of class $\mathcal{C}$ of $\mathcal{I}_1$ there exists a scheduler $\mathcal{D}'$ of class $\mathcal{C}$ of $\mathcal{I}_2$ such that for all outcomes/timed traces, i.e., (σ, θ), we have $P^{trace}_{\mathcal{I}_1,\mathcal{D}}(\sigma, \theta) = P^{trace}_{\mathcal{I}_2,\mathcal{D}'}(\sigma, \theta)$ and vice versa. Here, $P^{trace}_{\mathcal{I}_1,\mathcal{D}}(\sigma, \theta)$ denote the probability of all timed paths that are compatible with the outcome/timed trace (σ, θ) in $\mathcal{I}_1$ under scheduler $\mathcal{D}$. More specifically, we define six variants of trace equivalence on the basis of increasing power of schedulers, namely stationary deterministic (SD), stationary randomized (SR), history-dependent deterministic (HD), history-dependent randomized (HR), timed history-dependent deterministic (THD) and timed history-dependent randomized (THR) trace equivalence. We study the connections among these trace equivalences and also compare them with strong and weak bisimulation for IMCs [19,20].

In the weak setting, we define several variants of stutter trace equivalence for closed IMC models. We perform button pushing experiments on stutter insensitive stochastic trace machines to obtain these equivalences. We define six variants of stutter trace equivalence on the basis of increasing power of schedulers, namely stationary deterministic (SD), stationary randomized (SR), history-dependent deterministic (HD), history-dependent randomized (HR), timed history-dependent deterministic (THD) and timed history-dependent randomized (THR) stutter trace equivalence. We study the connections among these equivalences and their relationship with (strong) trace equivalence for IMCs. We also relate these stutter trace equivalences with strong and weak bisimulation for IMCs. Put in a nutshell, the major contributions of this paper are as follows:

[1] An IMC is said to be closed if it is not subject to any further synchronization.

- We define several variants of trace equivalence by experimenting with stochastic trace machines, establish connections among these equivalences and compare them with bisimulation.
- We define several variants of stutter trace equivalence by experimenting with stutter insensitive stochastic trace machines, establish connections among these equivalences and compare them with weak bisimulation.
- We also investigate the connections between strong equivalences and weak/stutter equivalences defined in this paper. Finally, we use these results to sketch the linear time-branching time spectrum of equivalences for closed IMCs.

We believe that our work is an important step forward in understanding the linear time-branching time spectrum of equivalences for models that support both non-deterministic and stochastic behavior.

1.1 Related Work

Branching Time: For continuous-time Markov chains (CTMCs), several variants of weak and strong bisimulation equivalence and simulation pre-orders have been defined in [5]. Their compatibility to (fragments of) stochastic variants of computation tree logic (CTL) has been thoroughly investigated, cf. [5]. In [26], authors have defined strong bisimulation relation for continuous-time Markov decision processes (CTMDPs). This paper also proves that continuous stochastic logic (CSL) properties are preserved under bisimulation for CTMDPs. Strong and weak bisimulation relations for IMCs have been defined in [19,20]. Both strong and weak bisimulation preserve time-bounded (as well as unbounded) reachability probabilities. For Markov automata (MAs), strong and weak bisimulation relations have been defined in [15–17]. In [17], weak bisimulation has been defined over state probability distributions rather than over individual ones. In [15,16], it has been shown that weak bisimulation provides a sound and complete proof methodology for a touchstone equivalence called reduction barbed congruence. Notions of early and late semantics for MAs have been proposed in [33]. Using these semantics, early and late weak bisimulations have been defined and it has been proved that late weak bisimulation is weaker than all of the other variants defined in [15–17].

Linear Time: In [6], Bernardo considered Markovian testing equivalence over sequential Markovian process calculus (SMPC), and coined the term T-lumpability [7] for the induced state-level aggregation where T stands for testing. His testing equivalence is a congruence w.r.t. parallel composition, and preserves transient as well as steady-state probabilities. Bernardo's T-lumpability has been reconsidered in [32] where weighted lumpability (WL) is defined as a structural notion on CTMCs. Note that deterministic timed automaton (DTA) [1] and metric temporal logic (MTL) [12,27] specifications are preserved under WL [32]. In [35], several linear time equivalences (Markovian trace equivalence, failure and ready trace equivalence) for CTMCs have been investigated. Testing scenarios based on push-button experiments have been used for defining these equivalences. Trace semantics for CTMDPs have been defined in [29]. Similarly, trace semantics for open interactive Markov chains (IMCs) have been defined in [36]. In this paper testing scenarios using button pushing experiments have been used to define several variants of trace equivalences that arise by varying the type of schedulers. Recently, trace equivalences have been defined for MA models [30,31]. This paper also uses button pushing experiments to define several variants of trace equivalence for MA models. Note that the trace machine used in [36] for open IMCs does not display Markovian transitions and τ actions. Additionally, it uses a timer to count down from a certain value that is set by an external observer at the beginning of the experiment. Moreover, in [36], trace denotes an ordered sequence of visible actions on a path fragment. Due to these differences, the precise relation of our work to the equivalences defined in [36] is not yet clear to us. Note that, our definitions of (stutter) trace equivalences allow investigating the preservation of linear real-time objectives, e.g., DTA specifications [1] and MTL formulas [12,27].

Organisation of the Paper. Section 2 briefly recalls the main concepts of IMCs. Section 3 defines trace equivalences. Section 4 compares trace equivalence with bisimulation. Section 5 defines stutter trace equivalences. Section 6 relates stutter trace equivalence with weak bisimulation. Section 7 sketches the linear time-branching time spectrum. Finally, Sect. 8 concludes the paper and provides pointers for future research.

2 Preliminaries

This section presents the necessary definitions and basic concepts related to interactive Markov chains (IMCs) that are needed for the understanding of the rest of this paper.

Definition 1 (IMC). *An* interactive Markov chain *(IMC) is a tuple* $\mathcal{I} = (S, s_0, Act, AP, \rightarrow, \Rightarrow, L)$ *where:*

- S *is a finite set of states,*
- s_0 *is the initial state,*
- Act *is a finite set of actions,*
- AP *is a finite set of atomic propositions,*
- $\rightarrow \subseteq S \times Act \times S$ *is a set of interactive transitions,*
- $\Rightarrow \subseteq S \times \mathbb{R}_{>0} \times S$ *is a set of Markovian transitions, and*
- $L : S \rightarrow 2^{AP}$ *is a labeling function.*

We abbreviate $(s, a, s') \in \rightarrow$ as $s \xrightarrow{a} s'$ and similarly, $(s, \lambda, s') \in \Rightarrow$ by $s \overset{\lambda}{\Rightarrow} s'$. Let $IT(s)$ and $MT(s)$ denote the set of interactive and Markovian transitions that leave state s. A state s is *Markovian* iff $MT(s) \neq \varnothing$ and $IT(s) = \varnothing$; it is *interactive* iff $MT(s) = \varnothing$ and $IT(s) \neq \varnothing$. Further, s is a *hybrid* state iff $MT(s) \neq \varnothing$ and $IT(s) \neq \varnothing$; finally s is a *deadlock* state iff $MT(s) = \varnothing$ and $IT(s) = \varnothing$. W.l.o.g. in this paper we only consider IMCs that do not have any deadlock states. Let $MS \subseteq S$ and $IS \subseteq S$ denote the set of Markovian and interactive states in an IMC $\mathcal{I}$. For any Markovian state $s \in MS$, let $R(s, s') = \sum\{\lambda | s \overset{\lambda}{\Rightarrow} s'\}$ be the rate to move from state s to state s'. For $C \subseteq S$, let $R(s, C) = \sum_{s' \in C} R(s, s')$ be the rate to move from state s to a set of states C. The exit rate for a state s is defined by: $E(s) = \sum_{s' \in S} R(s, s')$.

It is easy to see that an IMC where $MT(s) = \varnothing$ for any state s is an LTS [4]. An IMC where $IT(s) = \varnothing$ for any state s is a CTMC [5]. The semantics of IMCs can thus be given in terms of the semantics of CTMCs (for Markovian transitions) and LTSs (for interactive transitions).

The meaning of a Markovian transition $s \overset{\lambda}{\Rightarrow} s'$ is that the IMC moves from state s to s' within t time units with probability $1 - e^{-\lambda \cdot t}$. If s has multiple outgoing Markovian transitions to different successors, then we speak of a race between these transitions, known as the *race condition*. In this case, the probability to move from s to s' within t time units is $\frac{R(s,s')}{E(s)} \cdot (1 - e^{-E(s) \cdot t})$.

Example 1. Consider the IMC $\mathcal{I}$ shown in Fig. 1, where $S = \{s_0, s_1, s_2, s_3, s_4, s_5, s_6, s_7, s_8, s_9\}$, $AP = \{a, b, c\}$, $Act = \{\alpha, \beta, \gamma\}$ and s_0 is the initial state. The set of interactive states is $IS = \{s_0, s_1, s_2\}$; MS contains all the other states. Note that there is no hybrid state in IMC $\mathcal{I}$. Non-determinism between action transitions appears in state s_0. Similarly, race condition due to multiple Markovian transitions appears in s_3 and s_4.

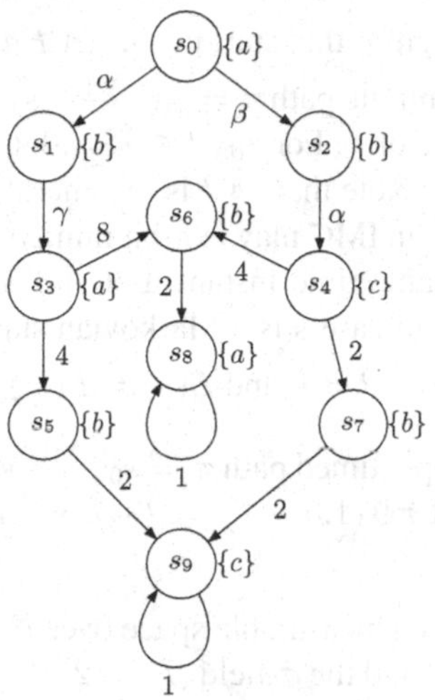

Fig. 1. An example IMC $\mathcal{I}$

We assume that in closed IMCs all outgoing interactive transitions from every state $s \in S$ are labeled with $\tau \in Act$ (internal action).

Definition 2 *(Maximal progress* [20]*). In any closed IMC, interactive transitions take precedence over Markovian transitions.*

Intuitively, the maximal progress assumption states that in closed IMCs, τ labeled transitions are not subject to interaction and thus can happen immediately[2], whereas the probability of a Markovian transition to happen immediately is zero. Accordingly, we assume that each state s has either only outgoing τ transitions or outgoing Markovian

[2] We restrict to models without zeneness. In simple words, this means that τ cycles are not allowed.

transitions. In other words, a closed IMC only has interactive and Markovian states. We use a distinguished action $\perp \notin Act$ to indicate Markovian transitions and extend the set of actions to $Act_\perp = Act \cup \{\perp\}$.

Definition 3 (IMC timed paths). *Let $\mathcal{I} = (S, s_0, Act, AP, \rightarrow, \Rightarrow, L)$ be an IMC. An* infinite *path π in $\mathcal{I}$ is a sequence $s_0 \xrightarrow{\sigma_0,t_0} s_1 \xrightarrow{\sigma_1,t_1} s_2 \ldots s_{n-1} \xrightarrow{\sigma_{n-1},t_{n-1}} s_n \ldots$ where $s_i \in S$, $\sigma_i \in Act$ or $\sigma_i = \perp$, and $t_i \in \mathbb{R}_{\geq 0}$ is the sojourn time in state s_i. For $\sigma_i \in Act$, $s_i \xrightarrow{\sigma_i,t_i} s_{i+1}$ denotes that after residing t_i time units in s_i, the IMC $\mathcal{I}$ has moved via action σ_i to s_{i+1}. Instead, $s_i \xrightarrow{\perp,t_i} s_{i+1}$ denotes that after residing t_i time units in s_i, a Markovian transition led to s_{i+1}. A finite path π is a finite prefix of an infinite path. The length of an infinite path π, denoted $|\pi|$ is ∞; the length of a finite path π with $n+1$ states is n.*

Let $Paths^{\mathcal{I}} = Paths^{\mathcal{I}}_{fin} \cup Paths^{\mathcal{I}}_{\omega}$ denote the set of all paths in $\mathcal{I}$ that start in s_0, where $Paths^{\mathcal{I}}_{fin} = \bigcup_{n \in \mathbb{N}} Paths^{\mathcal{I}}_n$ is the set of all finite paths in $\mathcal{I}$ and $Paths^{\mathcal{I}}_n$ denote the set of all finite paths of length n that start in s_0. Let $Paths^{\mathcal{I}}_{\omega}$ is the set of all infinite paths in $\mathcal{I}$ that start in s_0. For infinite path $\pi = s_0 \xrightarrow{\sigma_0,t_0} s_1 \xrightarrow{\sigma_1,t_1} s_2 \ldots$ and any $i \in \mathbb{N}$, let $\pi[i] = s_i$, the $(i+1)$st state of π. For any $t \in \mathbb{R}_{\geq 0}$, let $\pi@t$ denote the sequence of states that π occupies at time t. Note that $\pi@t$ is in general not a single state, but rather a sequence of several states, as an IMC may exhibit immediate transitions and thus may occupy various states at the same time instant. Let $Act(s)$ denote the set of enabled actions from state s. Note that in case s is a Markovian state then $Act(s) = \{\perp\}$. Step $s \xrightarrow{\sigma,t} s'$ is a *stutter step* if $\sigma = \tau$, $t = 0$ and $L(s) = L(s')$.

Example 2. Consider an example timed path $\pi = s_0 \xrightarrow{\alpha,0} s_1 \xrightarrow{\gamma,0} s_3 \xrightarrow{\perp,1.5} s_2 \xrightarrow{\gamma,0} s_5$. Here we have $\pi[2] = s_3$ and $\pi@(1.5 - \epsilon) = \langle s_3 \rangle$, where $0 < \epsilon < 1.5$. Similarly, $\pi@1.5 = \langle s_2 s_5 \rangle$.

σ-algebra. In order to construct a measurable space over $Paths^{\mathcal{I}}_{\omega}$, we define the following sets: $\Omega = Act_\perp \times \mathbb{R}_{\geq 0} \times S$ and the σ-field $\mathcal{J} = (2^{Act_\perp} \times \mathcal{J}_R \times 2^S)$, where $Act_\perp = Act \cup \{\perp\}$, $\mathcal{J}_R$ is the Borel σ-field over $\mathbb{R}_{\geq 0}$ [2,3]. The σ-field over $Paths^{\mathcal{I}}_n$ is defined as $\mathcal{J}_{Paths^{\mathcal{I}}_n} = \sigma(\{S_0 \times M_0 \times \ldots \times M_{n-1} | S_0 \in 2^S, M_i \in \mathcal{J}, 0 \leq i \leq n-1\})$. A set $B \in \mathcal{J}_{Paths^{\mathcal{I}}_n}$ is a base of a cylinder set C if $C = Cyl(B) = \{\pi \in Paths^{\mathcal{I}}_{\omega} | \pi[0 \ldots n] \in B\}$, where $\pi[0 \ldots n]$ is the prefix of length n of the path π. The σ-field $\mathcal{J}_{Paths^{\mathcal{I}}_{\omega}}$ of measurable subsets of $Paths^{\mathcal{I}}_{\omega}$ is defined as $\mathcal{J}_{Paths^{\mathcal{I}}_{\omega}} = \sigma(\cup_{n=0}^{\infty} \{Cyl(B) | B \in \mathcal{J}_{Paths^{\mathcal{I}}_n}\})$.

2.1 Schedulers

Non-determinism in an IMC is resolved by a scheduler. Schedulers are also known as adversaries or policies. More formally, schedulers are defined as follows:

Definition 4 (Scheduler). *A scheduler for an IMC $\mathcal{I} = (S, s_0, Act, AP, \rightarrow, \Rightarrow, L)$ is a measurable function $\mathcal{D} : Paths^{\mathcal{I}}_{fin} \rightarrow Distr(Act)$, such that for $n \in \mathbb{N}$,*

$$\mathcal{D}(s_0 \xrightarrow{\sigma_0,t_0} s_1 \xrightarrow{\sigma_1,t_1} \ldots \xrightarrow{\sigma_{n-1},t_{n-1}} s_n)(\alpha) > 0 \text{ implies } \alpha \in Act(s_n)$$

where Distr(Act) denotes the set of all distributions on Act.

Schedulers can be classified[3] according to the way they resolve non-determinism and the information on the basis of which a decision is taken. For example, the next action can be selected with probability one (deterministic schedulers) or at random according to a specific probability distribution (randomized schedulers). Similarly, non-determinism can be resolved by only considering the current state (stationary schedulers) or complete (time-abstract/timed) history. More formally, schedulers can be classified as follows:

Definition 5 (Classes of schedulers). *A scheduler $\mathcal{D}$ for an IMC $\mathcal{I}$ is*

- *stationary deterministic (SD) if* $\mathcal{D} : S \longrightarrow Act$ *such that* $\mathcal{D}(s) \in Act(s)$
- *stationary randomized (SR) if* $\mathcal{D} : S \rightarrow Distr(Act)$ *such that* $\mathcal{D}(s)(\alpha) > 0$ *implies* $\alpha \in Act(s)$
- *history-dependent deterministic (HD) if* $\mathcal{D} : (S \times Act)^* \times S \longrightarrow Act$ *such that we have* $\mathcal{D}(\underbrace{s_0 \xrightarrow{\sigma_0} s_1 \xrightarrow{\sigma_1} \ldots \xrightarrow{\sigma_{n-1}} s_n}_{\text{time-abstract history}}) \in Act(s_n)$
- *history-dependent randomized (HR) if* $\mathcal{D} : (S \times Act)^* \times S \rightarrow Distr(Act)$ *such that* $\mathcal{D}(\underbrace{s_0 \xrightarrow{\sigma_0} s_1 \xrightarrow{\sigma_1} \ldots \xrightarrow{\sigma_{n-1}} s_n}_{\text{time-abstract history}})(\alpha) > 0$ *implies* $\alpha \in Act(s_n)$
- *timed history-dependent deterministic (THD) if* $\mathcal{D} : (S \times Act \times \mathbb{R}_{>0})^* \times S \rightarrow Act$ *such that* $\mathcal{D}(\underbrace{s_0 \xrightarrow{\sigma_0, t_0} s_1 \xrightarrow{\sigma_1, t_1} \ldots \xrightarrow{\sigma_{n-1}, t_{n-1}} s_n}_{\text{timed history}}) \in Act(s_n)$
- *timed history-dependent randomized (THR) scheduler has been already defined in Definition 4*

Let $Adv(\mathcal{I})$ denotes the set of all schedulers of $\mathcal{I}$. Let $Adv_{\mathcal{C}}(\mathcal{I})$ denotes the set of all schedulers of class $\mathcal{C}$, e.g., $Adv_{THD}(\mathcal{I})$ denotes the set of all THD schedulers of IMC $\mathcal{I}$. Let $Paths^{\mathcal{I}}_{\mathcal{D}}$ denotes the set of all infinite paths of $\mathcal{I}$ under $\mathcal{D} \in Adv(\mathcal{I})$ that start in s_0. Once the non-deterministic choices of a IMC $\mathcal{I}$ have been resolved by a scheduler, say $\mathcal{D}$, the induced model obtained is purely stochastic. To that end the unique probability measure for probability space $(Paths^{\mathcal{I}}_{\omega}, \mathcal{J}_{Paths^{\mathcal{I}}_{\omega}})$ can be defined [25].

Remark 1. If the alternating sequences of state labels and actions are same for two time-abstract histories $\mathcal{H}$ and $\mathcal{H}'$, then for both the histories non-determinism will be resolved in the same way from state s. In other words, if two histories have the same traces then the decision taken by a history-dependent scheduler $\mathcal{D}$ from s is going to be the same for both the histories. The same holds true for timed histories where sequences of state labels, actions and timing information are taken into account.

3 Stochastic Trace Machines

This section proposes several variants of trace equivalence for closed IMCs. These equivalences are obtained by performing push-button experiments with a stochastic

[3] We only consider schedulers that make a decision as soon as a state is entered. Such schedulers are called early schedulers.

trace machine $\mathcal{I}$. The machine is equipped with an action display, a state label display, a timer and a reset button. Action display shows the last action that has been executed by the trace machine. For Markovian states, action display shows the distinguished action $\perp$. Note that this display is blank at the beginning of the experiment. The state label display shows the set of atomic propositions that are true in the current state of the machine $\mathcal{I}$. The timer display shows the absolute time. The reset button is used to restart the machine for another run starting from the initial state. Consider a run of the machine (under scheduler $\mathcal{D}$ of class $\mathcal{C}$) which always starts from the initial state. The state label display shows the label of the current state and action display shows the last action that has been executed. Note that the action display remains unchanged until the next action is executed by the machine. An external observer records the sequence of state labels, actions and time checks where each time check[4] is recorded at an arbitrary time instant between the occurrence of two successive actions. The observer can press the reset button to stop the current run. Once the reset button is pressed, the action display will be blank and the state label display shows the set of atomic propositions that are true in the initial state. The machine then starts for another run and the observer again records the sequence of actions, state labels and time checks. Note that the machine needs to be executed for infinitely many runs to complete the whole experiment. It is assumed that the observer can distinguish between two successive actions that are equal. For a sequence of τ actions, state labels can be different but the recorded time checks are going to stay the same. This is because τ actions are executed immediately. An outcome of this machine is $(\sigma, \theta) = (\langle L(s_0)\sigma_0 L(s_1)\sigma_1 \ldots L(s_{n-1})\sigma_{n-1} L(s_n)\rangle, \langle t'_0, t'_1, \ldots, t'_n\rangle)$, where $\sigma_0, \ldots, \sigma_{n-1} \in \{\tau, \perp\}$. This outcome can be interpreted as follows: for $0 \leq m < n$, action σ_m of machine is performed in the time interval $(y_m, y_{m+1}]$ where $y_m = \Sigma_{i=0}^{m} t'_i$.

Definition 6. *Let* $(\sigma, \theta) = (\langle L(s_0)\sigma_0 L(s_1)\sigma_1 \ldots L(s_{n-1})\sigma_{n-1} L(s_n)\rangle, \langle t'_0, t'_1, \ldots, t'_n\rangle)$ *be an outcome of* $\mathcal{I}$ *under* $\mathcal{D} \in Adv(\mathcal{I})$*, then a path* $\pi = s_0 \xrightarrow{\sigma_0, t_0} s_1 \xrightarrow{\sigma_1, t_1} s_2 \ldots s_{n-1} \xrightarrow{\sigma_{n-1}, t_{n-1}} s_n \ldots \in Paths_{\mathcal{D}}^{\mathcal{I}}$ *is said to be compatible with* (σ, θ)*, denoted* $\pi \triangleright (\sigma, \theta)$*, if the following holds:*

$$Trace(\pi[0 \ldots n]) = \sigma \text{ and } \Sigma_{j=0}^{i} t_j \in (y_i, y_{i+1}] \text{ for } 0 \leq i < n$$

where $y_i = \Sigma_{j=0}^{i} t'_j$.

Remark 2. Time check t'_i recorded by the external observer should not be confused with t_i in Definition 6. Here, t_i denote the time spent in state s_i of a path π. If s_i is a Markovian state then $t_i > 0$ otherwise $t_i = 0$.

Definition 7. *Let* (σ, θ) *be an outcome of trace machine* $\mathcal{I}$ *under* $\mathcal{D} \in Adv(\mathcal{I})$*. Then the probability of all the paths compatible with* (σ, θ) *is defined as follows:*

$$P_{\mathcal{I},\mathcal{D}}^{trace}(\sigma, \theta) = Pr_{\mathcal{D}}(\{\pi \in Paths_{\mathcal{D}}^{\mathcal{I}} | \pi \triangleright (\sigma, \theta)\})$$

[4] Time check should not be confused with the absolute time displayed by the timer.

Definition 8. *Let $P^{trace}_{\mathcal{I},\mathcal{D}}$ be an observation of machine $\mathcal{I}$ under $\mathcal{D} \in Adv(\mathcal{I})$. Then the set of observations for scheduler class $\mathcal{C}$, denoted $O_{\mathcal{C}}(\mathcal{I})$, is defined as follows:*

$$O_{\mathcal{C}}(\mathcal{I}) = \{P^{trace}_{\mathcal{I},\mathcal{D}} | \mathcal{D} \in Adv_{\mathcal{C}}(\mathcal{I})\}$$

Definition 9. *Two IMCs $\mathcal{I}_1$, $\mathcal{I}_2$ are trace equivalent w.r.t. scheduler class $\mathcal{C}$ denoted $\mathcal{I}_1 \equiv_{\mathcal{C}} \mathcal{I}_2$ iff $O_{\mathcal{C}}(\mathcal{I}_1) = O_{\mathcal{C}}(\mathcal{I}_2)$.*

Example 3. Consider the IMCs $\mathcal{I}$ and $\mathcal{I}'$ shown in Fig. 2. These two systems are $\equiv_{SD}$, $\equiv_{SR}$, $\equiv_{HD}$, $\equiv_{HR}$, $\equiv_{THD}$ and $\equiv_{THR}$.

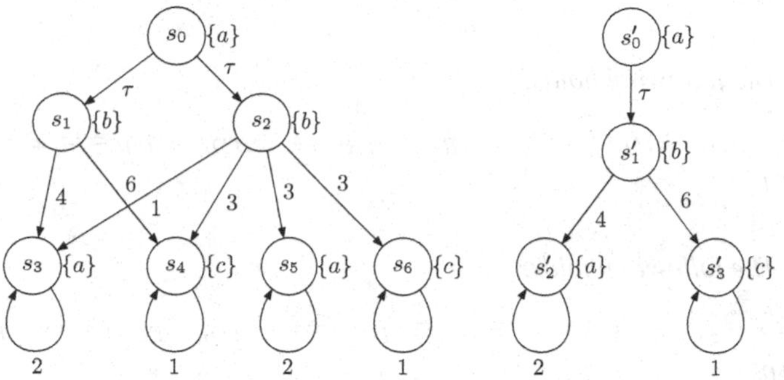

Fig. 2. IMCs $\mathcal{I}$ (left) and $\mathcal{I}'$ (right)

Next, we study the relationship between several variants of trace equivalence defined in this section. Connections among these equivalences can be understood from Fig. 3. Here, a directed edge from node labeled with, say $\equiv_{\mathcal{C}_1}$, to node labeled with $\equiv_{\mathcal{C}_2}$ denotes implication, i.e., $\equiv_{\mathcal{C}_1} \implies \equiv_{\mathcal{C}_2}$. Similarly, an edge that connects two nodes in both the directions denote bi-implication, i.e., coincidence. To avoid clutter, we have omitted the directed edge between $\equiv_{\mathcal{C}_1} \implies \equiv_{\mathcal{C}_3}$ whenever we have $\equiv_{\mathcal{C}_1} \implies \equiv_{\mathcal{C}_2}$ and $\equiv_{\mathcal{C}_2} \implies \equiv_{\mathcal{C}_3}$.

Theorem 1. *The following holds:*

- $\equiv_{SD} \implies \equiv_{SR}$, $\equiv_{SD} \implies \equiv_{HD}$, $\equiv_{SD} \implies \equiv_{HR}$, $\equiv_{SD} \not\Rightarrow \equiv_{THD}$, $\equiv_{SD} \not\Rightarrow \equiv_{THR}$

Theorem 2. *The following holds:*

- $\equiv_{SR} \not\Rightarrow \equiv_{SD}$, $\equiv_{SR} \not\Rightarrow \equiv_{HD}$, $\equiv_{SR} \implies \equiv_{HR}$, $\equiv_{SR} \not\Rightarrow \equiv_{THD}$, $\equiv_{SR} \not\Rightarrow \equiv_{THR}$

Theorem 3. *The following holds:*

- $\equiv_{HD} \implies \equiv_{SD}$, $\equiv_{HD} \implies \equiv_{SR}$, $\equiv_{HD} \not\Rightarrow \equiv_{THD}$, $\equiv_{HD} \implies \equiv_{HR}$, $\equiv_{HD} \not\Rightarrow \equiv_{THR}$

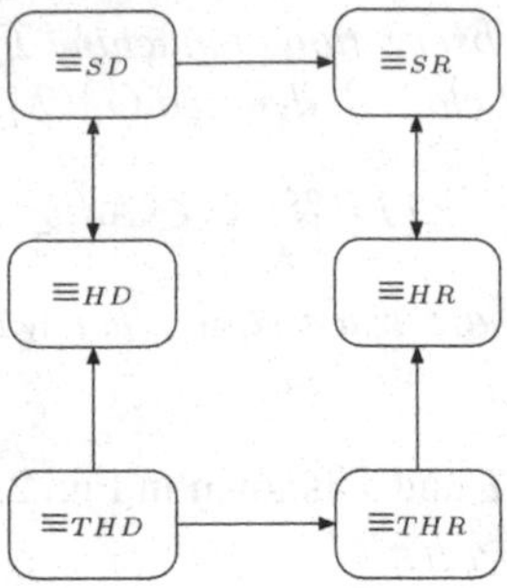

Fig. 3. Connections among six trace equivalences

Theorem 4. *The following holds:*

- $\equiv_{HR} \not\Longrightarrow \equiv_{SD}$, $\equiv_{HR} \Longrightarrow \equiv_{SR}$, $\equiv_{HR} \not\Longrightarrow \equiv_{HD}$, $\equiv_{HR} \not\Longrightarrow \equiv_{THD}$, $\equiv_{HR} \not\Longrightarrow \equiv_{THR}$

Theorem 5. *The following holds:*

- $\equiv_{THD} \Longrightarrow \equiv_{SD}$, $\equiv_{THD} \Longrightarrow \equiv_{SR}$, $\equiv_{THD} \Longrightarrow \equiv_{HD}$, $\equiv_{THD} \Longrightarrow \equiv_{HR}$, $\equiv_{THD} \Longrightarrow \equiv_{THR}$

Theorem 6. *The following holds:*

- $\equiv_{THR} \not\Longrightarrow \equiv_{SD}$, $\equiv_{THR} \Longrightarrow \equiv_{SR}$, $\equiv_{THR} \not\Longrightarrow \equiv_{HD}$, $\equiv_{THR} \Longrightarrow \equiv_{HR}$, $\equiv_{THR} \not\Longrightarrow \equiv_{THD}$

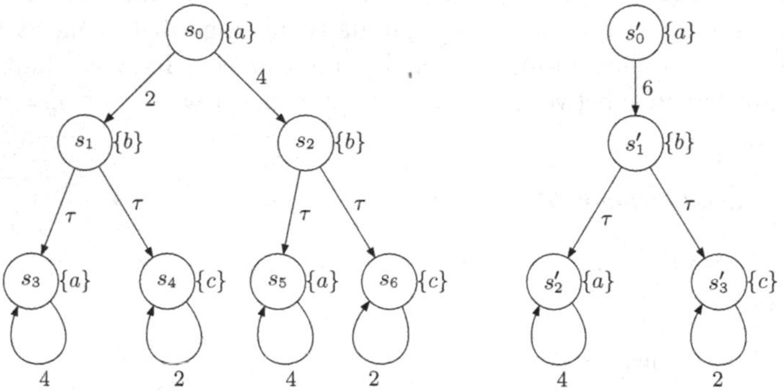

Fig. 4. IMCs $\mathcal{I}$ (left) and $\mathcal{I}'$ (right)

Example 4. Consider the two IMCs $\mathcal{I}$ and $\mathcal{I}'$ shown in Fig. 4. These two systems are $\equiv_{SR}$, $\equiv_{HR}$ and $\equiv_{THR}$. Note that these two systems are $\not\equiv_{SD}$, $\not\equiv_{HD}$ and $\not\equiv_{THD}$.

4 Trace Equivalences Versus Bisimulation

This section investigates the relationship of bisimulation to trace equivalences. We first recall the definition of bisimulation for IMCs [20]. Let $Post(s,\tau,C) = \{s' \in C | s \xrightarrow{\tau} s'\}$.

Definition 10. *(Strong bisimulation* [19,20]*) Let $\mathcal{I} = (S, s_0, Act, AP, \rightarrow, \Rightarrow, L)$ be a closed IMC. An equivalence relation $\mathcal{R} \subseteq S \times S$ is a strong bisimulation on $\mathcal{I}$ if for any $(s_1, s_2) \in \mathcal{R}$ and equivalence class $C \in S/_{\mathcal{R}}$ the following holds:*

- $L(s_1) = L(s_2)$,
- $R(s_1, C) = R(s_2, C)$,
- $Post(s_1, \tau, C) \neq \varnothing \Leftrightarrow Post(s_2, \tau, C) \neq \varnothing$.

States s_1 and s_2 are strongly bisimilar, denoted $s_1 \sim s_2$, if $(s_1, s_2) \in \mathcal{R}$ for some strong bisimulation[5] *$\mathcal{R}$.*

Strong bisimulation is rigid as it requires that each individual step should be mimicked.

Example 5. Consider the two IMCs shown in Fig. 2. Here, $s_0 \sim s_0'$.

Theorem 7. *The following holds:*

- $\sim \not\Longrightarrow \equiv_{SD}$ *and* $\equiv_{SD} \not\Longrightarrow \sim$
- $\sim \Longrightarrow \equiv_{SR}$ *and* $\equiv_{SR} \not\Longrightarrow \sim$
- $\sim \not\Longrightarrow \equiv_{HD}$ *and* $\equiv_{HD} \not\Longrightarrow \sim$
- $\sim \Longrightarrow \equiv_{HR}$ *and* $\equiv_{HR} \not\Longrightarrow \sim$
- $\sim \not\Longrightarrow \equiv_{THD}$ *and* $\equiv_{THD} \not\Longrightarrow \sim$
- $\sim \Longrightarrow \equiv_{THR}$ *and* $\equiv_{THR} \not\Longrightarrow \sim$

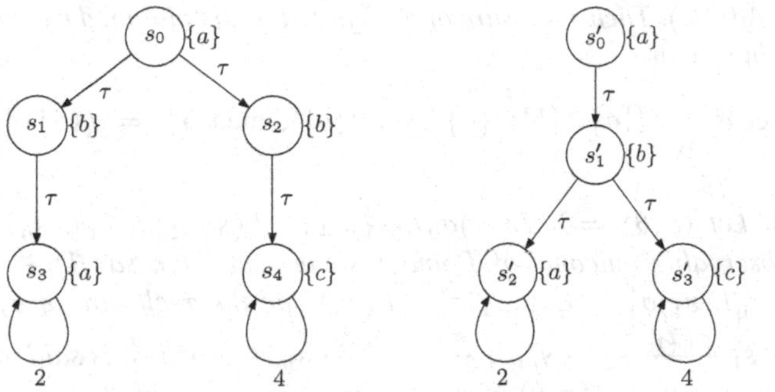

Fig. 5. IMCs $\mathcal{I}$ (left) and $\mathcal{I}'$ (right)

Example 6. Consider the two IMCs shown in Fig. 5. These two IMCs are $\equiv_{SD}$, $\equiv_{HD}$, $\equiv_{THD}$, $\equiv_{SR}$, $\equiv_{HR}$ and $\equiv_{THR}$ trace equivalent but not bisimilar, i.e., $\not\sim$.

[5] Note that the definition of strong bisimulation has been slightly modified to take into account the state labels.

5 Stutter Insensitive Stochastic Trace Machines

This section proposes several variants of stutter trace equivalence for closed IMCs. We perform experiments on stutter insensitive trace machine $\mathcal{I}$ to obtain these equivalences. Since our machine is insensitive to stutter steps, action display remains blank if the last action executed is an internal, i.e., τ action and shows $\perp$[6] otherwise. In other words, for interactive states no action is displayed on the action display. Note that when the display goes blank, the only information that an observer can deduce is that a τ action is being executed. From this information alone, it is not possible to find out if this τ action is part of a (sequence of) stutter step(s) or a non-stuttering step, i.e., $s \xrightarrow{\tau,0} s'$ such that $L(s) \neq L(s')$. As in the case of trace machine presented in Sect. 3, an observer records the sequence of actions, state labels and time checks. The machine needs to be executed for infinitely many runs to complete the whole experiment. An observable outcome of this machine is $(\sigma, \theta) = (\langle L(s_0)\sigma_0 L(s_1)\sigma_1 \ldots L(s_{n-1})\sigma_{n-1}L(s_n)\rangle, \langle t'_0, t'_1, \ldots, t'_n\rangle)$, where $\sigma_i = \perp$ or $\sigma_i = \varnothing$. Here, $\varnothing$ denotes that the action display is blank.

Example 7. Consider a path $\pi = s_0 \xrightarrow{\tau,0} s_1 \xrightarrow{\tau,0} s_2 \xrightarrow{\tau,0} s_3 \xrightarrow{\perp,4} s_4$. Let $L(s_0) = L(s_1) = L(s_2) = \{a\}$, $L(s_3) = \{b\}$ and $L(s_4) = \{c\}$. Here, the alternating sequence of state labels and actions recorded by an external observer is $\sigma = \langle\{a\}\varnothing\{b\}\perp\{c\}\rangle$. In this example we have two stutter steps.

Example 8. Consider another path $\pi' = s_0 \xrightarrow{\perp,2} s_1 \xrightarrow{\perp,3} s_2 \xrightarrow{\perp,1} s_3$. Let $L(s_0) = L(s_1) = \{a\}$, $L(s_2) = L(s_3) = \{b\}$. Here, the alternating sequence of state labels and actions recorded by an external observer is $\sigma = \langle\{a\}\perp\{a\}\perp\{b\}\perp\{b\}\rangle$. This is because the observer can distinguish between two successive $\perp$ labeled transitions.

Definition 11 **(τ-closure).** *Let (σ, θ) be an observable outcome of $\mathcal{I}$ under some scheduler, say $\mathcal{D} \in Adv(\mathcal{I})$. Then τ-closure of σ, denoted σ^τ, is obtained by replacing every instance of $\varnothing$ in σ with τ.*

Example 9. Let $\sigma = \langle\{a\}\varnothing\{b\}\varnothing\{a\}\perp\{c\}\varnothing\{b\}\rangle$, then $\sigma^\tau = \langle\{a\}\tau\{b\}\tau\{a\}\perp\{c\}\tau\{b\}\rangle$.

Definition 12. *Let $(\sigma, \theta) = (\langle L(s_0)\sigma_0 L(s_1)\sigma_1 \ldots L(s_{n-1})\sigma_{n-1}L(s_n)\rangle, \langle t'_0, t'_1, \ldots, t'_n\rangle)$ be an observable outcome of $\mathcal{I}$ under some scheduler, say $\mathcal{D} \in Adv(\mathcal{I})$ and $\sigma^\tau = \langle L(s_0)\sigma'_0 L(s_1)\sigma'_1 \ldots L(s_{n-1})\sigma'_{n-1}L(s_n)\rangle$ be the τ-closure of σ, then a path $\pi = s_0 \xrightarrow{\sigma_0,t_0} s_1 \xrightarrow{\sigma_1,t_1} s_2 \ldots s_{m-1} \xrightarrow{\sigma_{m-1},t_{m-1}} s_m \in Paths^{\mathcal{I}}_{\mathcal{D}}$ is said to be compatible with (σ, θ), denoted $\pi \rhd (\sigma, \theta)$, if exactly one of the following holds:*

- $n = m$ *and* $Trace(\pi[0 \ldots m]) = \sigma^\tau$ *and* $\Sigma^i_{j=0} t_j \in (y_i, y_{i+1}]$ *for* $0 \leq i < n$
- $m > n$ *and* $Trace(\pi[0 \ldots m]) = L(s_0) \underbrace{\tau L(s_0)}_{n_0-times} \sigma'_0 L(s_1) \underbrace{\tau L(s_1)}_{n_1-times} \sigma'_1 L(s_2) \ldots L(s_{n-1}) \underbrace{\tau L(s_{n-1})}_{n_{n-1}-times} \sigma'_{n-1} L(s_n)$ *and* $\Sigma^{n_0}_{k=0} t_k \in (y_0, y_1]$, $\Sigma^{n_0+n_1+1}_{k=0} t_k \in (y_1, y_2], \ldots, \Sigma^{m+n-1}_{k=0} t_k \in (y_{n-1}, y_n]$

[6] Recall that we use a distinguished action $\perp$ to indicate Markovian transitions.

where $y_i = \Sigma_{j=0}^{i} t'_j$, $n_k \geq 0$ *for* $0 \leq k \leq n-1$ *and* $m = n_0 + n_1 + \ldots + n_{n-1}$.

The first condition corresponds to the case where stuttering is absent in path π. The second condition corresponds to the case where stuttering is present in π. Since our machine is insensitive to stutter steps, it is possible that multiple paths of different lengths are compatible with the same outcome of trace machine $\mathcal{I}$. In other words, our machine does not distinguish between paths that have different number of stutter steps as long as one of the conditions of Definition 12 is satisfied. The probability of all the paths compatible with an observable outcome is defined as follows:

Definition 13. *Let* (σ, θ) *be an observable outcome of trace machine* $\mathcal{I}$ *under* $\mathcal{D} \in Adv(\mathcal{I})$. *Let* $|\sigma| = n$. *Then the probability of all the paths compatible with* (σ, θ) *is defined as follows:*

$$P_{\mathcal{I},\mathcal{D}}^{trace}(\sigma, \theta) = \sum_{i=n}^{m} Pr_{\mathcal{D}}(\{\pi \in Paths_{\mathcal{D}}^{\mathcal{I}} \mid |\pi| = i \wedge \pi \triangleright (\sigma, \theta)\})$$

where m *is the length of the longest path that is compatible with* (σ, θ).

Informally, $P_{\mathcal{I},\mathcal{D}}^{trace}$ is a function that gives the probability to observe (σ, θ) in machine $\mathcal{I}$ under scheduler $\mathcal{D}$.

Definition 14 (Set of observations). *Let* $P_{\mathcal{I},\mathcal{D}}^{trace}$ *be an observation of machine* $\mathcal{I}$ *under* $\mathcal{D} \in Adv(\mathcal{I})$. *Then the set of observations for scheduler class* $\mathcal{C}$, *denoted* $O_{\mathcal{C}}(\mathcal{I})$, *is defined as follows:*

$$O_{\mathcal{C}}(\mathcal{I}) = \{P_{\mathcal{I},\mathcal{D}}^{trace} \mid \mathcal{D} \in Adv_{\mathcal{C}}(\mathcal{I})\}$$

Informally, $O_{\mathcal{C}}(\mathcal{I})$ denote a set of functions where each function assigns a probability value to every possible observable outcome of the trace machine, i.e., (σ, θ).

Definition 15 (Stutter trace equivalence). *Two IMCs* $\mathcal{I}_1$, $\mathcal{I}_2$ *are stutter trace equivalent w.r.t. scheduler class* $\mathcal{C}$ *denoted* $\mathcal{I}_1 \simeq_{\mathcal{C}} \mathcal{I}_2$ *iff* $O_{\mathcal{C}}(\mathcal{I}_1) = O_{\mathcal{C}}(\mathcal{I}_2)$.

This definition says that for every $\mathcal{D} \in Adv_{\mathcal{C}}(\mathcal{I}_1)$ there exists a scheduler $\mathcal{D}' \in Adv_{\mathcal{C}}(\mathcal{I}_2)$ such that for all outcomes (σ, θ) we have $P_{\mathcal{I}_1,\mathcal{D}}^{trace}(\sigma, \theta) = P_{\mathcal{I}_2,\mathcal{D}'}^{trace}(\sigma, \theta)$ and vice versa. Next, we investigate the connections between trace equivalence ($\equiv_{\mathcal{C}}$) and stutter trace equivalence ($\simeq_{\mathcal{C}}$).

Theorem 8. *The following holds:*

- $\equiv_{SD} \Longrightarrow \simeq_{SD}$ *and* $\equiv_{HD} \Longrightarrow \simeq_{HD}$ *and* $\equiv_{THD} \Longrightarrow \simeq_{THD}$ *and* $\equiv_{SR} \Longrightarrow \simeq_{SR}$ *and* $\equiv_{HR} \Longrightarrow \simeq_{HR}$ *and* $\equiv_{THR} \Longrightarrow \simeq_{THR}$
- $\simeq_{SD} \not\Longrightarrow \equiv_{SD}$ *and* $\simeq_{HD} \not\Longrightarrow \equiv_{HD}$ *and* $\simeq_{THD} \not\Longrightarrow \equiv_{THD}$ *and* $\simeq_{SR} \not\Longrightarrow \equiv_{SR}$ *and* $\simeq_{HR} \not\Longrightarrow \equiv_{HR}$ *and* $\simeq_{THR} \not\Longrightarrow \equiv_{THR}$

This theorem says that $\equiv_{\mathcal{C}}$ is strictly finer than $\simeq_{\mathcal{C}}$. Next, we study the relationship between several variants of stutter trace equivalence defined in this section. Connections among these equivalences can be understood from Fig. 6.

Theorem 9. *The following holds:*

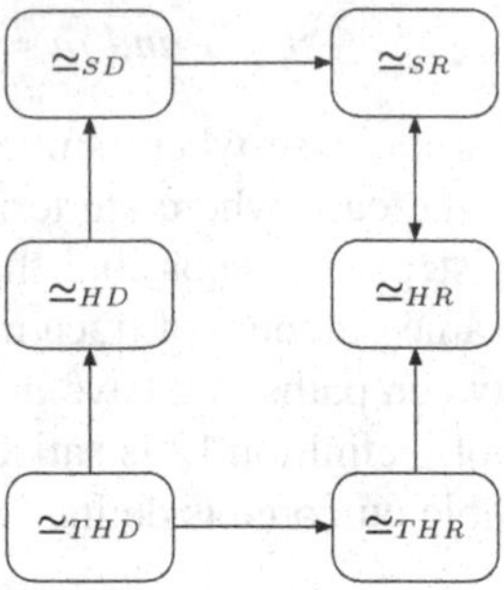

Fig. 6. Connections among six stutter trace equivalences

- $\simeq_{SD} \implies \simeq_{SR}$, $\simeq_{SD} \not\Rightarrow \simeq_{HD}$, $\simeq_{SD} \implies \simeq_{HR}$, $\simeq_{SD} \not\Rightarrow \simeq_{THD}$, $\simeq_{SD} \not\Rightarrow \simeq_{THR}$

Theorem 10. *The following holds:*

- $\simeq_{SR} \not\Rightarrow \simeq_{SD}$, $\simeq_{SR} \not\Rightarrow \simeq_{HD}$, $\simeq_{SR} \implies \simeq_{HR}$, $\simeq_{SR} \not\Rightarrow \simeq_{THD}$, $\simeq_{SR} \not\Rightarrow \simeq_{THR}$

Theorem 11. *The following holds:*

- $\simeq_{HD} \implies \simeq_{SD}$, $\simeq_{HD} \implies \simeq_{SR}$, $\simeq_{HD} \implies \simeq_{HR}$, $\simeq_{HD} \not\Rightarrow \simeq_{THD}$, $\simeq_{HD} \not\Rightarrow \simeq_{THR}$

Theorem 12. *The following holds:*

- $\simeq_{HR} \not\Rightarrow \simeq_{SD}$, $\simeq_{HR} \implies \simeq_{SR}$, $\simeq_{HR} \not\Rightarrow \simeq_{HD}$, $\simeq_{HR} \not\Rightarrow \simeq_{THD}$, $\simeq_{HR} \not\Rightarrow \simeq_{THR}$

Theorem 13. *The following holds:*

- $\simeq_{THD} \implies \simeq_{SD}$, $\simeq_{THD} \implies \simeq_{SR}$, $\simeq_{THD} \implies \simeq_{HD}$, $\simeq_{THD} \implies \simeq_{HR}$, $\simeq_{THD} \implies \simeq_{THR}$

Theorem 14. *The following holds:*

- $\simeq_{THR} \not\Rightarrow \simeq_{SD}$, $\simeq_{THR} \implies \simeq_{SR}$, $\simeq_{THR} \not\Rightarrow \simeq_{HD}$, $\simeq_{THR} \implies \simeq_{HR}$, $\simeq_{THR} \not\Rightarrow \simeq_{THD}$

6 Stutter Trace Equivalence Versus Weak Bisimulation

This section investigates the relationship between weak bisimulation and stutter trace equivalence. First, we fix some notations. A sequence of zero or more τ-labeled transitions is denoted by $s \xrightarrow{\tau^*} s'$. Similarly, a sequence of one or more τ-labeled transitions is denoted by $s \xrightarrow{\tau^+} s'$.

Definition 16. *(Weak bisimulation [20]) Let $\mathcal{I} = (S, s_0, Act, AP, \rightarrow, \Rightarrow, L)$ be a closed IMC. An equivalence relation $\mathcal{R} \subseteq S \times S$ is a weak bisimulation on $\mathcal{I}$ if for any $(s_1, s_2) \in \mathcal{R}$ and equivalence class $C \in S/_{\mathcal{R}}$ the following holds:*

- $L(s_1) = L(s_2)$,
- $\exists s' \in C : s_1 \xrightarrow{\tau^+} s' \Leftrightarrow \exists s'' \in C : s_2 \xrightarrow{\tau^+} s''$,
- $s_1 \xrightarrow{\tau^*} s' \wedge s' \in MS \Rightarrow s_2 \xrightarrow{\tau^*} s'' \wedge s'' \in MS \wedge R(s', C) = R(s'', C)$ *for some* $s'' \in S$.

States s_1 and s_2 are weakly bisimilar, denoted $s_1 \approx s_2$, if $(s_1, s_2) \in \mathcal{R}$ for some weak bisimulation[7] *$\mathcal{R}$.*

The first condition asserts that s_1 and s_2 are equally labeled. The second condition asserts that if s_1 can reach some equivalence class C solely via one or more τ-steps then s_2 can also do so and vice versa. Similarly, third condition requires that if s_1 can reach a Markovian state s' solely via zero or more τ-steps then s_2 can also reach a Markovian state s'' such that both these Markovian states have the same rate of moving to any equivalence class C. Note that in both these conditions all the extra steps are taken within the equivalence class of s_1, i.e., $[s_1]$ before reaching C. Similarly, for s_2 all the extra steps are taken within $[s_2]$.

Theorem 15. *The following holds:*

- $\approx \;\not\Leftrightarrow\; \simeq_{SD}$ *and* $\approx \;\not\Leftrightarrow\; \simeq_{HD}$ *and* $\approx \;\not\Leftrightarrow\; \simeq_{THD}$ *and* $\simeq_{SD} \;\not\Rightarrow\; \approx$ *and* $\simeq_{HD} \;\not\Rightarrow\; \approx$ *and* $\simeq_{THD} \;\not\Rightarrow\; \approx$
- $\approx \;\Longrightarrow\; \simeq_{SR}$ *and* $\approx \;\Longrightarrow\; \simeq_{HR}$ *and* $\approx \;\Longrightarrow\; \simeq_{THR}$ *and* $\simeq_{SR} \;\not\Rightarrow\; \approx$ *and* $\simeq_{HR} \;\not\Rightarrow\; \approx$ *and* $\simeq_{THR} \;\not\Rightarrow\; \approx$

7 Linear Time-Branching Time Spectrum

This section sketches the linear time-branching time spectrum of equivalences for closed IMCs. We first relate weak bisimulation to trace equivalences defined in Sect. 3. Next, we study the connections between stutter trace equivalences and bisimulation for closed IMCs.

Theorem 16. *The following holds:*

- $\approx \;\not\Leftrightarrow\; \equiv_{SD}$ *and* $\approx \;\not\Leftrightarrow\; \equiv_{HD}$ *and* $\approx \;\not\Leftrightarrow\; \equiv_{THD}$ *and* $\equiv_{SD} \;\not\Rightarrow\; \approx$ *and* $\equiv_{HD} \;\not\Rightarrow\; \approx$ *and* $\equiv_{THD} \;\not\Rightarrow\; \approx$
- $\approx \;\not\Leftrightarrow\; \equiv_{SR}$ *and* $\approx \;\not\Leftrightarrow\; \equiv_{HR}$ *and* $\approx \;\not\Leftrightarrow\; \equiv_{THR}$ *and* $\equiv_{SR} \;\not\Rightarrow\; \approx$ *and* $\equiv_{HR} \;\not\Rightarrow\; \approx$ *and* $\equiv_{THR} \;\not\Rightarrow\; \approx$

Theorem 17. *The following holds:*

- $\sim \;\not\Leftrightarrow\; \simeq_{SD}$ *and* $\sim \;\not\Leftrightarrow\; \simeq_{HD}$ *and* $\sim \;\not\Leftrightarrow\; \simeq_{THD}$ *and* $\simeq_{SD} \;\not\Rightarrow\; \sim$ *and* $\simeq_{HD} \;\not\Rightarrow\; \sim$ *and* $\simeq_{THD} \;\not\Rightarrow\; \sim$

[7] Note that the definition of weak bisimulation has been slightly modified to take into account the state labels.

– $\sim \implies \simeq_{SR}$ *and* $\sim \implies \simeq_{HR}$ *and* $\sim \implies \simeq_{THR}$ *and* $\simeq_{SR} \not\Longrightarrow \sim$ *and* $\simeq_{HR} \not\Longrightarrow \sim$ *and* $\simeq_{THR} \not\Longrightarrow \sim$

Connections between trace relations, stutter trace relations, strong bisimulation and weak bisimulation have been depicted in Fig. 7. A total of 121 connections have been investigated in this paper.

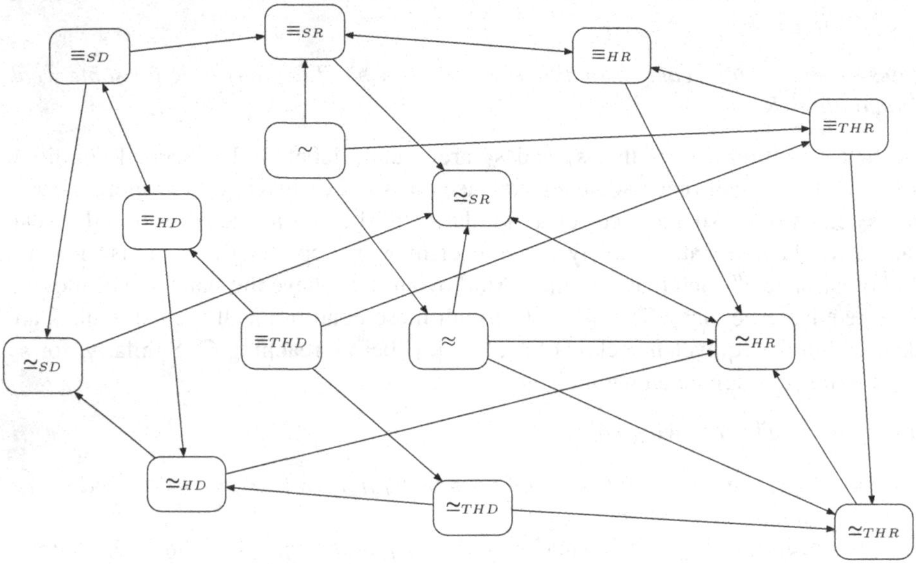

Fig. 7. Linear time-branching time spectrum for IMCs

8 Conclusions and Future Work

This paper presented several variants of trace and stutter trace equivalence for closed IMCs. These equivalences were obtained as a result of button pushing experiments performed on stochastic trace machines. We investigated the relationship among these trace equivalences and also compared them with strong bisimulation for IMCs. In the weak setting, we compared stutter trace equivalences with trace equivalences, bisimulation and weak bisimulation for IMCs. Additionally, we also established the relationship among these stutter trace equivalences. Finally, we used these results to sketch the linear time-branching time spectrum of equivalences for IMCs. Some interesting directions for future research are as follows:

– Investigate the connections between (stutter) trace equivalence and (weak) interactive Markovian equivalence (IME) [28].
– Define approximate (stutter) trace equivalence for IMCs.
– Study ready trace and failure trace semantics for IMCs and update the linear time-branching time spectrum of equivalences accordingly.
– Investigate the preservation of linear real-time objectives, e.g., MTL formulas [12, 27] and DTA specifications [1].

References

1. Alur, R., Dill, D.L.: A theory of timed automata. Theor. Comput. Sci. **126**(2), 183–235 (1994)
2. Ash, R.B., Doleans-Dade, C.A.: Probability and Measure Theory. Academic Press, San Diego (2000)
3. Baier, C., Haverkort, B.R., Hermanns, H., Katoen, J.-P.: Model-checking algorithms for continuous-time Markov chains. IEEE Trans. Software Eng. **29**(6), 524–541 (2003)
4. Baier, C., Katoen, J.-P.: Principles of Model Checking. MIT Press, Cambridge (2008)
5. Baier, C., Katoen, J.-P., Hermanns, H., Wolf, V.: Comparative branching-time semantics for Markov chains. Inf. Comput. **200**(2), 149–214 (2005)
6. Bernardo, M.: Non-bisimulation-based Markovian behavioral equivalences. J. Log. Algebraic Program. **72**(1), 3–49 (2007)
7. Bernardo, M.: Towards state space reduction based on T-lumpability-consistent relations. In: Thomas, N., Juiz, C. (eds.) EPEW 2008. LNCS, vol. 5261, pp. 64–78. Springer, Heidelberg (2008). https://doi.org/10.1007/978-3-540-87412-6_6
8. Böde, E., et al.: Compositional dependability evaluation for STATEMATE. IEEE Trans. Softw. Eng. **35**(2), 274–292 (2009)
9. Boudali, H., Crouzen, P., Haverkort, B.R., Kuntz, M., Stoelinga, M.: Architectural dependability evaluation with arcade. In: DSN, pp. 512–521. IEEE Computer Society (2008)
10. Boudali, H., Crouzen, P., Stoelinga, M.: A compositional semantics for dynamic fault trees in terms of interactive Markov chains. In: Namjoshi, K.S., Yoneda, T., Higashino, T., Okamura, Y. (eds.) ATVA 2007. LNCS, vol. 4762, pp. 441–456. Springer, Heidelberg (2007). https://doi.org/10.1007/978-3-540-75596-8_31
11. Boudali, H., Crouzen, P., Stoelinga, M.: Dynamic fault tree analysis using input/output interactive Markov chains. In: DSN, pp. 708–717. IEEE Computer Society (2007)
12. Bouyer, P.: From qualitative to quantitative analysis of timed systems. Mémoire d'habilitation, Université Paris 7, Paris, France, January 2009
13. Bozzano, M., Cimatti, A., Katoen, J., Nguyen, V.Y., Noll, T., Roveri, M.: Safety, dependability and performance analysis of extended AADL models. Comput. J. **54**(5), 754–775 (2011)
14. Coste, N., Hermanns, H., Lantreibecq, E., Serwe, W.: Towards performance prediction of compositional models in industrial GALS designs. In: Bouajjani, A., Maler, O. (eds.) CAV 2009. LNCS, vol. 5643, pp. 204–218. Springer, Heidelberg (2009). https://doi.org/10.1007/978-3-642-02658-4_18
15. Deng, Y., Hennessy, M.: On the semantics of Markov automata. In: Aceto, L., Henzinger, M., Sgall, J. (eds.) ICALP 2011. LNCS, vol. 6756, pp. 307–318. Springer, Heidelberg (2011). https://doi.org/10.1007/978-3-642-22012-8_24
16. Deng, Y., Hennessy, M.: On the semantics of Markov automata. Inf. Comput. **222**, 139–168 (2013)
17. Eisentraut, C., Hermanns, H., Zhang, L.: On probabilistic automata in continuous time. In: LICS, pp. 342–351 (2010)
18. Guck, D., Han, T., Katoen, J.-P., Neuhäußer, M.R.: Quantitative timed analysis of interactive Markov chains. In: Goodloe, A.E., Person, S. (eds.) NFM 2012. LNCS, vol. 7226, pp. 8–23. Springer, Heidelberg (2012). https://doi.org/10.1007/978-3-642-28891-3_4
19. Hermanns, H.: Interactive Markov Chains: And the Quest for Quantified Quality. LNCS, vol. 2428. Springer, Heidelberg (2002). https://doi.org/10.1007/3-540-45804-2
20. Hermanns, H., Katoen, J.-P.: The how and why of interactive Markov chains. In: de Boer, F.S., Bonsangue, M.M., Hallerstede, S., Leuschel, M. (eds.) FMCO 2009. LNCS, vol. 6286, pp. 311–337. Springer, Heidelberg (2010). https://doi.org/10.1007/978-3-642-17071-3_16

21. Hermanns, H., Katoen, J., Neuhäußer, M.R., Zhang, L.: GSPN model checking despite confusion. Technical report, RWTH Aachen UNIVERSITY (2010)
22. Katoen, J.-P., Kemna, T., Zapreev, I., Jansen, D.N.: Bisimulation minimisation mostly speeds up probabilistic model checking. In: Grumberg, O., Huth, M. (eds.) TACAS 2007. LNCS, vol. 4424, pp. 87–101. Springer, Heidelberg (2007). https://doi.org/10.1007/978-3-540-71209-1_9
23. Larsen, K.G., Skou, A.: Bisimulation through probabilistic testing. In: POPL, pp. 344–352 (1989)
24. Mateescu, R., Serwe, W.: A study of shared-memory mutual exclusion protocols using CADP. In: Kowalewski, S., Roveri, M. (eds.) FMICS 2010. LNCS, vol. 6371, pp. 180–197. Springer, Heidelberg (2010). https://doi.org/10.1007/978-3-642-15898-8_12
25. Neuhäußer, M.R.: Model checking non-deterministic and randomly timed systems. Ph.D. thesis, RWTH Aachen University (2010)
26. Neuhäußer, M.R., Katoen, J.-P.: Bisimulation and logical preservation for continuous-time Markov decision processes. In: Caires, L., Vasconcelos, V.T. (eds.) CONCUR 2007. LNCS, vol. 4703, pp. 412–427. Springer, Heidelberg (2007). https://doi.org/10.1007/978-3-540-74407-8_28
27. Ouaknine, J., Worrell, J.: Some recent results in metric temporal logic. In: Cassez, F., Jard, C. (eds.) FORMATS 2008. LNCS, vol. 5215, pp. 1–13. Springer, Heidelberg (2008). https://doi.org/10.1007/978-3-540-85778-5_1
28. Sharma, A.: Interactive Markovian equivalence. In: Reinecke, P., Di Marco, A. (eds.) EPEW 2017. LNCS, vol. 10497, pp. 33–49. Springer, Cham (2017). https://doi.org/10.1007/978-3-319-66583-2_3
29. Sharma, A.: Trace relations and logical preservation for continuous-time Markov decision processes. In: Hung, D., Kapur, D. (eds.) ICTAC 2017. LNCS, vol. 10580, pp. 192–209. Springer, Cham (2017)
30. Sharma, A.: Trace relations and logical preservation for Markov automata. In: Jansen, D.N., Prabhakar, P. (eds.) FORMATS 2018. LNCS, vol. 11022, pp. 162–178. Springer, Cham (2018). https://doi.org/10.1007/978-3-030-00151-3_10
31. Sharma, A.: Stuttering for Markov automata. In: TASE. IEEE Computer Society (2019)
32. Sharma, A., Katoen, J.-P.: Weighted lumpability on Markov chains. In: Clarke, E., Virbitskaite, I., Voronkov, A. (eds.) PSI 2011. LNCS, vol. 7162, pp. 322–339. Springer, Heidelberg (2012). https://doi.org/10.1007/978-3-642-29709-0_28
33. Song, L., Zhang, L., Godskesen, J.C.: Late weak bisimulation for Markov automata. CoRR, abs/1202.4116 (2012)
34. Glabbeek, R.J.: The linear time - branching time spectrum (extended abstract). In: Baeten, J.C.M., Klop, J.W. (eds.) CONCUR 1990. LNCS, vol. 458, pp. 278–297. Springer, Heidelberg (1990). https://doi.org/10.1007/BFb0039066
35. Wolf, V., Baier, C., Majster-Cederbaum, M.E.: Trace machines for observing continuous-time Markov chains. ENTCS **153**(2), 259–277 (2006)
36. Wolf, V., Baier, C., Majster-Cederbaum, M.E.: Trace semantics for stochastic systems with nondeterminism. Electr. Notes Theor. Comput. Sci. **164**(3), 187–204 (2006)

Computing Branching Distances Using Quantitative Games

Uli Fahrenberg[1(✉)], Axel Legay[2,3], and Karin Quaas[4]

[1] École polytechnique, Palaiseau, France
uli@lix.polytechnique.fr
[2] Université catholique de Louvain, Louvain-la-Neuve, Belgium
[3] Aalborg University, Aalborg, Denmark
[4] Universität Leipzig, Leipzig, Germany

Abstract. We lay out a general method for computing branching distances between labeled transition systems. We translate the quantitative games used for defining these distances to other, path-building games which are amenable to methods from the theory of quantitative games. We then show for all common types of branching distances how the resulting path-building games can be solved. In the end, we achieve a method which can be used to compute all branching distances in the linear-time–branching-time spectrum.

Keywords: Quantitative verification · Branching distance · Quantitative game · Path-building game

1 Introduction

During the last decade, formal verification has seen a trend towards modeling and analyzing systems which contain quantitative information. This is motivated by applications in real-time systems, hybrid systems, embedded systems and others. Quantitative information can thus be a variety of things: probabilities, time, tank pressure, energy intake, *etc.*

A number of quantitative models have hence been developed: probabilistic automata [43], stochastic process algebras [36], timed automata [2], hybrid automata [1], timed variants of Petri nets [30,42], continuous-time Markov chains [44], *etc.* Similarly, there is a number of specification formalisms for expressing quantitative properties: timed computation tree logic [35], probabilistic computation tree logic [31], metric temporal logic [37], stochastic continuous logic [3], *etc.*

Quantitative verification, *i.e.*, the checking of quantitative properties for quantitative systems, has also seen rapid development: for probabilistic systems in PRISM [38] and PEPA [27], for real-time systems in Uppaal [40],

U. Fahrenberg—This author's work is supported by the *Chaire ISC: Engineering Complex Systems* – École polytechnique – Thales – FX – DGA – Dassault Aviation – DCNS Research – ENSTA ParisTech – Télécom ParisTech.

R. M. Hierons and M. Mosbah (Eds.): ICTAC 2019, LNCS 11884, pp. 59–75, 2019.
https://doi.org/10.1007/978-3-030-32505-3_4

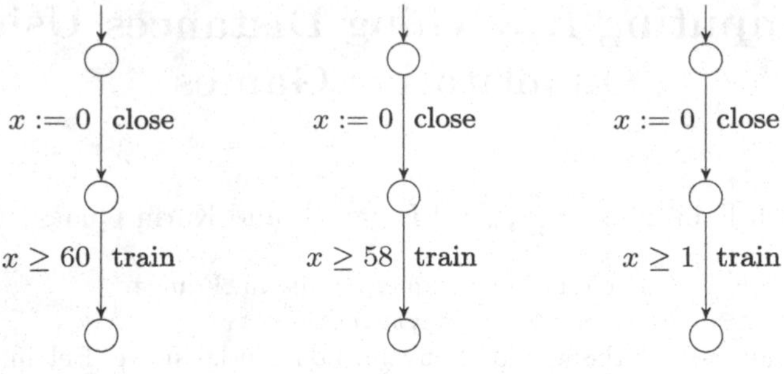

Fig. 1. Three timed automata modeling a train crossing.

RED [51], TAPAAL [5] and Romeo [26], and for hybrid systems in HyTech [33], SpaceEx [25] and HySAT [24], to name but a few.

Quantitative verification has, however, a problem of *robustness*. When the answers to model checking problems are Boolean—either a system meets its specification or it does not—then small perturbations in the system's parameters may invalidate the result. This means that, from a model checking point of view, small, perhaps unimportant, deviations in quantities are indistinguishable from larger ones which may be critical.

As an example, Fig. 1 shows three simple timed-automaton models of a train crossing, each modeling that once the gates are closed, some time will pass before the train arrives. Now assume that the specification of the system is

The gates have to be closed 60 seconds before the train arrives.

Model A does guarantee this property, hence satisfies the specification. Model B only guarantees that the gates are closed 58 s before the train arrives, and in model C, only one second may pass between the gates closing and the train.

Neither of models B and C satisfies the specification, so this is the result which a model checker like for example Uppaal would output. What this does not tell us, however, is that model C is dangerously far away from the specification, whereas model B only violates it slightly (and may be acceptable from a practical point of view given other constraints on the system which we have not modeled here).

In order to address the robustness problem, one approach is to replace the Boolean yes-no answers of standard verification with distances. That is, the Boolean co-domain of model checking is replaced by the non-negative real numbers. In this setting, the Boolean `true` corresponds to a distance of zero and `false` to the non-zero numbers, so that quantitative model checking can now tell us not only that a specification is violated, but also *how much* it is violated, or *how far* the system is from corresponding to its specification.

In the example of Fig. 1, and depending on precisely how one wishes to measure distances, the distance from A to our specification would be 0, whereas the distances from B and C to the specification may be 2 and 59, for example. The precise interpretation of distance values will be application-dependent; but in any case, it is clear that C is much farther away from the specification than B is.

The distance-based approach to quantitative verification has been developed in [8,11,28,34,46–48] and many other papers. Common to all these approaches is that they introduce distances between systems, or between systems and specifications, and then employ these for approximate or quantitative verification. However, depending on the application context, a plethora of different distances are being used. Consequently, there is a need for a general theory of quantitative verification which depends as little as possible on the concrete distances being used.

Different applications foster different types of quantitative verification, but it turns out that most of these essentially measure some type of distances between labeled transition systems. We have in [21] laid out a unifying framework which allows one to reason about such distance-based quantitative verification independently of the precise distance. This is essentially a general metric theory of labeled transition systems, with infinite quantitative games as its main theoretical ingredient and general fixed-point equations for linear and branching distances as one of its main results.

The work in [21] generalizes the linear-time–branching-time spectrum of preorders and equivalences from van Glabbeek's [50] to a quantitative linear-time–branching-time spectrum of distances, all parameterized on a given distance on traces, or executions; *cf.* Fig. 2. This is done by generalizing Stirling's bisimulation game [45] along two directions, both to cover all other preorders and equivalences in the linear-time–branching-time spectrum and into a game with quantitative (instead of Boolean) objectives.

What is missing in [21] are actual *algorithms* for computing the different types of distances. (The fixed-point equations mentioned above are generally defined over infinite lattices, hence Tarski's fixed-point theorem does not help here.) In this paper, we take a different route to compute them. We translate the general quantitative games used in [21] to other, path-building games. We show that under mild conditions, this translation can always be effectuated, and that for all common trace distances, the resulting path-building games can be solved using various methods which we develop.

We start the paper by reviewing the quantitative games used to define linear and branching distances in [21] in Sect. 2. Then we show the reduction to path-building games in Sect. 3 and apply this to show how to compute all common branching distances in Sect. 4. We collect our results in the concluding Sect. 5. The contributions of this paper are the following:

(1) A general method to reduce quantitative bisimulation-type games to path-building games. The former can be posed as *double* path-building games, where the players alternate to build *two* paths; we show how to transform such games into a form where the players instead build *one* common path.

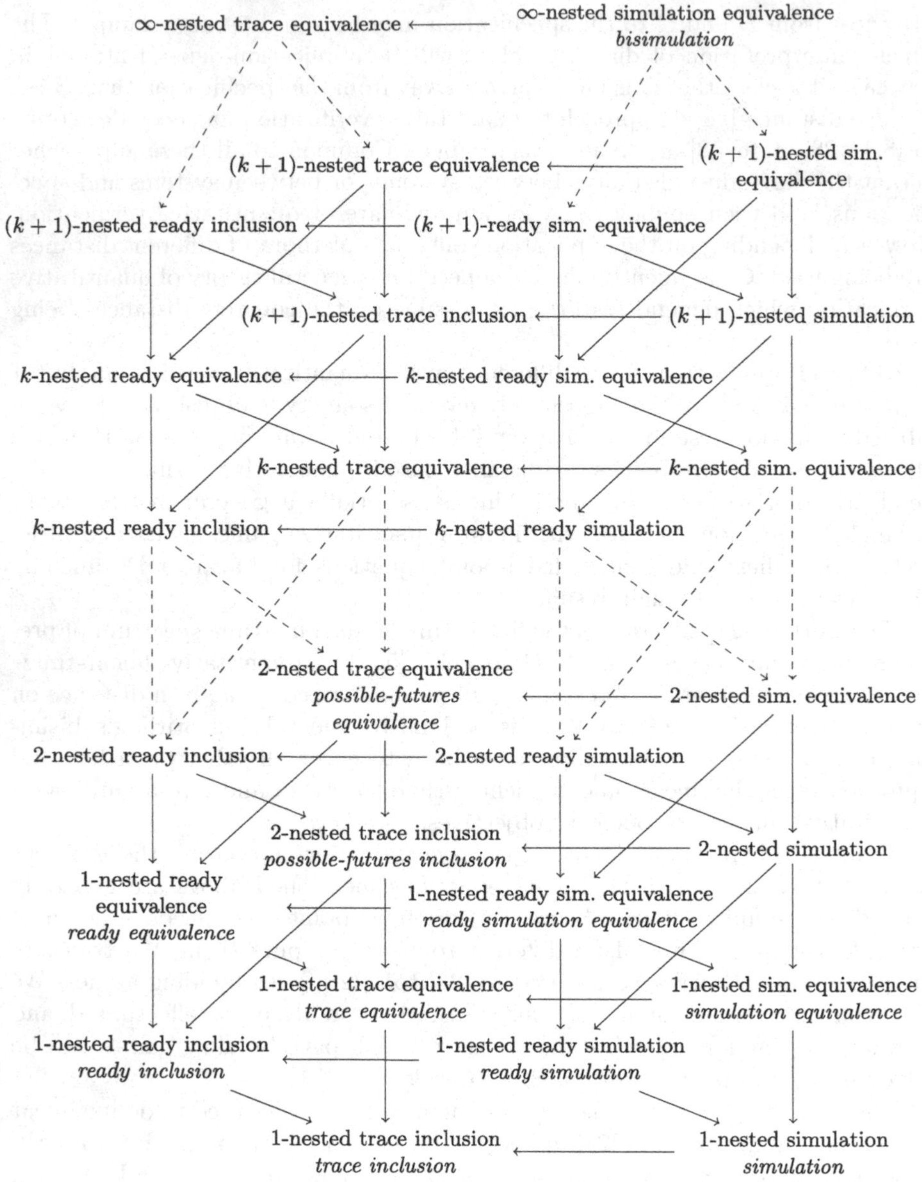

Fig. 2. The quantitative linear-time–branching-time spectrum from [21]. The nodes are different system distances, and an edge $d_1 \longrightarrow d_2$ or $d_1 \dashrightarrow d_2$ indicates that $d_1(s,t) \geq d_2(s,t)$ for all states s, t, and that d_1 and d_2 in general are topologically inequivalent.

(2) A collection of methods for solving different types of path-building games. Standard methods are available for solving discounted games and mean-payoff games; for other types we develop new methods.

(3) The application of the methods in (2) to compute various types of distances between labeled transition systems defined by the games of (1).

2 Linear and Branching Distances

Let Σ be a set of labels. Σ^ω denotes the set of infinite traces over Σ. We generally count sequences from index 0, so that $\sigma = (\sigma_0, \sigma_1, \dots)$. Let $\mathbb{R}_* = \mathbb{R}_{\geq 0} \cup \{\infty\}$ denote the extended non-negative real numbers.

2.1 Trace Distances

A *trace distance* is a hemimetric $D : \Sigma^\omega \times \Sigma^\omega \to \mathbb{R}_*$, *i.e.*, a function which satisfies $D(\sigma, \sigma) = 0$ and $D(\sigma, \tau) + D(\tau, \upsilon) \geq D(\sigma, \upsilon)$ for all $\sigma, \tau, \upsilon \in \Sigma^\omega$.

The following is an exhaustive list of different trace distances which have been used in different applications. We refer to [21] for more details and motivation.

The Discrete Trace Distance: $D_{\text{disc}}(\sigma, \tau) = 0$ if $\sigma = \tau$ and ∞ otherwise. This is equivalent to the standard Boolean setting: traces are either equal (distance 0) or not (distance ∞).

The Point-Wise Trace Distance: $D_{\sup}(\sigma, \tau) = \sup_{n \geq 0} d(\sigma_n, \tau_n)$, for any given label distance $d : \Sigma \times \Sigma \to \mathbb{R}_*$. This measures the greatest individual symbol distance in the traces and has been used for quantitative verification in, among others, [9,10,12,19,39,46].

The Discounted Trace Distance: $D_+(\sigma, \tau) = \sum_{n=0}^{\infty} \lambda^n d(\sigma_n, \tau_n)$, for any given *discounting factor* $\lambda \in [0, 1[$. Sometimes also called *accumulating* trace distance, this accumulates individual symbol distances along traces, using discounting to adjust the values of distances further off. It has been used in, for example, [6,19,39,46].

The Limit-Average Trace Distance: $D_{\text{lavg}}(\sigma, \tau) = \liminf_{n \geq 1} \frac{1}{n} \sum_{i=0}^{n-1} d(\sigma_i, \tau_i)$. This again accumulates individual symbol distances along traces and has been used in, among others, [6,7]. Both discounted and limit-average distances are well-known from the theory of discounted and mean-payoff games [16,52].

The Cantor Trace Distance: $D_{\text{C}}(\sigma, \tau) = \frac{1}{1 + \inf\{n \mid \sigma_n \neq \tau_n\}}$. This measures the (inverse of the) length of the common prefix of the traces and has been used for verification in [14].

The Maximum-Lead Trace Distance: $D_\pm(\sigma, \tau) = \sup_{n \geq 0} \left| \sum_{i=0}^{n} (\sigma_i - \tau_i) \right|$. Here it is assumed that Σ admits arithmetic operations of $+$ and $-$, for instance $\Sigma \subseteq \mathbb{R}$. As this measures differences of accumulated labels along runs, it is especially useful for real-time systems, *cf.* [20,34,46].

2.2 Labeled Transition Systems

A *labeled transition system* (LTS) over Σ is a tuple (S, i, T) consisting of a set of states S, with initial state $i \in S$, and a set of transitions $T \subseteq S \times \Sigma \times S$. We often write $s \xrightarrow{a} t$ to mean $(s, a, t) \in T$. We say that (S, i, T) is *finite* if S and T are finite. We assume our LTS to be *non-blocking* in the sense that for every state $s \in S$ there is a transition $(s, a, t) \in T$.

We have shown in [21] how any given trace distance D can be lifted to a quantitative linear-time–branching-time spectrum of distances on LTS. This is done via quantitative games as we shall review below. The point of [21] was that if the given trace distance has a recursive formulation, which, as we show in [21], every commonly used trace distance has, then the corresponding linear and branching distances can be formulated as fixed points for certain monotone functionals.

The fixed-point formulation of [21] does not, however, give rise to actual algorithms for computing linear and branching distances, as it happens more often than not that the mentioned monotone functionals are defined over infinite lattices. Concretely, this is the case for all but the point-wise trace distances in Sect. 2.1. Hence other methods are required for computing them; developing these is the purpose of this paper.

2.3 Quantitative Ehrenfeucht-Fraïssé Games

We review the quantitative games used in [21] to define different types of linear and branching distances for any given trace distance D. For conciseness, we only introduce *simulation games* and *bisimulation games* here, but similar definitions may be given for all equivalences and preorders in the linear-time–branching-time spectrum [50].

Quantitative Simulation Games. Let $\mathcal{S} = (S, i, T)$ and $\mathcal{S}' = (S', i', T')$ be LTS and $D : \Sigma^\omega \times \Sigma^\omega \to \mathbb{R}_*$ a trace distance. The *simulation game* from $\mathcal{S}$ to $\mathcal{S}'$ is played by two players, the maximizer and the minimizer. A play begins with the maximizer choosing a transition $(s_0, a_0, s_1) \in T$ with $s_0 = i$. Then the minimizer chooses a transition $(s_0', a_0', s_1') \in T'$ with $s_0' = i'$. Now the maximizer chooses a transition $(s_1, a_1, s_2) \in T$, then the minimizer chooses a transition $(s_1', a_1', s_2') \in T'$, and so on indefinitely. Hence this is what should be called a *double path-building game*: the players each build, independently, an infinite path in their respective LTS.

A *play* hence consists of two infinite paths, π starting from i, and π' starting from i'. The *utility* of this play is the distance $D(\sigma, \sigma')$ between the traces σ, σ' of the paths π and π', which the maximizer wants to maximize and the minimizer wants to minimize. The *value* of the game is, then, the utility of the play which results when both maximizer and minimizer are playing optimally.

To formalize the above intuition, we define a *configuration* for the maximizer to be a pair (π, π') of finite paths of equal length, π in $\mathcal{S}$ and starting in i, π' in $\mathcal{S}'$ starting in i'. The intuition is that this covers the *history* of a play; the choices both players have made up to a certain point in the game. Hence a configuration for the minimizer is a similar pair (π, π') of finite paths, but now π is one step longer than π'.

A *strategy* for the maximizer is a mapping from maximizer configurations to transitions in $\mathcal{S}$, fixing the maximizer's choice of a move in the given configuration. Denoting the set of maximizer configurations by Conf, such a strategy is hence a mapping $\theta : \mathsf{Conf} \to T$ such that for all $(\pi, \pi') \in \mathsf{Conf}$ with $\theta(\pi, \pi') = (s, a, t)$, we have $\mathsf{end}(\pi) = s$. Here $\mathsf{end}(\pi)$ denotes the last state of π. Similarly, and denoting the set of minimizer configurations by Conf', a strategy for the minimizer is a mapping $\theta' : \mathsf{Conf}' \to T'$ such that for all $(\pi, \pi') \in \mathsf{Conf}'$ with $\theta'(\pi, \pi') = (s', a', t')$, $\mathsf{end}(\pi') = s'$.

Denoting the sets of these strategies by Θ and Θ', respectively, we can now define the *simulation distance* from $\mathcal{S}$ to $\mathcal{S}'$ induced by the trace distance D, denoted $D^{\text{sim}}(\mathcal{S}, \mathcal{S}')$, by

$$D^{\text{sim}}(\mathcal{S}, \mathcal{S}') = \sup_{\theta \in \Theta} \inf_{\theta' \in \Theta'} D(\sigma(\theta, \theta'), \sigma'(\theta, \theta')),$$

where $\sigma(\theta, \theta')$ and $\sigma'(\theta, \theta')$ are the traces of the paths $\pi(\theta, \theta')$ and $\pi'(\theta, \theta')$ induced by the pair of strategies (θ, θ').

Remark 1. If the trace distance D is discrete, *i.e.*, $D = D_{\text{disc}}$ as in Sect. 2.1, then the quantitative game described above reduces to the well-known *simulation game* [45]: The only choice the minimizer has for minimizing the value of the game is to always choose a transition with the same label as the one just chosen by the maximizer; similarly, the maximizer needs to try to force the game into states where she can choose a transition which the minimizer cannot match. Hence the value of the game will be 0 if the minimizer always can match the maximizer's labels, that is, iff $\mathcal{S}$ is simulated by $\mathcal{S}'$.

Quantitative Bisimulation Games. There is a similar game for computing the *bisimulation distance* between LTS $\mathcal{S}$ and $\mathcal{S}'$. Here we give the maximizer the choice, at each step, to either choose a transition from s_k as before, or to "switch sides" and choose a transition from s'_k instead; the minimizer then has to answer with a transition on the other side.

Hence the players are still building two paths, one in each LTS, but now they are *both* contributing to *both* paths. The utility of such a play is still the distance between these two paths, which the maximizer wants to maximize and the minimizer wants to minimize. The *bisimulation distance* between $\mathcal{S}$ and $\mathcal{S}'$, denoted $D^{\text{bisim}}(\mathcal{S}, \mathcal{S}')$, is then defined to be the value of this quantitative bisimulation game.

Remark 2. If the trace distance $D = D_{\text{disc}}$ is discrete, then using the same arguments as in Remark 1, we see that $D^{\text{bisim}}_{\text{disc}}(\mathcal{S}, \mathcal{S}') = 0$ iff $\mathcal{S}$ and $\mathcal{S}'$ are *bisimilar*. The game which results being played is precisely the bisimulation game of [45], which also has been introduced by Fraïssé [23] and Ehrenfeucht [15] in other contexts.

The Quantitative Linear-Time–Branching-Time Spectrum. The above-defined quantitative simulation and bisimulation games can be generalized using different methods. One is to introduce a *switch counter* sc into the game which counts how often the maximizer has switched sides during an ongoing game. Then one can limit the maximizer's capabilities by imposing limits on sc: if the limit is $\mathsf{sc} = 0$, then the players are playing a simulation game; if there is no limit ($\mathsf{sc} \le \infty$), they are playing a bisimulation game. Other limits $\mathsf{sc} \le k$, for $k \in \mathbb{N}$, can be used to define *k-nested simulation distances*, generalizing the equivalences and preorders from [29,32].

Another method of generalization is to introduce *ready moves* into the game. These consist of the maximizer challenging her opponent by switching sides, but only requiring that the minimizer match the chosen transition; afterwards the game finishes. This can be employed to introduce the *ready simulation distance* of [41] and, combined with the switch counter method above, the *ready k-nested simulation distance*. We refer to [21] for further details on these and other variants of quantitative (bi)simulation games.

For reasons of exposition, we will below introduce our reduction to path-building games only for the quantitative simulation and bisimulation games; but all our work can easily be transferred to the general setting of [21].

3 Reduction

In order to compute simulation and bisimulation distances, we translate the games of the previous section to path-building games à la Ehrenfeucht-Mycielski [16]. Let $D : \Sigma^\omega \times \Sigma^\omega \to \mathbb{R}_*$ be a trace distance, and assume that there are functions $\mathsf{val}_D : \mathbb{R}_*^\omega \to \mathbb{R}_*$ and $f_D : \Sigma \times \Sigma \to \mathbb{R}_*$ for which it holds, for all $\sigma, \tau \in \Sigma^\infty$, that

$$D(\sigma, \tau) = \mathsf{val}_D(0, f_D(\sigma_0, \tau_0), 0, f_D(\sigma_1, \tau_1), 0, \dots). \tag{1}$$

We will need these functions in our translation, and we show in Sect. 3.2 below that they exist for all common trace distances.

3.1 Simulation Distance

Let $\mathcal{S} = (S, i, T)$ and $\mathcal{S}' = (S', i', T')$ be LTS. We construct a turn-based game $\mathcal{U} = \mathcal{U}(\mathcal{S}, \mathcal{S}') = (U, u_0, \dashrightarrow)$ as follows, with $U = U_1 \cup U_2$:

$$U_1 = S \times S' \qquad U_2 = S \times S' \times \Sigma \qquad u_0 = (i, i')$$
$$\dashrightarrow = \{(s, s') \overset{0}{\dashrightarrow} (t, s', a) \mid (s, a, t) \in T\}$$
$$\cup \{(t, s', a) \xrightarrow{f_D(a, a')} (t, t') \mid (s', a', t') \in T'\}$$

This is a two-player game. We again call the players maximizer and minimizer, with the maximizer controlling the states in U_1 and the minimizer the ones in

U_2. Transitions are labeled with extended real numbers, but as the image of f_D in $\mathbb{R}_*$ is finite, the set of transition labels in U is finite.

The game on $\mathcal{U}$ is played as follows. A play begins with the maximizer choosing a transition $(u_0, a_0, u_1) \in \twoheadrightarrow$ with $u_0 = i$. Then the minimizer chooses a transition $(u_1, a_1, u_2) \in \twoheadrightarrow$. Then the maximizer chooses a transition $(u_2, a_2, u_3) \in \twoheadrightarrow$, and so on indefinitely (note that $\mathcal{U}$ is non-blocking). A play thus induces an infinite path $\pi = (u_0, a_0, u_1), (u_1, a_1, u_2), \ldots$ in $\mathcal{U}$ with $u_0 = i$. The goal of the maximizer is to maximize the value $\mathsf{val}_D(\mathcal{U}) := \mathsf{val}_D(a_0, a_1, \ldots)$ of the trace of π; the goal of the minimizer is to minimize this value.

This is hence a path-building game, variations of which (for different valuation functions) have been studied widely in both economics and computer science since Ehrenfeucht-Mycielski's [16]. Formally, configurations and strategies are given as follows. A configuration of the maximizer is a path π_1 in $\mathcal{U}$ with $\mathsf{end}(\pi_1) \in U_1$, and a configuration of the minimizer is a path π_2 in $\mathcal{U}$ with $\mathsf{end}(\pi_2) \in U_2$. Denote the sets of these configurations by Conf_1 and Conf_2, respectively. A strategy for the maximizer is, then, a mapping $\theta_1 : \mathsf{Conf}_1 \to \twoheadrightarrow$ such that for all $\pi_1 \in \mathsf{Conf}_1$ with $\theta_1(\pi_1) = (u, x, v)$, $\mathsf{end}(\pi_1) = u$. A strategy for the minimizer is a mapping $\theta_2 : \mathsf{Conf}_2 \to \twoheadrightarrow$ such that for all $\pi_2 \in \mathsf{Conf}_2$ with $\theta_2(\pi_2) = (u, x, v)$, $\mathsf{end}(\pi_2) = u$. Denoting the sets of these strategies by Θ_1 and Θ_2, respectively, we can now define

$$\mathsf{val}_D(\mathcal{U}) = \sup_{\theta_1 \in \Theta_1} \inf_{\theta_2 \in \Theta_2} \mathsf{val}_D(\sigma(\theta_1, \theta_2)),$$

where $\sigma(\theta_1, \theta_2)$ is the trace of the path $\pi(\theta_1, \theta_2)$ induced by the pair of strategies (θ_1, θ_2).

By the next theorem, the value of $\mathcal{U}$ is precisely the simulation distance from $\mathcal{S}$ to $\mathcal{S}'$.

Theorem 3. *For all LTS $\mathcal{S}$, $\mathcal{S}'$, $D^{sim}(\mathcal{S}, \mathcal{S}') = \mathsf{val}_D(\mathcal{U}(\mathcal{S}, \mathcal{S}'))$.*

Proof. Write $\mathcal{S} = (S, i, T)$ and $\mathcal{S}' = (S', i', T')$. Informally, the reason for the equality is that any move $(s, a, t) \in T$ of the maximizer in the simulation distance game can be copied to a move $(s, s') \xrightarrow{0} (t, s', a)$, regardless of s', in $\mathcal{U}$. Similarly, any move (s', a', t') of the minimizer can be copied to a move $(t, s', a) \xrightarrow{f_D(a,a')} (t, t')$, and all the moves in $\mathcal{U}$ are of this form.

To turn this idea into a formal proof, we show that there are bijections between configurations and strategies in the two games, and that under these bijections, the utilities of the two games are equal. For $(\pi, \pi') \in \mathsf{Conf}$ in the simulation distance game, with $\pi = (s_0, a_0, s_1), \ldots, (s_{n-1}, a_{n-1}, s_n)$ and $\pi' = (s'_0, a'_0, s'_1), \ldots, (s'_{n-1}, a'_{n-1}, s'_n)$, define

$$\begin{aligned}\phi_1(\pi, \pi') =& ((s_0, s'_0), 0, (s_1, s'_0, a_0)), ((s_1, s'_0, a_0), f_D(a_0, a'_0), (s_1, s'_1)), \ldots, \\ & ((s_n, s'_{n-1}, a_{n-1}), f_D(a_{n-1}, a'_{n-1}), (s_n, s'_n)).\end{aligned}$$

It is clear that this defines a bijection $\phi_1 : \mathsf{Conf} \to \mathsf{Conf}_1$, and that one can similarly define a bijection $\phi_2 : \mathsf{Conf}' \to \mathsf{Conf}_2$.

Now for every strategy $\theta : \mathsf{Conf} \to T$ in the simulation distance game, define a strategy $\psi_1(\theta) = \theta_1 \in \Theta_1$ as follows. For $\pi_1 \in \mathsf{Conf}_1$, let $(\pi, \pi') = \phi_1^{-1}(\pi_1)$ and $s' = \mathsf{end}(\pi')$. Let $\theta(\pi, \pi') = (s, a, t)$ and define $\theta_1(\pi_1) = ((s, s'), 0, (t, s', a))$. Similarly we define a mapping $\psi_2 : \Theta' \to \Theta_2$ as follows. For $\theta' : \mathsf{Conf}' \to T'$ and $\pi_2 \in \mathsf{Conf}_2$, let $(\pi, \pi') = \phi_2^{-1}(\pi_2)$ with $\pi = (s_0, a_0, s_1), \dots, (s_n, a_n, s_{n+1})$. Let $\theta'(\pi, \pi') = (s', a', t')$ and define $\psi_2(\theta')(\pi_2) = ((s_{n+1}, s', a_n), f_D(a_n, a'), (s_{n+1}, t'))$.

It is clear that ψ_1 and ψ_2 indeed map strategies in the simulation distance game to strategies in $\mathcal{U}$ and that both are bijections. Also, for each pair $(\theta, \theta') \in \Theta \times \Theta'$, $D(\sigma(\theta, \theta'), \sigma'(\theta, \theta')) = \mathsf{val}_D(\sigma(\psi_1(\theta), \psi_2(\theta')))$ by construction. But then

$$\begin{aligned} D^{\text{sim}}(\mathcal{S}, \mathcal{S}') &= \sup_{\theta \in \Theta} \inf_{\theta' \in \Theta'} D(\sigma(\theta, \theta'), \sigma'(\theta, \theta')) \\ &= \sup_{\theta \in \Theta} \inf_{\theta' \in \Theta'} \mathsf{val}_D(\sigma(\psi_1(\theta), \psi_2(\theta'))) \\ &= \sup_{\theta_1 \in \Theta_1} \inf_{\theta_2 \in \Theta_2} \mathsf{val}_D(\sigma(\theta_1, \theta_2)) = \mathsf{val}_D(\mathcal{U}), \end{aligned}$$

the third equality because ψ_1 and ψ_2 are bijections. □

3.2 Examples

We show that the reduction applies to all trace distances from Sect. 2.1.

1. For the discrete trace distance $D = D_{\text{disc}}$, we let

$$\mathsf{val}_D(x) = \sum_{n=0}^{\infty} x_n, \qquad f_D(a, b) = \begin{cases} 0 & \text{if } a = b, \\ \infty & \text{otherwise,} \end{cases}$$

   then (1) holds. In the game on $\mathcal{U}$, the minimizer needs to play 0-labeled transitions to keep the distance at 0.
2. For the point-wise trace distance $D = D_{\text{sup}}$, we can let

$$\mathsf{val}_D(x) = \sup_{n \ge 0} x_n, \qquad f_D(a, b) = d(a, b).$$

   Hence the game on $\mathcal{U}$ computes the sup of a trace.
3. For the discounted trace distance $D = D_+$, let

$$\mathsf{val}_D(x) = \sum_{n=0}^{\infty} \sqrt{\lambda}^n x_n, \qquad f_D(a, b) = \sqrt{\lambda}\, d(a, b),$$

   then (1) holds. Hence the game on $\mathcal{U}$ is a standard discounted game [52].
4. For the limit-average trace distance $D = D_{\text{lavg}}$, we can let

$$\mathsf{val}_D(x) = \liminf_{n \ge 1} \frac{1}{n} \sum_{i=0}^{n-1} x_i, \qquad f_D(a, b) = 2d(a, b)\,;$$

   we will show below that (1) holds. Hence the game on $\mathcal{U}$ is a mean-payoff game [52].

5. For the Cantor trace distance $D = D_{\mathrm{C}}$, let

$$\mathsf{val}_D(x) = \frac{2}{1 + \inf\{n \mid x_n \neq 0\}}, \qquad f_D(a,b) = \begin{cases} 0 & \text{if } a = b, \\ 1 & \text{otherwise.} \end{cases}$$

The objective of the maximizer in this game is to reach a transition with weight 1 *as soon as possible.*

6. For the maximum-lead trace distance $D = D_{\pm}$, we can let

$$\mathsf{val}_D(x) = \sup_{n \geq 0} \Big| \sum_{i=0}^{n} x_i \Big|, \qquad f_D(a,b) = a - b,$$

then (1) holds.

3.3 Bisimulation Distance

We can construct a similar turn-based game to compute the bisimulation distance. Let $\mathcal{S} = (S, i, T)$ and $\mathcal{S}' = (S', i', T')$ be LTS and define $\mathcal{V} = \mathcal{V}(\mathcal{S}, \mathcal{S}') = (V, v_0, \dashrightarrow)$ as follows, with $V = V_1 \cup V_2$:

$$V_1 = S \times S' \qquad V_2 = S \times S' \times \Sigma \times \{1,2\} \qquad v_0 = (i, i')$$

$$\begin{aligned} \dashrightarrow = &\{(s,s') \xrightarrow{0} (t, s', a, 1) \mid (s,a,t) \in T\} \\ \cup &\{(s,s') \xrightarrow{0} (s, t', a', 2) \mid (s', a', t') \in T'\} \\ \cup &\{(t, s', a, 1) \xrightarrow{f_D(a,a')} (t, t') \mid (s', a', t') \in T'\} \\ \cup &\{(s, t', a', 2) \xrightarrow{f_D(a,a')} (t, t') \mid (s, a, t) \in T\} \end{aligned}$$

Here we have used the minimizer's states to both remember the label choice of the maximizer and which side of the bisimulation game she plays on. By suitable modifications, we can construct similar games for all distances in the spectrum of [21]. The next theorem states that the value of $\mathcal{V}$ is precisely the bisimulation distance between $\mathcal{S}$ and $\mathcal{S}'$.

Theorem 4. *For all LTS $\mathcal{S}$, $\mathcal{S}'$, $D^{bisim}(\mathcal{S}, \mathcal{S}') = \mathsf{val}_D(\mathcal{V}(\mathcal{S}, \mathcal{S}'))$.*

Proof. This proof is similar to the one of Theorem 3, only that now, we have to take into account that the maximizer may "switch sides". The intuition is that maximizer moves (s, a, t) in the $\mathcal{S}$ component of the bisimulation distance games are emulated by moves $(s, s') \xrightarrow{0} (t, s', a, 1)$, maximizer moves (s', a', t') in the $\mathcal{S}'$ component are emulated by moves $(s, s') \xrightarrow{0} (s, t', a', 2)$, and similarly for the minimizer. The values 1 and 2 in the last component of the V_2 states ensure that the minimizer only has moves available which correspond to playing in the correct component in the bisimulation distance game (*i.e.*, that ψ_2 is a bijection). □

4 Computing the Values of Path-Building Games

We show here how to compute the values of the different path-building games which we saw in the last section. This will give us algorithms to compute all simulation and bisimulation distances associated with the trace distances of Sect. 2.1.

We will generally only refer to the games $\mathcal{U}$ for computing simulation distance here, but the bisimulation distance games $\mathcal{V}$ are very similar, and everything we say also applies to them.

Discrete Distance: The game to compute the discrete simulation distances is a reachability game, in that the goal of the maximizer is to force the minimizer into a state from which she can only choose ∞-labeled transitions. We can hence solve them using the standard controllable-predecessor operator defined, for any set $S \subseteq U_1$ of maximizer states, by

$$\mathsf{cpre}(S) = \{u_1 \in U_1 \mid \exists u_1 \xrightarrow{0} u_2 : \forall u_2 \xrightarrow{x} u_3 : u_3 \in S\}.$$

Now let $S \subseteq U_1$ be the set of states from which the maximizer can force the game into a state from which the minimizer only has ∞-labeled transitions, *i.e.*,

$$S = \{u_1 \in U_1 \mid \exists u_1 \xrightarrow{0} u_2 : \forall u_2 \xrightarrow{x} u_3 : x = \infty\},$$

and compute $S^* = \mathsf{cpre}^*(S) = \bigcup_{n \geq 0} \mathsf{cpre}^n(S)$. By monotonicity of cpre and as the subset lattice of U_1 is complete and finite, this computation finishes in at most $|U_1|$ steps.

Lemma 5. $\mathsf{val}_D(\mathcal{U}) = 0$ *iff* $u_0 \notin S^*$.

Proof. As we are working with the discrete distance, we have either $\mathsf{val}_D(\mathcal{U}) = 0$ or $\mathsf{val}_D(\mathcal{U}) = \infty$. Now $u_o \in S^*$ iff the maximizer can force, using finitely many steps, the game into a state from which the minimizer only has ∞-labeled transitions, which is the same as $\mathsf{val}_D(\mathcal{U}) = \infty$. □

Point-Wise Distance: To compute the value of the point-wise simulation distance game, let $W = \{w_1, \ldots, w_m\}$ be the (finite) set of weights of the minimizer's transitions, ordered such that $w_1 < \cdots < w_m$. For each $i = 1, \ldots, m$, let $S_i = \{u_1 \in U_1 : \exists u_1 \xrightarrow{0} u_2 : \forall u_2 \xrightarrow{x} u_3 : x \geq w_i\}$ be the set of maximizer states from which the maximizer can force the minimizer into a transition with weight at least w_i; note that $S_m \subseteq S_{m-1} \subseteq \cdots \subseteq S_1 = U_1$. For each $i = 1, \ldots, m$, compute $S_i^* = \mathsf{cpre}^*(S_i)$, then $S_m^* \subseteq S_{m-1}^* \subseteq \cdots \subseteq S_1^* = U_1$.

Lemma 6. *Let p be the greatest index for which $u_0 \in S_p^*$, then $p = \mathsf{val}_D(\mathcal{U})$.*

Proof. For any k, we have $u_0 \in S_k^*$ iff the maximizer can force, using finitely many steps, the game into a state from which the minimizer only has transitions with weight at least w_k. Thus $u_0 \in S_p^*$ iff (1) the maximizer can force the minimizer into a w_p-weighted transition; (2) the maximizer *cannot* force the minimizer into a w_{p+1}-weighted transition. □

Discounted Distance: The game to compute the discounted simulation distance is a standard discounted game and can be solved by standard methods [52].

Limit-Average Distance: For the limit-average simulation distance game, let $(y_n)_{n\geq 1}$ be the sequence $(1, 1, \frac{3}{2}, 1, \frac{5}{4}, \dots)$ and note that $\lim_{n\to\infty} y_n = 1$. Then

$$\begin{aligned}\mathsf{val}_D(x) = \mathsf{val}_D(x) \lim_{n\to\infty} y_n &= \liminf_{n\geq 1} \frac{y_n}{n} \sum_{i=0}^{n-1} x_i \\ &= \liminf_{2k\geq 1} \frac{1}{2k} \sum_{i=0}^{k-1} f_D(\sigma_i, \tau_i) \\ &= \liminf_{k\geq 1} \frac{1}{k} \sum_{i=0}^{k-1} d(\sigma_i, \tau_i) = D_{\text{lavg}}(\sigma, \tau),\end{aligned}$$

so, indeed, (1) holds. The game is a standard mean-payoff game and can be solved by standard methods, see for example [13].

Cantor Distance: To compute the value of the Cantor simulation distance game, let $S_1 \subseteq U_1$ be the set of states from which the maximizer can force the game into a state from which the minimizer only has 1-labeled transitions, *i.e.*, $S_1 = \{u_1 \in U_1 \mid \exists u_1 \xrightarrow{0} u_2 : \forall u_2 \xrightarrow{x} u_3 : x = 1\}$. Now recursively compute $S_{i+1} = S_i \cup \mathsf{cpre}(S_i)$, for $i = 1, 2, \dots$, until $S_{i+1} = S_i$ (which, as $S_i \subseteq S_{i+1}$ for all i and U_1 is finite, will happen eventually). Then S_i is the set of states from which the maximizer can force the game to a 1-labeled minimizer transition which is at most $2i$ steps away. Hence $\mathsf{val}_D(\mathcal{U}) = 0$ if there is no p for which $u_0 \in S_p$, and otherwise $\mathsf{val}_D(\mathcal{U}) = \frac{1}{p}$, where p is the least index for which $u_0 \in S_p$.

Maximum-Lead Distance: For the maximum-lead simulation distance game, we note that the maximizer wants to maximize $\sup_{n\geq 0}\left|\sum_{i=0}^{n} x_i\right|$, *i.e.*, wants the accumulated values $\sum_{i=0}^{n} x_i$ or $-\sum_{i=0}^{n} x_i$ to exceed any prescribed bounds. A weighted game in which one player wants to keep accumulated values inside some given bounds, while the opponent wants to exceed these bounds, is called an *interval-bound energy game.* It is shown in [4] that solving general interval-bound energy games is EXPTIME-complete.

We can reduce the problem of computing maximum-lead simulation distance to an interval-bound energy game by first non-deterministically choosing a bound k and then checking whether player 1 wins the interval-bound energy game on $\mathcal{U}$ for bounds $[-k, k]$. (There is a slight problem in that in [4], energy games are defined only for *integer*-weighted transition systems, whereas we are dealing with real weights here. However, it is easily seen that the results of [4] also apply to *rational* weights and bounds; and as our transition systems are finite, one can always find a sound and complete rational approximation.)

We can thus compute maximum-lead simulation distance in non-deterministic exponential time; we leave open for now the question whether there is a more efficient algorithm.

5 Conclusion and Future Work

We sum up our results in the following corollary which gives the complexities of the decision problems associated with the respective distance computations. Note that the first part restates the well-known fact that simulation and bisimulation are decidable in polynomial time.

Corollary 7

1. *Discrete simulation and bisimulation distances are computable in PTIME.*
2. *Point-wise simulation and bisimulation distances are computable in PTIME.*
3. *Discounted simulation and bisimulation distances are computable in NP ∩ coNP.*
4. *Limit-average simulation and bisimulation distances are computable in NP ∩ coNP.*
5. *Cantor simulation and bisimulation distances are computable in PTIME.*
6. *Maximum-lead simulation and bisimulation distances are computable in NEXPTIME.*

In the future, we intend to expand our work to also cover *quantitative specification theories.* Together with several coauthors, we have in [17,18] developed a comprehensive setting for satisfaction and refinement distances in quantitative specification theories. Using our work in [22] on a qualitative linear-time–branching-time spectrum of specification theories, we plan to introduce a quantitative linear-time–branching-time spectrum of specification distances and to use the setting developed here to devise methods for computing them through path-building games.

Another possible extension of our work contains *probabilistic* systems, for example the probabilistic automata of [43]. A possible starting point for this is [49] which uses simple stochastic games to compute probabilistic bisimilarity.

References

1. Alur, R., et al.: The algorithmic analysis of hybrid systems. Theor. Comput. Sci. **138**(1), 3–34 (1995)
2. Alur, R., Dill, D.L.: A theory of timed automata. Theor. Comput. Sci. **126**(2), 183–235 (1994)
3. Aziz, A., Sanwal, K., Singhal, V., Brayton, R.K.: Model-checking continous-time Markov chains. ACM Trans. Comput. Log. **1**(1), 162–170 (2000)
4. Bouyer, P., Fahrenberg, U., Larsen, K.G., Markey, N., Srba, J.: Infinite runs in weighted timed automata with energy constraints. In: Cassez, F., Jard, C. (eds.) FORMATS 2008. LNCS, vol. 5215, pp. 33–47. Springer, Heidelberg (2008). https://doi.org/10.1007/978-3-540-85778-5_4
5. Byg, J., Jørgensen, K.Y., Srba, J.: TAPAAL: editor, simulator and verifier of timed-arc Petri nets. In: Liu, Z., Ravn, A.P. (eds.) ATVA 2009. LNCS, vol. 5799, pp. 84–89. Springer, Heidelberg (2009). https://doi.org/10.1007/978-3-642-04761-9_7
6. Černý, P., Henzinger, T.A., Radhakrishna, A.: Simulation distances. Theor. Comput. Sci. **413**(1), 21–35 (2012)

7. Chatterjee, K., Doyen, L., Henzinger, T.A.: Quantitative languages. ACM Trans. Comput. Log. **11**(4), 23:1–23:38 (2010)
8. de Alfaro, L., Faella, M., Henzinger, T.A., Majumdar, R., Stoelinga, M.: Model checking discounted temporal properties. Theor. Comput. Sci. **345**(1), 139–170 (2005)
9. de Alfaro, L., Faella, M., Stoelinga, M.: Linear and branching system metrics. IEEE Trans. Software Eng. **35**(2), 258–273 (2009)
10. de Alfaro, L., Henzinger, T.A., Majumdar, R.: Discounting the future in systems theory. In: Baeten, J.C.M., Lenstra, J.K., Parrow, J., Woeginger, G.J. (eds.) ICALP 2003. LNCS, vol. 2719, pp. 1022–1037. Springer, Heidelberg (2003). https://doi.org/10.1007/3-540-45061-0_79
11. Desharnais, J., Gupta, V., Jagadeesan, R., Panangaden, P.: Metrics for labelled Markov processes. Theor. Comput. Sci. **318**(3), 323–354 (2004)
12. Desharnais, J., Laviolette, F., Tracol, M.: Approximate analysis of probabilistic processes. In: QEST, pp. 264–273. IEEE Computer Society (2008)
13. Dhingra, V., Gaubert, S.: How to solve large scale deterministic games with mean payoff by policy iteration. In: Lenzini, L., Cruz, R.L. (eds.) VALUETOOLS. ACM International Conference Proceedings, vol. 180, p. 12. ACM (2006)
14. Doyen, L., Henzinger, T.A., Legay, A., Ničković, D.: Robustness of sequential circuits. In: Gomes, L., Khomenko, V., Fernandes, J.M. (eds.) ACSD, pp. 77–84. IEEE Computer Society, Washington, D.C. (2010)
15. Ehrenfeucht, A.: An application of games to the completeness problem for formalized theories. Fund. Math. **49**, 129–141 (1961)
16. Ehrenfeucht, A., Mycielski, J.: Positional strategies for mean payoff games. Int. J. Game Theory **8**, 109–113 (1979)
17. Fahrenberg, U., Křetínský, J., Legay, A., Traonouez, L.-M.: Compositionality for quantitative specifications. In: Lanese, I., Madelaine, E. (eds.) FACS 2014. LNCS, vol. 8997, pp. 306–324. Springer, Cham (2015). https://doi.org/10.1007/978-3-319-15317-9_19
18. Fahrenberg, U., Křetínský, J., Legay, A., Traonouez, L.-M.: Compositionality for quantitative specifications. Soft. Comput. **22**(4), 1139–1158 (2018)
19. Fahrenberg, U., Larsen, K.G., Thrane, C.: A quantitative characterization of weighted Kripke structures in temporal logic. Comput. Inform. **29**(6+), 1311–1324 (2010)
20. Fahrenberg, U., Legay, A.: A robust specification theory for modal event-clock automata. In: Bauer, S.S., Raclet, J.-B. (eds.) Proceedings Fourth Workshop on Foundations of Interface Technologies, FIT 2012. EPTCS, Tallinn, Estonia, 25 March 2012, vol. 87, pp. 5–16 (2012)
21. Fahrenberg, U., Legay, A.: The quantitative linear-time-branching-time spectrum. Theor. Comput. Sci. **538**, 54–69 (2014)
22. Fahrenberg, U., Legay, A.: A linear-time–branching-time spectrum of behavioral specification theories. In: Steffen, B., Baier, C., van den Brand, M., Eder, J., Hinchey, M., Margaria, T. (eds.) SOFSEM 2017. LNCS, vol. 10139, pp. 49–61. Springer, Cham (2017). https://doi.org/10.1007/978-3-319-51963-0_5
23. Fraïssé, R.: Sur quelques classifications des systèmes de relations. Publ. Scient. de l'Univ. d'Alger Série A **1**, 35–182 (1954)
24. Fränzle, M., Herde, C.: HySAT: an efficient proof engine for bounded model checking of hybrid systems. Formal Meth. Syst. Design **30**(3), 179–198 (2007)
25. Frehse, G., et al.: SpaceEx: scalable verification of hybrid systems. In: Gopalakrishnan, G., Qadeer, S. (eds.) CAV 2011. LNCS, vol. 6806, pp. 379–395. Springer, Heidelberg (2011). https://doi.org/10.1007/978-3-642-22110-1_30

26. Gardey, G., Lime, D., Magnin, M., Roux, O.H.: Romeo: a tool for analyzing time Petri nets. In: Etessami, K., Rajamani, S.K. (eds.) CAV 2005. LNCS, vol. 3576, pp. 418–423. Springer, Heidelberg (2005). https://doi.org/10.1007/11513988_41
27. Gilmore, S., Hillston, J.: The PEPA workbench: a tool to support a process algebra-based approach to performance modelling. In: Haring, G., Kotsis, G. (eds.) TOOLS 1994. LNCS, vol. 794, pp. 353–368. Springer, Heidelberg (1994). https://doi.org/10.1007/3-540-58021-2_20
28. Girard, A., Pappas, G.J.: Approximation metrics for discrete and continuous systems. IEEE Trans. Automat. Contr. **52**(5), 782–798 (2007)
29. Groote, J.F., Vaandrager, F.W.: Structured operational semantics and bisimulation as a congruence. Inf. Comput. **100**(2), 202–260 (1992)
30. Hanisch, H.-M.: Analysis of place/transition nets with timed arcs and its application to batch process control. In: Ajmone Marsan, M. (ed.) ICATPN 1993. LNCS, vol. 691, pp. 282–299. Springer, Heidelberg (1993). https://doi.org/10.1007/3-540-56863-8_52
31. Hansson, H., Jonsson, B.: A logic for reasoning about time and reliability. Formal Aspects Comput. **6**(5), 512–535 (1994)
32. Hennessy, M., Milner, R.: Algebraic laws for nondeterminism and concurrency. J. ACM **32**(1), 137–161 (1985)
33. Henzinger, T.A., Ho, P.-H., Wong-Toi, H.: HYTECH: a model checker for hybrid systems. Int. J. Softw. Tools Techn. Trans. **1**(1–2), 110–122 (1997)
34. Henzinger, T.A., Majumdar, R., Prabhu, V.S.: Quantifying similarities between timed systems. In: Pettersson, P., Yi, W. (eds.) FORMATS 2005. LNCS, vol. 3829, pp. 226–241. Springer, Heidelberg (2005). https://doi.org/10.1007/11603009_18
35. Henzinger, T.A., Nicollin, X., Sifakis, J., Yovine, S.: Symbolic model checking for real-time systems. Inf. Comput. **111**(2), 193–244 (1994)
36. Hillston, J.: A Compositional Approach to Performance Modelling. Cambridge University Press, Cambridge (1996)
37. Koymans, R.: Specifying real-time properties with metric temporal logic. Real-Time Syst. **2**(4), 255–299 (1990)
38. Kwiatkowska, M., Norman, G., Parker, D.: Probabilistic symbolic model checking with PRISM: a hybrid approach. In: Katoen, J.-P., Stevens, P. (eds.) TACAS 2002. LNCS, vol. 2280, pp. 52–66. Springer, Heidelberg (2002). https://doi.org/10.1007/3-540-46002-0_5
39. Larsen, K.G., Fahrenberg, U., Thrane, C.: Metrics for weighted transition systems: axiomatization and complexity. Theor. Comput. Sci. **412**(28), 3358–3369 (2011)
40. Larsen, K.G., Pettersson, P., Yi, W.: UPPAAL in a nutshell. Int. J. Softw. Tools Techn. Trans. **1**(1–2), 134–152 (1997)
41. Larsen, K.G., Skou, A.: Bisimulation through probabilistic testing. In: POPL, pp. 344–352. ACM Press (1989)
42. Merlin, P.M., Farber, D.J.: Recoverability of communication protocols-implications of a theoretical study. IEEE Trans. Commun. **24**(9), 1036–1043 (1976)
43. Segala, R., Lynch, N.: Probabilistic simulations for probabilistic processes. In: Jonsson, B., Parrow, J. (eds.) CONCUR 1994. LNCS, vol. 836, pp. 481–496. Springer, Heidelberg (1994). https://doi.org/10.1007/978-3-540-48654-1_35
44. Stewart, W.J.: Introduction to the Numerical Solution of Markov Chains. Princeton University Press, Princeton (1994)
45. Stirling, C.: Modal and temporal logics for processes. In: Moller, F., Birtwistle, G. (eds.) Logics for Concurrency. LNCS, vol. 1043, pp. 149–237. Springer, Heidelberg (1996). https://doi.org/10.1007/3-540-60915-6_5

46. Thrane, C., Fahrenberg, U., Larsen, K.G.: Quantitative analysis of weighted transition systems. J. Log. Alg. Prog. **79**(7), 689–703 (2010)
47. van Breugel, F.: An introduction to metric semantics: operational and denotational models for programming and specification languages. Theor. Comput. Sci. **258**(1–2), 1–98 (2001)
48. van Breugel, F., Worrell, J.: A behavioural pseudometric for probabilistic transition systems. Theor. Comput. Sci. **331**(1), 115–142 (2005)
49. van Breugel, F., Worrell, J.: The complexity of computing a bisimilarity pseudometric on probabilistic automata. In: van Breugel, F., Kashefi, E., Palamidessi, C., Rutten, J. (eds.) Horizons of the Mind. A Tribute to Prakash Panangaden. LNCS, vol. 8464, pp. 191–213. Springer, Cham (2014). https://doi.org/10.1007/978-3-319-06880-0_10
50. van Glabbeek, R.J.: The linear time - branching time spectrum I. In: Bergstra, J.A., Ponse, A., Smolka, S.A. (eds.) Handbook of Process Algebra, pp. 3–99. Elsevier, Amsterdam (2001)
51. Wang, F., Mok, A., Emerson, E.A.: Symbolic model checking for distributed real-time systems. In: Woodcock, J.C.P., Larsen, P.G. (eds.) FME 1993. LNCS, vol. 670, pp. 632–651. Springer, Heidelberg (1993). https://doi.org/10.1007/BFb0024671
52. Zwick, U., Paterson, M.: The complexity of mean payoff games on graphs. Theor. Comput. Sci. **158**(1&2), 343–359 (1996)

Clinical Pathways Formal Modelling Using Bigraphical Reactive Systems

Fateh Latreche[1(✉)], Abdelkader Moudjari[2], and Hichem Talbi[2]

[1] LIRE Laboratory Constantine 2 University, Constantine, Algeria
fateh.latreche@univ-constatntine2.dz
[2] Misc Laboratory Constantine 2 University, Constantine, Algeria
moudjariabdelkader@gmail.com, hichem.talbi@univ-constatntine2.dz

Abstract. Clinical pathways are multidisciplinary structured care plans that aim at increasing the quality of healthcare delivery. Despite the wide-spread adoption of clinical pathways by governments and healthcare authorities, works that associate formal semantics to clinical pathways do not model clearly the roles of health professionals during the care process. This has motivated us to propose a formal modelling approach for clinical pathways, based on Bigraphical Reactive Systems. We concentrate on showing graphically relationships among the healthcare stakeholders. To meet the control flow requirement of clinical pathways, we apply the Maude strategies language over a Maude implementation of Bigraphical reactive Systems key concepts.

Keywords: Clinical pathways · Roles coordination · Bigraphical Reactive Systems · Maude strategies

1 Introduction

Over the few past decades, healthcare systems have experienced the challenges of population ageing and the increased number of chronic diseases. In order to improve outcomes of clinical practices, national and international organisations such as the World Health Organization (WHO) have defined medical guidelines to be implemented by governments and healthcare ministries.

Clinical Pathways (CPs) were established as a tool to implement paper-based medical guidelines for a particular disease. CPs ensure visualization, documentation and communication [2].

Clinical Pathways are a method for patient-care management for a well-defined group of patients during a well-defined period of time [7]. CPs are considered as a standardization of medical practices, which provides a comprehensive guide of treatment for a specific disease. One of the most important objectives of CPs is to facilitate the communication and coordination among the healthcare workers.

In fact, there is no single, widely accepted approach to implement CPs. However, modelling languages should genuinely describe activities and process flow,

R. M. Hierons and M. Mosbah (Eds.): ICTAC 2019, LNCS 11884, pp. 76–90, 2019.
https://doi.org/10.1007/978-3-030-32505-3_5

and permit the annotation of relevant information and knowledge at the appropriate level of abstraction [25].

Two families of languages have been widely used to model CPs. The first family concerns generic modelling languages such as UML activity diagrams, Event Process Chains (EPCs), and Business Process Model and Notation (BPMN). The second family is about the domain specific modelling languages [3,25]. The main drawback of these languages is the lack of formal semantics, thus there is no support for a formal analysis of CPs.

Formal methods have also been applied in the field of healthcare systems. However, deduced models of CPs-formal-based approaches did not show clearly the healthcare stakeholders and their relationships during the care process.

The contribution of the approach we propose is the use of Bigraphical Reactive Systems (BRSs) as a formal modelling and analysis framework for the Chronic Obstructive Pulmonary Disease (COPD) clinical pathway. We aim at providing an intuitive and understandable model of CPs. To meet the control flow requirement of CPs, we provide a Maude implementation of BRSs key concepts. By means of this implementation, we are able to order execution of reaction rules.

BRSs have been introduced by Robin Milner [19] in order to provide an intuitive and graphical notation for distributed systems, showing both locality and connectivity. In our work, we apply the BRSs sorting logic to depict the structure of COPD CP, and reaction rules are used to describe care activities.

This paper is organised as follow: after reviewing some related works on formal modelling and analysis of healthcare systems, we present elementary concepts of BRSs, rewriting logic and the strategies language of Maude. In Sect. 4, we show how to apply our proposal on the COPD clinical pathway. Section 5 concludes this paper and provides future directions for this work.

2 Related Work

Formal methods have been applied to describe and validate critical systems such as healthcare systems. In this section, we review relevant related works aimed at improving outcomes of healthcare practices.

In [21], the authors used Meta-ECATNets [14], multilevel algebraic Petri nets, to model CPs. Their work aimed at controlling flexibility and adjusting the preestablished healthcare processes. A Meta-ECATNet with two levels were introduced, the higher level controls and manages transitions of the lower-level net (treatment actions). The Meta-ECATNets clinical pathways were implemented using the Real-time Maude system and checked by applying the TCTL model-checking technique [15].

Modular Petri nets were used for modelling healthcare systems in [16]. The medical protocol Petri net module is composed of: input places, output places, activity places, resources places and two kinds of transitions (immediate and timed). Patients undergoing therapies are considered as tokens within activity places. The treatment duration is defined by the firing delay of the output transition. To obtain the global health Petri net model, basic Petri net modules are fused on common inputs and outputs places. At the analysis step, authors were interested in checking the following properties: conservation of resources, repetitiveness of a medical protocol and its violation.

Oueida et al. [22] proposed a modular approach based on Petri nets to describe activities of emergency units. More precisely, this work aimed at efficiently controlling the work-flow synchronization among activities using a new class of Petri nets, called Resource Preservation Nets (RPN). The main objective was to improve the patients performance indicators: length of stay, resources utilization rates and waiting time. In order validate the soundness of the obtained model, authors have applied structural analysis of Petri nets.

All aforementioned works are useful; they apply Petri nets to model and/or reason about important aspects of healthcare systems. However, none of them shows clearly the roles of health-care professionals during the care process. Thus, the implementation of these approaches will very likely lead to some communication and coordination inconsistencies among the members of the multidisciplinary care team.

The authors of [9] have proposed a timed extension of the process algebra Communicating Sequential Processes (CSP) to model and verify healthcare work-flows. The approach integrates real-time constraints by means of the real-time logic duration calculus and data aspects using first-order formulae. The authors have checked the consistency of time constraints and the absence of blocking behaviour. To do so, the process with timing constraints is translated into timed automaton. Then, this automaton is converted into a Transition Constraint System to be checked using the model checkers ARMC [23] and SLAB [23]. This work uses another formalism (CSP) to verify healthcare workflows. But, like the previous works, the deduced models do not model collaboration between stakeholders, and no graphical notation for timed CSP has been employed. A distinctive aspect of our proposal is the use of BRSs, which are considered as a meta-calculus tool, to define diverse calculi and models of concurrent systems such as Petri nets and CSP [20]. Furthermore, the understandable bigraphical models of CPs improve the coordination among the medical staff.

In [1], the authors proposed a framework to resolve conflicts in the treatments given to patients suffering from two or more chronic diseases. First, *BPMN* was used to capture medical guidelines. Then, the deduced models were translated to an intermediate formal model (the labelled event structure) capturing the flattened structure of the pathway. Next, correctness of the resulting model was checked using together the constraint solver *z3* [8] and the theorem prover *isabelle* [23]. This work tackled the important problem of combining multiple clinical pathways. In addition, the deduced BPMN models provided a suitable graphical representation of CPs that can be used by clinicians. However, since the work transforms semi-formal models to formal ones, an additional operation to check the transformation process is needed.

3 Preliminaries

3.1 Bigraphical Reactive Systems Review

Bigraphs are a formalism initially introduced by Milner in [19]. Bigraphs are devoted to model and analyse spatial evolution and concurrency of distributed systems.

Bigraphs have been considered as a unifying framework for several other formalisms such as: Petri nets, CCS and λ-calculus [20]. An important advantage of bigraphs is their graphical notation that simplifies the modelling task. Also, bigraphs are characterized by their rigorous mathematical basis.

A bigraph is composed of two sub-graphs: a place graph and a link graph. The place graph represents the embedding of nodes. It consists of the following elements: Regions, nodes and sites (hole). Regions (called also roots) are the outermost elements of a place graph, they may include nodes and/or sites. A node could contain other nodes and sites. It is worth to mention that sibling nodes are unordered, while roots and holes are indexed.

Figure 1 shows the graphical notation associated to bigraphs. The two regions of this figure are depicted with dashed rectangles and nodes are depicted using solid circles. Ports are connection points for nodes, they are shown as black dots.

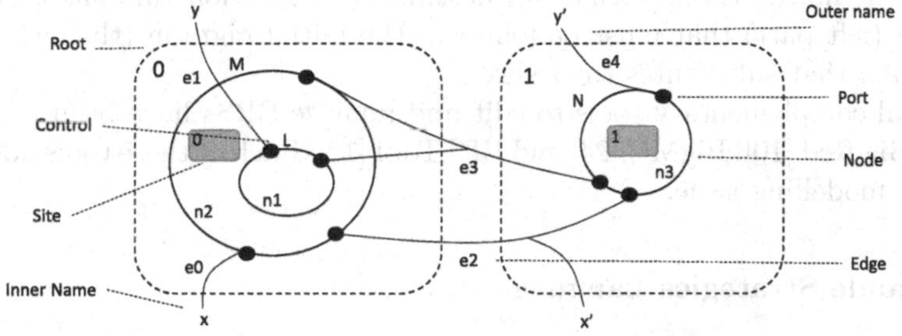

Fig. 1. Anatomy of bigraphs

The link-graph captures connectivity and communication between components of the system. A link graph connects bigraph nodes through their ports. Links and edges (closed links) are graphically depicted as continuous lines.

Definition 1 (Bigraphs). *A bigraph B is a tuple:* $B = (V, E, ctrl, prnt, link) : \langle m, X\rangle \rightarrow \langle n, Y\rangle$, *where:*

- V *is the set of nodes.*
- E *is a finite set of hyper-links.*
- $ctrl : V \rightarrow K$ *is a function that associates to each node* $v \in V$ *a control* $k \in K$ *defining its arities,* K *is the signature.*
- m *and* n *are the number of sites and the number of regions respectively.*
- X *and* Y *are the sets of inner names and the set of outer names respectively.*

- *The parent map prnt associates to each node and site its parent (node or region).*
- *The function link merges ports and inner names with edges and outer names.*
- $\langle m, X\rangle$ *and* $\langle n, Y\rangle$ *are the inner face and the outer face of* B.

The composition and tensor product operations are applied on basic bigraphs to obtain complex ones. The composition of two bigraphs is to insert the first bigraph regions' content inside sites of the second one having similar indexes, and merging links on common names. For the tensor product, it consists of juxtaposing the place graphs, constructing the union of the link graphs and increasing the indexes of sites and regions. Formal definition of composition and tensor product operations can be found in [20].

Nodes and links may be classified using controls. The latter show the behaviour of nodes through their status: *active*, *passive* and *atomic*. An atomic node cannot contain another node, active and passive status define whether reaction may occur within a node or not. The different controls used in a bigraph make its signature. To enhance the characterization of bigraphs, the sorting mechanism associates sorts to controls and uses the formation rules to express well-formedness conditions for bigraphs.

A Bigraphical Reactive system (BRS) consists of a bigraph endowed with a set of reaction rules. A reaction rule describes the evolution of a bigraph. It modifies the nodes' connectivity and nesting. The reaction rule has two parts: the *redex* (left part) that must be found in the initial bigraph (the agent), and the *reactum* that substitutes the redex.

Several complementary tools to edit and analyse BRSs have been proposed, such as: Big Red [10], BigMC [24] and BPLTool [12]. Each of these tools addresses a specific modelling issue.

3.2 Maude Strategies Language

Various formalisms have been introduced to improve reliability of information technology systems. Rewriting logic is one of the most commonly applied formalisms to ensure correctness of concurrent systems [18].

In rewriting logic, static and dynamic aspects of systems are modelled using membership equational and rewriting theories. The membership equational theory describing the system states is noted $(\Sigma, E \cup A)$, where Σ is the theory signature and $E \cup A$ is a set of equations and equational attributes defining properties of the different operators. The rewriting theory includes a set of rewriting rules having the form $l : t \rightarrow t'$ if *cond*, where t and t' represent the system's states and *cond* the rule condition.

The Maude system [6] is one of the most efficient implementations of rewriting logic. In fact, Maude provides two kind of modules: functional and system. The functional modules make concrete equational theories, they use equations to identify data and memberships to give type information for some data. System modules deal with rewriting aspects, they describe transitions between states.

Unlike equations, Maude rewriting rules could be divergent, leading to have many final states [5]. In order to look for states satisfying some properties, we could use the *search* command. In certain cases, users look only for some executions satisfying some constraints. In this case, the Maude system provides a strategy language that controls and guides non-deterministic behaviour when applying rewriting rules [17].

Maude strategies can be applied either at the *metalevel* using descent functions [6], or at the object level. At the metalevel, we take advantage of the reflective property of rewriting logic; rewrite rules are applied at a higher level.

Contrary to *metalevel* strategies, the object level strategies do not use the reflective capabilities, and they are based on a clear separation between the rewrite rules and the strategy expressions [17]. Consequently, different strategy modules could control rules of a single system module.

Basically, an object level strategy is to apply a rewriting rule (using its label) on a given term. Moreover, variables of this rule could be instantiated before its execution. Besides, conditions of conditional rewrite rules might be controlled using a search expression.

Strategies of Maude can be combined using the following combinators: ";" for concatenation, "|" for union, "*" for zero or more iterations and "+" for one or more iterations.

4 A BRSs Formal Modelling Approach of Clinical Pathways

In the present work, we propose a BRS based formalisation of COPD clinical pathway. The main objective is to provide an intuitive and understandable model, that improves the coordination of roles through the care process. In practical terms, we apply the BRSs sorting to determine the involved partners, and we use reaction rules to implement care activities. In addition, to respect the control flow of activities a Maude based implementation is proposed.

4.1 Case Study: COPD Clinical Pathway

To show the usefulness of our proposal, we apply it on the Chronic Obstructive Pulmonary Disease (COPD) clinical process. COPD is a chronic respiratory disease that causes blockage of airflow from the lungs. Shortness of breath and coughing are the main symptoms of COPD. COPD is frequently associated with obstructive lung changes caused by smoking and permanent exposure to harmful fumes or dust. According to [4], COPD is a major cause of morbidity and mortality in adulthood, it will become the third leading cause of death worldwide by 2020.

To diagnose COPD, the ratio of Forced Expiratory Volume in one second to Forced Vital Capacity (FEV1/FVC) is calculated. According to this ratio four stages of COPD can be identified [26]:

- Stage 1 (Early COPD) when FEV1/FVC $< 70\%$ and FVC $\geq 80\%$.
- Stage 2 (Mild COPD) when FEV1/FVC $< 70\%$ and $50\% \leq$ FVC $< 80\%$.
- Stage 3 (Severe COPD) when FEV1/FVC$<70\%$ and $30\% \leq$ FVC $< 50\%$.
- Stage 4 (Very severe COPD) when FEV1/FVC $< 70\%$ and FVC $< 30\%$.

The COPD patients treatment procedure often involves a whole medical staff. To reduce symptoms and slow down the progression of the disease, patients must take various medications and should follow a healthy lifestyle.

The COPD care pathway published by the French National Authority for Health (FNAH) [11] recommends the following plan of medical examinations: up to the stage 2, the monitoring is exclusively done by the general practitioner twice a year. The patient at the stage 3 or 4 who does not require supplemental oxygen, should consult his/her treating doctor every three months and the pulmonologist once a year. The patients at the stage 4 and requiring supplemental oxygen should consult the general practitioner every month and the pulmonologist twice a year.

4.2 A BRSs Formal Model for COPD Clinical Pathway

In this work, we address the formal modelling of COPD clinical pathway using BRSs. The BRSs formalism enables describing the collaborative plan of intervention which involves a multidisciplinary care team.

Figure 2 shows the bigraphical model associated with COPD healthcare system. Each actor involved in the cure of COPD is modelled by a node having well-defined control.

The set of controls associated to a bigraph depicts the type of modelled entities. It shows also how nodes behave dynamically. Thus, we have four kinds of nodes: P, G, L and S modelling respectively the following actors: the patient, the general practitioner, the laboratory technician and the pulmonologist.

The bigraph regions indexed 0, 1 and 2 depict respectively: the general practitioner office, the medical laboratory and the pulmonologist office. The sites represent placeholders for hosting further patients nodes. They indicate if there are patients in doctors' offices or in the laboratory.

The dynamics of the cure process is achieved using BRSs reaction rules. Table 1 shows the reaction rules that implement COPD care activities.

A set of tools for editing, executing and checking BRSs was proposed. But, these tools are not yet sufficiently mature, they do not allow expressing all concepts of BRSs. Also, the provided checking techniques only allow the verification of limited reachability properties, such as the deadlock freedom property [13,24].

Another limit of BRS tools is that they are unable to make an order among reaction rules. To overcome this limitation, we provide a Maude implementation of BRSs that enables ordering treatment tasks within CPs (the control flow requirement of CPs). More concretely, the Maude strategy expressions guide the execution of reaction rules.

Table 1. Reaction rules of COPD treatment process

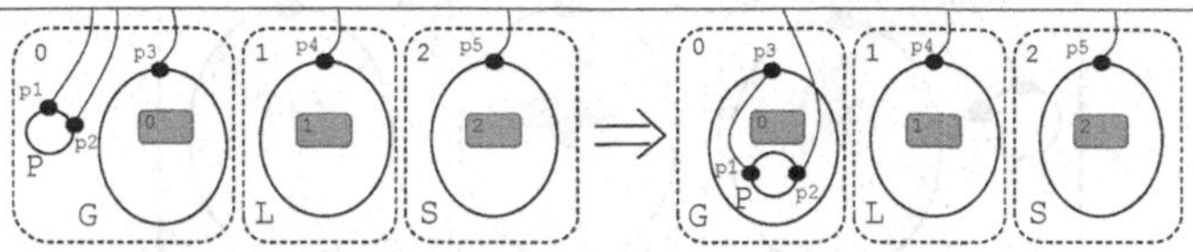 The first contact between the patient and the general practitioner.
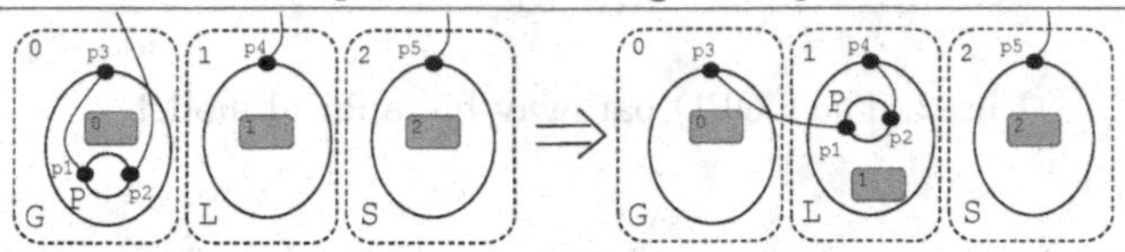 The general practitioner needs additional laboratory tests to confirm COPD.
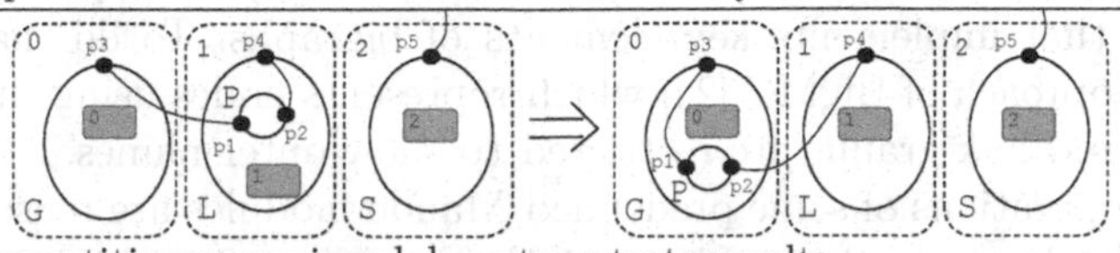 The general practitioner receives laboratory tests results.
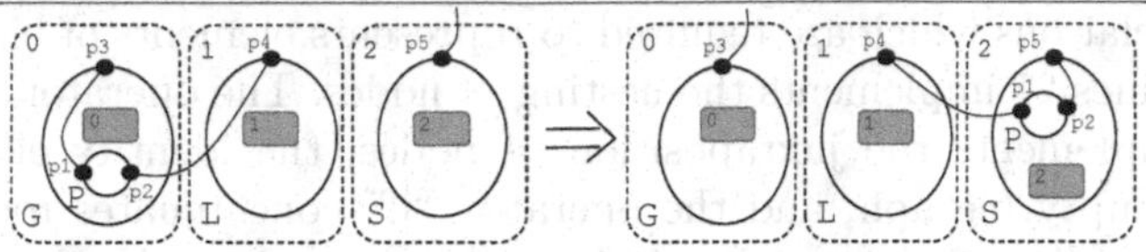 The diagnosis is not confirmed. Thus, the general practitioner refers the patient to the pulmonologist.
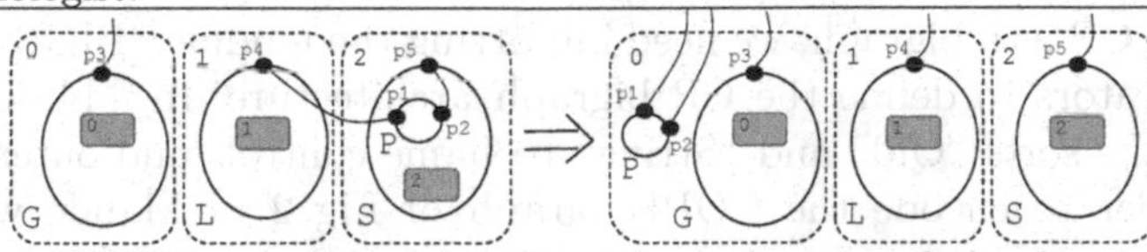 The stage of COPD is confirmed by the pulmonologist and the disease monitoring within a stage is started.
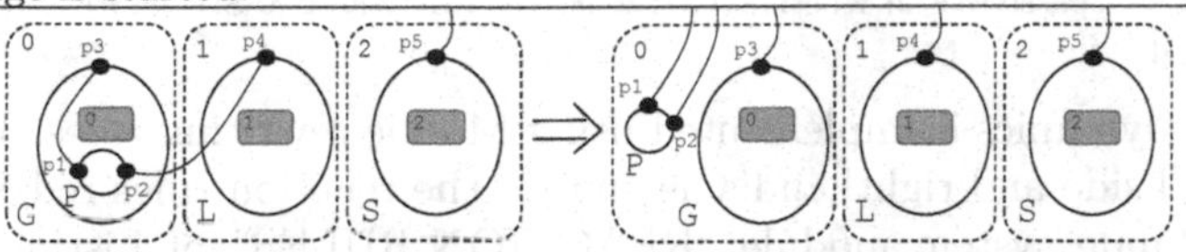 The stage of COPD is confirmed by the general practitioner and the disease monitoring within a stage is started.
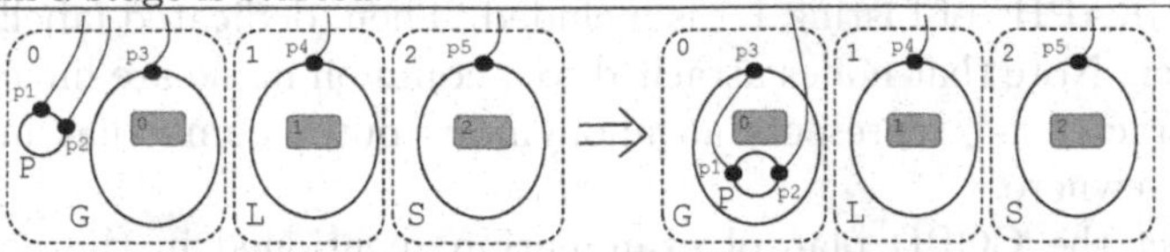 The monitoring of a patient suffering from COPDby the general practitioner, within a stage.
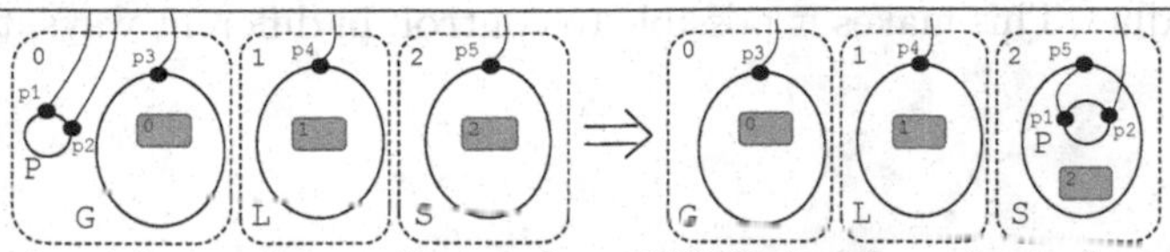 The monitoring of a patient suffering from COPD by the pulmonologist, within a stage.

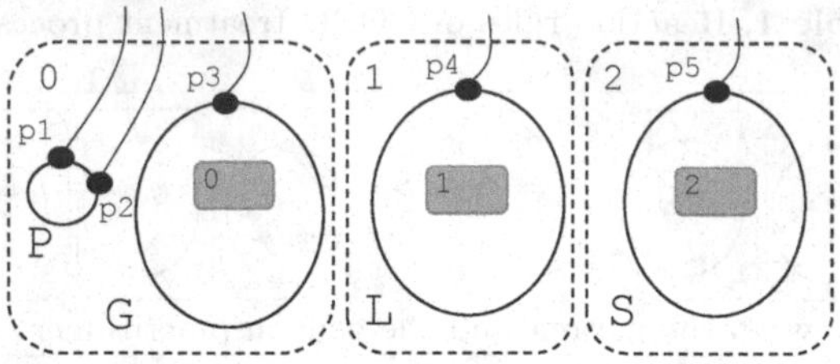

Fig. 2. The COPD pathway bigraphical model

Listing 1.1 shows the functional Maude module (the equational theory achievement) that implements key elements of bigraphs. To do that, we have followed the approach of BigMC [24] which represents nodes using their controls (their kinds) and link graphs are restricted to show outer names.

Data and operations of some predefined Maude modules are reused using the import statements of line 2 (see Listing 1.1). From lines 3 to 6, we declare sorts and subsort relations which are required to represents elements of bigraphs. The operation of lines 21 implements the nesting of nodes. The operator "|" declared on line 23 implements the juxtaposition of nodes, the identity element "*nil*" denotes the empty bigraph, and the operator "||" concatenates roots. Finally, the operation of line 27 defines sites indexed by natural numbers.

The "BIGRAPH" module of listing 1.1 constitutes a generic model of CPs. To model any CP as a bigraph, we need importing the module "BIGRAPH" and using its operators to define the CP bigraph architecture. In addition, we use constants of the sorts "Qid" and "String" to name controls and outer names.

So, in order to encode the COPD biraph of Fig. 2 in Maude we use the following equation:

```
eq system = %passive 'P:2; %active 'G:1; %active 'L:1; %active 'S:1 & 'P[-,-] | 'G[-] ||
    'L[-] || 'S[-] .
```

Bigraphs dynamics is implemented using Maude rewriting rules. The rewrite rules lefthand side and righthand side encode the reaction rules redex and reactum. The Maude system module "REACTION-RULES" of Listing 1.2 shows rewriting rules that implement reaction rules of Table 1. At the first line, the module "BIGRAPH" of Listing 1.1 is included. Then, dedicated labelled rewrite rules are given. Note that nodes attached to a common name are linked together and the character "-" represents an anonymous outer name that may not be referenced elsewhere.

To respect the COPD plan of examination mentioned in the Sect. 4.1, we make use of the Maude object level strategies. We use a strategy provided in a separate module. This makes it possible to control, in different ways, treatments

Listing 1.1. The Maude implmentation of the term language representation of Bigraphs

```
(fmod BIGRAPH is
protecting NAT . protecting STRING . protecting QID .
sorts Bigraph Cntrl Cntrls BigTerms BigTerm Nams Nam CtrlNm .
subsort Cntrl < Cntrls .        subsort Nam < Nams .
subsort Qid < CtrlNm .      subsort String < Nam .
subsort BigTerm < BigTerms .
*** Associate controls to nodes
op %passive_:_ : CtrlNm Nat -> Cntrl [ctor prec 25] .
op %active_:_ : CtrlNm Nat -> Cntrl [ctor prec 25] .
op _;_ : Cntrls Cntrls -> Cntrls [ctor assoc comm prec 26] .
op _&_ : Cntrls BigTerms -> Bigraph [ctor prec 27] .
*** Declaring outer names
op _‘,_ : Nams Nams -> Nams [ctor assoc comm prec 26] .
op - : -> Nam [ctor] .
op <_> : BigTerm -> BigTerm [ctor] .
*** Declaring the empty bigraph
op nil : -> BigTerm [ctor] .
*** Nodes structure
op _‘[_‘] : CtrlNm Nams -> BigTerm [ctor prec 22] .
*** Nesting of nodes
op _‘[_‘]._ : CtrlNm Nams BigTerm -> BigTerm [ctor prec 24 ] .
*** Juxtaposition of nodes
op _|_ : BigTerm BigTerm -> BigTerm [ctor assoc comm id: nil prec 24 ] .
*** Juxtaposition of roots
op _||_ : BigTerm BigTerm -> BigTerm [ctor assoc prec 25] .
*** numbered sites
op $_ : Nat -> BigTerm [ctor prec 21] .
endfm)
```

of the COPD. We Apply the strategies combinators: ";", "|", "*" and "+" over COPD reaction rules (identified by the corresponding rules labels) to implement the following types of treatments controls: sequential, conditional and iteration.

In the module COPD-STRAT of Listing 1.3 the strategy expression "pathWay" controls rewriting rules of the system module "REACTION-RULES".

The strategy "pathWay" begins by the sub strategy "rl1 ; rl2 ; rl3" that executes sequentially the clinical examination and the laboratory tests. Then, the strategy "(rl4 ; rl5 | rl6)" reflects either referral to a pulmonologist to confirm diagnosis or the launch of the monitoring within a stage. Finally, the last three iterative strategy expressions implement the three main treatment procedures.

Note that the strategy expression uses the three copies: *rl7'*, *rl7"* and *rl7'''* of the rule *rl7* to start differently the three main treatment procedures of the COPD.

Listing 1.2. COPD CP reaction rules

```
( mod REACTION-RULES is
including BIGRAPH .
*** The first appointment with the general practitioner
rl [rl1] : ctrls & 'P[-,-] | 'G[-] || 'L[-] || 'S[-] X ls => ctrls & 'G["cl-ex"] . <'P["
    cl-ex",-] > || 'L[-] || 'S[-] X ls .'gp-examination .
*** Performing medical analysis to identify COPD
rl [rl2] : ctrls & 'G["cl-ex"] . < 'P["cl-ex",-] > || 'L[-] || 'S[-] X ls => ctrls & 'G
    ["cl-ex"] || 'L["med-an"] . < 'P[ "cl-ex", "med-an"] > || 'S[-] X ls.'lab-analysis .
*** Receiving medical analysis results
rl [rl3] : ctrls & 'G["cl-ex"] || 'L ["med-an"] . <'P["cl-ex", "med-an"] > || 'S[-] X
    ls => ctrls & 'G["cl-ex"] . < 'P["cl-ex", "med-an"] > || 'L["med-an"] || 'S[-] X ls.'
    rcv-analysis .
*** The plumonologist performs further examinations to confirm COPD
rl [rl4] : ctrls & 'G["cl-ex"].< 'P["cl-ex", "med-an"] > || 'L ["med-an"] || 'S[-] X ls
    => ctrls & 'G[-] || 'L["med-an"] || 'S["cl-ex"] . <'P ["cl-ex", "med-an"] > X ls . '
    diag-diff .
*** The start of a new medical stage
rl [rl5] : ctrls & 'G[-] || 'L["med-an"] || 'S ["cl-ex"] . < 'P[ "cl-ex", "med-an"] >
    X ls => ctrls & 'P[-, -] | 'G [-] || 'L[-] || 'S[-] X ls .
*** The start of a new medical stage
rl [rl6] : ctrls & 'G["cl-ex"] . < 'P["cl-ex", "med-an"] > || 'L["med-an"] || 'S[-] X
    ls => ctrls & 'P[-, -] | 'G [-] || 'L[-] || 'S[-] X ls .
*** Monitoring by the general practitioner during a stage
rl [rl7] : ctrls & 'P[-,-] | 'G[-] || 'L[-] || 'S[-] X ls => ctrls & 'G["md-suprvs"] .
    < 'P["md-suprvs",-] > || 'L[-] || 'S[-] X ls .
*** Monitoring by the plumonologist during a stage
rl [rl8] : ctrls & 'P[-,-] | 'G[-] || 'L[-] || 'S[-] X ls => ctrls & 'G[-] || 'L[-] | 'S
    ["md-suprvs"] . < 'P ["md-suprvs", -] > X ls .
endm)
```

4.3 Execution and Analysis of COPD Clinical Pathway

In addition to the ease of its understanding, the bigraphical model of COPD CP can be analysed formally using the Maude System. The application of strategy expressions to guide execution of medical treatments reduces the size of the generated state space and minimizes the execution time for checking.

To ask for an execution of the COPD CP according to the strategy expression "pathWay" we use the command "(srew system using pathWay .)". The command "(next .)" can be used to generate the next solution (see Listing 1.5).

Listing 1.3. COPD CP control flow

```
(smod COPD-STRAT is
strat pathWay : @ Bigraph .
var B : Bigraph .
sd pathWay := rl1 ; rl2 ; rl3 ; ( rl4 ; rl5 | rl6 ) ; ( rl7' ; rl8 ; rl7 ; rl8 ) * ; (
    rl7'' ; rl8 ; rl7 ; rl8 ; rl7 ; rl8 ; rl7 ; rl9 ; rl10 ) * ; ( rl7''' ; rl8 ; rl7 ;
    rl8 ; rl7 ; rl8 ; rl7 ; rl8 ; rl7 ; rl8 ; rl7 ; rl9 ; rl10 ; rl7 ; rl8 ; rl7 ; rl8 ;
    rl7 ; rl8 ; rl7 ; rl8 ; rl7 ; rl8 ; rl7 ; rl9 ; rl10 ) * .
endsm)
```

Listing 1.4. COPD CP test strategies

```
including SYSTEM . var B : Bigraph .
rl [test1] : BX((('gp−examination.'lab−analysis). 'rcv−analysis).'diag−diff).'stage3 =>
   BX((('gp−examination . 'lab−analysis).'rcv−analysis).'diag−diff).'stage3 .
rl [test2] : BX((((gp−examination.'lab−analysis). 'rcv−analysis).'diag−diff).'stage2).'
   stage1 => BX((((gp−examination.'lab−analysis).'rcv−analysis). 'diag−diff).'stage2 ).'
   stage1 .
rl [test3] : BX((((gp−examination.'lab−analysis). 'rcv−analysis).'diag−diff).'stage1).'
   stage1 => BX((((gp−examination.'lab−analysis).'rcv−analysis). 'diag−diff).'stage1 ).'
   stage1 .
```

Listing 1.5. Execution and analysis of the COPD CP BRS model

```
Maude > (srew system using pathWay . )
rewrites: 463 in 6296478266ms cpu (431ms real) (0 rewrites/second)
rewrite with strategy :
result System :
%active 'L:1 ; %active 'G:1 ; %active 'S:1 ; %passive 'P : 2 & 'G[−] |'P[−,−] || 'L[−] ||
   'S[−]X((('gp−examination). 'lab−analysis).'rcv−analysis).'stage1
=================
Maude > (next .)
rewrites: 545 in 6296478266ms cpu (35ms real) (0 rewrites/second)
next solution rewriting with strategy :
result System :
%active 'L:1; %active 'G:1; %passive 'S:1; %passive 'P:2 & 'G[−]| 'P[−,−]|| 'L[−]|| 'S[−]
   X((('gp−examination). 'lab−analysis).'rcv−analysis).'stage2
=================
Maude > (srew system using pathWay ; test1 .)
rewrites: 2679 in 6296478266ms cpu (172ms real) (0 rewrites/second)
rewrite with strategy :
result System :
%active 'L:1; %active 'G:1; %active 'S:1; %passive 'P:2 & 'G[−]| 'P[−,−]|| 'L[−]|| 'S[−]
   X((('gp−examination. 'lab−analysis).'rcv−analysis).'diag−diff).'stage3
=================
Maude > (srew system using pathWay ; test2 .)
rewrites: 3262 in 6296478266ms cpu (176ms real) (0 rewrites/second)
rewrite with strategy :
No possible rewriting .
=================
Maude > (srew system using pathWay ; test3 .)
rewrites: 1491 in 6296478266ms cpu (111ms real) (0 rewrites/second)
rewrite with strategy :
result System :
%active 'LA : 1 ; %active 'MG : 1 ; %passive 'MP : 1 ; %passive 'P : 2 & 'MG[−]| 'P
   [−,−]|| 'LA[−]|| 'MP[−]X(((('gp−examination . 'lab−analysis). 'rcv−analysis). 'diag−
   diff). 'stage1). 'stage1
```

To be able to check some useful properties on the COPD CP, we have expanded the bigraphical model that represents the COPD CP by an ordered list of quoted identifiers. Each element of this list represents important examinations and the beginning of the three main treatment procedures.

The two rewriting rules "test1" and "test2" (see Listing 1.4, lines 3 and 4) are then combined with the strategy expression "pathWay" to check the properties:

- A patient who suffers from the COPD can be at a severe stage after the first diagnosis,
- Can early stages of COPD be reversed? More precisely, the property checks whether or not the second treatment procedure could come first.

The strategy command associated to the first property succeeds and returns a Maude term that represents a patient at the last treatment procedure after the first diagnostic (see the commands of line 15, Listing 1.5).

Concerning the second property, its strategy command fails by returning as results the message "No possible rewriting" (see line 22, Listing 1.5). This result is due to the fact that COPD is a chronic disease, a patient suffering from the COPD cannot be cured by medication.

Finally, the last strategy rewrite command of the Listing 1.5 checks repetitiveness of medical examinations of the first stage. The command succeeds because COPD patients can remain at the same stage for an indefinite period.

5 Conclusion

In this paper, we have proposed a BRSs approach for modelling both structural and behavioural aspects of clinical pathways. The sorting mechanism of BRSs is used to distinguish between the healthcare practitioners and to identify their roles in the healthcare process. Reaction rules are applied to simulate the different care activities. To respect medical dependencies among treatment steps, we have used the strategies language of Maude over a Maude implementation of key elements of bigraphs.

In this paper, we have focused more specifically on modelling coordination between the healthcare professionals for a single CP. Our proposal can be used to resolve contradictions and conflicts that can arise when combining several CPs.

As future extension of this work, we plan to enlarge the proposed formalisation to tackle the resources management using bigraphs with sharing. We intend also to tackle the exception and deviation issue of CPs.

References

1. Bowles, J., Caminati, M.B., Cha, S.: An integrated framework for verifying multiple care pathways. In: 2017 International Symposium on Theoretical Aspects of Software Engineering (TASE), pp. 1–8. IEEE (2017)
2. Braun, R., Schlieter, H., Burwitz, M., Esswein, W.: BPMN4CP: design and implementation of a BPMN extension for clinical pathways. In: 2014 IEEE International Conference on Bioinformatics and Biomedicine (BIBM), pp. 9–16. IEEE (2014)

3. Burwitz, M., Schlieter, H., Esswein, W.: Modeling clinical pathways-design and application of a domain-specific modeling language. Wirtschaftsinformatik Proceedings, pp. 1325–1339 (2013)
4. Celli, B.R.: Predictors of mortality in COPD. Respir. Med. **104**(6), 773–779 (2010)
5. Clavel, M., et al.: Maude manual (version 2.7), March 2015
6. Clavel, M., et al.: All About Maude - A High-Performance Logical Framework: How to Specify, Program, and Verify Systems in Rewriting Logic. LNCS, vol. 4350. Springer, Heidelberg (2007). https://doi.org/10.1007/978-3-540-71999-1
7. De Bleser, L., Depreitere, R., Waele, K.D., Vanhaecht, K., Vlayen, J., Sermeus, W.: Defining pathways. J. Nurs. Manag. **14**(7), 553–563 (2006)
8. de Moura, L., Bjørner, N.: Z3: an efficient SMT solver. In: Ramakrishnan, C.R., Rehof, J. (eds.) TACAS 2008. LNCS, vol. 4963, pp. 337–340. Springer, Heidelberg (2008). https://doi.org/10.1007/978-3-540-78800-3_24
9. Faber, J.: A timed model for healthcare workflows based on CSP. In: Proceedings of the 4th International Workshop on Software Engineering in Health Care, pp. 1–7. IEEE Press (2012)
10. Faithfull, A.J., Perrone, G., Hildebrandt, T.T.: Big red: a development environment for bigraphs. Electron. Commun. EASST **61** (2013)
11. for Health, F.N.A.: Guide du parcours de soins: Bronchopneumopathie chronique obstructive (2014). https://www.has-sante.fr/portail/upload/docs/application/pdf/2012-04/guide_parcours_de_soins_bpco_finale.pdf
12. Højsgaard, E.: Bigraphical Languages and their Simulation. Ph.D. thesis, IT University of Copenhagen (2012)
13. Hu, K., Lei, L., Tsai, W.T.: Multi-tenant verification-as-a-service (VaaS) in a cloud. Simul. Model. Pract. Theory **60**, 122–143 (2016)
14. Latreche, F., Belala, F.: A layered petri net model to formally analyse time critical web service composition. IJCCBS **7**(2), 119–137 (2017)
15. Lepri, D., Ábrahám, E., Ölveczky, P.C.: Timed CTL model checking in real-time maude. In: Durán, F. (ed.) WRLA 2012. LNCS, vol. 7571, pp. 182–200. Springer, Heidelberg (2012). https://doi.org/10.1007/978-3-642-34005-5_10
16. Mahulea, C., Mahulea, L., Soriano, J.M.G., Colom, J.M.: Modular Petri net modeling of healthcare systems. Flex. Serv. Manuf. J. **30**(1–2), 329–357 (2018)
17. Martí-Oliet, N., Meseguer, J., Verdejo, A.: Towards a strategy language for maude. Electron. Notes Theor. Comput. Sci. **117**, 417–441 (2005)
18. Meseguer, J.: Twenty years of rewriting logic. J. Log. Algebraic Program. **81**(7–8), 721–781 (2012)
19. Milner, R.: Bigraphs and their algebra. Electron. Notes Theor. Comput. Sci. **209**, 5–19 (2008)
20. Milner, R.: The Space and Motion of Communicating Agents. Cambridge University Press, Cambridge (2009)
21. Moudjari, A., Latreche, F., Talbi, H.: Meta-ECATNets for modelling and analyzing clinical pathways. In: Chikhi, S., Amine, A., Chaoui, A., Saidouni, D.E. (eds.) MISC 2018. LNNS, vol. 64, pp. 289–298. Springer, Cham (2019). https://doi.org/10.1007/978-3-030-05481-6_22
22. Oueida, S., Kotb, Y., Kadry, S., Ionescu, S.: Healthcare operation improvement based on simulation of cooperative resource preservation nets for none-consumable resources. Complexity **2018** (2018)
23. Paulson, L.C.: Isabelle: A Generic Theorem Prover, vol. 828. Springer, Heidelberg (1994). https://doi.org/10.1007/BFb0030541
24. Perrone, G., Debois, S., Hildebrandt, T.T.: A verification environment for bigraphs. Innov. Syst. Softw. Eng. **9**(2), 95–104 (2013)

25. Shitkova, M., Taratukhin, V., Becker, J.: Towards a methodology and a tool for modeling clinical pathways. Procedia Comput. Sci. **63**, 205–212 (2015)
26. Siafakas, N., Bizymi, N., Mathioudakis, A., Corlateanu, A.: Early versus mild chronic obstructive pulmonary disease (COPD). Respir. Med. **140**, 127–131 (2018)

Optimal Run Problem for Weighted Register Automata

Hiroyuki Seki[1(✉)], Reo Yoshimura[1], and Yoshiaki Takata[2]

[1] Nagoya University, Nagoya, Japan
seki@i.nagoya-u.ac.jp, yoshimura.reo@sqlab.jp
[2] Kochi University of Technology, Kami, Japan
takata.yoshiaki@kochi-tech.ac.jp

Abstract. Register automata (RA) are a computational model that can handle data values by adding registers to finite automata. Recently, weighted register automata (WRA) were proposed by extending RA so that weights can be specified for transitions. In this paper, we first investigate decidability and complexity of decision problems on the weights of runs in WRA. We then propose an algorithm for the optimum run problem related to the above decision problems. For this purpose, we use a register type as an abstraction of the contents of registers, which is determined by binary relations (such as =, <, etc.) handled by WRA. Also, we introduce a subclass where both the applicability of transition rules and the weights of transitions are determined only by a register type. We present a method of transforming a given WRA satisfying the assumption to a weighted directed graph such that the optimal run of WRA and the minimum weight path of the graph correspond to each other. Lastly, we discuss the optimal run problem for weighted timed automata as an example.

1 Introduction

There have been many extensions of finite automata that can manipulate data values. Among them, register automata (abbreviated as RA) introduced in [12] have the advantages that important decision problems including membership and emptiness are decidable and the class of languages accepted by RA is closed under standard language operations except complementation. In a k-RA, k registers are associated with each state. An input is a finite sequence of pairs of a symbol from a finite alphabet and a data value from an infinite set. Each transition can compare the contents of the registers and the current input data value and if this test succeeds, the input data value is loaded to the registers specified by the transition and the state is changed. The complexity of decision problems has been analyzed [11,18]. Also, [13] points out that RA is a good formal model for querying structured data such as XML documents. Recently, weighted RA was proposed in [5] by incorporating weights into RA so that various quantities such as time, information flow and costs needed for transitions and/or data manipulations can be formally represented as weights. A k-WRA is

R. M. Hierons and M. Mosbah (Eds.): ICTAC 2019, LNCS 11884, pp. 91–110, 2019.
https://doi.org/10.1007/978-3-030-32505-3_6

a k-RA equipped with weight functions for transitions and data manipulations. The weight function for data manipulations can represent weights depending on data values such as the cost depending on the elapsed time in timed automata. A semiring is assumed to represent weights and to assign a weight to a switch (one step move), a run (accepting sequence of switches), a data word in a systematic way. Closure properties of the data series recognized by WRA are discussed and an MSO logical counterpart of WRA is proposed and studied in depth in [5]. However, decidability and complexity of basic problems on WRA were not discussed. Timed automata (abbreviated as TA) are well-known extensions of finite automata that can deal with time by clock variables [3]. TA was extended to weighted TA (WTA) and the optimal-reachability problems have been investigated [4,15]. In [5], TA and WTA are shown to be regarded as subclasses of RA and WRA, respectively.

In this paper, we discuss optimal run problems and related decision problems for WRA, motivated by [3]. First, we clarify the decidability and complexity of the decision problems on weight computation and weight realizability. More concretely, we show that the problem to decide whether there is a run of a given data word whose weight takes a given value in a given WRA is NP-complete, and the problem to compute the weight of a given data word, which is the sum of all runs of the data word, in a given WRA is in PSPACE and #P-hard. We also show that the following two weight realizability problems are both undecidable: the problem to decide whether there is a run in a given WRA whose weight takes a given value and the problem to decide whether there is a data word whose weight in a given WRA equals to a given value. Note that the former two problems and the latter two problems can be regarded as extensions of the membership and emptiness problems for RA, which are known to be NP-complete and PSPACE-complete, respectively.

Next, we utilize register type, which was introduced in [20] as an abstraction of the contents of registers, by identifying the data values indistinguishable by comparisons allowed in the guards of transitions. We show an equivalence transformation from a given k-WRA to a k-WRA such that the exact register type is annotated to each state by associating register types with states before and after a transition. A WRA obtained by this transition decomposition by register type is called a normal form WRA.

Then, we move to the main topic, the optimal run problem for WRA, which is a problem to compute a run whose weight takes the infimum among all the runs in a given WRA. The idea is simple and similar to the one in [4]: A given WRA is translated into a directed graph where a node stands for a state and an edge between two nodes stands for switches between them where the weight of the edge is the infimum of the weights of those switches. In order to determine the weight of each edge, the infimum of the weights must be independent of the contents of registers. However, this does not hold in general, unlike for WTA. To overcome this issue, we introduce two reasonable assumptions: for each transition, the infimum of the weights of switches realized by the transition is uniquely determined independent of the contents of registers (weighted simulation); and the above infimum can be computed when weighted simulation holds (weight computability). These two assumptions are a weighted version of simulation and progress proposed in [20]. For a given WRA satisfying the above two properties, we can construct a directed graph as intended, and we can obtain an optimal run by an existing graph algorithm that computes the minimum-weight path in the constructed graph.

Finally, we discuss the optimal run problem for weighted timed automata (WTA) as an example of the application of the proposed method. We focus on the subclass of WRA obtained from WTA by the translation of [5]. Intuitively, a register type corresponds to a clock region of TA [3]. Moreover, [4] shows that there always exists an optimal (minimum weight) path that visits only boundary regions and limit regions because all clock constrains of TA are linear. If we restrict the register types to those corresponding to boundary regions and limit regions, weighted simulation and weight computability hold where the directed graph constructed in our paper corresponds to the subregion graph in [4].

Related Work. Register automata (RA) were proposed by Kaminsky and Francez [12] as finite-memory automata where they show that the membership and emptiness problems are decidable, and the class of languages recognized by RA are closed under union, concatenation and Kleene-star. Later, the computational complexity of the above two problems are analyzed in [11,18]. In [10], register context-free grammars (RCFG) as well as pushdown automata over an infinite alphabet were introduced as extensions of RA and the equivalence of the two models were shown. Properties of RCFG such as closure and complexity of decision problems are investigated in depth in [10,19,20].

As extensions of finite automata other than RA, data automata [9], pebble automata (PA) [16] and nominal automata (NA) [8] are known. Libkin and Vrgoč [14] argue that RA is the only model that has efficient data complexity for membership among the above mentioned formalisms. Neven et al. consider variations of RA and PA, which are either one way or two ways, deterministic, nondeterministic or alternating. They show inclusion and separation relationships among these automata, FO($\sim, <$) and EMSO($\sim, <$), and give the answer to some open problems including the undecidability of the universality problem for RA [17].

Time-optimal reachability and the related and generalized problems for weighed timed automata (WTA) have been investigated. The single-source optimal reachability problem for WTA is solved by a branch-and-bound algorithm in [7]. Alur et al. [4] solved the optimal reachability problem for TA, which is more general than the single-source one, by introducing limited regions and transforming a WTA to a weighted graph. The decision version of the optimal reachability problem is shown to be PSPACE-complete in [15].

The existing study most related to this paper is Babari et al.'s [5,6], where RA is extended to weighted RA (WRA), and properties including closure and MSO logical characterizations are studied in depth as mentioned in the beginning of this section. Note that WRA is different from cost register automata [2] where data values and weights are not separated and the basic problems are undecidable even for very restricted subclass such as copyless cost register automata (CRA) [1]. This paper partially answers to open problems and conjectures raised in [5] about the decidability of the optimal run problem for WRA under reasonable assumptions as well as the complexity of decision problems for WRA which are counterparts of the membership and emptiness problems for models without weights.

2 Definitions

Let $\mathbb{B} = \{0, 1\}$ be the set of truth values, $\mathbb{N} = \{0, 1, \ldots\}$ be the set of natural numbers and $\mathbb{R}_{\geq 0}$ be the set of nonnegative reals. For a natural number $k \in \mathbb{N}$, let $[k] = \{1, \ldots, k\}$. By $|\beta|$, we mean the cardinality of β if β is a set and the length of β if β is a finite sequence. Let Σ be a finite alphabet and D be an infinite set of data values. We call $w \in (\Sigma \times D)^+$ a *data word* (over Σ and D). For a finite collection $\mathcal{R}$ of binary relations over D, $\mathbb{D} = \langle D, \mathcal{R} \rangle$ is called a *data structure*.

Intuitively, an automaton is equipped with a certain number of registers that can store a data value. Formally, an assignment of data values to k registers (abbreviated as k-register assignment or just assignment if k is irrelevant) is a mapping $\theta : [k] \to D$. The collection of k-register assignments is denoted as Θ_k. For a k-register assignment θ, $\theta(i) (i \in [k])$ is the data value assigned to the i-th register by θ. Let F_k denote the set of guard formulas (or simply, guards) defined by $\varphi := tt \mid x_i^R \mid x_i^{R^{-1}} \mid \mathrm{in}^R \mid \varphi \wedge \varphi \mid \neg\varphi \quad (i \in [k], R \in \mathcal{R})$. For an assignment θ, a data value $d \in D$ and a guard φ, the satisfaction relation $(\theta, d) \models \varphi$ is defined inductively on the structure of φ as $(\theta, d) \models x_i^R$ iff $(\theta(i), d) \in R$, $(\theta, d) \models x_i^{R^{-1}}$ iff $(d, \theta(i)) \in R$, $(\theta, d) \models \mathrm{in}^R$ iff $(d, d) \in R$ and the meaning of tt, $\wedge$ and $\neg$ are defined in the usual way. Define $\mathit{ff} \equiv \neg tt$, $\varphi_1 \vee \varphi_2 \equiv \neg(\neg\varphi_1 \wedge \neg\varphi_2)$.

Definition 1 ([12,13]). *A k-register automaton (k-RA) over a finite alphabet Σ and a data structure $\mathbb{D}$ is a tuple $A = (Q, Q_0, T, Q_f)$ where*

- *Q is a finite set of states,*
- *$Q_0, Q_f \subseteq Q$ are sets of initial and final states, respectively,*
- *$T \subseteq Q \times \Sigma \times F_k \times 2^{[k]} \times Q$ is a set of state transitions.* □

Let $A = (Q, Q_0, T, Q_f)$ be a k-RA over Σ and $\langle D, \mathcal{R} \rangle$. A state transition (or transition) $t = (q, a, \varphi, \Lambda, q') \in T$ where $q, q' \in Q, a \in \Sigma, \varphi \in F_k, \Lambda \in 2^{[k]}$ is written as $q \to^a_{\varphi,\Lambda} q'$ and we denote by label(t) the second component a of t. The description length of a k-RA $A = (Q, Q_0, T, Q_f)$ is defined as $\|A\| = |Q| + |T| \max\{(\log |Q| + k) + \|\varphi\| \mid q \to^a_{\varphi,\Lambda} q' \in T\}$, where $\|\varphi\|$ is the description length of φ, defined in a usual way.

For an assignment $\theta \in \Theta_k$, $\Lambda \in 2^{[k]}$ and a data value $d \in D$, the updated assignment $\theta[\Lambda \leftarrow d] \in \Theta_k$ is $\theta[\Lambda \leftarrow d](i) = d$ if $i \in \Lambda$ and $\theta[\Lambda \leftarrow d](i) = \theta(i)$ otherwise. For a state $q \in Q$ and an assignment $\theta \in \Theta_k$, (q, θ) is called an *instantaneous description (ID)*. For two IDs $c = (q, \theta)$ and $c' = (q', \theta')$, if there are $d \in D$, $t = q \to^a_{\varphi,\Lambda} q' \in T$ such that $(\theta, d) \models \varphi$ and $\theta' = \theta[\Lambda \leftarrow d]$, then

$c \vdash_{t,d} c'$ is called a *switch* from c to c' by t and d in A. The initial value of any register is $\bot$ ($\bot \in D$). The initial ID and an accepting ID are $c_0 \in Q_0 \times \bot^k$ and $c_f \in Q_f \times \Theta_k$, respectively. A *run* in A is a finite sequence of switches from the initial ID to an accepting ID $\rho = c_0 \vdash_{t_1,d_1} c_1 \vdash_{t_2,d_2} c_2 \cdots \vdash_{t_n,d_n} c_n$. The label of a run ρ is $\text{label}(\rho) = (\text{label}(t_1), d_1) \dots (\text{label}(t_n), d_n)$ and ρ is called a run of $\text{label}(\rho)$ in A. For $w \in (\Sigma \times D)^+$, $\text{Run}_A(w)$ is the set of all runs of w in A.

We define $L(A) = \{w \mid \text{Run}_A(w) \neq \emptyset\}$, called the *data language* recognized by A. A data language $L \subseteq (\Sigma \times D)^+$ is recognizable if there is an RA A such that $L = L(A)$.

Example 1. Let $\Sigma = \{a\}$, $\mathcal{R} = \{<, =, >\}$. An example of 2-RA A_1 is shown in Fig. 1 where $\bot = 0$. For an input data word w, A_1 loads any data value, say d_i, in w to the first register nondeterministically by t_2. After that, every time a data value not equal to d_i comes, A_1 stays at q_1 by t_3 or t_4 until the same value d_i comes, at which A_1 moves to q_2 by t_5. In this way, A_1 nondeterministically chooses two positions having an identical data value d_i from the input data word, and the data values between them are not equal to d_i. We have $L(A_1) = \{(a, d_1) \dots (a, d_n) \in (D \times \Sigma)^+ \mid i, j \in [n], i < j, d_i = d_j \text{ and for } k = i+1, ..., j-1, d_i \neq d_k\}$.

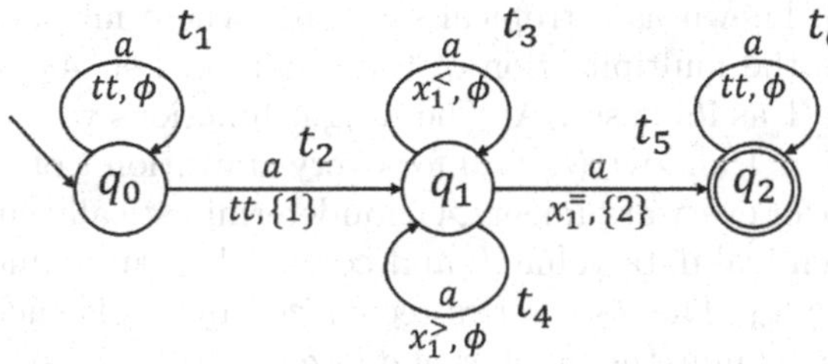

Fig. 1. RA A_1

We will use notations Σ, $\mathbb{D} = \langle D, \mathcal{R} \rangle$ and $\mathcal{S} = (S, +, \cdot, 0, 1)$ to implicitly denote a finite alphabet, a data structure and a semiring, respectively.

Definition 2 ([5]). *A k-register weighted automaton (k-WRA) over $\Sigma, \mathbb{D}, \mathcal{S}$ is a tuple $\mathcal{A} = (Q, Q_0, T, Q_f, \text{wt})$ where*

- *(Q, Q_0, T, Q_f) is a k-RA over $\Sigma, \mathbb{D}$, called the base RA of $\mathcal{A}$,*
- *$\text{wt} = (\text{wtt}, \text{wtd})$ where $\text{wtt} : T \to S$ and $\text{wtd} : (T \times [k]) \to ((D \times D) \to S)$.* □

Let $\mathcal{A} = (Q, Q_0, T, Q_f, \text{wt})$ be a k-WRA as above. $\text{wtt}(t)$ represents the weight of a transition $t \in T$. $\text{wtd}(t, j)$ is the weight of the j-th register at a transition $t \in T$. More precisely, $\text{wtd}(t, j)(\theta(j), d)$ represents the weight needed for manipulating the j-th register for a switch $(q, \theta) \vdash_{t,d} c'$. The weight of a switch $c \vdash_{t,d} c'$ is defined as

$$\mathrm{wt}((q,\theta) \vdash_{t,d} c') = \prod_{j=1}^{k} \mathrm{wtd}(t,j)(\theta(j), d) \cdot \mathrm{wtt}(t).$$

A run in $\mathcal{A}$ is just a run in the base RA of $\mathcal{A}$. The weight of a run $\rho = c_0 \vdash_{t_1,d_1} c_1 \vdash_{t_2,d_2} c_2 \cdots \vdash_{t_n,d_n} c_n$ in $\mathcal{A}$ is defined as

$$\mathrm{wt}(\rho) = \prod_{i=1}^{n} \mathrm{wt}(c_{i-1} \vdash_{t_i,d_i} c_i).$$

We assume that there are constants $W_1, W_2 \in \mathbb{R}_{\geq 0}$ such that for any $t \in T$, $\mathrm{wtt}(t)$ can be computed in W_1 time and for $t \in T, j \in [k], d_1, d_2 \in D$, $\mathrm{wtd}(t,j)(d_1,d_2)$ can be computed in W_2 time. We define the description length of a k-WRA $\mathcal{A}$ as $\|\mathcal{A}\| = \|A_b\|$ where A_b is the base RA of $\mathcal{A}$.[1]

A data series over Σ, D and $\mathcal{S}$ is a mapping $U : (\Sigma \times D)^+ \to S$. The data series recognized by a WRA $\mathcal{A}$ is the data series $[\![\mathcal{A}]\!]$ defined as $[\![\mathcal{A}]\!](w) = \sum_{\rho \in \mathrm{Run}_{\mathcal{A}}(w)} \mathrm{wt}(\rho)$ for each $w \in (\Sigma \times D)^+$. A data series $U : (\Sigma \times D)^+ \to S$ is recognizable if there is a WRA that recognizes U.

Example 2. Let $\Sigma = \{a\}$, $\mathbb{D} = \langle \mathbb{N}, \{<, =, >\}\rangle$, and the semiring $\mathcal{R}_{\mathrm{trpc}} = (\mathbb{R}_{\geq 0} \cup \{\infty\}, \min, +, \infty, 0)$, known as a tropical semiring, where min acts as the addition and $+$ acts as the multiplication of the semiring. Let $\mathcal{A}_2$ be 2-WRA that has A_1 of Example 1 as its base RA. The weight functions $\mathrm{wt} = (\mathrm{wtt}, \mathrm{wtd})$ are defined as: $\mathrm{wtt}(t_3) = 1$ and $\mathrm{wtt}(t) = 0$ for every transition t other than t_3, and $\mathrm{wtd}(t,j)(d,d') = 0$ for every argument. $\mathcal{A}_2$ nondeterministically chooses two positions having an identical data value d_i and counts the data values greater than d_i between them by t_3. The data series recognized by $\mathcal{A}_2$ is such that for $w \in (\Sigma \times D)^+$, $[\![\mathcal{A}_2]\!](w) = \min\{$the number of d in $d_{i+1}, \cdots, d_{j-1}$ such that $d > d_i \mid w = (a, d_1) \ldots (a, d_n), i, j \in [n], i < j, d_i = d_j$ and for $k = i+1, ..., j-1, d_i \neq d_k\}$.

3 Decision Problems

In this section, we analyze the computational complexity of the following problems for WRA. The results are summarized in Table 1.

Definition 3 (The weight computation problems)
Input: a k-WRA $\mathcal{A}$ over Σ, $\mathbb{D}$, $\mathcal{S}$ and a data word $w \in (\Sigma \times D)^+$. For the run weight computation problem, a weight $s \in S$ is also given.
(The run weight computation problem) $\exists \rho \in \mathrm{Run}_{\mathcal{A}}(w).\mathrm{wt}(\rho) = s$?
(The data word weight computation problem) Compute $[\![\mathcal{A}]\!](w)$.
The input size of both problems is $\|\mathcal{A}\| + |w|$.

[1] We do not include the size of the weight part because of the assumption that the computation of the weights of a single transition and a single register can be done in constant time.

Table 1. Complexity results

Problem	Complexity
Run weight computation	NP-complete
Data word weight computation	PSPACE-solvable, #P-hard (#P-complete when a weight is a natural number, a transition weight function is bounded and every register manipulation weight is 1)
Run weight realizability	Undecidable
Data word weight realizability	Undecidable

Definition 4 (The weight realizability problems)
Input: a k-WRA $\mathcal{A}$ over Σ, $\mathbb{D}$, $\mathcal{S}$ and a weight $s \in S$
(The run weight realizability problem) $\exists w.\exists \rho \in \mathrm{Run}_{\mathcal{A}}(w).\mathrm{wt}(\rho) = s$?
(The data word weight realizability problem) $\exists w.[\![\mathcal{A}]\!](w) = s$?
The input size of both problems is $\|\mathcal{A}\|$.

Theorem 1. *The run weight computation problem is NP-complete.*

Proof. Assume we are given a k-WRA $\mathcal{A} = (Q, Q_0, T, Q_f, \mathrm{wt})$ over Σ, $\langle D, \mathcal{R}\rangle$, $\mathcal{S} = (S, +, \cdot, 0, 1)$, a data word $w \in (\Sigma \times D)^+$ and $s \in S$.

(NP solvability). By the assumption on complexity of computing weights of WRA, $\mathrm{wt}(c \vdash_{t,d} c')$ can be computed in $O(W_1 + W_2 k)$ time. Thus, for any run $\rho \in \mathrm{Run}_{\mathcal{A}}(w)$, the weight $\mathrm{wt}(\rho)$ can be computed in $O((W_1 + W_2 k)|w|)$ time. Hence, we can nondeterministically choose a run of w and test whether $\mathrm{wt}(\rho) = s$ in polynomial time.

(NP-hardness). We restrict the problem as:

> For every transition $t \in T$, $j \in [k]$ and $d_1, d_2 \in D$, $\mathrm{wtt}(t) = \mathrm{wtd}(t, j)(d_1, d_2) = 1$. Also $s = 1$.

Then, for any switch $c \vdash_{t,d} c'$, we have $\mathrm{wt}(c \vdash_{t,d} c') = 1$. This implies that for every run $\rho \in \mathrm{Run}_{\mathcal{A}}(w)$, we have $\mathrm{wt}(\rho) = 1 = s$. Therefore, the problem restricted in this way asks for an input k-WRA $\mathcal{A}$ and a data word w, whether $\exists \rho \in \mathrm{Run}_{\mathcal{A}}(w)$. The k-WRA in this setting can be regarded as a RA (standard register automata without weight) and the above problem is equivalent to the membership problem that asks whether a given data word w is accepted by $\mathcal{A}$ regarded as an RA. Hence the run weight computation problem is NP-hard because the membership problem for RA is NP-complete [13]. □

To discuss the complexity of the data word computation problem, we use the complexity class #P, the class of function problems that can be solved by counting the number of accepting runs of a polynomial-time non-deterministic Turing machine. An example of #P-complete problem is #SAT: How many different variable assignments will satisfy a given general boolean formula?

Let $\mathcal{N} = (\mathbb{N}, +, \cdot, 0, 1)$ be the semiring of natural numbers.

Lemma 1. *The data word weight computation problem of k-WRA $\mathcal{A} = (Q, Q_0, T, Q_f, (\mathrm{wtt}, \mathrm{wtd}))$ over Σ, $\langle D, \mathcal{R}\rangle$ and $\mathcal{N}$ is #P-hard even if* $\mathrm{wtt}(t) = \mathrm{wtd}(t, j)(d, d') = 1$ *for every* $t \in T, j \in [k], d, d' \in D$.

Proof. We reduce #2SAT problem, which is known to be #P-complete, to the data word weight computation problem. Let $\phi = c_1 \wedge c_2 \wedge \cdots \wedge c_m$ be a given 2-CNF, where each c_i ($i \in [m]$) is a clause consisting of two literals and $z_1, \ldots, z_n$ are Boolean variables appearing in ϕ. We construct n-WRA $\mathcal{A}_\phi = (Q, Q_0, T, Q_f, (\mathrm{wtt}, \mathrm{wtd}))$ over $\Sigma, \langle D, \mathcal{R}\rangle, \mathcal{N}$ and input data word w from ϕ as follows. Let $\Sigma = \{a\}$, D be an infinite set containing $\top$ and $\bot$, and let $\mathcal{R} = \{=, \neq\}$ where $=$ and $\neq$ are (an extension of) the equality on Boolean values (logical equivalence) and its negation, respectively. Note that $\bot$ is the initial value. The values of all weight functions wtt and wtd are defined as $1 \in \mathbb{N}$. Let $Q = \{q_i \mid i \in [n]\} \cup \{q_{k,l} \mid k \in [m], l \in [2]\} \cup \{q'_k \mid k \in [m]\} \cup \{q_{\mathrm{r}}, q_f\}$, $Q_0 = \{q_1\}$ and $Q_f = \{q_f\}$. The input word is $w = (a, \top) \cdots (a, \top)$ of length $|w| = n + 2m$. We construct the following transitions and add them to T: The first group of transitions nondeterministically simulates an assignment of a Boolean value to each z_i ($i \in [n]$). If x_i is updated to be $\top$, it means z_i is assigned *tt*, and otherwise, it means z_i is assigned *ff*.

$$q_1 \xrightarrow{a}_{tt,\{1\}} q_2,\ q_1 \xrightarrow{a}_{tt,\emptyset} q_2,\ \ldots,\ q_n \xrightarrow{a}_{tt,\{n\}} q_{1,1},\ q_n \xrightarrow{a}_{tt,\emptyset} q_{1,1}.$$

The second group of transitions deterministically evaluates the truth value of each clause $c_k = y_{k,1} \vee y_{k,2}$ ($k \in [m]$).

$$\begin{array}{lll}
q_{k,1} \xrightarrow{a}_{x_i^=,\emptyset} q'_k, & q_{k,1} \xrightarrow{a}_{x_i^{\neq},\emptyset} q_{k,2} & \text{if } y_{k,1} = z_i, \\
q_{k,1} \xrightarrow{a}_{x_i^{\neq},\emptyset} q'_k, & q_{k,1} \xrightarrow{a}_{x_i^=,\emptyset} q_{k,2}, & \text{if } y_{k,1} = \overline{z_i}, \\
q_{k,2} \xrightarrow{a}_{x_i^=,\emptyset} q_{k+1,1}, & q_{k,2} \xrightarrow{a}_{x_i^{\neq},\emptyset} q_{\mathrm{r}}, & \text{if } y_{k,2} = z_i, \\
q_{k,2} \xrightarrow{a}_{x_i^{\neq},\emptyset} q_{k+1,1}, & q_{k,2} \xrightarrow{a}_{x_i^=,\emptyset} q_{\mathrm{r}}, & \text{if } y_{k,2} = \overline{z_i}, \\
q'_k \xrightarrow{a}_{tt,\emptyset} q_{k+1,1} & &
\end{array}$$

where $q_{m+1,1}$ is the final state q_f. The state q_{r} is a dead state with no outgoing transition. The states q'_k are used to skip the evaluation of literals when a preceding literal evaluates to $\top$ in the clause.

For a truth-value assignment $\alpha : \{z_1, \ldots, z_n\} \to \{tt, ff\}$, let $\theta_\alpha \in \Theta_n$ be $\theta_\alpha(x_i) = \top$ if $\alpha(z_i) = tt$ and $\theta_\alpha(x_i) = \bot$ otherwise. Assume $\mathcal{A}_\phi$ is fed with the input data word $w = (a, \top) \ldots (a, \top)$ of length $n + 2m$. After conducting the first group of transitions, the assignment of $\mathcal{A}_\phi$ becomes θ_α for some truth-value assignment α. Because the second group of transitions deterministically verifies whether ϕ evaluates to *tt* without register update, that part of the run is uniquely determined. In other words, there is a one-to-one correspondence between the set of maximal sequences of switches of w in $\mathcal{A}_\phi$ and the set of assignments. Therefore, a maximal sequence of switches of w in $\mathcal{A}_\phi$ is a run ρ of w if and only if ϕ is satisfied by the truth-value assignment α corresponding to the assignment θ_α obtained by ρ. □

Lemma 2. *The data word weight computation problem for k-WRA is PSPACE-solvable. When the semiring is $\mathcal{N}$,* wtt *is bounded and* $\mathrm{wtd}(t, j)(d_1, d_2) = 1$ *for every $t \in T$ and $j \in [k]$ and $d_1, d_2 \in D$ for a given k-WRA $\mathcal{A} = (Q, Q_0, T, Q_f, (\mathrm{wtt}, \mathrm{wtd}))$, the problem becomes #P-solvable.*

Proof. PSPACE-solvability is easy to show. The weight of a run of an input data word can be calculated in polynomial time by the proof of Theorem 1, and we need additional polynomial space to store the sum of the weights of all runs of the input data word.

Next, we discuss #P-solvability. From a k-WRA $\mathcal{A} = (Q, Q_0, T, Q_f, (\mathrm{wtt}, \mathrm{wtd}))$ over Σ, $\langle D, \mathcal{R}\rangle$, $\mathcal{N}$, we construct k-WRA $\mathcal{A}' = (Q', Q_0', T', Q_f', (\mathrm{wtt}', \mathrm{wtd}'))$ such that $[\![\mathcal{A}']\!] = [\![\mathcal{A}]\!]$ by dividing each run in $\mathcal{A}$ into several runs whose weights are 1. For $M = \max\{\mathrm{wtt}(t) \mid t \in T\}$, we introduce new states $q_1, \ldots, q_M$ not included in Q. Note that M is a constant by the assumption. The set of the states of $\mathcal{A}'$ is $Q' = \{(q, q_i) \mid q \in Q, i \in [M]\}$, and the set of transitions is $T' = \{(q, q_i) \to^a_{\varphi,\Lambda} (q', q_j) \mid t = q \to^a_{\varphi,\Lambda} q' \in T, \mathrm{wtt}(t) = m, i \in [M], j \in [m]\}$. Also, let $Q_0' = \{(q_I, q_1) \mid q_I \in Q_0\}$, and $Q_f' = \{(q_f, q_i) \mid q_f \in Q_f, i \in [M]\}$. This construction of $\mathcal{A}'$ can be done in polynomial time. Therefore, the data word weight computation problem is in #P under the given condition.

Theorem 2. *Let $\mathcal{A} = (Q, Q_0, T, Q_f, (\mathrm{wtt}, \mathrm{wtd}))$ be a k-WRA over Σ, $\langle D, \mathcal{R}\rangle$, $\mathcal{N}$. If* $\max\{\mathrm{wtt}(t) \mid t \in T\}$ *is uniformly bounded and* $\mathrm{wtd}(t, j)(d_1, d_2) = 1$ *for every $t \in T$, $j \in [k]$ and $d_1, d_2 \in D$, then the data word weight computation problem is #P-complete.*

Proof. By Lemmas 1 and 2.

Theorem 3. *The run weight realizability problem for k-WRA is undecidable even if $k = 1$, all the values of weight functions are one and every relation of the data structure is decidable.*

Proof. We prove the theorem by a reduction from the Post correspondence problem (PCP). Let $I = \langle(u_1, \ldots, u_m), (v_1, \ldots, v_m)\rangle$ be a given instance of PCP over Σ where $u_i, v_i \in \Sigma^*$ for $i \in [m]$. From I, we construct a 1-WRA $\mathcal{A}_I = (\{q_0, q, q_f\}, \{q_0\}, T, \{q_f\}, \mathrm{wt})$ over $\{a\}$, $\langle D, \mathcal{R}\rangle$, $\mathcal{N}$ where the data structure $\langle D, \mathcal{R}\rangle$, the set T of transitions and the weight functions $\mathrm{wt} = (\mathrm{wtt}, \mathrm{wtd})$ are defined as follows.

- $D = \Sigma^* \times \Sigma^*$ with $\perp = (\varepsilon, \varepsilon) \in D$ as the initial value and $\mathcal{R} = \{R_i \mid i \in [m]\} \cup \{\mathrm{EQ}\}$ where for $x, y, x', y' \in \Sigma^*$, $(x, y)R_i(x', y') \Leftrightarrow (x' = xu_i$ and $y' = yv_i)$ for $i \in [m]$ and $(x, y)\mathrm{EQ}(x', y') \Leftrightarrow (x = y)$.
- $T = \{q_0 \to^a_{x_1^{R_i},\{1\}} q,\ q \to^a_{x_1^{R_i},\{1\}} q \mid i \in [m]\} \cup \{q \to^a_{x_1^{\mathrm{EQ}},\emptyset} q_f\}$.
- $\mathrm{wtt}(t) = \mathrm{wtd}(t, 1)(d_1, d_2) = 1$ for every $t \in T$, $d_1, d_2 \in D$.

It is easy to see that I has a solution of PCP if and only if there is a run ρ of some $w \in (\{a\} \times D)^+$ in $\mathcal{A}_I$ such that $\mathrm{wt}(\rho) = 1$.

Corollary 1. *The data word weight realizability problem of k-WRA is undecidable even if $k = 1$, all the values of weight functions are one and every relation of the data structure is decidable.* □

The above results imply that the realizability problems are already undecidable for ordinary RA (w/o weights). This motivates us to introduce a subclass of WRA for which the realizability problems and related optimization problems are solvable while the weights make sense, which are given in Sect. 5.2.

4 Transition Decomposition by Register Type

In this section, we will define a *normal form* WRA. First, we introduce a *register type* as a finite abstraction of assignments with respect to the relations in $\mathcal{R}$ of a given data structure $\langle D, \mathcal{R}\rangle$.

Definition 5 ([20]). *A register type (of k registers) for a data structure $\langle D, \mathcal{R}\rangle$ is an arbitrary function $\gamma : ([k] \times [k]) \to (\mathcal{R} \to \mathbb{B})$. Let Γ_k denote the collection of all register types of k registers. For an assignment $\theta \in \Theta_k$ and a register type $\gamma \in \Gamma_k$, if $\forall i, j \in [k] \forall R \in \mathcal{R}.(\gamma(i,j)(R) = 1 \Leftrightarrow (\theta(i), \theta(j)) \in R)$ holds, we write $\theta : \gamma$ and we say that the type of θ is γ.* □

Let $\mathcal{A} = (Q, Q_0, T, Q_f, \mathrm{wt})$ be an arbitrary k-WRA. From $\mathcal{A}$, we define k-WRA $\mathcal{A}' = (Q', Q_0', T', Q_f', \mathrm{wt}')$ as follows: $Q' = Q \times \Gamma_k$. $Q_0' = Q_0 \times \{\gamma_0\}$ where γ_0 is defined as $\forall R \in \mathcal{R}\big[(\forall i, j \in [k].\gamma_0(i,j)(R) = 1) \Leftrightarrow ((\bot, \bot) \in R)\big]$. $Q_f' = Q_f \times \Gamma_k$. T' is the smallest set of transitions $t' = (p, \gamma) \to^a_{\varphi', \Lambda} (q, \gamma')$ where

$t = p \to^a_{\varphi,\Lambda} q \in T$, $\gamma, \gamma' \in \Gamma_k$, $\varphi' = \varphi \wedge \prod_{R\in\mathcal{R}}(\prod_{i=1}^k \alpha_i^R \wedge \beta_i^R) \wedge \delta^R$, $\alpha_i^R \in \{x_i^R, \neg x_i^R\}, \beta_i^R \in \{x_i^{R^{-1}}, \neg x_i^{R^{-1}}\}, \delta^R \in \{\mathrm{in}^R, \neg \mathrm{in}^R\}$, $\varphi' \not\equiv \mathit{ff}$ and if $\theta \in \Theta_k$, $d \in D$, $\theta : \gamma$, $(\theta, d) \models \varphi'$, $(p, \theta) \vdash_{t,d} (q, \theta[\Lambda \leftarrow d])$ then $\theta[\Lambda \leftarrow d] : \gamma'$.

In the above definition, φ' says that in addition to φ, whether the contents of the i-th register and an input data value d satisfies R (resp. d is reflexive on R) is exactly determined by α_i^R and β_i^R (resp. by δ^R). Furthermore, an input data value is loaded to the registers specified by Λ when t' is applied. Therefore, if $t \in T$, $\gamma \in \Gamma_k$ and φ' are given, the transition belonging to T' is uniquely determined. We write that transition as $s_{t,\gamma,\varphi'}$. Finally, define $\mathrm{wt}' = (\mathrm{wtt}', \mathrm{wtd}')$ where for each $t' = s_{t,\gamma,\varphi'} \in T'$, $\mathrm{wtt}'(t') = \mathrm{wtt}(t)$ and for $j \in [k]$, $\mathrm{wtd}'(t', j) = \mathrm{wtd}(t, j)$. This completes the definition of k-WRA $\mathcal{A}'$.

Example 3. Let $k = 2$, $\mathcal{R} = \{R\}$ and consider transition $t = p \to^{a}_{x_1^R,\{2\}} q$ and register type γ such that $\gamma(i,j)(R) = 1$ ($(i,j) = (1,1),(1,2),(2,2)$), $\gamma(2,1)(R) = 0$. If we merge transitions whose target states are the same, then we have the following four transitions:

$$(p,\gamma) \to^{a}_{x_1^R \wedge x_1^{R^{-1}} \wedge \mathrm{in}^R,\{2\}} (q,\gamma^{(1)}), \quad (p,\gamma) \to^{a}_{x_1^R \wedge x_1^{R^{-1}} \wedge \neg\mathrm{in}^R,\{2\}} (q,\gamma^{(2)}),$$
$$(p,\gamma) \to^{a}_{x_1^R \wedge \neg x_1^{R^{-1}} \wedge \mathrm{in}^R,\{2\}} (q,\gamma^{(3)}), \quad (p,\gamma) \to^{a}_{x_1^R \wedge \neg x_1^{R^{-1}} \wedge \neg\mathrm{in}^R,\{2\}} (q,\gamma^{(4)})$$

where

$$\gamma^{(i)}(1,2)(R) = 1, \gamma^{(i)}(2,1)(R) = 1 \ (i = 1,2),$$
$$\gamma^{(i)}(1,2)(R) = 1, \gamma^{(i)}(2,1)(R) = 0 \ (i = 3,4),$$
$$\gamma^{(i)}(2,2)(R) = 0 \ (i = 2,4), \gamma^{(i)}(j,j)(R) = 1 \ (\textit{otherwise}).$$

Theorem 4. *Let $\mathcal{A} = (Q, Q_0, T, Q_f, \mathrm{wt})$ be an arbitrary k-WRA and $\mathcal{A}' = (Q', Q_0', T', Q_f', \mathrm{wt}')$ be the k-WRA obtained from $\mathcal{A}$ by the transition decomposition by register type. Also let $w = (a_1,d_1)\cdots(a_n,d_n) \in (\Sigma \times D)^+$ be an arbitrary data word. For a run $\rho = c_0 \vdash_{t_1,d_1} c_1 \vdash_{t_2,d_2} \cdots \vdash_{t_n,d_n} c_n \in \mathrm{Run}_{\mathcal{A}}(w)$, there exists a run $\rho' = c_0' \vdash_{s_{t_1,\gamma_0,\varphi_1'},d_1} c_1' \vdash_{s_{t_2,\gamma_1,\varphi_2'},d_2} \cdots \vdash_{s_{t_n,\gamma_{n-1},\varphi_n'},d_n} c_n' \in \mathrm{Run}_{\mathcal{A}'}(w)$ such that $\mathrm{wt}(\rho') = \mathrm{wt}(\rho)$. Conversely, for a run $\rho' \in \mathrm{Run}_{\mathcal{A}'}(w)$, there exists a run $\rho \in \mathrm{Run}_{\mathcal{A}}(w)$ such that $\mathrm{wt}(\rho) = \mathrm{wt}(\rho')$.*

Proof. Consider a data word w and a run ρ stated in the lemma and assume $c_i = (q_i, \theta_i), \theta_i : \gamma_i$ for $i \in \{0\} \cup [n]$. By the construction of T', there exists a unique transition $s_{t_i,\gamma_{i-1},\varphi_i'} = (q_{i-1},\gamma_{i-1}) \to^{a_i}_{\varphi_i',\Lambda} (q_i,\gamma_i) \in T'$ such that $((q_{i-1},\gamma_{i-1}),\theta_{i-1}) \vdash_{s_{t_i,\gamma_{i-1},\varphi_i'},d_i} ((q_i,\gamma_i),\theta_i)$ in $\mathcal{A}'$ where φ_i' is determined by whether $(\theta_{i-1},d_i) \models x_j^R$, $(\theta_{i-1},d_i) \models x_j^{R^{-1}}$ and $(\theta_{i-1},d_i) \models \mathrm{in}^R$ hold or not for $j \in [k]$ and $R \in \mathcal{R}$. If we concatenate the above switches, we obtain a run ρ' of w in $\mathcal{A}'$ and $\mathrm{wt}(\rho') = \mathrm{wt}(\rho)$.

Conversely, for $i \in [n]$, let $c_{i-1}' \vdash_{s_{t_i,\gamma_{i-1},\varphi_i'},d_i} c_i'$ be a switch in $\mathcal{A}'$ where $s_{t_i,\gamma_{i-1},\varphi_i'} = (q_{i-1},\gamma_{i-1}) \to^{a_i}_{\varphi_i',\Lambda} (q_i,\gamma_i) \in T'$. The transition of $\mathcal{A}$ corresponding to $s_{t_i,\gamma_{i-1},\varphi_i'} \in T'$ is exactly $t_i \in T$. By the construction of T', $(\theta_{i-1},d_i) \models \varphi_i'$ implies $(\theta_{i-1},d_i) \models \varphi_i$. Therefore, $c_{i-1} \vdash_{t_i,d_i} c_i$ is a switch in $\mathcal{A}$. The rest of the proof is similar to the former case; we lift the obtained switches to the run. □

A WRA obtained by the above transformation is called a *normal form WRA*.

5 The Optimal Run Problem

5.1 Definition of the Problem

We introduce the problem of computing the optimal (infimum) weight of the runs from the initial ID to an accepting ID of a given WRA. We assume the tropical semiring $\mathcal{R}_{\mathrm{trpc}}$ (see Example 2) because by $\mathcal{R}_{\mathrm{trpc}}$ we can represent the minimum weight by the addition of the semiring. Of course, we could use the max-tropical semiring $(\mathbb{R} \cup \{-\infty\}, \max, +, -\infty, 0)$ instead.

Definition 6 (The optimal run problem)
Input: a k-WRA $\mathcal{A}$ over $\Sigma, \langle D, \mathcal{R}\rangle, \mathcal{R}_{\mathrm{trpc}}$
Output: The infimum of $\{\mathrm{wt}(\rho) \mid \exists w \in (\Sigma \times D)^{+}.\rho \in \mathrm{Run}_{\mathcal{A}}(w)\}$ □

By Theorem 4 and the definition of the problem, the following property holds.

Corollary 2. *Let $\mathcal{A} = (Q, Q_0, T, Q_f, \mathrm{wt})$ be an arbitrary k-WRA and $\mathcal{A}' = (Q', Q'_0, T', Q'_f, \mathrm{wt}')$ be the normal form k-WRA obtained from $\mathcal{A}$. The solutions to the optimal run problem for $\mathcal{A}$ and $\mathcal{A}'$ are the same.* □

Let $\mathcal{A}$ and $\mathcal{A}'$ be as assumed in the above corollary. We will transform $\mathcal{A}'$ to an edge-weighted directed graph $G = \langle V, E\rangle$ such that the solution of the optimal run problem is equal to the weight of the minimum-weight path of G. The difficulty lies in the requirement that we must construct G without knowing an input data word w to $\mathcal{A}'$ or assignments appearing in a run of w in $\mathcal{A}'$. To overcome this problem, we introduce two properties in the next subsection.

5.2 Weighted Simulation and Weight Computability

Let $\mathcal{A} = (Q, Q_0, T, Q_f, \mathrm{wt})$ be an arbitrary k-WRA and $\mathcal{A}' = (Q', Q'_0, T', Q'_f, \mathrm{wt}')$ be the normal form k-WRA obtained from $\mathcal{A}$. We say that k-WRA $\mathcal{A}'$ has *weighted simulation property* if every $t' = (p, \gamma) \to^{a}_{\varphi', \Lambda} (q, \gamma') \in T'$ satisfies the following condition: for every $\theta \in \Theta_k$ such that $\theta : \gamma$, $\inf\{\mathrm{wt}(((p, \gamma), \theta) \vdash_{t',d} ((q, \gamma'), \theta[\Lambda \leftarrow d])) \mid d \in D, \theta[\Lambda \leftarrow d] : \gamma'\}$ takes a same value[2]. Also, we say that k-WRA $\mathcal{A}'$ has *weight computability* if the above infimum, denoted as $\mathrm{wt}(t')$, can be computed in polynomial time of $\|\mathcal{A}\|$. Weighted simulation is a natural extension of the property of TA and WTA that the infinite set of IDs can be divided into finite sets called clock regions such that any IDs belonging to a same clock region are indistinguishable. The above two properties are undecidable in general because a binary relation appearing in the guard of a transition may be undecidable. Weighted simulation says that if two assignments θ_1, θ_2 have a same register type γ, the infimum of the weights of switches from (p, θ_1) to (q, θ'_1) by t' is the same as that from (p, θ_2) to (q, θ'_2) by t'. This property, together with weight computability, enables us to compute the infimum of the weights from (p, γ) to (q, γ') without knowing an assignment or an input data value.

These assumptions also make the following two problems related to the weight realizability problems decidable.

Definition 7 (The weight bounding problems)
Input: a k-WRA $\mathcal{A}$ over Σ, $\langle D, \mathcal{R}\rangle$, $\mathcal{R}_{\mathrm{trpc}}$ and a weight $s \in \mathbb{R}_{\geq 0}$
(The run weight bounding problem) $\exists w.\exists \rho \in \mathrm{Run}_{\mathcal{A}}(w).\mathrm{wt}(\rho) \leq s$?
(The data word weight bounding problem) $\exists w.[\![\mathcal{A}]\!](w) \leq s$?
The input size for both problems is $\|\mathcal{A}\|$.

---

[2] If there is no such a switch $(p, \theta) \vdash_{t',d} (q, \theta[\Lambda \leftarrow d])$ for any $d \in D$, we define the infimum as ∞.

Theorem 5. *The run weight bounding problem for k-WRA over Σ, $\langle D, \mathcal{R}\rangle$, $\mathcal{R}_{\mathrm{trpc}}$ is PSPACE-complete if weighted simulation and weight computability hold.*

Proof (PSPACE-solvability). Let $\mathcal{A} = (Q, Q_0, T, Q_f, \mathrm{wt})$ be a k-WRA over Σ, $\langle D, \mathcal{R}\rangle$ and $\mathcal{S}$ and $s \in \mathbb{R}_{\geq 0}$. Let $\mathcal{A}' = (Q', Q'_0, T', Q'_f, \mathrm{wt}')$ be the normal form k-WRA obtained from $\mathcal{A}$. By weighted simulation, the number of IDs of $\mathcal{A}'$ that must be examined is not more than $|Q'|(k+1)^k$ by a similar reason discussed in [13]. Therefore, it is enough to check whether $\mathrm{wt}(\rho) \leq s$ for every run ρ of w whose length is at most $|Q'|(k+1)^k$. By the proof of Theorem 1, computing the weight of a run can be done in polynomial time. Also, the space needed to simulate a run of an input data word of length $|Q'|(k+1)^k$ is $\log(|Q'|(k+1)^k) = k\log(k+1) + \log|Q'|$. Because $Q' = Q \times \Gamma_k$ holds, $|Q'| \in O(|Q|2^{k^2|\mathcal{R}|})$. Hence, the space complexity is $O(k\log k + \log|Q| + k^2|\mathcal{R}|)$, which is a polynomial order of k, $|Q|$ and $|\mathcal{R}|$. Consequently, this problem can be solved in PSPACE.

(PSPACE-Hardness). As in the proof of NP-hardness in Theorem 1, we assume the value of every weight function is 0. Then, for every data word w and every run $\rho \in \mathrm{Run}_{\mathcal{A}}(w)$, $\mathrm{wt}(\rho) = 0$. When the given semiring value is $s = 0$, the run weight realizability problem is expressed as: for a given k-WRA $\mathcal{A}$, $\exists w, \exists \rho \in \mathrm{Run}_{\mathcal{A}}(w)$?, which is equivalent to the emptiness problem for k-RA. Because the emptiness problem for RA is PSPACE-complete [13], the run weight bounding problem is PSPACE-hard.

Theorem 6. *The data word weight bounding problem of k-WRA over Σ, $\langle D, \mathcal{R}\rangle$, $\mathcal{R}_{\mathrm{trpc}}$ is PSPACE-complete if weighted simulation and weight computability hold.*

Proof. Let $\mathcal{A}$ be a given k-WRA and s be a semiring value.

(PSPACE-solvability). We only need to take the sum of the weights of all the runs of w in $\mathcal{A}$ to compute $[\![\mathcal{A}]\!](w)$, which needs only $O(1)$ additional space. Therefore, this problem can be solved in PSPACE.

(PSPACE-hardness). The run weight bounding problem is PSPACE-complete by Theorem 5, and so this problem is PSPACE-hard.

5.3 Transformation to a Directed Graph

We will present a transformation from a given k-WRA to an edge-weighted directed graph when weighted simulation and weight computability hold. Let $\mathcal{A}$ be a k-WRA over Σ, $\langle D, \mathcal{R}\rangle$ and $\mathcal{R}_{\mathrm{trpc}}$ that satisfies weighted simulation and weight computability and $\mathcal{A}' = (Q', Q'_0, T', Q'_f, \mathrm{wt}')$ be the normal form k-WRA obtained from $\mathcal{A}$.

Construct the edge-weighted directed graph $G = \langle V, E\rangle$ where V and E are the sets of nodes and edges respectively, where $V = Q'$ and $E \subseteq V \times V \times T' \times \mathbb{R}_{\geq 0}$ is defined as follows: For each transition $s_{t,\gamma,\varphi'} = (p, \gamma) \to^{a}_{\varphi',\mathcal{A}} (q, \gamma') \in T'$ of $\mathcal{A}'$, compute $\mathrm{wt}(s_{t,\gamma,\varphi'})$, which is possible by weighted simulation and weight computability. If $\mathrm{wt}(s_{t,\gamma,\varphi'}) < \infty$, add $((p,\gamma), (q,\gamma'), s_{t,\gamma,\varphi'}, \mathrm{wt}(s_{t,\gamma,\varphi'}))$ to E.

For a path π in an edge-weighted directed graph, the weight of π is the sum of the weights of the edges in π, denoted by $\mathrm{wt}(\pi)$.

Theorem 7. *Let $\mathcal{A}$ and $\mathcal{A}'$ be the WRA above, and $\mathcal{A}'$ have weighted simulation property and weight computability. Let $G = \langle V, E\rangle$ be the directed graph obtained from $\mathcal{A}'$ by the above construction. For a path π in G starting with the initial state and ending with a final state of $\mathcal{A}'$, there is a run ρ in $\mathcal{A}'$ such that $\mathrm{wt}(\rho) = \mathrm{wt}(\pi)$. Conversely, for a run ρ in $\mathcal{A}'$, there is a path π in G such that $\mathrm{wt}(\pi) = \mathrm{wt}(\rho)$.*

Proof. Let $\pi = e_1 e_2 \cdots e_n$ be a path in G starting with the initial state and ending with a final state of $\mathcal{A}'$ where $e_i \in E$ $(i \in [n])$. By the construction of G, for the i-th edge $e_i = (v_{i-1}, v_i, s_i, m_i)$ of π $(i \in [n])$, the third component s_i can be written as $s_i = s_{t_i,\gamma_{i-1},\varphi'_i} = (p_{i-1}, \gamma_{i-1}) \to^{a_i}_{\varphi'_i,\Lambda} (p_i, \gamma_i) \in T'$ (p_0 is the initial state and p_n is a final state) and $m_i = \mathrm{wt}(s_{t_i,\gamma_{i-1},\varphi'_i}) = \inf\{\mathrm{wt}((p_{i-1}, \theta_{i-1}) \vdash_{s_{t_i,\gamma_{i-1},\varphi'_i}, d_i} (p_i, \theta_{i-1}[\Lambda \leftarrow d_i])) \mid d_i \in D\}$ for some $\theta_{i-1} \in \Theta_k$ such that $\theta_{i-1}[\Lambda \leftarrow d_i] : \gamma_i$. Note that (p_0, θ_0) is the initial ID. By weighted simulation, there is a run $\rho = (p_0, \theta'_0) \vdash_{s_{t_1,\gamma_0,\varphi'_1}, d'_1} (p_1, \theta'_1) \vdash_{s_{t_2,\gamma_1,\varphi'_2}, d'_2} \cdots \vdash_{s_{t_n,\gamma_{n-1},\varphi'_n}, d'_n} (p_n, \theta'_n)$ of some data word $(a_1, d'_1) \cdots (a_n, d'_n)$ where $\theta'_0 = \theta_0$ and a_i is the second component of t_i. Also, it is easy to see $\mathrm{wt}(\pi) = \mathrm{wt}(\rho)$. The converse direction holds by the construction of G. □

By Theorems 4 and 7, the optimal run problem for a given k-WRA $\mathcal{A}$ can be solved by solving the minimum weight path problem for the directed graph G obtained from $\mathcal{A}$ via the normal form $\mathcal{A}'$ if $\mathcal{A}'$ satisfies weighted simulation and weight computability. Furthermore, we can find the original transition $t \in T$ of $\mathcal{A}$ from a given transition $s_{t,\gamma,\varphi'} \in T'$ as described in the proof of Theorem 4. In this way, we can easily reconstruct the run in $\mathcal{A}$ that provide the infimum weight from a minimum path found in G.

The description length $\|\mathcal{A}'\|$ of k-WRA $\mathcal{A}' = (Q', Q'_0, T', Q'_f, \mathrm{wt}')$ can be represented by the following relationship between the sizes of the corresponding components of $\mathcal{A}'$ and $\mathcal{A}$: $|\Gamma_k| = 2^{k^2|\mathcal{R}|}$, $|Q'| = |Q| \times |\Gamma_k|$, $|Q'_0| = |Q_0|$, $|Q'_f| = |Q_f| \times |\Gamma_k|$, $|T'| = (|Q| \times |\Gamma_k|) \times |\Sigma| \times 2^{2k|\mathcal{R}|+|\mathcal{R}|} \times 2^k \times (|Q| \times 1)$.

Theorem 8. *When the normal form k-WRA $\mathcal{A}'$ constructed from k-WRA $\mathcal{A} = (Q, Q_0, T, Q_f, \mathrm{wt})$ has weighted simulation property and weight computability, the time complexity of the optimal run problem for k-WRA $\mathcal{A}$ is $O(2^{k^2|\mathcal{R}|}|Q|(4^{k|\mathcal{R}|}2^{|\mathcal{R}|+k}|\Sigma||Q| + k^2|\mathcal{R}|))$.*

Proof. The above complexity is derived from the time complexity $O(|E| + |V|\log|V|)$ of Dijkstra algorithm by $|V| = |Q'|$, $|E| = |T'|$.

Example 4. Consider the WRA $\mathcal{A}_2$ of Example 2 again. Let $\mathcal{A}'_2$ be the normal form WRA obtained from $\mathcal{A}_2$. $\mathcal{A}'_2$ satisfies weighted simulation and weight computability. We show the directed graph G_1 for $\mathcal{A}'_2$. For a label (t', w) of an edge, t' represents the applied transition and w represents the infimum of the weights of switches corresponding to the edge. The register types $\gamma_0, \gamma_1, \gamma_2$ in the node labels are as follows where γ_0 is the initial register type:

$$\gamma_0(1,2)(<) = 0, \quad \gamma_0(2,1)(<) = 0, \quad \gamma_0(1,2)(=) = \gamma_0(2,1)(=) = 1,$$
$$\gamma_1(1,2)(<) = 0, \quad \gamma_1(2,1)(<) = 1, \quad \gamma_1(1,2)(=) = \gamma_1(2,1)(=) = 0,$$
$$\gamma_2(1,2)(<) = 1, \quad \gamma_2(2,1)(<) = 0, \quad \gamma_2(1,2)(=) = \gamma_2(2,1)(=) = 0,$$
$$\gamma_m(j,j)(<) = 0, \quad \gamma_m(j,j)(=) = 1, \quad \text{for } m \in \{0\} \cup [2],\ j \in [2],$$
$$\gamma_m(i,j)(>) = \gamma_m(j,i)(<) \quad \text{for } m \in \{0\} \cup [2],\ i,j \in [2].$$

The edge with $(s_{t_1,\gamma_0,tt}, 0)$ represents the three edges generated from t_1 in $\mathcal{A}_2$. The optimal paths of G_1 are the simple paths from (q_0, γ_0) to (q_2, γ_0), and the weight infimum is 0 (Fig. 2).

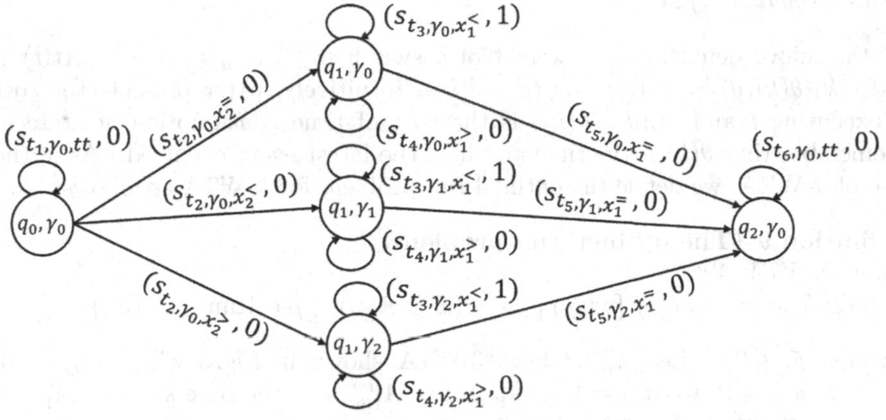

Fig. 2. The directed graph G_1 for $\mathcal{A}_2$ of Example 2.

6 Weighted Timed Automata

Weighted timed automata (WTA) are an extension of timed automata (TA) by introducing the weight to TA. In this subsection, we directly define WTA as a subclass of WRA based on Lemma 5.1 of [5] that every k-WTA can be simulated by a $(k+1)$-WRA by using one extra register to keep the current time instant (in particular, a clock reset can be simulated by loading the current time to the corresponding register). An input data word $w = (a_1, d_1)(a_2, d_2) \ldots (a_n, d_n)$ to a WTA means a_i occurs at time instant d_i ($i \in [n]$). In every switch, an input data value is loaded to the last register x_{k+1} so that x_{k+1} remembers when the latest symbol a_i occurred. The guard formula of every transition requires that an input data value is always not less than x_{k+1} to guarantee that $d_1 \leq d_2 \leq \ldots \leq d_n$.

For a binary relation $\bowtie$ over $\mathbb{R}_{\geq 0}$ and $c \in \mathbb{N}$, let $\bowtie c$ be the binary relation defined as $\bowtie c = \{(r, r') \mid r, r' \in \mathbb{R}_{\geq 0},\ r' - r \bowtie c\}$. Note that $(\theta, d) \models x_i^{\bowtie c}$ means $d - \theta(i) \bowtie c$, not $\theta(i) - d \bowtie c$. We let the data structure $\mathbb{D}^{\text{timed}} = \langle \mathbb{R}_{\geq 0}, \{\bowtie c \mid \bowtie \in \{<, =, >\}, c \in \mathbb{N}\} \rangle$ with the initial value $\bot = 0$.

Definition 8 ([5]). *A k-register weighted timed automaton (abbreviated as k-WTA) over Σ is a k-WRA $\mathcal{A}^{\mathrm{timed}} = (Q, Q_0, T, Q_f, (\mathrm{wtt}, \mathrm{wtd}))$ over Σ, $\mathbb{D}^{\mathrm{timed}}$ and $\mathcal{R}_{\mathrm{trpc}}$ where*

- $A_b^{\mathrm{timed}} = (Q, Q_0, T, Q_f)$ *is a k-RA (called the base k-TA of $\mathcal{A}^{\mathrm{timed}}$) such that for each transition $q \to^a_{\varphi,\Lambda} q' \in T$, $\varphi = \varphi' \wedge x_k^{\geq 0}$ for some $\varphi' \in F_k$,*
- wtt *is a function from T to $\mathbb{N}$, and*
- *for each $q \in Q$, a constant natural number $w_q \in \mathbb{N}$ is specified and for each transition $t = q \to^a_{\varphi,\Lambda} q' \in T$ and $d, d' \in \mathbb{R}_{\geq 0}$,*

$$\mathrm{wtd}(t, j)(d, d') = 0 \quad (j \in [k-1]), \tag{1}$$
$$\mathrm{wtd}(t, k)(d, d') = w_q(d' - d). \tag{2}$$

$L(A_b^{\mathrm{timed}})$ is the timed language recognized by A_b^{timed} and $[\![\mathcal{A}^{\mathrm{timed}}]\!]$ is the timed series recognized by $\mathcal{A}^{\mathrm{timed}}$. □

By the above definition, the weight of a switch $(q, \theta) \vdash_{t,d} (q', \theta')$ is $\mathrm{wtt}(t) + \mathrm{wtd}(t, k)(\theta(k), d) = \mathrm{wtt}(t) + w_q(d - \theta(k))$. Intuitively, wtt represents the cost of executing t and $w_q(d - \theta(k))$ is the cost of time consumption at state q. (Remember that $\theta(k)$ is the time at which the latest event occurred.) As in the case of k-WRA, we define the optimal run problem for k-WTA as follows.

Definition 9 (The optimal run problem)
Input: k-WTA $\mathcal{A}^{\mathrm{timed}}$
Output: The infimum of $\{\mathrm{wt}(\rho) \mid \exists w \in (\Sigma \times \mathbb{R}_{\geq 0})^+.\ \rho \in \mathrm{Run}_{\mathcal{A}^{\mathrm{timed}}}(w)\}$

Example 5. ([4]). Let $\mathcal{A}_1^{\mathrm{timed}}$ be a 3-WTA shown in Fig. 3 where $w_{q_0} = 3$, $w_{q_1} = 1$, $w_{q_2} = 0$, $\mathrm{wtt}(t_j) = 1$ ($j \in [3]$). Let $A_{1,b}^{\mathrm{timed}}$ be the base k-TA of $\mathcal{A}_1^{\mathrm{timed}}$. Then, $L(A_{1,b}^{\mathrm{timed}}) = \{(a, 2)\} \cup \{(a, d)(a, 2) \mid 0 \leq d < 2\}$. $\rho_1 \in \mathrm{Run}_{\mathcal{A}_1^{\mathrm{timed}}}((a, 2))$ is unique and $\mathrm{wt}(\rho_1) = \mathrm{wtd}(t_1, 3)(0, 2) + \mathrm{wtt}(t_1) = 3 \cdot 2 + 1 = 7$. For each $w_d = (a, d)(a, 2)$ where $0 \leq d < 2$, $\rho_d \in \mathrm{Run}_{\mathcal{A}_1^{\mathrm{timed}}}(w_d)$ is unique and $\mathrm{wt}(\rho_d) = \mathrm{wtd}(t_2, 3)(0, d) + \mathrm{wtt}(t_2) + \mathrm{wtd}(t_3, 3)(d, 2) + \mathrm{wtt}(t_3) = 3d + 1 + (2 - d) + 1 = 4 + 2d$. We have $\inf\{\mathrm{wt}(\rho) \mid \exists w \in (\Sigma \times D)^+.\ \rho \in \mathrm{Run}_{\mathcal{A}_1^{\mathrm{timed}}}(w)\} = 4$.

Example 6. ([4]). Let $\mathcal{A}_2^{\mathrm{timed}}$ be a 2-WTA shown in Fig. 4 where $w_{q_0} = 1$, $w_{q_1} = 2$, $w_{q_2} = 0$, $\mathrm{wtt}(t_1) = \mathrm{wtt}(t_2) = 1$. Let $A_{2,b}^{\mathrm{timed}}$ be the base k-TA of $\mathcal{A}_2^{\mathrm{timed}}$. Then, $L(A_{2,b}^{\mathrm{timed}}) = \{(a, 2 - \xi)(a, 2) \mid 0 < \xi \leq 2\}$. For each $w_\xi = (a, 2 - \xi)(a, 2)$ where $0 < \xi \leq 2$, $\rho_\xi \in \mathrm{Run}_{\mathcal{A}_2^{\mathrm{timed}}}(w_\xi)$ is unique and

$$\begin{aligned} \mathrm{wt}(\rho_\xi) &= \mathrm{wtd}(t_1, 2)(0, 2 - \xi) + \mathrm{wtt}(t_1) + \mathrm{wtd}(t_2, 2)(2 - \xi, 2) + \mathrm{wtt}(t_2) \\ &= (1 \cdot (2 - \xi)) + 1 + (2 \cdot \xi) + 1 = 4 + \xi. \end{aligned}$$

Hence, $\inf\{\mathrm{wt}(\rho) \mid \exists w \in (\Sigma \times D)^+.\ \rho \in \mathrm{Run}_{\mathcal{A}_2^{\mathrm{timed}}}(w)\} = 4$. □

In [4], an algorithm that solves the optimal run problem for WTA is proposed by extending the region construction for TA. Region construction is a well-known method to divide the infinite set of IDs of TA into a finite set of regions where

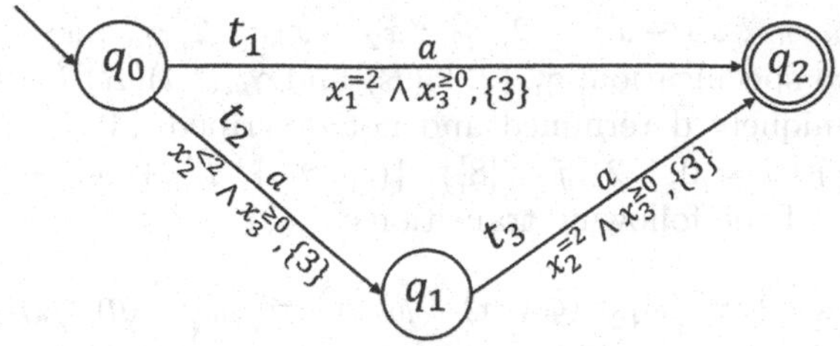

Fig. 3. WTA $\mathcal{A}_1^{\text{timed}}$

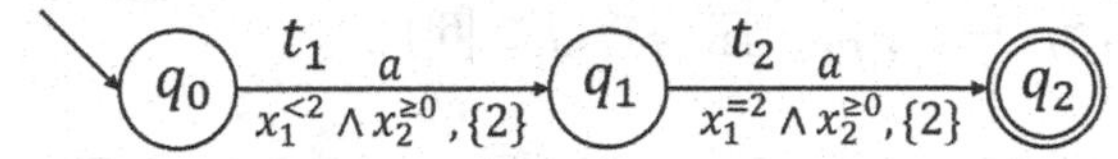

Fig. 4. WTA $\mathcal{A}_2^{\text{timed}}$

two IDs in a same region are indistinguishable (or bisimilar) with respect to any transition and time progress. In [4], a sub-region is defined as a refinement of a region by distinguishing $x < y$ and $x \lesssim y$ where the distance of x and y is large in the former case while the distance is very (arbitrarily) small in the latter case. This distinction is needed because it may happen that there is no run that has the minimum weight but there are infinite number of runs whose weights has the infimum as shown in Example 5 (see [4] for details). An edge-weighted directed graph, called the sub-region graph G is constructed from a given WTA and a minimum weight path of G is computed by any existing graph algorithm, which corresponds to a solution to the optimal run problem for the WTA.

The method proposed in this paper can also compute an optimal run of a WTA by distinguishing $<$ and $\lesssim$ as follows. Let $c_{\max}$ be the largest natural number appearing in the guard formula of some transition in a given WTA and let $\mathcal{R}^{\text{BL}}$ be the collection of relations $\{\bowtie c \mid \bowtie \in \{\lesssim, =, \gtrsim\}, c \in \mathbb{N}, c \leq c_{\max}\} \backslash \{\lesssim 0\}$. We redefine the data structure for WTA as $\mathbb{D}^{\text{timed,BL}} = \langle \mathbb{R}_{\geq 0}, \mathcal{R}^{\text{BL}} \rangle$. A boundary region is a region specified by at least one constraints using $=$ and no constraints using $\lesssim$ or $\gtrsim$. A limit region is a region specified by at least one constraints using $\lesssim$ or $\gtrsim$. Since the guard formula of any transition of WTA is a linear constraint on the contents of registers, it suffices to consider only the boundary regions and limit regions to compute the solution of the optimal run problem for WTA (see [4] for example). This implies weighted simulation and weight computability if we replace every $<$ and $>$ with $\lesssim$ and $\gtrsim$, respectively, and use $\mathbb{D}^{\text{timed,BL}}$ instead of $\mathbb{D}^{\text{timed}}$.

Example 7. Let us revisit Example 6. First, we replace every $<$ and $>$ with $\lesssim$ and $\gtrsim$, respectively, and consider its normal form. Since $c_{\max} = 2$, $\mathcal{R}^{\text{BL}} = \{= 0, \gtrsim 0, \lesssim 1, = 1, \gtrsim 1, \lesssim 2, = 2, \gtrsim 2\}$. After simplifications by using properties of the total order on $\mathbb{N}$, we have the following eight register types to be considered in this example: $\gamma_1 : x_2 - x_1 = 0$, $\gamma_2 : x_2 - x_1 \gtrsim 0$, $\gamma_3 : x_2 - x_1 \lesssim 1$, $\gamma_4 : x_2 - x_1 = 1$,

$\gamma_5 : x_2 - x_1 \gtrsim 1$, $\gamma_6 : x_2 - x_1 \lesssim 2$, $\gamma_7 : x_2 - x_1 = 2$, $\gamma_8 : x_2 - x_1 \gtrsim 2$. Note that by the above specification, $\gamma_m(2,1)(R)$ and $\gamma_m(i,i)(R)$ ($m \in [8]$, $i = 1,2$, $R \in \mathcal{R}^{\mathrm{BL}}$) are uniquely determined and not described. $\mathcal{A}_2^{\mathrm{timed}}$ is transformed to $\mathcal{A}_2' = (\{(q_i, \gamma_j) \mid i = 0,1,2,\ j \in [8]\},\ \{(q_0,\gamma_1)\},\ T',\ \{(q_2,\gamma_7)\},\ (\mathrm{wtt}', \mathrm{wtd}'))$ where T' consists of the following transitions:

$$\begin{aligned}
&(q_0,\gamma_1) \rightarrow^a_{x_1^{=0},\{2\}} (q_1,\gamma_1),\ (q_0,\gamma_1) \rightarrow^a_{x_1^{\gtrsim 0},\{2\}} (q_1,\gamma_2),\\
&(q_0,\gamma_1) \rightarrow^a_{x_1^{\lesssim 1},\{2\}} (q_1,\gamma_3),\ (q_0,\gamma_1) \rightarrow^a_{x_1^{=1},\{2\}} (q_1,\gamma_4),\\
&(q_0,\gamma_1) \rightarrow^a_{x_1^{\gtrsim 1},\{2\}} (q_1,\gamma_5),\ (q_0,\gamma_1) \rightarrow^a_{x_1^{\lesssim 2},\{2\}} (q_1,\gamma_6),\\
&(q_1,\gamma_j) \rightarrow^a_{x_1^{=2},\{2\}} (q_2,\gamma_7)\ \ (j \in [6])
\end{aligned}$$

and wtt', wtd' are defined accordingly. Note that $(\theta, d) \models \mathrm{in}^{=0} \wedge \neg\mathrm{in}^{R} \wedge \neg\mathrm{in}^{R^{-1}}$ for $R \in \mathcal{R}^{\mathrm{BL}} \backslash \{=0\}$ and an input data value is always loaded to x_2 (the previous data in x_2 is overwritten), and hence constraints on x_2 and an input data value are not needed in the guard formulas. We have the following six kinds of runs, each of which corresponds to one of the above six transitions from (q_0, γ) followed by the last transition, which have the following weights: $\mathrm{wt}(\rho_1) = 6$, $\mathrm{wt}(\rho_1) = 6 - \xi$, $\mathrm{wt}(\rho_1) = 5 + \xi$, $\mathrm{wt}(\rho_1) = 5$, $\mathrm{wt}(\rho_1) = 5 - \xi$, $\mathrm{wt}(\rho_1) = 4 + \xi$ for small $\xi > 0$. Hence, the solution of the optimal run problem for this example is 4, which is realized by ρ_6 by $\xi \to 0$. □

7 Conclusion

In this paper, we discussed the optimal run problem for weighted register automata (WRA). We first introduced register type to WRA and provided a transformation from a given WRA into a normal form such that the register types before and after each transition are uniquely determined. Because the decision problem related to the optimal run problem is undecidable, we proposed a sufficient condition called weighted simulation and weight computability for the problem to become decidable. Lastly, we illustrated computing the optimal run of weighted timed automata as an example. Investigating the problem for semirings other than the tropical reals is an interesting future study.

Acknowledgements. The authors thank the reviewers for providing valuable comments to the paper. This work was supported by JSPS KAKENHI Grant Number JP19H04083.

References

1. Almagor, S., Cadilhac, M., Mazowiecki, F., Pérez, G.A.: Weak cost register automata are still powerful. In: Hoshi, M., Seki, S. (eds.) DLT 2018. LNCS, vol. 11088, pp. 83–95. Springer, Cham (2018). https://doi.org/10.1007/978-3-319-98654-8_7

2. Alur, R., D'Antoni, L., Deshmukh, J.V., Raghothaman, M., Yuan, Y.: Regular functions and cost register automata. In: 28th Annual ACM/IEEE Symposium on Logic in Computer Science, LICS 2013, New Orleans, 25–28 June 2013, pp. 13–22 (2013). https://doi.org/10.1109/LICS.2013.65
3. Alur, R., Dill, D.L.: A theory of timed automata. Theor. Comput. Sci. **126**(2), 183–235 (1994). https://doi.org/10.1016/0304-3975(94)90010-8
4. Alur, R., La Torre, S., Pappas, G.J.: Optimal paths in weighted timed automata. In: Di Benedetto, M.D., Sangiovanni-Vincentelli, A. (eds.) HSCC 2001. LNCS, vol. 2034, pp. 49–62. Springer, Heidelberg (2001). https://doi.org/10.1007/3-540-45351-2_8
5. Babari, P., Droste, M., Perevoshchikov, V.: Weighted register automata and weighted logic on data words. In: Sampaio, A., Wang, F. (eds.) ICTAC 2016. LNCS, vol. 9965, pp. 370–384. Springer, Cham (2016). https://doi.org/10.1007/978-3-319-46750-4_21
6. Babari, P., Droste, M., Perevoshchikov, V.: Weighted register automata and weighted logic on data words. Theor. Comput. Sci. **744**, 3–21 (2018). https://doi.org/10.1016/j.tcs.2018.01.004
7. Behrmann, G., Fehnker, A., Hune, T., Larsen, K., Pettersson, P., Romijn, J., Vaandrager, F.: Minimum-cost reachability for priced time automata. In: Di Benedetto, M.D., Sangiovanni-Vincentelli, A. (eds.) HSCC 2001. LNCS, vol. 2034, pp. 147–161. Springer, Heidelberg (2001). https://doi.org/10.1007/3-540-45351-2_15
8. Bojańczyk, M., Klin, B., Lasota, S.: Automata theory in nominal sets. Logical Methods Comput. Sci. **10**(3) (2014). https://doi.org/10.2168/LMCS-10(3:4)2014
9. Bouyer, P.: A logical characterization of data languages. Inf. Process. Lett. **84**(2), 75–85 (2002). https://doi.org/10.1016/S0020-0190(02)00229-6
10. Cheng, E.Y., Kaminski, M.: Context-free languages over infinite alphabets. Acta Informatica **35**(3), 245–267 (1998). https://doi.org/10.1007/s002360050120
11. Demri, S., Lazić, R.: LTL with the freeze quantifier and register automata. ACM Trans. Comput. Log. **10**(3), 16:1–16:30 (2009). https://doi.org/10.1145/1507244.1507246
12. Kaminski, M., Francez, N.: Finite-memory automata. Theoret. Comput. Sci. **134**(2), 329–363 (1994). https://doi.org/10.1016/0304-3975(94)90242-9
13. Libkin, L., Tan, T., Vrgoč, D.: Regular expressions for data words. J. Comput. Syst. Sci. **81**(7), 1278–1297 (2015). https://doi.org/10.1016/j.jcss.2015.03.005
14. Libkin, L., Vrgoč, D.: Regular path queries on graphs with data. In: 15th International Conference on Database Theory (ICDT 2012), pp. 74–85 (2012). https://doi.org/10.1145/2274576.2274585
15. Lygeros, J., Tomlin, C., Sastry, S.: Controllers for reachability specifications for hybrid systems. Automatica **35**(3), 349–370 (1999). https://doi.org/10.1016/S0005-1098(98)00193-9
16. Milo, T., Suciu, D., Vianu, V.: Typechecking for XML transformers. In: 19th ACM Symposium on Principles of Database Systems (PODS 2000), pp. 11–22 (2000). https://doi.org/10.1145/335168.335171
17. Neven, F., Schwentick, T., Vianu, V.: Finite state machines for strings over infinite alphabets. ACM Trans. Comput. Log. **5**(3), 403–435 (2004). https://doi.org/10.1145/1013560.1013562
18. Sakamoto, H., Ikeda, D.: Intractability of decision problems for finite-memory automata. Theor. Comput. Sci. **231**(2), 297–308 (2000). https://doi.org/10.1016/S0304-3975(99)00105-X

19. Senda, R., Takata, Y., Seki, H.: Complexity results on register context-free grammars and register tree automata. In: Fischer, B., Uustalu, T. (eds.) ICTAC 2018. LNCS, vol. 11187, pp. 415–434. Springer, Cham (2018). https://doi.org/10.1007/978-3-030-02508-3_22
20. Senda, R., Takata, Y., Seki, H.: Generalized register context-free grammars. In: Martín-Vide, C., Okhotin, A., Shapira, D. (eds.) LATA 2019. LNCS, vol. 11417, pp. 259–271. Springer, Cham (2019). https://doi.org/10.1007/978-3-030-13435-8_19

Real-Time and Temporal Logics

Time4sys2imi: A Tool to Formalize Real-Time System Models Under Uncertainty

Étienne André[1,2,3], Jawher Jerray[1(✉)], and Sahar Mhiri[1]

[1] Université Paris 13, LIPN, CNRS, UMR 7030, 93430 Villetaneuse, France
jerray@lipn.univ-paris13.fr
[2] JFLI, CNRS, Tokyo, Japan
[3] National Institute of Informatics, Tokyo, Japan

Abstract. Time4sys is a formalism developed by Thales Group, realizing a graphical specification for real-time systems. However, this formalism does not allow to perform formal analyses for real-time systems. So a translation of this tool to a formalism equipped with a formal semantics is needed. We present here Time4sys2imi, a tool translating Time4sys models into parametric timed automata in the input language of IMITATOR. This translation allows not only to check the schedulability of real-time systems, but also to infer some timing constraints (deadlines, offsets...) guaranteeing schedulability. We successfully applied Time4sys2imi to various examples.

Keywords: Real-time systems · Scheduling · Model checking · Parametric timed automata · Parameter synthesis

1 Introduction

Due to the increasing complexity in real-time systems, designing and analyzing such systems is an important challenge, especially for safety-critical real-time systems, for which the correctness is crucial. The scheduling problem for real-time systems consists in deciding which task the processor runs at each moment by taking into consideration the needs of urgency, importance and reactivity in the execution of the tasks. Systems can feature one processor ("uniprocessor") or several processors ("multiprocessor"). Each processor features a scheduling policy, according to which it schedules new task instances. Tasks are usually characterized by a best and worst case execution times (BCET and WCET), and are assigned a deadline and often a priority. Tasks can be activated periodically ("periodic task"), sporadically ("sporadic tasks"), or be activated upon

This work is supported by the ASTREI project funded by the Paris Île-de-France Region, with the additional support of the ANR national research program PACS (ANR-14-CE28-0002) and ERATO HASUO Metamathematics for Systems Design Project (No. JPMJER1603), JST.

R. M. Hierons and M. Mosbah (Eds.): ICTAC 2019, LNCS 11884, pp. 113–123, 2019.
https://doi.org/10.1007/978-3-030-32505-3_7

completion of another task—to which we refer to "dependency" or "task chain". This latter feature is often harder to encode using traditional scheduling models. Periodic tasks may be subject to a "jitter", i.e., a variation in the period; all tasks can be subject to an "offset", i.e., a constant time from the system start to the first activation of the task. The schedulability problem consists in verifying that all tasks can finish their computation before their relative deadline, for a given scheduling policy. This problem is a very delicate task: The origin of complexity arises from a large number of parameters to consider (BCET and WCET, tasks priorities, deadlines, periodic and sporadic tasks, tasks chains, etc.). The schedulability problem becomes even more complicated when periods, deadlines or execution times become uncertain or completely unknown: we refer to this problem as schedulability under uncertainty.

Thales Group, a large multinational company specialized in aerospace, defense, transportation and security, developed a graphical formalism Time4sys[1] to allow interoperability between timed verification tools. Time4sys responds to a need to unify the approaches within Thales Group: This formalism is being rolled out at TSA (Thales Airborne Systems) and studies are underway at TAS (Thales Alenia Space). Time4sys is now an open source framework, offering many features to represent real-time systems. However, Time4sys lacks for a formalization: it does not perform any verification nor simulation, nor can it assess the schedulability of the depicted systems.

Since Time4sys does not allow to perform formal analyzes for real-time systems, a translation to a well-grounded formalism is needed to verify and analyze real-time systems. In this paper, we present a tool Time4sys2imi which allows to translate Time4sys into parametric timed automata (PTAs) [2] described in the input language of IMITATOR. PTAs extend finite-state automata with clocks (i.e., real-valued variables evolving at the same rate) and parameters (unknown timing constants). PTAs are a formalism well-suited to verify systems where some timing delays are known with uncertainty, or completely unknown. IMITATOR [5] is the *de-facto* standard tool to analyze models represented using PTAs. This translation allows not only to assess the schedulability of systems modeled using Time4sys, but only to *synthesize* some timing constants guaranteeing schedulability.

In [4], we presented a set of rules translating Time4sys to PTAs. We introduce here the tool performing this translation, with its practical description, as well as a set of case studies, absent from [4].

Related Works. Scheduling using (extensions) of timed automata was proposed in the past (e.g., [1]). For uniprocessor real-time systems only, (parametric) *task automata* offer a more compact representation than (parametric) timed automata [3,8,12]; however [8,12] do not offer an automated translation and, while [3] comes with a script translating some parametric task automata to parametric timed automata, the case of multiprocessor is not addressed. Schedulability analysis under uncertainty was also tackled in the past, e.g., in [7,9,15].

[1] https://github.com/polarsys/time4sys.

The main difference with our tool is that we allow here a systematic translation from an industrial formalism.

An export from Time4sys is available to Cheddar [14]. However, while Cheddar is able to deduce schedulability of real-time systems, it suffers from two main limitations:

1. it does not allow task dependencies; and
2. all timing constants must be fixed in order to study the schedulability.

In contrast, our translation in Time4sys2imi allows for both.

A model represented with Time4sys can also be exported to MAST [10] which is an open-source suite of tools to perform schedulability analysis of real-time distributed systems. However, the effectiveness of this tool is limited: it does not allow us to have a complete solution to our problem since it only works with instantiated systems, so we can not perform a real-time system with unknown parameters.

Outline. Section 2 describes Time4sys, and states the problem. Section 3 exposes the architecture of Time4sys2imi. As a proof of concept, Sect. 4 gives the results obtained on some examples. We discuss future works in Sect. 5.

2 Time4sys in a Nutshell

We review here Time4sys, and make a few (minor) assumptions to ease our translation.

Time4sys is a formalism that provides an environment to prepare the design phase of a system through the graphical visualization developed. Time4sys contains two modes: Design and Analysis. In our translation, we use the Time4sys Design mode which uses a subset of the OMG MARTE standard [13] as a basis for displaying a synthetic view to the real-time system. This graphical representation encompasses all the elements and properties that can define a real-time system.

The Time4sys Design tool allows users to define the following elements:

- **Hardware Resource**: a hardware resource in Time4sys is a processor, and it contains a set of tasks; it is also assigned a scheduling policy.
- **Software Resource**: a software resource in Time4sys is a task, and it features a (relative) deadline.
- **Execution Step**: an execution step can be seen as a subtask. It is characterized by a BCET, a WCET, and a priority. In our translation, we assume that each software resource contains exactly one execution step. That is, we do not encompass for subtasks.
- **Event**: an event can be seen as an activation policy for tasks. There are two main types of Events:
  - **PeriodicEvent**: defined by its period, its jitter and its phase (i.e., offset).
  - **SporadicEvent**: defined by its minimum and maximum interarrival times, its jitter and its phase.

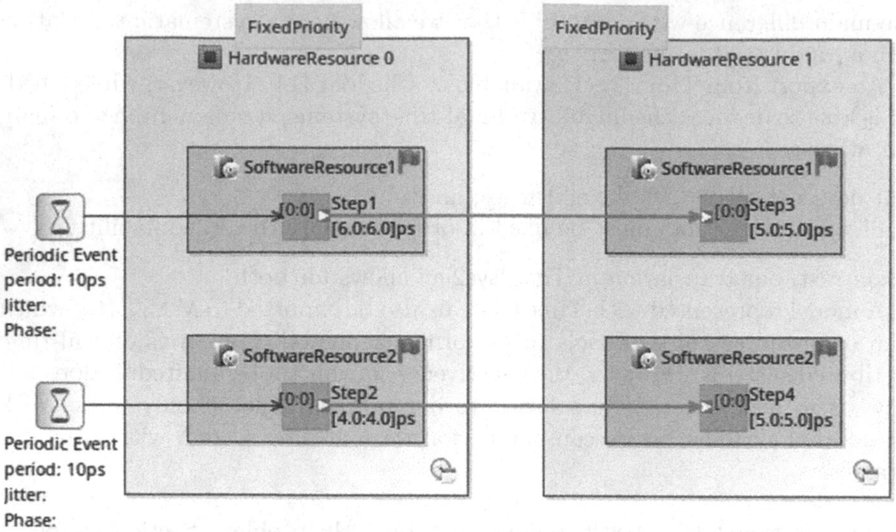

Fig. 1. Example of a Time4sys design

Example 1. Figure 1 shows an example of a real-time system designed with Time4sys. In this example, we have two hardware resources (HardwareResource0, HardwareResource1) both using fixed priority as a scheduling policy, two software resources (SoftwareResource1, SoftwareResource2) in each hardware resource, and four execution tasks, with the following timing constraints:

- Step1: $WCET = BCET = 6\,ps$
- Step2: $WCET = BCET = 4\,ps$
- Step3: $WCET = BCET = 5\,ps$
- Step4: $WCET = BCET = 5\,ps$

Finally, this example features two periodic events, both characterized by a 10 ps period, a 0 ps jitter and a 0 ps phase ("offset").

In this example, we start executing with Step1 in the CPU HardwareResource0. After 6 ps, the execution of Step1 ends so Step2 takes its place. At the same time, Step3 in the CPU HardwareResource1 starts performing. At $t = 10\,ps$, the execution of Step2 finishes and a new period of Step1 starts, however at that time Step3 is still executing. So this real-time system is not schedulable i.e., the period of StepT1 is strictly less than the WCET of Step1 plus the WCET of Step3.

Objective

The main objective of Time4sys2imi is as follows: given a real-time system with some unknown timing constants (period, jitter, deadlines...), *synthesize* the timing constants for which the system is schedulable. Note that, when all timing constants are known precisely, this problem is schedulability analysis.

3 Architecture and Principle

The main purpose of Time4sys2imi is to perform the translation of Time4sys models into the input language of IMITATOR. The schedulability analysis itself is done by IMITATOR, using reachability synthesis.

3.1 Targeted User

The application is intended primarily for the designer of real-time systems, aiming to verify the schedulability of her/his system, or synthesize the timing constants ensuring schedulability.

Time4sys2imi can automatically analyze a graphical representation of a real-time system realized by Time4sys using IMITATOR. The end-user does not need to have skills on PTAs nor on model checking.

Time4sys2imi allows the user to:

- Use the GUI of Time4sys2imi (cf. Fig. 2) and configure the options of both the translation and IMITATOR.
- Import an XML file generated by Time4sys. This file contains the data that describes the real-time system to be analyzed.
- Generate an `.imi` model analyzable by IMITATOR.

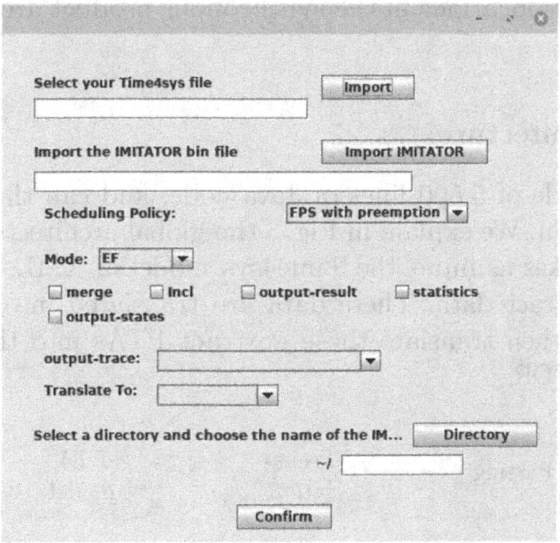

Fig. 2. GUI of Time4sys2imi

3.2 User Workflow

The analysis of real-time systems, using the proposed translation, can be summed up in three main parts:

1. Graphical modeling of a real-time system containing all its components with Time4sys. This part allows us to have a complete architecture of the system on the one hand. The architecture is encoded in an XML file automatically generated by Time4sys. This file contains all the data needed to describe the system.
2. The second part is the automatic translation of the XML file to the input language of IMITATOR, and is performed by Time4sys2imi. Time4sys2imi creates an `.imi` file that is analyzable with IMITATOR.
3. Finally, the user can run IMITATOR from Time4sys2imi to get the answer to the schedulability problem.

The translation rules are described in [4]. In short, we translate each task, each task chain and each processor scheduling policy (earliest deadline first, rate monotonic, shortest job first...) into a PTA; most of these PTAs feature a special location corresponding to a deadline miss (i.e., this location is reachable iff a deadline miss occurs). Timing constants are encoded either as constants (if they are known) or as timing parameters (if they are unknown). Then, we build (on-the-fly) the synchronous product of these PTAs. Finally, the set of valuations for which the system is schedulable is exactly those for which the special deadline miss locations in the synchronous product are unreachable. See [4] for details.

3.3 Global Architecture

Time4sys2imi is made of 5,500 lines of Java code, and can therefore run under any operating system. We explain in Fig. 3 the global architecture of the system.

Time4sys2imi takes as input the Time4sys model in XML, then we used the DOM parser to extract data. These data are translated into an abstract syntax for PTAs. We then translate these abstract PTAs into the concrete input language of IMITATOR.

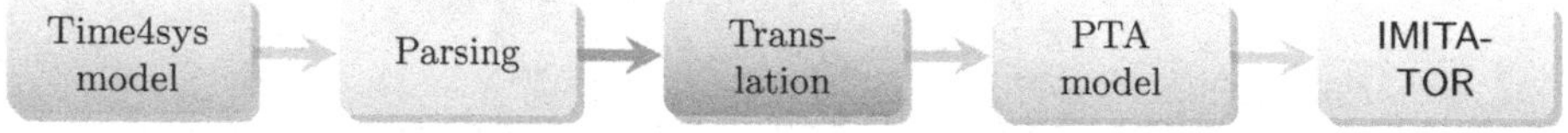

Fig. 3. Workflow of Time4sys2imi

3.4 Detailed Architecture

The global process is in Fig. 4.

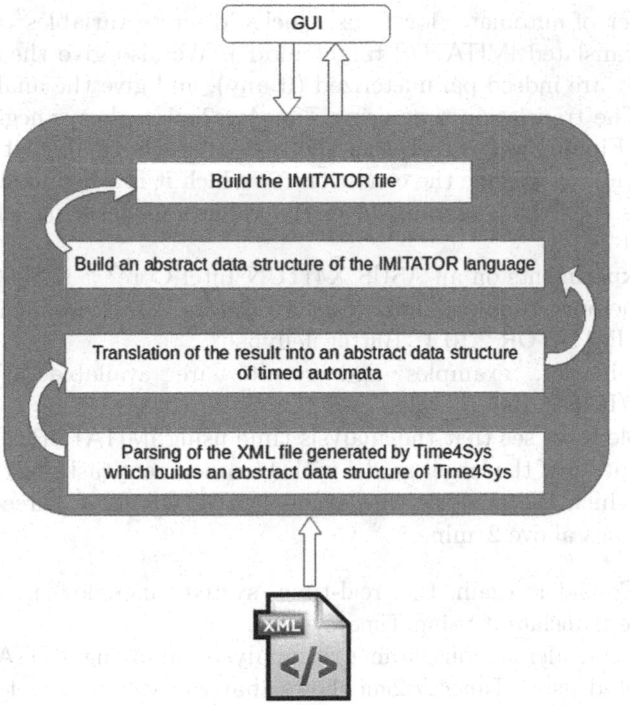

Fig. 4. Detailed architecture

Level 1. This level is the interface between the translation tool and the user: It allows the user to import the XML file to be translated, to choose the name of the IMITATOR model and to confirm the translation request.

Level 2. This level is loaded by the translation of the XML file through the following steps:

1. Parsing the XML file that Time4sys generates in order to get an abstract data structure from Time4sys.
2. Translation of the result into an abstract data structure of PTAs.
3. Construct an IMITATOR file from the PTAs abstract data structure.

Level 3. This level shows the XML files generated by Time4sys when designing a real-time system.

4 Proof of Concept

As a proof of concept to show the applicability of our translation tool, we modeled some real-time systems with Time4sys, then we translated those models to PTAs using with Time4sys2imi and analyzed them using IMITATOR.

We give in Table 1 a list of four case studies with, from top to bottom, the number of CPU, of tasks and task chains in the original Time4sys model, followed by the number of automata, locations, clocks, discrete variables[2] and parameters in the translated IMITATOR target model. We also give the name of the constants that are indeed parameterized (if any), and give the analysis time by IMITATOR. The translation time using Time4sys2imi is always negligible in our experiments. Finally, we give whether the system is schedulable (if it is entirely non-parametric), or we give the condition for which it is schedulable. The parametric results (i.e., the constraints over the valuations ensuring schedulability) are given in [6].

We ran experiments on an ASUS X411UN Intel Core™ i7-8550U 1.80 GHz with 8 GiB memory running Linux Mint 19 64 bits. All experiments were conducted using IMITATOR 2.10.4 "Butter Jellyfish".

Source, binaries, examples and results are available at www.imitator.fr/static/ICTAC19.

From Table 1, we see that the analysis time using IMITATOR remains small, with the exception of the larger model with 11 concurrent tasks featuring dependencies, for which the analysis time using IMITATOR for a three-dimensional analysis becomes above 2 min.

Example 2. Consider again the real-time system modeled in Fig. 1 using Time4sys. We translate it using Time4sys2imi.

First, we consider a non-parametric analysis: applying IMITATOR to the PTAs translated using Time4sys2imi shows that the system is not schedulable, as it was expected from Example 1.

Second, we parameterize the BCET and WCET of Step1. The result of the schedulability synthesis using IMITATOR yields the following constraint: $0 \leq BCETStep1 \leq WCETStep1 < 5$.

This constraint explains why this real-time system was not schedulable when $WCET = BCET = 6$ i.e., the values taken for $WCET$ and $BCET$ are not in the interval for which the system is schedulable.

Additional examples with models and translated PTAs are given in [6].

5 Perspectives

A short term future work will be to optimize our translation: while we followed the rules developed in [4], it is likely that varying the rules in order to optimize the size of the automata or reducing the clocks, may help to make the model more compact and the analysis more efficient.

Second, when the model is entirely non-parametric, we believe that using the UPPAAL model checker [11] instead of IMITATOR may be more efficient; for that purpose, we plan to develop a translator to the input language of UPPAAL too; this implies to modify only the last step of our translation (from the abstract (P)TAs into the concrete input language of the target model checker).

[2] Discrete variables are global rational-valued variables that can be read and modified by the PTAs.

Table 1. Summary of experiments

Case study	Example 1 [Fig. 1]		Example 2 [[6]]		Example 3 [[6]]		Example 4 [[6]]	
# CPU	2		1		1		4	
# tasks	4		4		3		11	
# tasks chains	2		0		1		4	
# number of automata	6		9		3		12	
# total number of locations	22		26		14		53	
# clocks	8		8		6		22	
# discrete	4		4		3		11	
# parameters	0	2	0	1	0	2	0	3
Parameters	-	WCETStep1 BCETStep1	-	DeadlineStep2	-	DeadlineStep1	-	WCETStep5 BCETStep5 DealineStep11
Execution time (seconds)	0.040	0.112	0.263	0.289	0.042	0.045	2.276	144.627
Schedulable?	×		√		√		√	

Third, so far the analysis using IMITATOR is exact, i.e., sound and complete; however, it may sometimes be interesting to get only *some* ranges of parameter valuations for which the system is schedulable. Such optimizations (on the IMITATOR side) should help to make the analysis faster.

Seeing from our experiments, it is unlikely that the toolkit made of Time4sys, Time4sys2imi and IMITATOR can analyze models with hundreds of processors and thousands of tasks, especially with unknown timing constants. However, we believe that our approach can give first useful guarantees at the preliminary stage of system design and verification, notably to help designers to exhibit suitable ranges of timing parameters guaranteeing schedulability. Finally, real-time systems with uncertain timing constants were recently proved useful when Thales Group published an open challenge[3] for a system (actually modeled using Time4sys) with periods known with a limited precision only; while this problem was not strictly speaking a schedulability problem (but rather a computation of minimum/maximum execution times), it shed light on the practical need for methods to formally analyze real-time systems under uncertainty in the industry.

Acknowledgements. We thank Romain Soulat and Laurent Rioux from Thales R&D for useful help concerning Time4sys.

References

1. Adbeddaïm, Y., Asarin, E., Maler, O.: Scheduling with timed automata. Theor. Comput. Sci. **354**(2), 272–300 (2006). https://doi.org/10.1016/j.tcs.2005.11.018
2. Alur, R., Henzinger, T.A., Vardi, M.Y.: Parametric real-time reasoning. In: STOC, pp. 592–601. ACM, New York, NY, USA (1993). https://doi.org/10.1145/167088.167242
3. André, É.: A unified formalism for monoprocessor schedulability analysis under uncertainty. In: Petrucci, L., Seceleanu, C., Cavalcanti, A. (eds.) FMICS/AVoCS -2017. LNCS, vol. 10471, pp. 100–115. Springer, Cham (2017). https://doi.org/10.1007/978-3-319-67113-0_7
4. André, É.: Formalizing Time4sys using parametric timed automata. In: Méry, D., Qin, S. (eds.) TASE, pp. 176–183. IEEE (2019). https://doi.org/10.1109/TASE.2019.00031
5. André, É., Fribourg, L., Kühne, U., Soulat, R.: IMITATOR 2.5: a tool for analyzing robustness in scheduling problems. In: Giannakopoulou, D., Méry, D. (eds.) FM 2012. LNCS, vol. 7436, pp. 33–36. Springer, Heidelberg (2012). https://doi.org/10.1007/978-3-642-32759-9_6
6. André, É., Jerray, J., Mhiri, S.: Time4sys2imi: a tool to formalize real-time system models under uncertainty (long version). arXiv (2019). https://arxiv.org/pdf/1907.13447.pdf
7. Cimatti, A., Palopoli, L., Ramadian, Y.: Symbolic computation of schedulability regions using parametric timed automata. In: RTSS, pp. 80–89. IEEE Computer Society (2008). https://doi.org/10.1109/RTSS.2008.36

[3] "Formal Methods for Timing Verification Challenge", in the WATERS workshop: http://waters2015.inria.fr/challenge/.

8. Fersman, E., Krcál, P., Pettersson, P., Yi, W.: Task automata: schedulability, decidability and undecidability. Inf. Comput. **205**(8), 1149–1172 (2007). https://doi.org/10.1016/j.ic.2007.01.009
9. Fribourg, L., Lesens, D., Moro, P., Soulat, R.: Robustness analysis for scheduling problems using the inverse method. In: TIME, pp. 73–80. IEEE Computer Society Press (2012). https://doi.org/10.1109/TIME.2012.10
10. González Harbour, M., Gutiérrez García, J.J., Palencia Gutiérrez, J.C., Drake Moyano, J.M.: MAST: modeling and analysis suite for real time applications. In: ECRTS, pp. 125–134. IEEE Computer Society (2001). https://doi.org/10.1109/EMRTS.2001.934015
11. Larsen, K.G., Pettersson, P., Yi, W.: UPPAAL in a nutshell. Int. J. Softw. Tools Technol. Transf. **1**(1–2), 134–152 (1997). https://doi.org/10.1007/s100090050010
12. Norström, C., Wall, A., Yi, W.: Timed automata as task models for event-driven systems. In: RTCSA, pp. 182–189. IEEE Computer Society (1999). https://doi.org/10.1109/RTCSA.1999.811218
13. OMG: Modeling and analysis of real-time and embedded systems (MARTE) (2008). http://www.omg.org/omgmarte/
14. Singhoff, F., Legrand, J., Nana, L., Marcé, L.: Cheddar: a flexible real time scheduling framework. In: SIGAda, pp. 1–8. ACM (2004). https://doi.org/10.1145/1032297.1032298
15. Sun, Y., Soulat, R., Lipari, G., André, É., Fribourg, L.: Parametric schedulability analysis of fixed priority real-time distributed systems. In: Artho, C., Ölveczky, P.C. (eds.) FTSCS 2013. CCIS, vol. 419, pp. 212–228. Springer, Cham (2014). https://doi.org/10.1007/978-3-319-05416-2_14

Testing Real-Time Systems Using Determinization Techniques for Automata over Timed Domains

Moez Krichen[1,2](✉)

[1] Faculty of Computer Science and Information Technology, Al-Baha University, Al Bahah, Kingdom of Saudi Arabia
[2] ReDCAD Research Laboratory, University of Sfax, Sfax, Tunisia
moez.krichen@redcad.org

Abstract. In this work, we are interested in Model-Based Testing for Real-Time Systems. The proposed approach is based on the use of the model of Automata over Timed Domains (ATD) which corresponds to an extension of the classical Timed Automaton Model. First, we explain the main advantages of adopting this new formalism. Then, we propose a testing framework based on ATD and which is an extension of our initial framework presented in previous contributions. We extend the notion of correctness requirements (soundness and completeness) along with the notion of timed input-output conformance relation (tioco) used to compare between implementations and specifications. Moreover we propose a determinization technique used to generate test cases. Finally, several possible extensions of the present work are proposed.

Keywords: Model-Based Testing · Real-time systems · Automaton over timed domains · Correctness · Conformance relation · Determinization

1 Introduction

In general MBT (model based testing) [12] consists in describing the behavior of the SUT (system under test) using a particular adequate formalism and then generating automatically test scenarios from the considered description with respect to some coverage criteria adopting some selection methods. The following step consists in executing the obtained case studies on the SUT and collecting the corresponding verdicts in order to check whether the implementation conforms to its specification or not.

This paper extends some of our previous contributions [5–9] about MBT for real-time systems. Theses works were mainly built on the classical timed automaton model [1]. Our new proposed approach is mainly inspired by [3,4]. We adopt a new variant of timed automata called automata over timed domains (ATD). This new variant allows to model a much wider class of timed systems and it is equipped with a determinization technique which can be used for test generation.

R. M. Hierons and M. Mosbah (Eds.): ICTAC 2019, LNCS 11884, pp. 124–133, 2019.
https://doi.org/10.1007/978-3-030-32505-3_8

Next in Sect. 2 we propose some definitions related to the proposed formalism. In Sect. 3 we give details about the determinization procedure for ATD. Section 4 introduces the adopted testing framework. Finally Sect. 5 proposes some directions for future work.

2 Definitions

2.1 Timed Domains and Updates

Let $\mathbb{N}$ (respectively $\mathbb{R}\geq 0$) be the set of natural numbers (respectively the set of non-negative real-numbers). *Timed domains* are introduced to represent the progression of continuous entities. A timed domain $\mathcal{D}om$ is made of a set of values denoted by $\mathcal{V}al$ and a time transition function denoted by $\rightsquigarrow$ encoding the progression of those values when time evolves.

Definition 1. *A* timed domain *is a tuple* $\mathcal{D}om = \langle \mathcal{V}al, \rightsquigarrow \rangle$ *such that:*

- $\mathcal{V}al$ *is the set of values;*
- $\rightsquigarrow: \mathcal{V}al \times \mathbb{R}_{\geq 0} \rightarrow \mathcal{V}al$ *is the time transition function such that for all* $val \in \mathcal{V}al$ *and all* $t, t' \in \mathbb{R}_{\geq 0}$ *we have* $\rightsquigarrow (\rightsquigarrow (val, t), t') = \rightsquigarrow (val, t + t')$.

For simplicity we will write $val \overset{t}{\rightsquigarrow} val'$ instead of $\rightsquigarrow (val, t) = val'$. Moreover we consider the particular symbol $\perp$ which is assigned to the considered resource as soon as it becomes inactive and no longer evolves over time (that is for each $t \in \mathbb{R}_{\geq 0}$ we have $\perp \overset{t}{\rightsquigarrow} \perp$).

For the timed domains $\mathcal{D}om = \langle \mathcal{V}al, \rightsquigarrow_{\mathcal{V}al} \rangle$ and $\mathcal{D}om' = \langle \mathcal{V}al', \rightsquigarrow_{\mathcal{V}al'} \rangle$ we associate the *product*

$$\mathcal{D}om \times \mathcal{D}om' = \langle \mathcal{V}al_{\mathsf{Prod}}, \rightsquigarrow_{\mathsf{Prod}} \rangle$$

such that

$$\mathcal{V}al_{\mathsf{Prod}} = (\mathcal{V}al \cup \{\perp\}) \times (\mathcal{V}al' \cup \{\perp\})$$

and for $(val, val') \in \mathcal{V}al_{\mathsf{Prod}}$ and $t \in \mathbb{R}_{\geq 0}$ we have

$$\rightsquigarrow_{\mathsf{Prod}} \big((val, val'), t\big) = \big(\rightsquigarrow_{\mathcal{V}al} (val, t), \rightsquigarrow_{\mathcal{V}al'} (val', t)\big).$$

For $n \in \mathbb{N}_{>1}$, we define the timed domain $\mathcal{D}om^n$ inductively as

$$\mathcal{D}om^1 = \mathcal{D}om$$

and

$$\mathcal{D}om^{n+1} = \mathcal{D}om \times \mathcal{D}om^n.$$

Moreover, for $\mathcal{D}om = \langle \mathcal{V}al, \rightsquigarrow_{\mathcal{V}al} \rangle$ we define the timed domain

$$\mathcal{P}(\mathcal{D}om) = \langle \mathcal{V}al_{\mathcal{P}}, \rightsquigarrow_{\mathcal{P}} \rangle$$

such that

$$\mathcal{V}al_{\mathcal{P}} = \mathcal{P}(\mathcal{V}al)$$

and

$$\forall V \in \mathcal{V}al_{\mathcal{P}} \cdot \forall t \in \mathbb{R}_{\geq 0} : \rightsquigarrow_{\mathcal{P}} (V, t) = \{\rightsquigarrow_{\mathcal{V}al} (val, t) \mid val \in V\}.$$

Definition 2. *Consider a timed domain $\mathcal{D}om = \langle \mathcal{V}al, \rightsquigarrow \rangle$ and an alphabet Δ. An* update set *for $\mathcal{D}om$ and Δ is a set $\mathcal{U} \subseteq \Delta \times \mathcal{V}al^{\mathcal{V}al}$.*

For an update set $\mathcal{U}$ and a symbol $\delta \in \Delta$, we define the set

$$\mathcal{U}_\delta = \{y \in \mathcal{V}al^{\mathcal{V}al} \mid (\delta, y) \in \mathcal{U}\}.$$

An element from $\mathcal{U}_\delta$ is called a δ-update.[1]

Let $\mathcal{D}om$ and $\mathcal{D}om'$ be two timed domains. Moreover consider two update sets $\mathcal{U}$ (for $\mathcal{D}om$ and Δ) and $\mathcal{U}'$ (for $\mathcal{D}om'$ and Δ). We then define the update set $\mathcal{U} \times \mathcal{U}'$ with respect to $\mathcal{D}om \times \mathcal{D}om'$ and Δ such that

$$\mathcal{U} \times \mathcal{U}' = \big\{(\delta, (y, y')) \mid \delta \in \Delta \wedge (y, y') \in \mathcal{U}_\delta \times \mathcal{U}'_\delta\big\}.$$

For $n \in \mathbb{N}_{>1}$, we define the update set $\mathcal{U}^n$ for $\mathcal{D}om^n$ and Δ inductively as

$$\mathcal{U}^1 = \mathcal{U}$$

and

$$\mathcal{U}^{n+1} = \mathcal{U} \times \mathcal{U}^n.$$

We also define the update set $\mathcal{P}(\mathcal{U})$ with respect to $\mathcal{P}(\mathcal{D}om)$ and Δ such that

$$\mathcal{P}(\mathcal{U}) = \big\{(\delta, \mathcal{Y}) \in \Delta \times \mathcal{P}(\mathcal{V}al)^{\mathcal{P}(\mathcal{V}al)} \mid \forall V \subseteq \mathcal{V}al \cdot \mathcal{Y}(V) = \bigcup_{(y, val) \in \mathcal{U}_\delta \times V} \{y(val)\}\big\}.$$

2.2 Automata over Timed Domains (ATD)

Definition 3. *Consider a timed domain $\mathcal{D}om = \langle \mathcal{V}al, \rightsquigarrow \rangle$ and an update set $\mathcal{U}$ for $\mathcal{D}om$ over Δ. An* automaton *on $\mathcal{D}om$ and $\mathcal{U}$ is a tuple $\mathcal{A} = \langle S, s_{ini}, v_{ini}, E, T \rangle$ where:*

- *S is a finite set of states;*
- *$s_{ini} \in S$ is the initial state;*
- *$val_{ini} \in \mathcal{V}al$ is the initial value;*
- *$E \subseteq S \times \mathcal{V}al \times \mathcal{U} \times S$ is the set of edges;*
- *$T \subseteq S$ is the set of terminal states.*

For the automaton over timed domains (*ATD*) $\mathcal{A}$ over $\mathcal{D}om$ and $\mathcal{U}$, we consider the set $C_\mathcal{A} = S \times \mathcal{V}al$ called the set of *configurations* of $\mathcal{A}$. The ATD $\mathcal{A}$ yiels a *timed labeled transition system* (TLTS)

$$\mathcal{L}_\mathcal{A} = \langle C_\mathcal{A}, c_\mathcal{A}^{ini}, \rightarrow_\mathcal{A} \rangle$$

where $c_\mathcal{A}^{ini}$ is the initial configuration of $\mathcal{A}$ such that

$$c_\mathcal{A}^{ini} = (s_{ini}, val_{ini})$$

and

$$\rightarrow_\mathcal{A} = (\xrightarrow{t}_\mathcal{A})_{t \in \mathbb{R}_{\geq 0}} \uplus (\xrightarrow{\delta, y}_\mathcal{A})_{(\delta, y) \in \mathcal{U}}$$

[1] Or simply an update.

such that

- $(s, val) \xrightarrow{t}_{\mathcal{A}} (s', val')$ if and only if $s = s'$ and $val \overset{t}{\rightsquigarrow} val'$;
- $(s, val) \xrightarrow{\delta,y}_{\mathcal{A}} (s', val')$ if and only if $(s, val, (\delta, y), s') \in E$ and $val' = y(val)$.

Similarly, the ATD $\mathcal{A}$ yields an *observable timed labeled transition system* (OTLTS)

$$\mathcal{OL}_{\mathcal{A}} = \langle C_{\mathcal{A}}, c_{\mathcal{A}}^{ini}, \twoheadrightarrow_{\mathcal{A}} \rangle$$

where

$$\twoheadrightarrow_{\mathcal{A}} = (\overset{t}{\twoheadrightarrow}_{\mathcal{A}})_{t \in \mathbb{R}_{\geq 0}} \uplus (\overset{\delta}{\twoheadrightarrow}_{\mathcal{A}})_{\delta \in \Delta}$$

such that

- $(s, val) \overset{t}{\twoheadrightarrow}_{\mathcal{A}} (s', val')$ if and only if $(s, val) \xrightarrow{t}_{\mathcal{A}} (s', val')$;
- $(s, val) \overset{\delta}{\twoheadrightarrow}_{\mathcal{A}} (s', val')$ if and only if $\exists y \in \mathcal{V}al^{\mathcal{V}al} : (s, val) \xrightarrow{\delta,y}_{\mathcal{A}} (s', val')$.

The first type of transitions is called *timed transitions* and the second type *discrete transitions*. For $(s, val) \in C_{\mathcal{A}}$ and $\mu \in \mathbb{R}_{\geq 0} \uplus \Delta$, we write $(s, val) \overset{\mu}{\twoheadrightarrow}_{\mathcal{A}}$ if there exists $(s', val') \in C_{\mathcal{A}}$ such that $(s, val) \overset{\mu}{\twoheadrightarrow}_{\mathcal{A}} (s', val')$.

2.3 Finitely Representable ATD

A *set of guards* is a set $\mathcal{G} \subseteq \mathcal{P}(\mathcal{V}al)$. For $\delta \in \Delta$, a $\mathcal{G}$-guarded update for δ is a couple $(G, \Gamma) \in \mathcal{G} \times \mathcal{P}(\mathcal{U}_\delta)$. For $I \subseteq \mathbb{N}$, consider $\Psi_I = \{(G_i, \Gamma_i) \mid i \in I\}$ a set of $\mathcal{G}$-guarded updates for δ. Also consider $\mathcal{A} = \langle S, s_{ini}, val_{ini}, E, T \rangle$ an automaton on $\mathcal{D}om$ and $\mathcal{U}$. A pair of states (s, s') of $\mathcal{A}$ is said to be *compatible* with Ψ_I if the two following conditions hold:

- $\forall i \in I. \forall\, val \in G_i. \forall\, y \in \Gamma_i : (s, val, (\delta, y), s') \in E$;
- $\forall\, (s, val, (\delta, y), s') \in E. \exists\, i \in I : v \in G_i \wedge y \in \Gamma_i$.

Definition 4. *The ATD $\mathcal{A}$ is said to be* finitely representable *using $\mathcal{G}$ if for every pair of states (s, s') of $\mathcal{A}$ and for every $\delta \in \Delta$, there is a finite set Ψ of $\mathcal{G}$-guarded updates for δ, such that (s, s') is compatible with Ψ.*

2.4 Deterministic ATD (DATD)

Consider the ATD $\mathcal{A} = \langle S, s_{ini}, val_{ini}, E, T \rangle$ on $\mathcal{D}om$ and $\mathcal{U}$. Let (s, v) be a possible configuration of $\mathcal{A}$. A *timed trace* from (s, val) is a sequence $ttr = (s_i, val_i)_{0 \leq i \leq n}$ such that:

- $(s_0, val_0) = (s, val)$;
- $\forall 1 \leq i \leq n. \exists w_i \in \mathbb{R}_{\geq 0} \uplus \Delta : (s_i, val_i) \overset{w_i}{\twoheadrightarrow}_{\mathcal{A}} (s_{i+1}, val_{i+1})$.

The timed trace ttr is said to be *produced* by the *timed word* $tw = (w_i)_{1 \leq i \leq n}$. In this case we write $(s, val) \overset{tw}{\twoheadrightarrow}_{\mathcal{A}} (s_n, val_n)$ and $(s, val) \overset{tw}{\twoheadrightarrow}_{\mathcal{A}}$. The duration of tw is defined as follows

$$\mathcal{D}uration(tw) = \sum_{1 \leq i \leq n} |w_i|$$

where

$$|w_i| = \begin{cases} w_i & \text{if} w_i \in \mathbb{R}_{\geq 0} \\ 0 & \text{otherwise} \end{cases}$$

that is $\mathcal{D}uration(tw)$ denotes the complete amount of time consumed during the execution of the timed word tw.

In general given a timed word $tw \in (\mathbb{R}_{\geq 0} \uplus \Delta)^*$, the set $TTr_{\mathcal{A}}((s, val), tw)$ stands for the set of timed traces produced by tw starting from (s, val).

Definition 5. *The ATD $\mathcal{A}$ is said to be* deterministic *if for any timed word $tw \in (\mathbb{R}_{\geq 0} \uplus \Delta)^*$ the cardinality of $TTr_{\mathcal{A}}((s_{ini}, val_{ini}), tw)$ is less or equal to one.*

For a positive integer n, we say that the timed word $tw \in (\mathbb{R}_{\geq 0} \uplus \Delta)^n$ is *accepted* by the ATD $\mathcal{A}$ if there is a timed trace $ttr = (s_i, val_i)_{0 \leq i \leq n}$ such that $ttr \in TTr_{\mathcal{A}}((s_{ini}, val_{ini}), tw)$ and $s_{n+1} \in T$ (i.e. s_{n+1} is a terminal state of $\mathcal{A}$). In this case all the configurations (s_i, val_i) are said to be *reachable*. The set of accepted timed words by $\mathcal{A}$ is denoted $\mathcal{L}ang(\mathcal{A})$ and the set of reachable configurations is denoted $\mathcal{R}each(\mathcal{A})$.

3 Determinization of Non-Deterministic ATD (NDATD)

Consider the (possibly) non-deterministic ATD $\mathcal{A} = \langle S, s_{ini}, val_{ini}, E, T \rangle$ on $\mathcal{D}$ and $\mathcal{U}$ and let Δ be the alphabet corresponding to the update set $\mathcal{U}$. For simplicity we assume that

$$S = \{s_1, \cdots, s_p\}$$

such that $p \in \mathbb{N}_{>0}$ and $s_{ini} = s_1$. For every state $s \in S$, we let $\mathsf{index}(s)$ denote the index of s. That is if $s = s_i$ then $\mathsf{index}(s) = i$. For each $1 \leq i \leq p$, we consider the set $Val_i \subseteq \mathcal{V}al$ which corresponds to the set of values corresponding to the state s_i.

For $\mathbf{V} = (V_i)_{1 \leq i \leq p} \in \mathcal{P}(\mathcal{V}al)^p$ we consider the set

$$S_{\mathbf{V}} = \{s \in S \mid V_{\mathsf{index}(s)} \neq \emptyset\}$$

which is the group of states the system may occupy when the values for the different states are given by $\mathbf{V}$.

For $\delta \in \Delta$ and $1 \leq i \leq j \leq p$, we associate the set

$$\lambda_\delta^{i \to j} = \{(val, y) \in \mathcal{V}al \times \mathcal{U}_\delta \mid (s_i, val, (\delta, y), s_j) \in E\}$$

which corresponds the different ways allowing to move from state s_i to state s_j. We also define

$$\lambda_\delta^{\to j} = (\lambda_\delta^{i \to j})_{1 \leq i \leq p}$$

which in turn records all the ways which allow to reach state s_j starting from any other state of the ATD $\mathcal{A}$.

For the considered letter δ and each $\lambda_\delta^{i\to j}$, we associate the *successor* function

$$\mathsf{succ}_{\delta,\lambda_\delta^{i\to j}} : \mathcal{V}al \to \mathcal{P}(\mathcal{V}al)$$

such that for $v \in \mathcal{V}al$ we have

$$\mathsf{succ}_{\delta,\lambda_\delta^{i\to j}}(val) = \{y(val) \mid (val, y) \in \lambda_\delta^{i\to j}\}$$

which collects all possible obtained values of *val* after executing instructions in $\lambda_\delta^{i\to j}$.

In a natural way we extend $\mathsf{succ}_{\delta,\lambda_\delta^{i\to j}}$ to elements from $\mathcal{P}(\mathcal{V})al$ and we define the function

$$\mathsf{Succ}_{\delta,\lambda_\delta^{i\to j}} : \mathcal{P}(\mathcal{V}al) \to \mathcal{P}(\mathcal{V}al)$$

such that for $V \subseteq \mathcal{V}al$ we have

$$\mathsf{Succ}_{\delta,\lambda_\delta^{i\to j}}(V) = \bigcup_{val \in V} \mathsf{succ}_{\delta,\lambda_\delta^{i\to j}}(val)$$

which this time collects the possible obtained values of all elements in V after executing instructions in $\lambda_\delta^{i\to j}$.

Moreover we define the function

$$\mathsf{Succ}_{\delta,\lambda_\delta^{\to j}} : \mathcal{P}(\mathcal{V}al)^p \to \mathcal{P}(\mathcal{V}al)$$

such that for $\mathbf{V} = (V_i)_{1\leq i\leq p} \in \mathcal{P}(\mathcal{V}al)^p$ we have

$$\mathsf{Succ}_{\delta,\lambda_\delta^{\to j}}(\mathbf{V}) = \bigcup_{1\leq i\leq p} \mathsf{Succ}_{\delta,\lambda_\delta^{i\to j}}(V_i)$$

which aggregates the possible updated values of $V_1, \cdots, V_p$ following respectively the instructions in $\delta^{1\to j}, \cdots, \delta^{p\to j}$.

Furthermore, we define the function

$$\mathsf{Succ}_{\delta,\mathcal{A}} : \mathcal{P}(\mathcal{V}al)^p \to \mathcal{P}(\mathcal{V}al)^p$$

such that for $\mathbf{V}al \in \mathcal{P}(\mathcal{V}al)^p$ we have

$$\mathsf{Succ}_{\delta,\mathcal{A}}(\mathbf{V}) = \left(\mathsf{Succ}_{\delta,\lambda_\delta^{\to 1}}(\mathbf{V}), \cdots, \mathsf{Succ}_{\delta,\lambda_\delta^{\to p}}(\mathbf{V})\right)$$

which can be seen as the successor of $\mathbf{V}$ after the execution of δ.

Lemma 1. *Let* $\mathbf{V} = (V_i)_{1\leq i\leq p} \in \mathcal{P}(\mathcal{V}al)^p$ *and* $\mathbf{V}' = (V_i')_{1\leq i\leq p} = \mathsf{Succ}_{\delta,\mathcal{A}}(\mathbf{V})$. *Then for every* $s' \in S$ *and* $v' \in \mathcal{V}$*:*

$$val' \in V'_{\mathsf{index}(s')} \Leftrightarrow \exists (s, val, (\delta, y), s') \in E \ s.t.\ v \in V_{\mathsf{index}(s)} \text{ and } v' = y(val).$$

Finally we define the update set $\mathcal{P}^p(\mathcal{U})$ with respect to $\mathcal{P}(\mathcal{D})^p$ and Δ such that[2]

$$\mathcal{P}^p(\mathcal{U}) = \{(\delta, \mathsf{Succ}_{\delta,\mathcal{A}}) \mid \delta \in \Delta\}.$$

We now have all the ingredients to define a deterministic ATD (*DATD*)

$$\mathcal{A}_{det} = \langle S_{det}, s_{ini}^{det}, val_{ini}^{det}, E_{det}, T_{det} \rangle$$

on $\mathcal{P}(\mathcal{D})^p$ and $\mathcal{P}^p(\mathcal{U})$ which is equivalent to the considered NDATD $\mathcal{A}$. The proposed DATD $\mathcal{A}_{det}$ is defined as follows:

- $S_{det} = \mathcal{P}(S)$;
- $s_{ini}^{det} = \{s_{ini}\}$;
- $val_{ini}^{det} = (\{val_{ini}\}, \emptyset, \cdots, \emptyset) \in \mathcal{P}(\mathcal{V})^p$;
- E_{det} is to the set of transitions $(S_{\mathbf{V}}, \mathbf{V}, (\delta, \mathsf{Succ}_{\delta,\mathcal{A}}), S')$ such that $\mathbf{V} \in \mathcal{P}(\mathcal{V}al)^p$ and $S' = S_{\mathbf{V}'}$ with $\mathbf{V}' = \mathsf{Succ}_{\delta,\mathcal{A}}(\mathbf{V})$;
- $T_{det} = \{S' \subseteq S \mid S' \cap T \neq \emptyset\}$.

4 Testing Framework

4.1 ATD with Inputs and Outputs (ATDIO)

Consider the ATD $\mathcal{A} = \langle S, s_{ini}, v_{ini}, E, T \rangle$ on $\mathcal{D}$ and $\mathcal{U}$ and let Δ be the alphabet corresponding to the update set $\mathcal{U}$. We assume that the alphabet Δ is split into two disjoint sets namely Δ_I a set of inputs and Δ_O a set of outputs (i.e., $\Delta = \Delta_I \sqcup \Delta_O$)[3]. In this case the ATD $\mathcal{A}$ is called an *ATD with inputs and outputs* (ATDIO). Moreover for we suppose that all the states of $\mathcal{A}$ are terminal (i.e., $T = E$).

The ATDIO $\mathcal{A}$ is said to be *input-enabled* if for any reachable configuration $conf \in \mathcal{R}each(\mathcal{A})$ and any input symbol $inp \in \Delta_I$ we have $conf \xrightarrow{inp}\!\!\!\!\rightarrow_{\mathcal{A}}$.

Moreover the considered ATDIO is called *non-blocking* if for any reachable configuration $conf \in \mathcal{R}each(\mathcal{A})$ and any duration $t \in \mathbb{R}_{\geq 0}$ there exists $tw \in (\mathbb{R}_{\geq 0} \uplus \Delta_O)^*$ such that $conf \xrightarrow{tw}\!\!\!\!\rightarrow_{\mathcal{A}}$ and $\mathcal{D}uration(tw) = t$.

Next we suppose that the specification of the system under test and the implementation are given as two non-blocking ATDIO $\mathcal{S}p$ and $\mathcal{I}m$ respectively.[4]

4.2 Parallel Composition of OTLTS with Inputs and Outputs

Given two OTLTS with inputs and outputs LTS_1 and LTS_2, we define the parallel product $LTS_1 || LTS_2$. For $i = 1, 2$, $LTS_i = (Q_i, q_0^i, \Delta_\mathsf{I}^i \cup \Delta_{(3-i)\to i}, \Delta_\mathsf{O}^i \cup \Delta_{i\to(3-i)}, T_d^i, T_t^i)$. The sets Δ_I^1, Δ_O^1, Δ_I^2, Δ_O^2, $\Delta_{1\to 2}$ and $\Delta_{2\to 1}$ are pairwise disjoint. The two OTLTS synchronize on shared common actions $\Delta_{1\leftrightarrow 2} = \Delta_{1\to 2} \cup \Delta_{2\to 1}$ and time delays. The parallel product of the two OTLTS is

$$LTS_1 || LTS_2 = (Q, (q_0^1, q_0^2), \Delta_\mathsf{I}, \Delta_\mathsf{O}, T_d, T_t)$$

[2] Note that $\mathcal{P}^p(\mathcal{U})$ is not the same as $\mathcal{P}(\mathcal{U})^p$.

[3] $\sqcup$ stands for the disjoint union symbol.

[4] We do not assume $\mathcal{I}m$ is known.

such that

$$\Delta_{\mathsf{I}} = \bigcup_{i=1,2} \Delta_{\mathsf{I}}^i,\ \Delta_{\mathsf{O}} = \bigcup_{i=1,2} \Delta_{\mathsf{O}}^i$$

and Q, T_d and T_t are the smallest groups of elements such that:

- $(q_0^1, q_0^2) \in Q$;
- For $(q_1, q_2) \in Q$ and $\delta \in \mathsf{R}$:

$$q_1 < aq_1' \in T_d^1 \Rightarrow (q_1', q_2) \in S \wedge (q_1, q_2) < a(q_1', q_2) \in T_d;$$

- For $(q_1, q_2) \in Q$ and $a \in \Delta_{\mathsf{I}}^1 \cup \Delta_{\mathsf{O}}^1 \cup \{\tau_1\}$:

$$q_1 < aq_1' \in T_d^1 \Rightarrow (q_1', q_2) \in S \wedge (q_1, q_2) < a(q_1', q_2) \in T_d;$$

- For $(q_1, q_2) \in Q$ and $a \in \Delta_{\mathsf{I}}^2 \cup \Delta_{\mathsf{O}}^2 \cup \{\tau_2\}$:

$$q_2 < aq_2' \in T_d^2 \Rightarrow (q_1, q_2') \in S \wedge (q_1, q_2) < a(q_1, q_2') \in T_d;$$

- For $(q_1, q_2) \in Q$ and $a \in \Delta_{1\leftrightarrow 2}$:

$$q_1 < aq_1' \in T_d^1 \wedge q_2 < aq_2' \in T_d^2 \Rightarrow (q_1', q_2') \in Q \wedge (q_1, q_2) < \tau_a(q_1', q_2') \in T_d.$$

4.3 Conformance Relation

Given a ATDIO $\mathcal{A}$ and a timed word $tw \in (\mathbb{R}_{\geq 0} \uplus \Delta)^*$, $\mathcal{A}$ after tw is the set of configurations of $\mathcal{A}$ that can be reached after the execution of tw. Formally:

$$\mathcal{A} \text{ after } tw = \{conf \in C_{\mathcal{A}} \mid c_{ini}^{\mathcal{A}} \xrightarrow{tw}\!\!\!\!\rightarrow_{\mathcal{A}} conf\}.$$

Given configuration $conf \in C_{\mathcal{A}}$, $\mathsf{out}(conf)$ is the set of all observations (either outputs or the elapsing of time) that may happen when the system is at configuration $conf$. The definition is extended in a natural way to a set of configurations $Conf$. Formally:

$$\mathsf{out}(conf) = \{\mu \in \mathbb{R}_{\geq 0} \uplus \Delta_O \mid conf \xrightarrow{\mu}\!\!\!\!\rightarrow_{\mathcal{A}}\},\ \mathsf{out}(Conf) = \bigcup_{conf \in Conf} \mathsf{out}(conf).$$

The definition of the relation tioco [10,11] is as follows:

$$\mathcal{I}m \text{ tioco } \mathcal{S}p \quad \textit{iff} \quad \forall tw \in \mathcal{L}ang(\mathcal{S}p) \ :\ \mathsf{out}(\mathcal{I}m \text{ after } tw) \subseteq \mathsf{out}(\mathcal{S}p \text{ after } tw).$$

The relation indicates that the implementation $\mathcal{I}m$ conforms to the specification $\mathcal{S}p$ if and only if for any timed word tw of $\mathcal{S}p$, the set of outputs (including time elapse) of $\mathcal{I}m$ after the execution of tw is a subset of the set of outputs that can be generated by $\mathcal{S}p$.

4.4 Timed Test Cases

A timed test case for the specification $\mathcal{S}p$ over Δ_τ is a total function

$$TTest : (\mathbb{R}_{\geq 0} \uplus \Delta)^* \rightarrow \Delta_\mathsf{I} \cup \{\mathsf{WT}, \mathsf{PS}, \mathsf{FL}\}.$$

$TTest(tw)$ indicates the action that must executed by the tester once it observes tw. If $TTest(tw) = inp \in \Delta_\mathsf{I}$ then the tester produces input inp. If $TTest(tw) = \mathsf{WT}$ then the tester lets time elapse (waits). If $TTest(tw) \in \{\mathsf{PS}, \mathsf{FL}\}$ then the tester emits a verdict and stops.

The execution of $TTest$ on $\mathcal{I}m$ may be seen as the *parallel composition* of the OTLTS with inputs and outputs defined by $TTest$ and $\mathcal{I}m$. The product TIOLTS is denoted by $\mathcal{I}m \| TTest$. In a formal fashion, we announce that the implementation $\mathcal{I}m$ *passes* $TTest$, denoted $\mathcal{I}m$ passes $TTest$, if state FL can not be reached in $\mathcal{I}m \| TTest$. We declare that the implementation passes (respectively fails) the test suite $\mathcal{TT}$ if it passes all tests (respectively fails at least one test) in $\mathcal{TT}$. $\mathcal{TT}$ is said to be *sound with respect to* $\mathcal{S}p$ if

$$\forall \mathcal{I}m \ : \ \mathcal{I}m \ \mathsf{tioco} \ \mathcal{S}p \Rightarrow \mathcal{I}m \ \mathsf{passes} \ \mathcal{TT}.$$

Similarly $\mathcal{TT}$ is said to be *complete with respect to* $\mathcal{S}p$ if

$$\forall \mathcal{I}m \ : \ \mathcal{I}m \ \mathsf{passes} \ \mathcal{TT} \Rightarrow \mathcal{I}m \ \mathsf{tioco} \ \mathcal{S}p.$$

Our goal it then to produce test suites which are both sound and complete. More precisely, we aim to generate timed tests in the form of deterministic ATDIO and which are finitely representable. For that purpose we need to use the determinization technique proposed in Sect. 3.

5 Future Work

The work proposed in this paper is at its beginning. In the future we need to work on many aspects:

- First, we need to extend the presented framework to the case where the specification of the SUT is given as a product of ATDIO and not simply one ATDIO. In this way we can deal with distributed and multi-components systems.
- Second, we should find a way which guarantees that the generated timed tests are finitely representable so that we can store them and execute them later on. For that, we need to use some approximation techniques based on game theory techniques like the ones proposed in [2].
- Third, we need to use some coverage and selection techniques which allow to reduce the size of generated test cases and to efficiently deal the state explosion problem usually encountered when following model-based approaches.

References

1. Alur, R., Dill, D.: The theory of timed automata. In: de Bakker, J.W., Huizing, C., de Roever, W.P., Rozenberg, G. (eds.) REX 1991. LNCS, vol. 600, pp. 45–73. Springer, Heidelberg (1992). https://doi.org/10.1007/BFb0031987
2. Bertrand, N., Stainer, A., Jéron, T., Krichen, M.: A game approach to determinize timed automata. Formal Methods Syst. Des. **46**(1), 42–80 (2015)
3. Bojańczyk, M., Lasota, S.: A machine-independent characterization of timed languages. In: Czumaj, A., Mehlhorn, K., Pitts, A., Wattenhofer, R. (eds.) ICALP 2012. LNCS, vol. 7392, pp. 92–103. Springer, Heidelberg (2012). https://doi.org/10.1007/978-3-642-31585-5_12
4. Bouyer, P., Jaziri, S., Markey, N.: On the determinization of timed systems. In: Abate, A., Geeraerts, G. (eds.) FORMATS 2017. LNCS, vol. 10419, pp. 25–41. Springer, Cham (2017). https://doi.org/10.1007/978-3-319-65765-3_2
5. Krichen, M.: A formal framework for conformance testing of distributed real-time systems. In: Lu, C., Masuzawa, T., Mosbah, M. (eds.) OPODIS 2010. LNCS, vol. 6490, pp. 139–142. Springer, Heidelberg (2010). https://doi.org/10.1007/978-3-642-17653-1_12
6. Krichen, M.: A formal framework for black-box conformance testing of distributed real-time systems. Int. J. Crit. Comput. Based Syst. **3**(1/2), 26–43 (2012). https://doi.org/10.1504/IJCCBS.2012.045075. http://dx.doi.org/10.1504/IJCCBS.2012.045075
7. Krichen, M., Tripakis, S.: Black-box conformance testing for real-time systems. In: Graf, S., Mounier, L. (eds.) SPIN 2004. LNCS, vol. 2989, pp. 109–126. Springer, Heidelberg (2004). https://doi.org/10.1007/978-3-540-24732-6_8
8. Krichen, M., Tripakis, S.: Real-time testing with timed automata testers and coverage criteria. In: Lakhnech, Y., Yovine, S. (eds.) FORMATS/FTRTFT -2004. LNCS, vol. 3253, pp. 134–151. Springer, Heidelberg (2004). https://doi.org/10.1007/978-3-540-30206-3_11
9. Krichen, M., Tripakis, S.: An expressive and implementable formal framework for testing real-time systems. In: Khendek, F., Dssouli, R. (eds.) TestCom 2005. LNCS, vol. 3502, pp. 209–225. Springer, Heidelberg (2005). https://doi.org/10.1007/11430230_15
10. Krichen, M., Tripakis, S.: Interesting properties of the real-time conformance relation tioco. In: Barkaoui, K., Cavalcanti, A., Cerone, A. (eds.) ICTAC 2006. LNCS, vol. 4281, pp. 317–331. Springer, Heidelberg (2006). https://doi.org/10.1007/11921240_22
11. Krichen, M., Tripakis, S.: Conformance testing for real-time systems. Formal Methods Syst. Des. **34**(3), 238–304 (2009). https://doi.org/10.1007/s10703-009-0065-1
12. Tretmans, J.: Testing concurrent systems: a formal approach. In: Baeten, J.C.M., Mauw, S. (eds.) CONCUR 1999. LNCS, vol. 1664, pp. 46–65. Springer, Heidelberg (1999). https://doi.org/10.1007/3-540-48320-9_6. http://dl.acm.org/citation.cfm?id=646734.701460

Verification of Multi-agent Systems with Timeouts for Migration and Communication

Bogdan Aman[1,2] and Gabriel Ciobanu[1,2](✉)

[1] Faculty of Computer Science,
Alexandru Ioan Cuza University, Iasi, Romania
[2] Institute of Computer Science, Romanian Academy, Iasi, Romania
{bogdan.aman,gabriel}@info.uaic.ro

Abstract. A prototyping high-level language is used to describe multi-agent systems using timeouts for migration between explicit locations and local communication in a distributed system. We translate such a high-level specification into the real-time Maude rewriting language. We prove that this translation is correct, and provide an operational correspondence between the evolutions of the mobile agents with timeouts and their rewriting translations. These results allow to analyze the multi-agent systems with timeouts for migration and communication by using the real-time Maude tools. A running example is used to illustrate the whole approach.

1 Introduction

Multi-agent systems are composed of a large number of agents that behave according to their timed actions. The mobility of agents and the communication between agents may lead to unexpected behaviours. Components can be highly heterogeneous, having individual objectives and using different temporal scales to achieve them. As multi-agent systems are getting more complex, automated verification of such systems is needed. Actually, the specification and analysis of multi-agent systems represent an active research direction in the last years. It is important to have modelling techniques able to describe easily such systems, as well as tools to simulate and verify some complex (qualitative and quantitative) properties of their behaviours. We take a step in this direction by developing a high-level specification language for specifying the mobile agents with timeouts, and providing a way to perform automated verification of some complex systems involving explicit locations and timeouts for migration and local communication in distributed networks.

There exist already some approaches to formalize timed systems, for instance timed automata [1]. Software platforms as UPPAAL [3] represent model checking tools used for the simulation and verification of real-time systems modelled as timed automata [10]. Logic-based models complement the timed automata models as they are able to capture other aspects of real-time systems: e.g., mobility of agents and the communication between agents. Rewriting logic is appropriate for

R. M. Hierons and M. Mosbah (Eds.): ICTAC 2019, LNCS 11884, pp. 134–151, 2019.
https://doi.org/10.1007/978-3-030-32505-3_9

providing a general semantic framework for various languages and models of concurrency [11]. Maude is a system that supports computations based on rewriting and equational logic, while real-time Maude [14] provides a specification formalism with several decidability results for many system properties. Also, in real-time Maude different types of communication used in process calculi can be modelled. The real-time Maude tool is developed using an extension of rewriting logic, and seem to be an appropriate tool for specification, validation and verification of real-time systems using features as migration of agents and communication between agents. This tool is useful in applications that use features not yet implemented in several existing model checkers for real-time systems [13].

For the specification of the multi-agent systems with timeouts for migration and local communication we use a real-time version of an existing high-level framework called TiMo, a framework able to describe easily interacting mobile agents in distributed systems. Then we translate this high-level specification into real-time Maude. There are some problems to overcome in order to obtain a fully executable specification in real-time Maude. Firstly, the transitions of a system need sometimes to use fresh names (to overcome binding problems); this is due to the fact that the communication of values takes place eventually after alpha-converting (to avoid clashes) in the high-level specification. Secondly, since infinite computations are not supported by real-time Maude, implementing an unbounded recursion operator is not possible; a solution to this problem is to consider a bounded recursion in which a process can be unfolded only a finite number of times during an execution. This restriction does not influence the results because we model real systems in which the recursive processes need to be unfolded only for a finite number of times.

2 Syntax and Semantics of the High-Level Specification

In the high-level specification language, the processes are allowed to migrate between explicit distributed locations and to communicate locally with other processes. The coordination of the processes in time and space is done by using timed migration and timed communication. The timeouts added to a migration action enforce the process to migrate to the target location after a period of time equal to the timeout constraint. Two processes are allowed to communicate only if they are both available into the same location at the same unit of time, and if the timeout restrictions of the active communication actions are non-negative. If a communication action cannot be executed before its timeout restriction expires, then the action is removed and the actions of an alternative process are executed as a continuation. The transitions involving either processes migration between locations, processes communication or unfolding of recursive processes are executed in a maximal parallel manner. This means that if a process can migrate, communicate or unfold, it has to do it. The transitions with timeouts are alternated with transitions involving the passage of time over all the processes; a global clock is used to model the passage of time. The operational semantics of the high-level specification is provided by using these two types of transitions:

a transition relation for timed migration and communication actions executed in the maximal parallel manner, and a transition relation used to model the passage of time.

A timeout restriction assigned to a migration action is given as a natural number t, while a timeout restriction assigned to a communication action is given as Δt, where $t \in \mathbb{N}$. The t notation means that migration action can be consumed exactly after t units of time, while the Δt notation means that the communication can be consumed at any moment in the next t units of time.

The syntax of the high-level language is given in Table 1, where the following notations are used:

- we use the set *Loc* of locations, set *Chan* of communication channels, and set *Id* of process identifiers;
- for each process identifier $id \in Id$ there exists a unique process definition $id(u_1, \ldots, u_{m_{id}}) \stackrel{def}{=} P_{id}$ in which the m_{id} parameters are identified by the distinct variables u_i;
- a, l, t denote a communication channel, a location or a location variable, and an action timeout, respectively; u and v denote a tuple of variables and a tuple of expressions built from values (e.g., strings, integers, bools), variables and allowed operations.

Table 1. The syntax of the high-level language

Processes	P, Q	$::=$	$a^{\Delta t}!\langle v\rangle$ then P else Q |	(output)
			$a^{\Delta t}?(u)$ then P else Q |	(input)
			$go^{t}l$ then P |	(move)
			0 |	(termination)
			$id(v)$ |	(recursion)
			$P \mid Q$	(parallel)
Located Processes	L	$::=$	$l[[P]]$	
Multi-Agent Systems	N	$::=$	L | $N \parallel N$ | $\mathbf{0}$	

An output communication process $a^{\Delta t}!\langle z\rangle$ then P else Q describes the fact that for t time units the process is available for sending on channel a the value z. Whenever the process succeeds in sending the value before the deadline, it continues its evolution according to process P; otherwise, it continues its evolution by the alternative process Q. The input communication process $a^{\Delta t}?(x)$ then P else Q describes the fact that for t time units the process is available to receive on channel a a value to instantiate the variable x. In a similar manner as for the output communication process, the continuation of an input communication process depends on the success of the communication.

A migration process $go^{t}l$ then P indicates a location change after t time units, namely after t units of time the process continues its execution as P at location l (and not at the current location). Since variables are instantiated through communication, this means that the location variables can be instantiated; this feature allows a flexible behaviour as processes can adapt their migration based on received information. The process 0 models an inactive process, while the process $P \mid Q$ models the parallel composition of the process P and Q that might also interact through communication. A process P currently located in location l is denoted by $l[[P]]$, while a system is composed of located processes composed by using the parallel operator.

There is only one binding operator in our calculus: in the input process $a^{\Delta t}?(u)$ then P else Q, the variable u is bound in process P. However, as process Q is an alternative process executed when the input action is not consumed, this means that variable u is not bound in process Q. Given a process P, we denote by $fv(P)$ its set of free variables. In case u_i are the m_{id} parameters of the process P_id, then the assumption $fv(P_{id}) \subseteq \{u_1, \ldots, u_{m_{id}}\}$ holds. As usually assumed in process calculi, we consider that processes are defined up to an alpha-conversion. Also, $P\{v/u, \ldots\}$ denotes a process P in which v replaces all the free occurrences of the variable u, possible after using alpha-conversion inside P to remove possible clashes. A system N is said to be well-formed if $fv(N) = \emptyset$.

Operational Semantics. The structural equivalence relation $\equiv$ represents an ingredient of the operational semantics; it is defined as the smallest congruence relation satisfying the equations of Table 2. The purpose of this relation $\equiv$ is to provide a way of rearranging the processes in a system such that they can evolve by using the operational semantics rules from Table 3.

Table 2. Structural congruence in high-level specification

(PNULL)	$P \mid 0 \equiv P$
(LNULL)	$N \parallel \mathbf{0} \equiv N$
(LCOMM)	$N \parallel N' \equiv N' \parallel N$
(LASSOC)	$(N \parallel N') \parallel N'' \equiv N \parallel (N' \parallel N'')$
(LSPLIT)	$l[[P \mid Q]] \equiv l[[P]] \parallel l[[Q]]$

The equalities of Table 2 are useful for transforming a system N into the system $l_1[[P_1]] \parallel \ldots \parallel l_n[[P_n]]$ composed of located process $l[[P_i]]$ such that there do not exist Q_i and R_i such that $P_i \equiv Q_i \mid R_i$. A located process that cannot be split into parallel located processes by using the rule (LSPLIT) is called a component of N, while the component decomposition of a system N is the system $l_1[[P_1]] \parallel \ldots \parallel l_n[[P_n]]$, where all $l_i[[P_i]]$ are components.

Table 3 presents the operational semantics rules. The transitions of the form $N \rightarrow N'$ indicate either processes migrating between locations, processes communicating locally or unfolding of processes, all these executed in parallel in one step. The passing of t time units is given by transitions of the form $N \overset{t}{\rightsquigarrow} N'$.

In rule (COM), two process $a^{\Delta t}!\langle v\rangle$ then P else Q and $a^{\Delta t}?(u)$ then P' else Q', both located at location l, are using channel a to communicate a tuple of values v to be used for the instantiation of the variable u. Applying the rule (COM) does not lead to a location change for any of the involved processes, but to a consumption of the output and input action. Upon a successful communication, the processes

Table 3. The operational semantics of the high-level language

(STOP)	$l[[0]] \not\rightarrow$ (DSTOP) $l[[0]] \overset{t}{\rightsquigarrow} l[[0]]$
(COM)	$l[[a^{\Delta t}!\langle v\rangle \text{ then } P \text{ else } Q]] \mid\mid l[[a^{\Delta t'}?(u) \text{ then } P' \text{ else } Q']] \rightarrow l[[P]] \mid\mid l[[P'\{v/u\}]]$
(DPUT)	$\dfrac{t \geq t' > 0}{l[[a^{\Delta t}!\langle v\rangle \text{ then } P \text{ else } Q]] \overset{t'}{\rightsquigarrow} l[[a^{\Delta t-t'}!\langle v\rangle \text{ then } P \text{ else } Q]]}$
(PUT0)	$l[[a^{\Delta 0}!\langle v\rangle \text{ then } P \text{ else } Q]] \rightarrow l[[Q]]$
(DGET)	$\dfrac{t \geq t' > 0}{l[[a^{\Delta t}?(u) \text{ then } P \text{ else } Q]] \overset{t'}{\rightsquigarrow} l[[a^{\Delta t-t'}?(u) \text{ then } P \text{ else } Q]]}$
(GET0)	$l[[a^{\Delta 0}?(u) \text{ then } P \text{ else } Q]] \rightarrow l[[Q]]$
(DMOVE)	$\dfrac{t \geq t'}{l[[go^{t} l' \text{ then } P]] \overset{t'}{\rightsquigarrow} l[[go^{t-t'} l' \text{ then } P]]}$
(MOVE0)	$l[[go^{0} l' \text{ then } P]] \rightarrow l'[[P]]$
(DCALL)	$\dfrac{l[[P_{id}\{v/x\}]] \overset{t}{\rightsquigarrow} l[[P'_{id}]] \qquad id(v) \overset{def}{=} P_{id}}{l[[id(v)]] \overset{t}{\rightsquigarrow} l[[P'_{id}]]}$
(CALL)	$\dfrac{l[[P_{id}\{v/x\}]] \rightarrow l[[P'_{id}]] \qquad id(v) \overset{def}{=} P_{id}}{l[[id(v)]] \rightarrow l[[P'_{id}]]}$
(DPAR)	$\dfrac{N_1 \overset{t}{\rightsquigarrow} N_1' \qquad N_2 \overset{t}{\rightsquigarrow} N_2' \qquad N_1 \mid\mid N_2 \not\rightarrow}{N_1 \mid\mid N_2 \overset{t}{\rightsquigarrow} N_1' \mid\mid N_2'}$
(PAR)	$\dfrac{N_1 \rightarrow N_1' \qquad N_2 \rightarrow N_2'}{N_1 \mid\mid N_2 \rightarrow N_1' \mid\mid N_2'}$
(DEQUIV)	$\dfrac{N_1 \equiv N_1' \qquad N_1' \overset{t}{\rightsquigarrow} N_2' \qquad N_2' \equiv N_2}{N_1 \overset{t}{\rightsquigarrow} N_2}$
(EQUIV)	$\dfrac{N_1 \equiv N_1' \qquad N_1' \rightarrow N_2' \qquad N_2' \equiv N_2}{N_1 \rightarrow N_2}$

$a^{\Delta t}!\langle v\rangle$ then P else Q and $a^{\Delta t}?(u)$ then P' else Q' continue their executions as processes P and $P'\{v/u\}$, respectively. If the process $a^{\Delta 0}!\langle v\rangle$ then P else Q exists in the system, then the communication action is discarded by using the rule (PUT0), and the execution continues as the alternative process Q. In a similar manner, by using the rule (GET0), the process $a^{\Delta 0}?(u)$ then P' else Q' continues its execution as the alternative process Q. In rule (MOVE0), a process $go^{0} l'$ then P is able to change its location by migrating from the current location l to the given location l' where it continues its execution as process P. The unfolding of recursive processes is performed by using the rule (CALL). In order to use the structural equivalence relation $\equiv$ to rearrange a system such that its components can interact for communication or migration, the rule (EQUIV) its used. Composing larger systems from smaller systems is done by using the rule (PAR) for the parallel composition operator.

The passage of time is described by the rules having their names starting with the capital letter D. The hypothesis $N_1 \mid\mid N_2 \not\rightarrow$ from the rule (DPAR) indicates the fact that placing the two systems N_1 and N_2 in parallel does not trigger the application of a rule (COM) that would modify these systems. The negative premises are essential to separate the steps based on the execution of actions by those based on time passing (i.e., time cannot pass when an action is executed).

A transition of the form $N \rightarrow N_1$ followed by a time passing transition of the form $N_1 \stackrel{t}{\rightsquigarrow} N'$ describe a complete step that can be written as:

$$N \rightarrow N_1 \stackrel{t}{\rightsquigarrow} N'.$$

Thus, a complete step indicates that a parallel execution of processes migrating between locations, processes communicating or unfolding is necessarily followed by a time step. We say that the system N' is directly reachable from N if a complete computational step $N \stackrel{\Lambda}{\rightarrow} N_1 \stackrel{t}{\rightsquigarrow} N'$ exists. If $N \not\rightarrow$, then only a time step $N \stackrel{t}{\rightsquigarrow} N'$ can be performed in the system N.

Theorem 1. *For all the systems N, N_1 and N_2,*

$$\text{if } N \stackrel{t}{\rightsquigarrow} N_1 \text{ and } N \stackrel{t}{\rightsquigarrow} N_2, \text{ then } N_1 \equiv N_2.$$

Theorem 1 claims that nondeterminism cannot be introduced upon executing a time transition in a system, namely the obtained system is unique up to structural congruence.

Theorem 2. *For all the systems N, N_1, N_2 and $0 < t' < t$, we have $N \stackrel{t}{\rightsquigarrow} N_2$ if and only if there is a N_1 such that $N \stackrel{t'}{\rightsquigarrow} N_1$ and $N_1 \stackrel{t-t'}{\rightsquigarrow} N_2$.*

Theorem 2 claims that whenever a time transition of length t can be performed in a system N leading to a system N_2, then always a time transition of length t' with $0 < t' < t$ can be performed in the same system N leading to a system N_1 followed by another time transition of length $t - t'$ in the systems N_1 leading to N_2, and vice versa. This result ensures that the passage of time in a system is continuous (no jumps).

Example 1. Let us consider an example in which a client wants to buy, at a good price, a flight ticket to a given location. The scenario is depicted in Fig. 1, where the names and values have the meanings given below (we explain the names and values in the order they appear, from left to right).

- The process *client* initially resides at location *home*. It has access to 130 cash units to be used for purchasing a flight ticket. Once the *client* reaches the *travelshop* location, an *agent* communicates to it the location of a *standard* offer. The *client* process goes to this location to receive the *standard* offer details. Here it also receives the location for a *special* offer. After receiving the information about the *special* offer, it goes to the bank for paying the cheaper offer between the *standard* and the *special* offers, and returns *home* (its initial location).
- The process *update* is able to migrate to the *special* location by starting from its initial location *travelshop* in order to communicate locally a reduction for the price *special* from 90 to 60 cash units.

- The process *agent* resides at the *travelshop* location, and has access to 100 cash units available in the cash register. Once a *client* reaches the *travelshop* location and the *agent* is available for communication, the client receives the location where the details of the standard offer are available. The *agent* has also the possibility to go to the *bank* to withdraw the available money from the *till*. Regardless of the amount of money taken from the *bank*, the *agent* always returns to *travelshop*, its initial location.
- The process *flightinfo* process residing at the *standard* location is able to do only local communications in order to provide to any interested client the details about the *standard* offer: the price of 110 cash units, and the location where the *special* offer resides.
- The process *saleinfo* process residing at the *special* location is able to do only local communications in order to provide (to any interested client) the details about the *standard* offer: the price of 90 cash units, and the location of the *bank* for the payment. The *saleinfo* process can also interact locally with the *update* process in order to modify the price of the *special*offer.
- The process *till* process owning 10 cash units and residing at the *bank* location is able to do only local communications: it can interact with a *client* to receive the payment for a flight ticket, and can interact with the *agent* in order to transfer the accumulated cash to it.

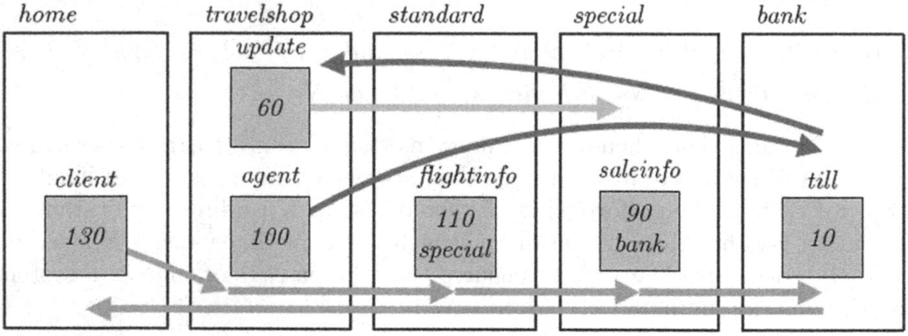

Fig. 1. Initial scenario

After all the interactions described in Fig. 1, the system looks like in Fig. 2.

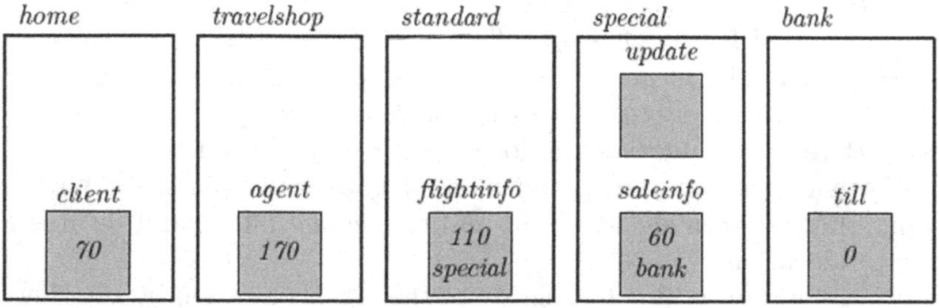

Fig. 2. A possible outcome

In the above example we have: (i) agents migrating in a distributed network with explicit locations; (ii) local communication of these agents (to get specific results); (iii) both migration and communication require certain time indicated by timeouts.

We show how this example can be easily described in our high-level language. First of all, in order to simplify the syntax, we consider that:
$a^{\Delta\infty}!\langle v\rangle$ then P else Q can be written as $a!\langle v\rangle \to P$,
$a^{\Delta\infty}?(u)$ then P else Q can be written as $a?(u) \to P$, and
$\mathsf{go}^t l$ then P can be written as $\mathsf{go}^t l \to P$.
This is because branch Q is ignored as it can never be executed.
The system presented in Fig. 1 is described in the high-level language as:

$$\begin{gathered} \mathit{TravelShop} = \mathit{home}\,[[\mathit{client}\,(130)]] \parallel \mathit{travelshop}\,[[\mathit{update}\,(60) \mid \mathit{agent}\,(100)]] \\ \parallel \mathit{standard}\,[[\mathit{flightinfo}\,(110, \mathit{special}\,)]] \parallel \mathit{special}\,[[\mathit{saleinfo}\,(90, \mathit{bank}\,)]] \\ \parallel \mathit{bank}\,[[\mathit{till}\,(10)]], \end{gathered}$$

where:

$$\begin{aligned}
&\mathit{client}\,(\mathit{init}\,) = \mathsf{go}^5 \mathit{travelshop} \to \mathit{flight}\,?(\mathit{standardoffer}\,) \\
&\quad \to \mathsf{go}^4 \mathit{standardoffer} \to \mathit{finfo2a}\,?(p1) \to \mathit{finfo2b}\,?(\mathit{specialoffer}\,) \\
&\quad \to \mathsf{go}^3 \mathit{specialoffer} \to \mathit{sinfo2a}\,?(p2) \to \mathit{sinfo2b}\,?(\mathit{paying}\,) \\
&\quad \to \mathsf{go}^6 \mathit{paying} \to \mathit{payc}\,!\langle \mathit{min}\{p1, p2\}\rangle \\
&\quad \to \mathsf{go}^4 \mathit{home} \to \mathit{client}\,(\mathit{init} - \mathit{min}\{p1, p2\})\ ; \\
&\mathit{update}\,(\mathit{saleprice}\,) = \mathsf{go}^1 \mathit{special} \to \mathit{info1}\,!\langle \mathit{saleprice}\,\rangle\ ; \\
&\mathit{agent}\,(\mathit{balance}\,) = \mathit{flight}\,!\langle \mathit{standard}\,\rangle \\
&\quad \to \mathsf{go}^{20} \mathit{bank} \to \mathit{paya}\,?(\mathit{profit}\,) \\
&\quad \to \mathsf{go}^{12} \mathit{travelshop} \to \mathit{agent}\,(\mathit{balance} + \mathit{profit}\,)\ ; \\
&\mathit{flightinfo}\,(\mathit{price}\,, \mathit{next}\,) = \mathit{finfo2a}\,!\langle \mathit{price}\,\rangle \to \mathit{finfo2b}\,!\langle \mathit{next}\,\rangle \\
&\quad \to \mathit{flightinfo}\,(\mathit{price}\,, \mathit{next}\,)\ ; \\
&\mathit{saleinfo}\,(\mathit{price}\,, \mathit{next}\,) = \mathit{info1}^{\,\Delta 2}?(\mathit{newprice}\,) \\
&\quad \mathsf{then}\ \mathit{sinfo2a}\,!\langle \mathit{newprice}\,\rangle \to \mathit{sinfo2a}\,!\langle \mathit{next}\,\rangle \to \mathit{saleinfo}\,(\mathit{newprice}\,, \mathit{next}\,) \\
&\quad \mathsf{else}\ \mathit{sinfo2a}\,!\langle \mathit{newprice}\,\rangle \to \mathit{sinfo2a}\,!\langle \mathit{next}\,\rangle \to \mathit{saleinfo}\,(\mathit{price}\,, \mathit{next}\,)\ ; \\
&\mathit{till}\,(\mathit{cash}\,) = \mathit{payc}^{\,\Delta 22}?(\mathit{newpayment}\,) \\
&\quad \mathsf{then}\ \mathit{paya}^{\,10}!\langle \mathit{cash} + \mathit{newpayment}\,\rangle\ \mathsf{then}\ \mathit{till}\,(0) \\
&\quad\qquad\qquad\qquad\qquad\qquad\qquad\qquad\ \mathsf{else}\ \mathit{till}\,(\mathit{cash} + \mathit{newpayment}\,)) \\
&\quad \mathsf{else}\ \mathit{paya}^{\,10}!\langle \mathit{cash}\,\rangle\ \mathsf{then}\ \mathit{till}\,(0) \\
&\quad\qquad\qquad\qquad\qquad\quad\ \mathsf{else}\ \mathit{till}\,(\mathit{cash}\,))\ .
\end{aligned}$$

3 Translating the High-Level Specification into Maude

In what follows we define a rewriting theory corresponding to the semantics of our high-level language defined in Table 3. The syntax used to give the rewriting theory is that of *real-time Maude*. A *rewrite theory* $\mathcal{R}$ is defined as a triple (Σ, E, R), where Σ stands for signature of function symbols, E and R are sets of Σ-equations and Σ-rewrite rules, respectively. The Σ-equations and Σ-rewrite rules can contain side conditions; for example, the conditions appearing in a rewrite rule can contain equations or other rewrite rules. Just like in [9], we use a typed setting given as an order-sorted equational logic (Σ, E) including sorts and an inclusion relation *subsort* between sorts. Given a rewrite theory $\mathcal{R}$, we write $\mathcal{R} \vdash t \Rightarrow t'$ if $t \Rightarrow t'$ is provable in $\mathcal{R}$ by using the rewrite rules of R. Rewriting logic is basically a computational logic that combines term rewriting with equational logic.

Let us discuss first the high-level recursion operator that is not directly encodable into real-time Maude (because infinite computations are not supported into this tool). Our solution is to use the construction $id(v, n)$ that is an extension of the constructions $id(v)$ of our language with a number n that limits the number of recursive calls to be executed during the evolution of the system.

In order to translate the high-level language (whose syntax is given in Table 1), we consider sorts corresponding to sets from our language: e.g., for the set *Chan* of channels, the sort `Channel` is created. Certain new aspects are provided by the sorts `AGuard` and `MGuard`. The sort `AGuard` contains the action parts $a^{\Delta t}!\langle v\rangle$ and $a^{\Delta t}?(u)$ of the communication processes $a^{\Delta t}!\langle v\rangle$ then P else Q and $a^{\Delta t}?(u)$ then P else Q, while the sort `MGuard` contains the action part $\mathsf{go}^{\mathsf{t}}l$ of the migration processes of the form $go^{t}l$ then P. The elements of the sorts `AGuard` and `MGuard` are essential in constructing the sequential processes of our language. Among the subsorting relations between the given sorts, we explain `subsorts Var < Location Channel Value` that illustrates the fact that location names, channel names or values can be used to instantiate variables. To work with multisets of values, we use the sort `MValue`.

```
sorts Location Channel Value MValue Var Process
            AGuard MGuard System .
subsorts Var < Location Channel Value < MValue .
subsort Location < Value .
subsorts System < GlobalSystem .
```

To each operator used in the syntax of Table 1 we attach the attribute `ctor` marking the fact that this operator is used to construct the system, and attribute `prec` followed by a number marking its applicability precedence with respect to other operators. Moreover, in order to encode properly into real-time Maude the parallel operators $|$ and $||$ from Table 1, we add to them the attributes `comm` and `assoc` to illustrate that they are commutative and associative constructors that respect the rules of Table 2.

```
op  _^_!‘<_>  :  Channel TimeInf Value -> AGuard [ctor prec 2] .
op  _^_?‘(_‘) : Channel TimeInf Var -> AGuard [ctor prec 2] .
op  go‘^__ : TimeInf Location -> MGuard [ctor prec 2] .
op _then‘(_‘)else‘(_‘) : AGuard Process Process -> Process
          [ctor prec 1] .
op _then‘(_‘) : MGuard Process -> Process [ctor prec 1] .
op _|_  :  Process Process -> Process [ctor prec 4 comm assoc] .
op stop : -> Process [ctor] .
op _‘[‘[_‘]‘] : Location Process -> System [ctor prec 3] .
op  _||_  :  System System -> System [ctor prec 5 comm assoc] .
op void : -> System [ctor] .
```

As already stated, most of the rules of the structural congruence (Table 2) are encoded by using the attributes `comm` and `assoc` when defining the previous operators. For the rest of the rules we provide the following equations:

```
eq P | stop = P .
eq M || void = M .
eq k[[P | Q]] = (k[[P]]) || (k[[Q]]) .
```

As communication between two processes by using rule (Com)) leads to a substitution of variables by the communicated values, we need to define this operation explicitly in real-time Maude. Such an operator acts only upon the free occurrences of a name, while leaving bound names as they are.

```
op _'{_/_'} : Process Value Var -> Process [prec 8] .

eq ((c ^ t ! < b + a > ) then (P) else (Q)) { V / b } =
        (( c ^ t ! < V + a > ) then  (P { V / b } ) else (Q { V / b } )) .
eq ((c ^ t ! < min(b , a) > ) then (P) else (Q)) { V / b } =
     (( c ^ t ! < min(V , a) > ) then (P { V / b } )
                                 else (Q { V / b } )) .

eq ((c ^ t ! < X > ) then (P) else (Q)) { V / X } =
     (( c ^ t ! < V > ) then (P { V / X } ) else (Q { V / X })) .
ceq ((c ^ t ! < W > ) then (P) else (Q)) { V / X } =
     (( c ^ t ! < W > ) then (P { V / X } )
                        else (Q { V / X })) if V =/= W .
eq ((c ^ t ? ( X )) then (P) else (Q)) {V / X} =
     ((c ^ t ? ( X )) then (P) else (Q)) .
ceq ((c ^ t ? ( Y )) then (P) else (Q)) {V / X} =
     ((c ^ t ? ( Y )) then (P { V / X } )
                      else (Q { V / X } )) if X =/= Y .
eq ((go ^ t X) then (P)) {V / X} = ((go ^ t V) then (P {V / X})) .
ceq ((go ^ t l) then (P)) {V / X} =
     ((go ^ t l) then (P {V / X})) if X =/= l .
eq (P | Q) {V / X} = ((P {V / X}) | (Q {V / X})) .
eq stop {V / X} = stop .
eq (P) { V / X } = (P) [owise] .
```

However, the above operator does not take into account the need for alpha-conversion in order to avoid name clashes once substitution takes place. To illustrate this issue, let us consider the process $P = a^t(b)$ then $(go^{t'}\, l$ then $X)$ else *stop*) in which the name b is bound inside the input prefix. If the substitution $\{b/X\}$ needs to be performed over this process, the obtained process would be $P\{b/X\} = a^t(b)$ then $(go^{t'}\, l$ then $b)$ else *stop*). This means that once variable X is replaced by the name b, name b would become bound not only in the input action. To avoid this, we define an operator able to perform alpha-conversion by using terms of the form $[X]$ that contain fresh names:

```
op '[_'] : Var -> System [ctor] .
```

The terms of the form $[X]$ containing fresh names are composed with the system by using the parallel operator $||$. Using the given fresh names, the renaming is done (when necessary) before substitution. This is provided by the operator:

```
op _'(_/_') : Process Value Var -> Process [prec 8] .
```

This operator has a definition similar with the substitution operator, except the case when we deal with bound names.

```
eq ((c ^ t ? ( X )) then (P) else (Q)) (V / X) =
     ((c ^ t ? ( V )) then (P { V / X } ) else (Q { V / X } )) .
```

It is worth noting that this is different from the substitution operator that does not allow the change of the bound name:

```
eq ((c ^ t ? ( X )) then (P) else (Q)) {V / X} =
     ((c ^ t ? ( X )) then (P) else (Q)) .
```

As most of the rules in Table 3 contain hypotheses, translating these rules in real-time Maude requires the use of conditional rewrite rules in which the conditions are the hypotheses of rules of Table 3. Notice that in what follows we do not directly implement the rules (PAR), (DEQUIV) and (EQUIV) as rewrite rules into real-time Maude, due to the fact that the commutativity, associativity and the congruence rewriting of the parallel operators $|$ and $||$ are already encoded into the matching mechanism of Maude. In order to identify for each of the below rewrite rule which rule from Table 3 it models, we consider simple intuitive names for these rewrite rules. More complicated names could be considered by using rewriting rules similar with the ones given for the executable specification of the π-calculus in Maude [15].

```
crl [Comm] : (k[[(c ^ t ! < V >) then (P) else (Q) ]])
      || (k[[(c ^ t' ? ( X )) then (P') else (Q') ]])
    => (k [[ P ]]) || (k [[ P' {V / X} ]]) if notin(V , bnP(P')) .
crl [Comm'] : (([Z]) || (k[[(c ^ t ! < V >) then (P) else (Q) ]]))
      || (k[[(c ^ t' ? ( X )) then (P') else (Q') ]])
    => (([X]) || (k [[ P ]])) || (k [[ (P' (Z / V)) { V / X} ]])
       if in(V , bnP(P'))  /\ (notin(Z , bnP(P'))) .
crl [Input0] : (k[[ (c ^ t ! < V >) then (P) else (Q) ]]) => k[[Q]]
       if t == 0 .
crl [Output0] : (k[[(c ^ t ? ( X )) then (P) else (Q) ]]) => k[[Q]]
       if t == 0 .
crl [Move] : k[[(go ^ t l) then (P)]] => l[[P]] if t == 0 .
```

It is also worth noting that there are two instances for the rule `[Comm]`. This is a consequence of the fact that after communication, before a substitution takes place, one may need to perform alpha-conversion to avoid name clashes. Rule `[Comm]` is applicable if the variable V is not bound inside process P (modelled by the condition `notin(V , bnP(L'))`), and so only a simple substitution is enough to complete the replacement of the variable X by name V. On the other hand, rule `[Comm']` is applicable if the variable V is bound inside process P (modelled by the condition `in(V , bnP(L'))`; in this case an alpha-conversion is needed to avoid the clash of name V. To be able to perform the alpha-conversion we also check before applying the rule `[Comm']` if a fresh name $[Z]$ exists in the system, name not present in process P' (modelled by the condition `notin(Z , bnP(Q))`).

The conditions of the rules [Comm] and [Comm'] make use of the functions in, notin and bnP for checking the membership of a name to the set of bound names for a given process.

A tick rewriting rule is used to model the passing of time in the encoded system by a positive amount of time that is at most equal with the maximal times that can elapse in the system. Such a tick rule has the form:

```
crl [tick] : {M} => {delta(M, t)} in time t if t <= mte(M) [nonexec] .
```

The [tick] rule uses the function delta to decrease all time constraints in a system by the same positive value. In order to correctly model the steps needed to obtain complete computational steps, namely the time cannot elapse if rewrite rules are applicable, we use the frozen attribute for the function delta. The attribute (1) marks the argument to be frozen (first one in this case).

```
op delta : System TimeInf -> System [frozen (1)] .
eq delta (k[[(c ^ t ! < V >) then (P) else (Q) ]] , t') =
     k[[((c ^ (t monus t') ! < V >) then (P) else (Q)) ]] .
eq delta (k[[(c ^ t ? ( X ) ) then (P) else (Q) ]] , t') =
     k[[((c ^ (t monus t') ? ( X )) then (P) else (Q)) ]] .
eq delta (k[[(go ^ t l) then (P)]] , t') =
     k[[(go ^ (t monus t') l) then (P)]] .
eq delta (k[[P | Q]] , t') = delta (k[[P]] , t')
     || delta (k[[Q]] , t') .
eq delta (M || N , t') = delta(M , t') || delta(N , t') .
eq delta (void , t') = void .
eq delta (l[[stop]] , t') = l[[stop]] .
eq delta (M , t') = M [owise] .
```

The function *mte* from the condition of rule [tick] is used to compute the maximal time that can be elapsed in a system, a time that is equal with the minimum time constraint of the applicable actions in the system.

```
op mte : System -> TimeInf [frozen (1)] .
eq mte (k[[(c ^ t ! < V >) then (P) else (Q) ]] ) = t .
eq mte (k[[(c ^ t ? ( X ) ) then (P) else (Q) ]]) = t .
eq mte (k[[(go ^ t l) then (P)]]) = t .
eq mte (k[[P | Q]] ) = min(mte (k[[P]]) , mte (k[[Q]] )) .
eq mte (M || N) = min(mte(M) , mte(N)) .
eq mte (void) = INF .
eq mte (l[[stop]]) = INF .
eq mte (M) = INF [owise] .
```

The full description of the translation into real-time Maude is available at https://profs.info.uaic.ro/~bogdan.aman/RTMaude/TiMoSpec.rtmaude .

In order to study the correspondence between the operational semantics of our high-level specification language and that of the real-time Maude, we inductively define a mapping ψ : TIMO $\rightarrow$ System as

$$\psi(M) = \begin{cases} l[[\varphi(P)]] & \text{if } M = l[[P]] \\ \psi(N_1)||\psi(N_2) & \text{if } M = N_1||N_2\,; \\ \texttt{void} & \text{if } M = \mathbf{0} \end{cases}$$

$$\varphi(P) = \begin{cases} a^{\Delta t}!\langle v\rangle \text{ then } \varphi(R) \text{ else } \varphi(Q) & \text{if } P = a^{\Delta t}!\langle v\rangle \text{ then } R \text{ else } Q \\ a^{\Delta t}?(X) \text{ then } \varphi(R) \text{ else } \varphi(Q) & \text{if } P = a^{\Delta t}?(X) \text{ then } R \text{ else } Q \\ (go^t\ l)\ .\ \varphi(R) & \text{if } P = go^t l \text{ then } R \\ \texttt{stop} & \text{if } P = 0 \\ \varphi(Q)\ \mid\ \varphi(R) & \text{if } P = Q \mid R \\ \varphi(R)\{v/u\} & \text{if } P = R\{v/u\} \text{ and } v \notin bn(R) \\ \varphi(R)(Z/v)\{v/u\} & \text{if } P = R\{v/u\} \text{ and } v \in bn(R) \\ & \quad \text{and } Z \notin bn(R). \end{cases}$$

By $\mathcal{R}_D$ we denote the rewrite theory defined previously in this section by the rewrite rules [Comm], [Comm'], [Input0], [Output0], [Move] and [tick], and also by the additional operators and equations appearing in these rewrite rules.

The next result relates the structural congruence of the high-level specification language with the equational equality of the rewrite theory.

Lemma 1. *$M \equiv N$ if and only if $\mathcal{R}_D \vdash \psi(M) = \psi(N)$.*

Proof. $\Rightarrow$: By induction on the congruence rules of our high-level language.
$\Leftarrow$: By induction on the equations of the rewrite theory $\mathcal{R}_D$.

The next result emphasizes the operational correspondence between the high-level systems M, N and their translations into a rewriting theory. We denote by $M \rightarrow N$ any rule of Table 3.

Theorem 3. *$M \rightarrow N$ if and only if $\mathcal{R}_D \vdash \psi(M) \Rightarrow \psi(N)$.*

Proof. $\Rightarrow$: By induction on the derivation $M \rightarrow N$.

- (Com): We have $M = l[[a^{\Delta t}!\langle v\rangle \text{ then } P \text{ else } Q]] \mid\mid l[[a^{\Delta t'}?(u) \text{ then } P' \text{ else } Q']]$ and $N = l[[P]] \mid\mid l[[P'\{v/u\}]]$. By definition of ψ, we obtain $\psi(M) = l[[a^{\Delta t}!\langle v\rangle$ then $\varphi(P)$ else $\varphi(Q)]] \mid\mid l[[a^{\Delta t'}?(u) \text{ then } \varphi(P') \text{ else } \varphi(Q')]]$. Depending on the fact that v appears or not as a bound name in P', we have two cases:
  - if $v \notin bn(P')$: By applying [Comm], we have $\mathcal{R}_D \vdash \psi(M) \Rightarrow l[[\varphi(P)]] \mid\mid l[[\varphi(P')\{v/u\}]] = N'$, and by the definition of ψ, we have $\psi(N) = N'$.
  - if $v \in bn(P')$: We should apply first an alpha-conversion before the value is communicated. This is done by using a fresh name $[Z]$ such that by applying the rule [Comm'] we get $\mathcal{R}_D \vdash [Z]||\psi(M) \Rightarrow [[v]]||l[[\varphi(P)]] \mid\mid l[[\varphi(P')(Z/v)\{v/u\}]] = N'$. By the definition of ψ, we have $\psi(N) = N'$.

- (Move0), (Put0) and (Get0): These cases are similar to the previous one, by using the rules [Move], [Input0] and [Output0], respectively.
- (DMove): We have that $M = l[[go^t l' \text{ then } P]]$ and $N = l[[go^{t-t'} l' \text{ then } P]]$. By definition of ψ, we obtain $\psi(M) = l[[(go^t\ l')\ .\ \varphi(P)]]$. By applying the rule [tick] we get $\mathcal{R}_D \vdash \psi(M) \Rightarrow l[[(go^{t-t'}\ l')\ .\ \varphi(P)]] = N'$. By definition of ψ, we have $\psi(N) = N'$.
- (DStop), (DPut) and (DGet): These cases are similar to the previous one, by using also the rule [tick].
- The rest of the rules are simulated using the implicit constructors of Maude.

$\Leftarrow$: By induction on the derivation $\mathcal{R}_D \vdash \psi(M) \Rightarrow \psi(N)$.

- [Comm]: We have $\psi(M) = l[[a^{\Delta t}!\langle v\rangle \text{ then } P \text{ else } Q]] \,||\, l[[a^{\Delta t'}?(u) \text{ then } P' \text{ else } Q']]$ and $\psi(N) = l[[P]] \,||\, l[[P'\{v/u\}]]$. According to the definition of ψ, we get $M = l[[a^{\Delta t}!\langle v\rangle \text{ then } P_1 \text{ else } Q_1]] \,||\, l[[a^{\Delta t'}?(u) \text{ then } P'_1 \text{ else } Q'_1]]$, where $P = \varphi(P_1)$ and $Q = \varphi(Q_1)$. By applying (Com), we get $M \rightarrow l[[P_1]] \,||\, l[[Q_1\{v/u\}]] = N'$. By definition of ψ, we have $N = N'$.
- The other rules are treated in a similar manner.

4 Analyzing Timed Mobile Agents by Using Maude Tools

We have the translation of the high-level specification of the multi-agent systems into real-time Maude rewriting system, and have also the operational correspondence between their semantics. The *TravelShop* system presented in Example 1 can now be described in real-time Maude. The entire system looks like this:

```
eq TravelShop = home[[client(130 , 1)]]
        || travelshop[[agent(100 , 1) | update(60 , 1)]]
        || standard[[flightinfo(110 , special , 1)]]
        || special[[saleinfo(90 , bank , 1)]]
        || bank[[till(10 , 1)]] .
```

where, e.g., the *client* syntax in real-time Maude is:

```
ceq client(init , applyC)=
 ((go ^ 5 travelshop)
  then ((flight ^ INF ? ( standardoffer ))
    then ((go ^ 4 standardoffer)
      then ((finfo2a ^ INF ? ( p1 ))
        then ((finfo2b ^ INF ? ( specialoffer ))
          then ((go ^ 3 specialoffer)
            then ((sinfo2a ^ INF ? ( p2 ))
              then ((sinfo2b ^ INF ? ( paying ))
                then ((go ^ 6 paying)
                  then ((payc ^ INF ! < min(p1 , p2) >)
                    then ((go ^ 4 home)
                      then (client(sd(init,min(p1,p2)),applyC monus 1)))
                    else (stop) ) )
```

```
              else (stop) )
            else (stop) ) )
          else (stop) )
        else (stop) ) )
      else (stop) ) )
    if applyC >= 1 .
```

Since the recursion operator cannot be directly encoded into real-time Maude, we include for each recursion process appearing in *TravelShop* system a second parameter saying how many times the process can be unfolded. For our example this is 1 (but it could be any finite value).

Before doing any verification, we have the possibility in real-time Maude to define the length of the time units performed by the whole system. For our example we choose a time unit of length 1 by using the following command:

```
(set tick def 1 .)
```

When using the rewrite command `(frew {TravelShop} in time < 38 .)`, the Maude platform executes `TravelShop` by using the equations and rewrite rules of $\mathcal{R}_D$ as given in the previous section, and outputs the following result:

```
Timed fair rewrite  {TravelShop} in Example with mode default time
    increase 1  in time < 38

Result ClockedSystem :
    {bank[[till(0,0)]]|| home[[client(70,0)]]|| special[[stop]]
    || special[[saleinfo(60,bank,0)]]
    || standard[[flightinfo(110,special,0)]]
    || travelshop[[agent(170,0)]]} in time 37

rewrites: 786514 in 404ms cpu (406ms real) (1946816 rewrites/second)
```

We use the real-time Maude platform to perform timed reachability tests, namely if starting from the initial configuration of a system one can reach a given configurations of the system before a time threshold. The real-time Maude is able to provide answers to such inquires by searching into the state space obtained into the given time framework for the given configuration. As we are interested in searching the appearance of the given configuration within a time-framework, the fact that multiple computational steps can be performed is marked by the use of the `=>*`. Also, the annotation `[n]` bounds the number of performed computational steps to n, thus reducing the possible state space.

```
(tsearch [2] {TravelShop} =>* {bank[[till(0,0)]]|| home[[client(70,0)]]
      || special[[stop]]||special[[saleinfo(60,bank,0)]]
      || standard[[flightinfo(110,special,0)]]
      || travelshop[[agent(170,0)]]} in time < 40 . )
```

The result of performing the above inquiry is:

```
Timed search [2] in Example
    {TravelShop} =>* {bank[[till(0,0)]]|| home[[client(70,0)]]
    || special[[stop]]|| special[[saleinfo(60,bank,0)]]
    || standard[[flightinfo(110,special,0)]]
    || travelshop[[agent(170,0)]]}
in time < 40 and with mode default time increase 1 :

Solution 1
TIME_ELAPSED:Time --> 37

Solution 2
TIME_ELAPSED:Time --> 38

rewrites: 3684 in 24ms cpu (25ms real) (153500 rewrites/second)
```

Instead of searching for the entire reachable system, we can also search only for certain parts of it: for instance, to check when the *client* remains with 70 cash units in a given interval of time. This can be done by using the command:

```
(tsearch {TravelShop} =>* {home[[client(70,0)]] || X:System}
in time-interval between >= 22 and < 40 . )
```

The answer returns that there exists such a situation at time 22.

```
Timed search [1] in Example
{TravelShop} =>* {home[[client(70,0)]]|| X:System}
in time between >= 22 and < 40 and with mode default time increase 1 :

Solution 1
TIME_ELAPSED:Time --> 22 ; X:System --> bank[[(paya ^ 6 ! < 70 >) then
(till(0,0))else(till(70,0))]]|| special[[stop]]|| special[[saleinfo(60,
bank,0)]]|| standard[[flightinfo(110,special,0)]]|| travelshop[[(go ^
3 bank) then ((paya ^ INF ?(profit)) then ((go ^ 12 travelshop) then
(agent(profit + 100,0)))else(stop))]]
```

The real-time Maude tool allows also the following command to find the shortest time to reach a desired configuration:

```
(find earliest {TravelShop} =>* {home[[client(70,0)]] || X:System} . )
```

It returns the same solution as the previous one, and tells that it was reached `in time 22`.
If time is not relevant for such a search, we can use the *untimed* search command:

```
(utsearch [1] {TravelShop} =>* {home[[client(70,0)]] || X:System} . )
```

5 Conclusion and Related Work

In the current paper we translated our high-level specifications of the multi-agent systems with timeouts for migration and communication into an existing rewriting engine able to execute and analyze timed systems. This translation satisfies an operational correspondence result. Thus, such a translation is suitable for analyzing complex multi-agent systems with timeouts in order to be sure that they have the expected behaviours and properties. We analyze the multi-agent systems with timeouts by using the real-time Maude software platform. The approach is illustrated by an example.

The used high-level specification is given in a language forthcoming realistic programming systems for multi-agent systems, a language with explicit locations and timeouts for migration and communication. It is essentially a simplified version of the timed distributed π-calculus [7]. It can be viewed as a prototyping language of the TiMo family for multi-agent systems in which the agents can migrate between explicit locations in order to perform local communications with other agents. The initial version of TiMo presented in [5] lead to some extensions; e.g., with access permissions in perTiMo [6], with real-time in rTiMo [2]. In [4] it was presented a Java-based software in which the agents are able to perform timed migration just like in TiMo. Using the model checker Process Analysis Toolkit (PAT), the tool TiMo@PAT [8] was created to verify timed systems. In [16], the authors consider an UTP semantics for rTiMo in order to provide a different understanding of this formalism. Maude is used in [17] to define a rewrite theory for the BigTiMo calculus, a calculus for structure-aware mobile systems combining TiMo and the bigraphs [12]. However, the authors of [17] do not tackle the fresh names and recursion problems presented in our current approach.

References

1. Alur, R., Dill, D.L.: A theory of timed automata. Theoret. Comput. Sci. **126**, 183–235 (1994)
2. Aman, B., Ciobanu, G.: Real-time migration properties of rTiMo verified in Uppaal. In: Hierons, R.M., Merayo, M.G., Bravetti, M. (eds.) SEFM 2013. LNCS, vol. 8137, pp. 31–45. Springer, Heidelberg (2013). https://doi.org/10.1007/978-3-642-40561-7_3
3. Behrmann, G., David, A., Larsen, K.G.: A tutorial on Uppaal. In: Bernardo, M., Corradini, F. (eds.) SFM-RT 2004. LNCS, vol. 3185, pp. 200–236. Springer, Heidelberg (2004). https://doi.org/10.1007/978-3-540-30080-9_7
4. Ciobanu, G., Juravle, C.: Flexible software architecture and language for mobile agents. Concurrency Comput. Pract. Experience **24**, 559–571 (2012)
5. Ciobanu, G., Koutny, M.: Timed mobility in process algebra and Petri nets. J. Logic Algebraic Program. **80**, 377–391 (2011)
6. Ciobanu, G., Koutny, M.: Timed migration and interaction with access permissions. In: Butler, M., Schulte, W. (eds.) FM 2011. LNCS, vol. 6664, pp. 293–307. Springer, Heidelberg (2011). https://doi.org/10.1007/978-3-642-21437-0_23

7. Ciobanu, G., Prisacariu, C.: Timers for distributed systems. Electron. Not. Theor. Comput. Sci. **164**, 81–99 (2006)
8. Ciobanu, G., Zheng, M.: Automatic analysis of TiMo systems in PAT. In: IEEE Computer Society Proceedings 18th Engineering of Complex Computer Systems (ICECCS), pp. 121–124 (2013)
9. Goguen, J.A., Meseguer, J.: Order-sorted algebra I: equational deduction for multiple inheritance, overloading, exceptions and partial operations. Theoret. Comput. Sci. **105**, 217–273 (1992)
10. Hessel, A., Larsen, K.G., Mikucionis, M., Nielsen, B., Pettersson, P., Skou, A.: testing real-time systems using UPPAAL. In: Hierons, R.M., Bowen, J.P., Harman, M. (eds.) Formal Methods and Testing. LNCS, vol. 4949, pp. 77–117. Springer, Heidelberg (2008). https://doi.org/10.1007/978-3-540-78917-8_3
11. Meseguer, J.: Twenty years of rewriting logic. In: Ölveczky, P.C. (ed.) WRLA 2010. LNCS, vol. 6381, pp. 15–17. Springer, Heidelberg (2010). https://doi.org/10.1007/978-3-642-16310-4_2
12. Milner, R.: The Space and Motion of Communicating Agents. Cambridge University Press, Cambridge (2009)
13. Ölveczky, P.C.: Real-time Maude and its applications. In: Escobar, S. (ed.) WRLA 2014. LNCS, vol. 8663, pp. 42–79. Springer, Cham (2014). https://doi.org/10.1007/978-3-319-12904-4_3
14. Ölveczky, P.C., Meseguer, J.: Semantics and pragmatics of real-time Maude. Higher-Order Symbolic Comput. **20**, 161–196 (2007)
15. Thati, P., Sen, K., Martí-Oliet, N.: An executable specification of asynchronous π-calculus semantics and may testing in Maude 2.0. Electron. Not. Theor. Comput. Sci. **71**, 261–281 (2004)
16. Xie, W., Xiang, S.: UTP semantics for rTiMo. In: Bowen, J.P., Zhu, H. (eds.) UTP 2016. LNCS, vol. 10134, pp. 176–196. Springer, Cham (2017). https://doi.org/10.1007/978-3-319-52228-9_9
17. Xie, W., Zhu, H., Zhang, M., Lu, G., Fang, Y.: Formalization and verification of mobile systems calculus using the rewriting engine Maude. In: IEEE 42nd Annual Computer Software and Applications Conference, pp. 213–218 (2018)

LTL to Smaller Self-Loop Alternating Automata and Back

František Blahoudek[1], Juraj Major[2(✉)], and Jan Strejček[2]

[1] University of Mons, Mons, Belgium
xblahoud@fi.muni.cz
[2] Masaryk University, Brno, Czech Republic
{major,strejcek}@fi.muni.cz

Abstract. Self-loop alternating automata (SLAA) with Büchi or co-Büchi acceptance are popular intermediate formalisms in translations of LTL to deterministic or nondeterministic automata. This paper considers SLAA with generic transition-based Emerson-Lei acceptance and presents translations of LTL to these automata and back. Importantly, the translation of LTL to SLAA with generic acceptance produces considerably smaller automata than previous translations of LTL to Büchi or co-Büchi SLAA. Our translation is already implemented in the tool LTL3TELA, where it helps to produce small deterministic or nondeterministic automata for given LTL formulae.

1 Introduction

Translation of *linear temporal logic (LTL)* [23] into equivalent automata over infinite words is an important part of many methods for model checking, controller synthesis, monitoring, etc. This paper presents improved translations of LTL to *self-loop alternating automata (SLAA)* [27], which are alternating automata that contain no cycles except self-loops. The SLAA class is studied for more than 20 years under several different names including *very weak* [13,25], *linear* [17], *linear weak* [14], or *1-weak* [22] *alternating automata.* The first publications showing that any LTL formula can be easily translated to an SLAA with only a linear number of states in the length of the formula are even older [19,28]. An LTL to SLAA translation forms the first step of many LTL to automata translations. For example, it is used in popular tools LTL2BA [13] and LTL3BA [3] translating LTL to nondeterministic automata, and also in the tool LTL3DRA [2] translating a fragment of LTL to deterministic automata.

A nice survey of various instances of LTL to SLAA translations can be found in Tauriainen's doctoral thesis [27], where he also presents another improved LTL to SLAA translation. To our best knowledge, the only new improvement since publication of the thesis has been presented by Babiak et al. [3]. All the LTL

F. Blahoudek has been supported by the F.R.S.-FNRS grant F.4520.18 (ManySynth). J. Major and J. Strejček have been supported by the Czech Science Foundation grant GA19-24397S.

R. M. Hierons and M. Mosbah (Eds.): ICTAC 2019, LNCS 11884, pp. 152–171, 2019.
https://doi.org/10.1007/978-3-030-32505-3_10

to SLAA translations considered so far produce SLAA with (state-based) Büchi or co-Büchi acceptance. The only exception is the translation by Tauriainen producing SLAA with transition-based co-Büchi acceptance.

In this paper, we follow a general trend of recent research and development in the field of automata over infinite words and their applications: consider a more general acceptance condition to construct smaller automata. In theory, this change usually does not decrease the upper bound on the size of constructed automata. Moreover, the complexity of algorithms processing automata with a more involved acceptance condition can be even higher. However, practical experiences show that achieved reduction of automata size often outweighs complications with a more general acceptance condition. This can be documented by observations of nondeterministic as well as deterministic automata.

Nondeterministic automata are traditionally considered with Büchi acceptance. However, all three most popular LTL to nondeterministic automata translators, namely LTL2BA [13], LTL3BA [3], and Spot [9], translate LTL formulae to *transition-based generalized Büchi automata (TGBA)*, which are further transformed to Büchi automata. When solving emptiness check, which is the central part of many model checking tools, algorithms designed for TGBA perform better than algorithms analyzing the corresponding Büchi automata [7,24].

Deterministic automata were typically considered with Rabin or Streett acceptance. Tools of the Rabinizer family [16] and the tool LTL3DRA [2] produce also deterministic automata with transition-based generalized Rabin acceptance. The equivalent Rabin automata are often dramatically larger. Direct processing of generalized Rabin automata can be substantially more efficient as shown by Chatterjee et al. [6] for probabilistic model checking and LTL synthesis.

All the previously mentioned acceptance conditions can be expressed by a generic acceptance condition originally introduced by Emerson and Lei [11] and recently reinvented in the *Hanoi omega-automata (HOA) format* [1]. Emerson-Lei acceptance condition is any positive boolean formula over terms of the form Inf● and Fin●, where ● is an *acceptance mark*. A run of a nondeterministic automaton (or an infinite branch of a run of an alternating automaton) satisfies Inf● or Fin● if it visits the acceptance mark ● infinitely often or finitely often, respectively. The acceptance marks placed on states denote traditional state-based acceptance, while marks placed on transitions correspond to transition-based acceptance.

Some tools that work with *transition-based Emerson-Lei automata (TELA)* already exist. For example, Delag [20] produces deterministic TELA and Spot is now able to produce both deterministic and nondeterministic TELA. The produced TELA are often smaller than the automata produced by the tools mentioned in the previous paragraphs. The development version of Spot provides also an emptiness check for TELA, and a probabilistic model checking algorithm working with deterministic Emerson-Lei automata has been implemented in PRISM. In both cases, an improved performance over previous solutions has been reported [4].

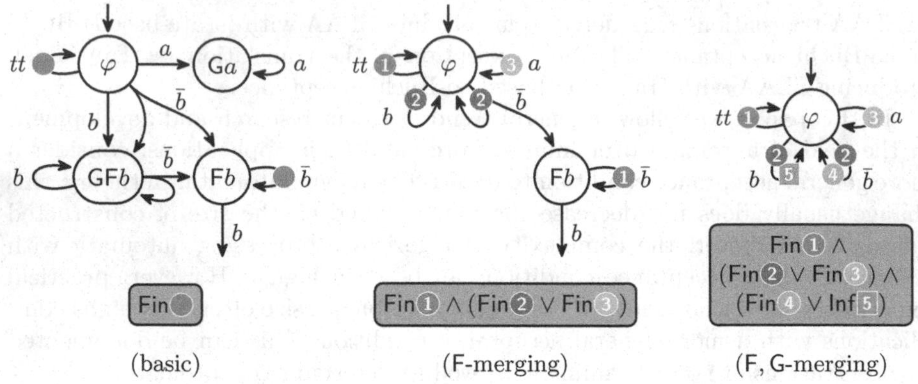

Fig. 1. Automata for the formula $\varphi = \mathsf{F}(\mathsf{G}a \vee \mathsf{GF}b)$: the co-Büchi SLAA produced by the basic translation (left), the Inf-less SLAA produced by F-merging (middle), and the SLAA produced by F, G-merging (right). Graphical notation is explained in Sect. 2.

This paper presents a translation of LTL to SLAA with transition-based Emerson-Lei acceptance. The translation aims to take advantage of the generic acceptance and produce SLAA with less states. We present it in three steps.

1. Section 3 recalls a basic translation producing co-Büchi SLAA. The description uses the same terminology and notation as the following modified translations. In particular, the acceptance marks are on transitions.
2. In Sect. 4, we modify the translation such that states for subformulae of the form $\mathsf{F}\psi$ are merged with states for ψ. The technique is called F-*merging*. The acceptance condition of constructed SLAA is a positive boolean combination of Fin-terms. We call such automata Inf-*less SLAA*.
3. Finally, we further modify the translation in Sect. 5, where states for some subformulae of the form $\mathsf{G}\psi$ are merged with states for ψ. The resulting technique is thus called F, G-*merging*. Constructed SLAA use acceptance condition containing both Inf- and Fin-terms.

The difference between these translations is illustrated by Fig. 1 showing three SLAA for the formula $\mathsf{F}(\mathsf{G}a \vee \mathsf{GF}b)$. One can observe that the initial state of the middle automaton is merged with the states for $\mathsf{G}a$ and $\mathsf{GF}b$ due to F-merging. In the automaton on the right, the state for $\mathsf{GF}b$ is merged with $\mathsf{F}b$ and the initial state is then merged with $\mathsf{G}a$ and $\mathsf{GF}b$. Hence, the resulting automaton contains only one state and the LTL to SLAA translation in this case produces directly a nondeterministic automaton.

LTL to SLAA translations are traditionally accompanied by automata simplification based on transition dominance [13]. Section 6 extends this idea to SLAA with generic acceptance.

Section 7 completes the theoretical part of the paper with a backward translation which takes an SLAA with transition-based Emerson-Lei acceptance and produces an equivalent LTL formula. Altogether, we get that SLAA with the generic acceptance have the same expressiveness as LTL.

The three presented LTL to SLAA translations have been implemented and Sect. 8 provides their experimental comparison extended with the results of the LTL to SLAA translation implemented in LTL3BA. On randomly generated formulae containing only temporal operators F and G, which are favourable to our translation improvements, the F, G-merging can save over 45% of states. This is a considerable reduction, especially with respect to the fact that even the simplest LTL to SLAA translations produce automata of linear size and thus the space for reduction is not big.

As we said at the beginning, SLAA are mainly used as an intermediate formalism in translations of LTL to other kinds of automata. We have already developed a *dealternation* algorithm transforming the produced SLAA to nondeterministic TELA. Both F, G-merging translation and dealternation are implemented in LTL3TELA 2.1 [18], which combines them with some heuristics and many functions of Spot in order to translate LTL to small deterministic and nondeterministic automata. Experiments[1] show that the tool produces, on average, smaller automata than established state-of-the-art translators.

2 Preliminaries

This section recalls the notion of *linear temporal logic* [23] and the definition of *self-loop alternating automata* [27]. We always use automata with transition-based acceptance condition given in the format of Emerson-Lei acceptance. For example, co-Büchi automaton is an automaton with acceptance condition Fin●.

2.1 Linear Temporal Logic (LTL)

We define the syntax of LTL formulae directly in the positive normal form as

$$\varphi ::= tt \mid ff \mid a \mid \neg a \mid \varphi \vee \varphi \mid \varphi \wedge \varphi \mid \mathsf{X}\varphi \mid \varphi \,\mathsf{U}\, \varphi \mid \varphi \,\mathsf{R}\, \varphi,$$

where *tt* stands for *true*, *ff* for *false*, *a* ranges over a set *AP* of *atomic propositions*, and X, U, R are temporal operators called *next*, *until*, and *release*, respectively. A *word* is an infinite sequence $u = u_0 u_1 u_2 \ldots \in \Sigma^\omega$, where $\Sigma = 2^{AP}$. By $u_{i..}$ we denote the suffix $u_{i..} = u_i u_{i+1} \ldots$. We define when a word u *satisfies* φ, written $u \models \varphi$, as follows:

$$\begin{array}{lll}
u \models tt & & \\
u \models a & \text{iff} & a \in u_0 \\
u \models \neg a & \text{iff} & a \notin u_0 \\
u \models \varphi_1 \vee \varphi_2 & \text{iff} & u \models \varphi_1 \text{ or } u \models \varphi_2 \\
u \models \varphi_1 \wedge \varphi_2 & \text{iff} & u \models \varphi_1 \text{ and } u \models \varphi_2 \\
u \models \mathsf{X}\varphi & \text{iff} & u_{1..} \models \varphi \\
u \models \varphi_1 \,\mathsf{U}\, \varphi_2 & \text{iff} & \exists i \geq 0 \text{ such that } u_{i..} \models \varphi_2 \text{ and } \forall 0 \leq j < i \,.\, u_{j..} \models \varphi_1 \\
u \models \varphi_1 \,\mathsf{R}\, \varphi_2 & \text{iff} & \exists i \geq 0 \text{ such that } u_{i..} \models \varphi_1 \text{ and } \forall 0 \leq j \leq i \,.\, u_{j..} \models \varphi_2, \\
& & \text{or } \forall i \geq 0 \,.\, u_{i..} \models \varphi_2
\end{array}$$

[1] https://github.com/jurajmajor/ltl3tela/blob/master/ATVA19.md.

A formula φ defines the language $L(\varphi) = \{u \in (2^{AP(\varphi)})^\omega \mid u \models \varphi\}$, where $AP(\varphi)$ denotes the set of atomic propositions occurring in φ. Further, we use derived operators *eventually* (F) and *always* (G) defined by $\mathsf{F}\varphi \equiv tt \,\mathsf{U}\, \varphi$ and $\mathsf{G}\varphi \equiv ff \,\mathsf{R}\, \varphi$. A *temporal formula* is a formula where the topmost operator is neither conjunction, nor disjunction. A formula without any temporal operator is called *state formula.* Formulae $tt, ff, a, \neg a$ are both temporal and state formulae.

2.2 Self-Loop Alternating Automata (SLAA)

An *alternating automaton* is a tuple $\mathcal{A} = (S, \Sigma, \mathcal{M}, \Delta, s_I, \Phi)$, where

- S is a finite set of *states*,
- Σ is a finite *alphabet*,
- $\mathcal{M}$ is a finite set of *acceptance marks*,
- $\Delta \subseteq S \times \Sigma \times 2^{\mathcal{M}} \times 2^S$ is an *alternating transition relation*,
- $s_I \in S$ is the *initial state*, and
- Φ is an *acceptance formula*, which is a positive boolean combination of terms Fin● and Inf●, where ● ranges over $\mathcal{M}$.

An alternating automaton is a *self-loop alternating automaton (SLAA)* if there exists a partial order on S such that for every $(s, \alpha, M, C) \in \Delta$, all states in C are lower or equal to s. In other words, SLAA contain no cycles except self-loops.

Subsets $C \subseteq S$ are called *configurations.* A quadruple $t = (s, \alpha, M, C) \in \Delta$ is called a *transition* from s to C under α (or labelled by α or α-transition) marked by elements of M. A transition $t = (s, \alpha, M, C) \in \Delta$ is *looping* (or simply a *loop*) if its *destination configuration* C contains its *source* s.

A *multitransition* T under α is a set of transitions under α such that the source states of the transitions are pairwise different. The *source configuration* $\mathrm{dom}(T)$ of a multitransition T is the set of source states of transitions in T. The *destination configuration* $\mathrm{range}(T)$ is the union of destination configurations of transitions in T. For an alternating automaton $\mathcal{A}$, $\Gamma^{\mathcal{A}}$ denotes the set of all multitransitions of $\mathcal{A}$ and $\Gamma^{\mathcal{A}}_\alpha$ denotes the set of all multitransitions of $\mathcal{A}$ under α.

A *run* ρ of an alternating automaton $\mathcal{A}$ over a word $u = u_0u_1 \ldots \in \Sigma^\omega$ is an infinite sequence $\rho = T_0T_1\ldots$ of multitransitions such that $\mathrm{dom}(T_0) = \{s_I\}$ and, for all $i \geq 0$, T_i is labelled by u_i and $\mathrm{range}(T_i) = \mathrm{dom}(T_{i+1})$. Each run ρ defines a directed acyclic edge-labelled graph $G_\rho = (V, E, \lambda)$, where

$$V = \bigcup_{i=0}^{\infty} V_i, \text{ where } V_i = \mathrm{dom}(T_i) \times \{i\},$$

$$E = \bigcup_{i=0}^{\infty} \left\{((s,i),(s',i+1)) \mid (s,\alpha,M,C) \in T_i, s' \in C\right\}, \text{ and}$$

the labeling function $\lambda : E \to 2^{\mathcal{M}}$ assigns to each edge $e = ((s,i),(s',i+1)) \in E$ the acceptance marks from the corresponding transition, i.e., $\lambda(e) = M$ where $(s, \alpha, M, C) \in T_i$. A *branch* of the run ρ is a maximal (finite or infinite) sequence $b = (v_0, v_1)(v_1, v_2)\ldots$ of consecutive edges in G_ρ such that $v_0 \in V_0$. For an infinite branch b, let $M(b)$ denote the set of marks that appear in infinitely many sets of the sequence $\lambda((v_0, v_1))\lambda((v_1, v_2))\dots$. An infinite branch b satisfies $\mathsf{Inf}\,\bullet$ if $\bullet \in M(b)$ and it satisfies $\mathsf{Fin}\,\bullet$ if $\bullet \notin M(b)$. An infinite branch is *accepting* if it satisfies the acceptance formula Φ. We say that a run ρ is *accepting* if all its infinite branches are accepting. The language of $\mathcal{A}$ is the set $L(\mathcal{A}) = \{u \in \Sigma^\omega \mid \mathcal{A}$ has an accepting run over $u\}$.

Several examples of SLAA are given in Fig. 1. Examples of SLAA with their runs can be found in Fig. 5. Note that a transition $(s, \alpha, M, C) \in \Delta$ of an alternating automaton is visualised as a branching edge leading from s to all states in C. In this paper, an automaton alphabet has always the form $\Sigma = 2^{AP'}$, where AP' is a finite set of atomic propositions. To keep the visual representation of automata concise, edges are labelled with boolean formulae over atomic propositions in a condensed notation: $\bar{a}$ denotes $\neg a$ and conjunctions are omitted. Hence, $ab\bar{c}$ would stand for $a \wedge b \wedge \neg c$. Every edge represents all transitions under combinations of atomic propositions satisfying its label.

3 Basic Translation

This section presents a basic translation of LTL to co-Büchi SLAA similar to the one implemented in LTL3BA [3]. To simplify the presentation, in contrast to the translation of LTL3BA we omit the optimization called *suspension*, we describe transitions for each $\alpha \in \Sigma$ separately, and we slightly modify the acceptance condition of the SLAA; in particular, we switch from state-based to transition-based acceptance.

Let φ be an LTL formula, where subformulae of the form $\mathsf{F}\psi$ and $\mathsf{G}\psi$ are seen only as abbreviations for $tt\ \mathsf{U}\ \varphi$ and $f\!f\ \mathsf{R}\,\varphi$, respectively. An equivalent SLAA is constructed as $\mathcal{A}_\varphi = (S, \Sigma, \{\bullet\}, \Delta, \varphi, \mathsf{Fin}\,\bullet)$, where states in S are subformulae of φ and $\Sigma = 2^{AP(\varphi)}$. The construction of the transition relation Δ treats it equivalently as a function $\Delta : S \times \Sigma \rightarrow 2^{\mathcal{P}}$ where $\mathcal{P} = 2^{\mathcal{M}} \times 2^{S}$. The construction of Δ is defined inductively and it directly corresponds to the semantics of LTL. The acceptance mark $\bullet$ is used to ensure that an accepting run cannot stay in a state $\psi_1\ \mathsf{U}\ \psi_2$ forever. In other words, it ensures that ψ_2 will eventually hold. The translation uses an auxiliary product operator $\otimes$ and a marks eraser me defined for each $P, P' \subseteq \mathcal{P}$ as:

$$P \otimes P' = \{(M \cup M', C \cup C') \mid (M, C) \in P, (M', C') \in P'\}$$
$$\mathsf{me}(P) = \{(\emptyset, C) \mid (M, C) \in P\}$$

The product operator is typically used to handle conjunction: to get successors of $\psi_1 \wedge \psi_2$, we compute the successors of ψ_1 and the successors of ψ_2 and combine them using the product operator $\otimes$. The marks eraser has two applications. First, it is used to remove unwanted acceptance marks on transitions looping on states of the form $\psi_1\ \mathsf{R}\ \psi_2$. Second, it is used to remove irrelevant accepting marks from the automaton, which are all marks not lying on loops. Indeed, only looping transition can appear infinitely often on some branch of an SLAA run and thus only marks on loops are relevant for acceptance.

$$\begin{aligned}
\Delta(tt,\alpha) &= \{(\emptyset,\emptyset)\}\\
\Delta(\mathit{ff},\alpha) &= \emptyset\\
\Delta(a,\alpha) &= \{(\emptyset,\emptyset)\} \text{ if } a\in\alpha,\ \emptyset \text{ otherwise}\\
\Delta(\neg a,\alpha) &= \{(\emptyset,\emptyset)\} \text{ if } a\notin\alpha,\ \emptyset \text{ otherwise}\\
\Delta(\psi_1\wedge\psi_2,\alpha) &= \mathsf{me}\big(\Delta(\psi_1,\alpha)\otimes\Delta(\psi_2,\alpha)\big)\\
\Delta(\psi_1\vee\psi_2,\alpha) &= \mathsf{me}\big(\Delta(\psi_1,\alpha)\cup\Delta(\psi_2,\alpha)\big)\\
\Delta(\mathsf{X}\psi,\alpha) &= \{(\emptyset,\{\psi\})\}\\
\Delta(\psi_1\mathsf{U}\psi_2,\alpha) &= \mathsf{me}\big(\Delta(\psi_2,\alpha)\big)\cup\Big(\big\{(\{\bullet\},\{\psi_1\mathsf{U}\psi_2\})\big\}\otimes\mathsf{me}\big(\Delta(\psi_1,\alpha)\big)\Big)\\
\Delta(\psi_1\mathsf{R}\psi_2,\alpha) &= \mathsf{me}\big(\Delta(\psi_1,\alpha)\otimes\Delta(\psi_2,\alpha)\big)\cup\mathsf{me}\Big(\big\{(\emptyset,\{\psi_1\mathsf{R}\psi_2\})\big\}\otimes\Delta(\psi_2,\alpha)\Big)
\end{aligned}$$

The automaton $\mathcal{A}_\varphi$ has at most $|\varphi|$ states as the states are subformulae of φ. To prove that the constructed automaton is a self-loop alternating automaton, it is enough to consider the partial order *'being a subformula of'* on states.

4 F-Merging Translation

Now we modify the basic translation on subformulae of the form $\mathsf{F}\psi$. The modified translation produces Inf*-less SLAA*, which are SLAA without $\mathsf{Inf}\blacksquare$ terms in acceptance formulae.

Before giving the formal translation, we discuss three examples to explain the ideas behind F-merging. We start with a formula $\mathsf{F}\psi$ where ψ is a temporal formula. Further, assume that the state ψ of the SLAA constructed by the basic translation has two types of transitions: non-looping labelled by α and loops labelled by β. The SLAA $\mathcal{A}$ for $\mathsf{F}\psi$ can be found in Fig. 2 (left). States $\mathsf{F}\psi$ and ψ can be merged into a single state that represents their disjunction (which is equivalent to $\mathsf{F}\psi$) as shown by the SLAA $\mathcal{A}_\mathsf{F}$ of Fig. 2 (right). The construction is still correct: (i) Clearly, each sequence of transitions that can be taken in $\mathcal{A}$ can be also taken in $\mathcal{A}_\mathsf{F}$. (ii) The sequences of transitions of $\mathcal{A}_\mathsf{F}$ that cannot be taken in $\mathcal{A}$ are those where the tt-loop is taken after some β-loop. However, every accepting run of $\mathcal{A}_\mathsf{F}$ use the tt-loop only finitely many times and thus we can find a corresponding accepting run of $\mathcal{A}$: instead of each β-loop that occurs before the last tt-loop we can use the tt-loop since β implies tt.

The second example deals with the formula $\mathsf{F}\psi$ where $\psi = (a\,\mathsf{R}\,b)\wedge\mathsf{G}c$. Figure 3 (left) depicts the SLAA $\mathcal{A}$ produced by the basic translation. The state ψ is dotted as it is unreachable. Hence, merging $\mathsf{F}\psi$ with ψ would not save any state. However, we can modify the translation rules to make ψ reachable and $a\,\mathsf{R}\,b$ unreachable at the same time. The modification is based on the following observation. Taking the red bc-edge in $\mathcal{A}$ would mean that both $a\,\mathsf{R}\,b$ and $\mathsf{G}c$ have to hold in the next step, which is equivalent to $(a\,\mathsf{R}\,b)\wedge\mathsf{G}c$. Thus we can replace the red edge by the red bc-loop as shown in the automaton $\mathcal{A}'$ of Fig. 3 (right).

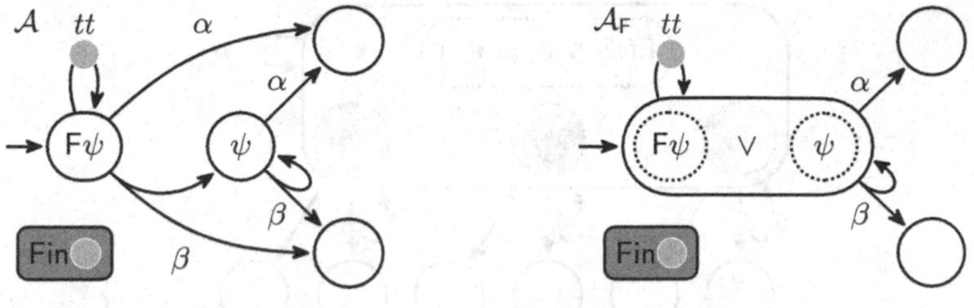

Fig. 2. Automata for $\mathsf{F}\psi$: the SLAA $\mathcal{A}$ built by the basic translation (left) and the SLAA $\mathcal{A}_{\mathsf{F}}$ built by the F-merging translation (right).

Because transitions leaving the state $\mathsf{F}\psi$ are computed from transitions of ψ, this replacement makes the state $(a \mathsf{R} b) \wedge \mathsf{G}c$ reachable and the state $a \mathsf{R} b$ becomes unreachable. The states $\mathsf{F}\psi$ and ψ of $\mathcal{A}'$ can be now merged for the same reason as in Fig. 2.

While the previous paragraph studied a formula $\mathsf{F}\psi$ where ψ is a conjunction, the third example focuses on disjunctions. Let us consider a formula $\mathsf{F}\psi$ where $\psi = \psi_1 \vee \psi_2 \vee \psi_3$ and each ψ_i is a temporal formula. As in the previous example, the state ψ is unreachable in the SLAA produced by the basic translation and thus merging $\mathsf{F}\psi$ with ψ does not make any sense. However, we can merge the state $\mathsf{F}\psi$ with states ψ_1, ψ_2, ψ_3 as indicated in Fig. 4. In contrast to the original SLAA, a single run of the merged SLAA can use a loop corresponding to a state ψ_i and subsequently a transition corresponding to a different ψ_j. Instead of every such a loop, the original SLAA can simply use the tt-loop of $\mathsf{F}\psi$. However, as the tt-loop is marked by ●, we can use it only finitely many times. In fact, the runs of the merged automaton that contain infinitely many loops corresponding to two or more different states ψ_i should be nonaccepting. Therefore we adjust the acceptance formula to Fin● ∧ (Fin❶ ∨ Fin❷ ∨ Fin❸) and place the new acceptance marks as shown in Fig. 4. Clearly, Fin❶ says that transitions of ψ_2 and ψ_3 are

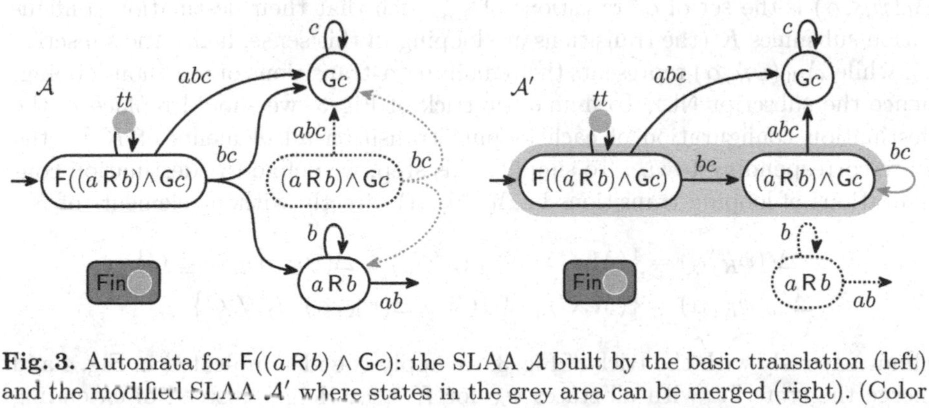

Fig. 3. Automata for $\mathsf{F}((a \mathsf{R} b) \wedge \mathsf{G}c)$: the SLAA $\mathcal{A}$ built by the basic translation (left) and the modified SLAA $\mathcal{A}'$ where states in the grey area can be merged (right). (Color figure online)

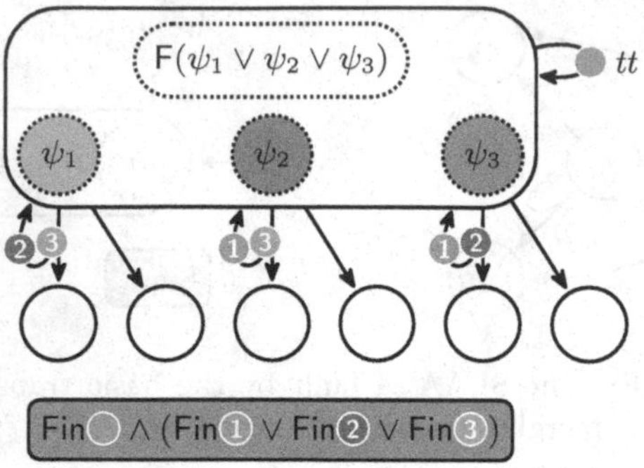

Fig. 4. Transitions of the state $\mathsf{F}(\psi_1 \vee \psi_2 \vee \psi_3)$ merged with states ψ_1, ψ_2, and ψ_3.

taken only finitely many times, Fin❷ does the same for transitions of ψ_1 and ψ_3, and Fin❸ for transitions of ψ_1 and ψ_2.

The F-merging translation combines the ideas presented above when processing any F-subformula, while other subformulae are handled as in the basic translation. Hence, we no longer treat $\mathsf{F}\psi$ as an abbreviation for $tt\,\mathsf{U}\,\psi$. Further, we think about formulae $\mathsf{F}\psi$ as formulae of the form $\mathsf{F} \bigvee_i \bigwedge_j \psi_{i,j}$, where $\psi_{i,j}$ are temporal formulae. Formally, we define formula decomposition into disjunctive normal form $\overline{\psi}$ as follows:

$$\begin{aligned} \overline{\psi} &= \{\{\psi\}\} \text{ if } \psi \text{ is a temporal formula} \\ \overline{\psi_1 \vee \psi_2} &= \overline{\psi_1} \cup \overline{\psi_2} \\ \overline{\psi_1 \wedge \psi_2} &= \{C_1 \cup C_2 \mid C_1 \in \overline{\psi_1} \text{ and } C_2 \in \overline{\psi_2}\}. \end{aligned}$$

Let $K \in \overline{\psi}$ be a set of temporal formulae. We use ψ_K to denote $\psi_K = \bigwedge_{\psi' \in K} \psi'$. Clearly, ψ is equivalent to $\bigvee_{K \in \overline{\psi}} \psi_K$. We define two auxiliary transition functions $\Delta_{\mathsf{L}}, \Delta_{\mathsf{NL}}$ to implement the trick illustrated with red edges in Fig. 3. Intuitively, $\Delta_{\mathsf{L}}(\psi_K, \alpha)$ is the set of α-transitions of ψ_K such that their destination configuration subsumes K (the transitions are looping in this sense, hence the subscript L), while $\Delta_{\mathsf{NL}}(\psi_K, \alpha)$ represents the remaining α-transitions of ψ_K (non-looping, hence the subscript NL). To mimic the trick of Fig. 3, we should replace in the destination configuration of each looping transition all elements of K by the state corresponding to ψ_K. To simplify this step, we define the destination configurations of looping transitions in $\Delta_{\mathsf{L}}(\psi_K, \alpha)$ directly without elements of K.

$$\begin{aligned} \Delta_{\mathsf{L}}(\psi_K, \alpha) &= \big\{(M, C \smallsetminus K) \mid (M, C) \in \Delta(\psi_K, \alpha),\ K \subseteq C\big\} \\ \Delta_{\mathsf{NL}}(\psi_K, \alpha) &= \big\{(M, C) \mid (M, C) \in \Delta(\psi_K, \alpha),\ K \nsubseteq C\big\} \end{aligned}$$

Simultaneously with the trick of Fig. 3, we apply the idea indicated in Fig. 4 and merge the state $\mathsf{F}\psi$ with all states ψ_K for $K \in \overline{\psi}$. Hence, instead of extending the looping transitions with states ψ_K, we extend it with the merged state called simply $\mathsf{F}\psi$. Altogether, we get

$$\begin{aligned}\Delta(\mathsf{F}\psi,\alpha) &= \{(\{\bullet\},\{\mathsf{F}\psi\})\} \cup \\ &\cup \bigcup_{K\in\overline{\psi}} \Big(\mathsf{me}(\Delta_{\mathsf{NL}}(\psi_K,\alpha)) \cup \{(M_K,\{\mathsf{F}\psi\})\} \otimes \Delta_{\mathsf{L}}(\psi_K,\alpha)\Big) \text{ where} \\ M_K &= \{\bullet^{K'} \mid K' \in \overline{\psi} \text{ and } K' \neq K\}.\end{aligned}$$

In other words, the merged state $\mathsf{F}\psi$ has three kinds of transitions: the *tt*-loop marked by $\bullet$, non-looping transitions of states ψ_K for each disjunct $K \in \overline{\psi}$, and the looping transitions of states ψ_K which are marked as shown in Fig. 4. Finally, we redefine the set of acceptance marks $\mathcal{M}$ and the acceptance formula Φ of the constructed SLAA as follows, where $\mathcal{F}_\varphi$ denotes the set of all subformulae of φ of the form $\mathsf{F}\psi$:

$$\mathcal{M} = \{\bullet\} \cup \{\bullet^K \mid \mathsf{F}\psi \in \mathcal{F}_\varphi \text{ and } K \in \overline{\psi}\}$$

$$\Phi = \mathsf{Fin}\bullet \wedge \bigwedge_{\mathsf{F}\psi\in\mathcal{F}_\varphi} \bigvee_{K\in\overline{\psi}} \mathsf{Fin}\bullet^K$$

In fact, we can reduce the number of orange marks (marks with an upper index) produced by F-merging translation. Let $n_\varphi = \max\{|\overline{\psi}| \mid \mathsf{F}\psi \in \mathcal{F}_\varphi\}$ be the maximal number of such marks corresponding to any subformula $\mathsf{F}\psi$ of φ. We can limit the total number of orange marks to n_φ by reusing them. We redefine $\mathcal{M}$ and Φ as

$$\mathcal{M} = \{\bullet\} \cup \{\bullet^i \mid 1 \leq i \leq n_\varphi\} \qquad \text{and} \qquad \Phi = \mathsf{Fin}\bullet \wedge \bigvee_{i=1}^{n_\varphi} \mathsf{Fin}\bullet^i$$

and alter the above definition of M_K. For every $\mathsf{F}\psi \in \mathcal{F}_\varphi$, we assign a unique index i_K, $1 \leq i_K \leq n_\varphi$, to each $K \in \overline{\psi}$. Sets M_K in transitions of $\mathsf{F}\psi$ are then defined to contain all orange marks except $\bullet^{i_K}$, formally $M_K = \mathcal{M} \setminus \{\bullet, \bullet^{i_K}\}$. This optimization is correct as any branch of an SLAA run cannot cycle between two states of the form $\mathsf{F}\psi$.

5 F, G-Merging Translation

We further improve the F-merging translation by adding a special rule also for subformulae of the form $\mathsf{G}\psi$. The resulting F, G-merging translation produces SLAA with an acceptance formula that is not Inf-less.

We start again with a simple example. Consider the formula $\mathsf{GF}a$. The basic translation produces the SLAA $\mathcal{A}$ from Fig. 5 (top). In general, each transition of the state $\mathsf{G}\psi$ is a transition of ψ extended with a loop back to $\mathsf{G}\psi$. The one-to-one correspondence between transitions of $\mathsf{G}\psi$ and ψ leads to the idea to merge the states into one that corresponds to their conjunction $(\mathsf{G}\psi) \wedge \psi$, which is equivalent to $\mathsf{G}\psi$. However, merging these states needs a special care.

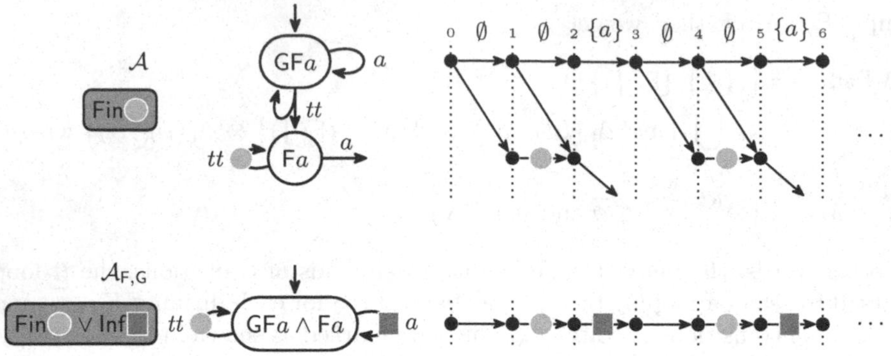

Fig. 5. An SLAA $\mathcal{A}$ for the formula $\mathsf{GF}a$ built by the basic translation (top) and an equivalent SLAA $\mathcal{A}_{\mathsf{F,G}}$ built by the F,G-merging translation (bottom) and their runs over the word $(\emptyset\emptyset\{a\})^\omega$.

Figure 5 (bottom) shows an SLAA where the states $\mathsf{GF}a$ and $\mathsf{F}a$ are merged. Consider now the word $u = (\emptyset\emptyset\{a\})^\omega$ and the runs of the two automata over u. The branches of the top run collapse into a single branch in the bottom run. While each branch of the top run has at most one occurrence of ●, the single branch in the bottom run contains infinitely many of these marks. The SLAA $\mathcal{A}_{\mathsf{F,G}}$ accepts u only because of the added ■ marks. The intuition for their placement is explained using the concept of *escaping multitransitions*.

A multitransition T of an SLAA $\mathcal{A}'$ is *s-escaping* for a state s if it contains a non-looping transition $(s, \alpha, M, C) \in T$. For an acceptance mark ●, we define its *owners* $\mathcal{O}(●) = \{s \in S \mid (s, \alpha, M, C) \in \Delta$ and $● \in M\}$ as the set of all states with outgoing transitions marked by ●. The following observation holds for every mark ● with a single owner s. *A run of $\mathcal{A}'$ satisfies* Fin● *(i.e., all its infinite branches satisfy* Fin●*) if and only if the run contains only a finite number of multitransitions T marked by ● or if it contains infinitely many s-escaping multitransitions.*

Transitions of $\mathcal{A}_{\mathsf{F,G}}$ correspond to multitransitions of $\mathcal{A}$ with source configuration $\{\mathsf{GF}a, \mathsf{F}a\}$. The observation implies that $\mathcal{A}_{\mathsf{F,G}}$ would be equivalent to $\mathcal{A}$ if we mark all transitions corresponding to $\mathsf{F}a$-escaping multitransitions with a new mark ■ and change the acceptance formula to Fin● ∨ Inf■ as shown in Fig. 5 (bottom).

This approach naturally extends to the case of $\mathsf{G}\psi$ where $\psi = \bigwedge_i \psi_i$ is a conjunction of temporal formulae. In this case, $\mathsf{G}\psi$ can be merged with all states ψ_i into a single state representing the conjunction $(\mathsf{G}\psi) \wedge \bigwedge_i \psi_i$. However, we have to pay a special attention to acceptance marks as the observation formulated above holds only for marks with a single owner. For every ψ_i with a ●-marked transition, we need to track ψ_i-escaping multitransitions separately. To implement this, we create a copy of ● for each ψ_i and we use a specific $■_{\psi_i}$ mark to label transitions corresponding to ψ_i-escaping multitransitions.

Unfortunately, due to the duality of the F and G operators and the fact that the transition relation of SLAA is naturally in disjunctive normal form (which is suitable for F), we did not find any way to improve the translation of $\mathsf{G}\psi$ if ψ is a disjunction. On the bright side, we can generalize the merging to $\mathsf{G}\bigwedge_i \psi_i$ where each ψ_i is a temporal or state formula (which can contain disjunctions). This is due to the fact that a state formula ψ_i affects only the labels of transitions with origin in the state $\mathsf{G}\bigwedge_i \psi_i$ and thus it does not create any reachable state or acceptance mark.

Formally, we first modify the translation rules introducing acceptance marks to work with marks of the form $\bullet_\psi$ as discussed above. More precisely, we change the rule for $\psi_1 \mathsf{U} \psi_2$ presented in the basic translation and the rule for $\mathsf{F}\psi$ of the F-merging translation to the following (note that the optimization reusing orange marks make no longer sense as we need to create their copies for each $\mathsf{F}\psi$ anyway).

$$\begin{aligned}\Delta(\psi_1\mathsf{U}\psi_2,\alpha) &= \mathsf{me}\big(\Delta(\psi_2,\alpha)\big)\cup\Big(\big\{(\{\bullet_{\psi_1\mathsf{U}\psi_2}\},\{\psi_1\mathsf{U}\psi_2\})\big\}\otimes\mathsf{me}\big(\Delta(\psi_1,\alpha)\big)\Big)\\ \Delta(\mathsf{F}\psi,\alpha) &= \big\{(\{\bullet_{\mathsf{F}\psi}\},\{\mathsf{F}\psi\})\big\}\cup\\ &\quad\cup\bigcup_{K\in\overline{\psi}}\Big(\mathsf{me}\big(\Delta_{\mathsf{NL}}(\psi_K,\alpha)\big)\cup\big\{(M_K,\{\mathsf{F}\psi\})\big\}\otimes\Delta_{\mathsf{L}}(\psi_K,\alpha)\Big),\\ \text{where } M_K &= \{\bullet^{K'}_{\mathsf{F}\psi}\mid K'\in\overline{\psi}\text{ and }K'\neq K\}\end{aligned}$$

Further, we add a specific rule for formulae $\mathsf{G}\psi$ where $\psi = \bigwedge_{\psi'\in K}\psi'$ for some set K of temporal and state formulae. Formulae $\mathsf{G}\psi$ of other forms are handled as $f\!f\,\mathsf{R}\,\psi$. On the top level, the rule simply defines transitions of $\mathsf{G}\psi$ as transitions of $\mathsf{G}\psi \;\wedge \bigwedge_{\psi'\in K}\psi'$:

$$\Delta(\mathsf{G}\psi,\alpha) = \big\{(\emptyset,\{\mathsf{G}\psi\})\big\}\otimes\bigotimes_{\psi'\in K}\Delta'(\psi',\alpha)$$

The definition of $\Delta'(\psi',\alpha)$ differs from $\Delta(\psi',\alpha)$ in two aspects. First, it removes ψ' from destination configurations because ψ' is merged with $\mathsf{G}\psi$ and the state $\mathsf{G}\psi$ is added to each destination configuration by the product on the top level. Second, it identifies all non-looping transitions of ψ' and marks them with $\blacksquare_{\psi'}$ as ψ'-escaping. We distinguish between looping and non-looping transitions only when ψ' has the form $\psi_1 \mathsf{U} \psi_2$ or $\mathsf{F}\psi_1$. All other ψ' have only looping transitions (e.g., G-formulae) or no marked transitions (e.g., state formulae or R- or X-formulae) and thus there are no ψ'-escaping transitions or we do not need to watch them. Similarly to $\mathcal{F}_\varphi$, we use $\mathcal{U}_\varphi$ for the set of all subformulae of φ of the form $\psi_1 \mathsf{U} \psi_2$. The function $\Delta'(\psi', \alpha)$ is defined as follows:

$$\Delta'(\psi', \alpha) = \begin{cases} \Delta'_{\mathsf{L}}(\psi', \alpha) \cup \Delta'_{\mathsf{NL}}(\psi', \alpha) & \text{if } \psi' \in \mathcal{F}_\varphi \text{ or } \psi' \in \mathcal{U}_\varphi \\ \{(M, C \smallsetminus \{\psi'\}) \mid (M, C) \in \Delta(\psi', \alpha)\} & \text{otherwise} \end{cases}$$

$$\Delta'_{\mathsf{L}}(\psi', \alpha) = \{(M, C \smallsetminus \{\psi'\}) \mid (M, C) \in \Delta(\psi', \alpha), \psi' \in C\}$$

$$\Delta'_{\mathsf{NL}}(\psi', \alpha) = \{(\{■_{\psi'}\}, C) \mid (M, C) \in \Delta(\psi', \alpha), \psi' \notin C\}$$

Finally, we redefine the set of marks $\mathcal{M}$ and the acceptance formula Φ. Now each subformula from $\mathcal{U}_\varphi$ and $\mathcal{F}_\varphi$ has its own set of marks, and the ■ marks are used to implement the intuition given using the Fig. 5.

$$\mathcal{M} = \{●_\psi, ■_\psi \mid \psi \in \mathcal{U}_\varphi\} \cup \left\{●_{\mathsf{F}\psi}, ■_{\mathsf{F}\psi}, ●^K_{\mathsf{F}\psi} \mid \mathsf{F}\psi \in \mathcal{F}_\varphi \text{ and } K \in \overline{\psi}\right\}$$

$$\Phi = \bigwedge_{\psi \in \mathcal{U}_\varphi} (\mathsf{Fin}●_\psi \vee \mathsf{Inf}■_\psi) \wedge \bigwedge_{\mathsf{F}\psi \in \mathcal{F}_\varphi} \left(\left(\mathsf{Fin}●_{\mathsf{F}\psi} \wedge \bigvee_{K \in \overline{\psi}} \mathsf{Fin}●^K_{\mathsf{F}\psi} \right) \vee \mathsf{Inf}■_{\mathsf{F}\psi} \right)$$

Theorem 1. *Let φ be an LTL formula and let $\mathcal{A}_\varphi$ be the corresponding SLAA built by the* F, G*-merging translation. Then $L(\mathcal{A}_\varphi) = L(\varphi)$. Moreover, the number of states of $\mathcal{A}_\varphi$ is linear to the size of φ and the number of acceptance marks is at most exponential to the size of φ.*

The proof can be found in the full version of this paper [5].

6 SLAA Simplification

Simplification by transition dominance is a basic technique improving various automata constructions [13]. The idea is that an automata construction does not add any transition that is dominated by some transition already present in the automaton. In SLAA, a transition (q, α, M_1, C_1) *dominates* a transition (q, α, M_2, C_2) if $C_1 \subseteq C_2$ and M_1 is "at least as helpful and at most as harmful for acceptance" as M_2. In the rest of this section, we focus on the precise formulation of the last condition.

In the classic case of co-Büchi SLAA with acceptance formula Fin●, the condition translates into $● \in M_1 \implies ● \in M_2$. For Büchi SLAA with acceptance formula Inf■, the condition has the form $■ \in M_2 \implies ■ \in M_1$.

Now consider an SLAA with an arbitrary acceptance formula Φ. Let $\mathsf{Fin}(\Phi)$ and $\mathsf{Inf}(\Phi)$ be the sets of all acceptance marks appearing in Φ in subformulae of the form Fin● and Inf■, respectively. A straightforward formulation of the condition is $M_1 \cap \mathsf{Fin}(\Phi) \subseteq M_2$ and $M_2 \cap \mathsf{Inf}(\Phi) \subseteq M_1$. This formulation is correct, but we can do better. For example, consider the case $\Phi = \mathsf{Fin}❶ \wedge (\mathsf{Fin}❷ \vee \mathsf{Inf}\boxed{3})$ and transitions $t_1 = (q, \alpha, \{❶, ❷\}, \{p\})$ and $t_2 = (q, \alpha, \{❶\}, \{p, p'\})$. Then t_1 does not dominate t_2 according to this formulation as $\{❶, ❷\} \cap \mathsf{Fin}(\Phi) = \{❶, ❷\} \not\subseteq \{❶\}$. However, any branch of an accepting run cannot take neither t_1 nor t_2 infinitely often and thus t_1 can be seen as dominating in this case.

To formalize this observation, we introduce *transition dominance with respect to acceptance formula*.

A *minimal model* O of an acceptance formula Φ is a subset of its terms satisfying Φ and such that no proper subset of O is a model of Φ. For example, the formula $\Phi = \mathsf{Fin}❶ \wedge (\mathsf{Fin}❷ \vee \mathsf{Inf}❸)$ has two minimal models: $\{\mathsf{Fin}❶, \mathsf{Fin}❷\}$ and $\{\mathsf{Fin}❶, \mathsf{Inf}❸\}$. For each minimal model O, by $\mathsf{Fin}(O)$ and $\mathsf{Inf}(O)$ we denote the sets of all acceptance marks appearing in O in terms of the form $\mathsf{Fin}●$ and $\mathsf{Inf}■$, respectively. We say that a transition (q, α, M_1, C_1) *dominates* a transition (q, α, M_2, C_2) *with respect to* Φ if $C_1 \subseteq C_2$ and for each minimal model O of Φ it holds $\mathsf{Fin}(O) \cap M_2 = \emptyset \implies \mathsf{Fin}(O) \cap M_1 = \emptyset$ and $\mathsf{Inf}(O) \cap M_1 = \emptyset \implies \mathsf{Inf}(O) \cap M_2 = \emptyset$. In other words, if t_2 can be used infinitely often without breaking some $\mathsf{Fin}●$ of a minimal model then t_1 can be used as well, and if an infinite use of t_1 does not satisfy any term $\mathsf{Inf}■$ of a minimal model then an infinite use of t_2 does not as well.

Note that our implementation of LTL to SLAA translation used later in experiments employs this simplification.

7 Translation of SLAA to LTL

This section presents a translation of an SLAA to an equivalent LTL formula. Let $\mathcal{A} = (S, \Sigma, \mathcal{M}, \Delta, s_I, \Phi)$ be an SLAA. We assume that the alphabet has the form $\Sigma = 2^{AP'}$ for some finite set of atomic propositions AP'. For each $\alpha \in \Sigma$, φ_α denotes the formula

$$\varphi_\alpha = \bigwedge_{a \in \alpha} a \quad \wedge \bigwedge_{a \in AP' \smallsetminus \{\alpha\}} \neg a.$$

For each state $s \in S$ we construct an LTL formula $\varphi(s)$ such that $u \in \Sigma^\omega$ satisfies $\varphi(s)$ if and only if u is accepted by $\mathcal{A}$ with its initial state replaced by s. Hence, $\mathcal{A}$ is equivalent to the formula $\varphi(s_I)$.

The formula construction is inductive. Let s be a state such that we have already constructed LTL formulae for all its successors. Further, let $Mod(\Phi)$ denote the set of all minimal models of Φ. Further, given a set of states C, by $\varphi(C)$ we denote the conjunction $\varphi(C) = \bigwedge_{s \in C} \varphi(s)$. We define $\varphi(s)$ as follows:

$$\varphi(s) = \varphi_1(s) \vee (\varphi_2(s) \wedge \varphi_3(s))$$

$$\varphi_1(s) = \bigvee_{\substack{(s,\alpha,M,C) \in \Delta \\ s \in C}} \varphi_\alpha \wedge \mathsf{X}\varphi(C \smallsetminus \{s\}) \quad \mathsf{U} \bigvee_{\substack{(s,\alpha,M,C) \in \Delta \\ s \notin C}} \varphi_\alpha \wedge \mathsf{X}\varphi(C)$$

$$\varphi_2(s) = \mathsf{G} \bigvee_{\substack{(s,\alpha,M,C) \in \Delta \\ s \in C}} \varphi_\alpha \wedge \mathsf{X}\varphi(C \smallsetminus \{s\})$$

$$\varphi_3(s) = \bigvee_{O \in Mod(\Phi)} \Bigg(\bigwedge_{■ \in \mathsf{Inf}(O)} \Big(\mathsf{GF} \bigvee_{\substack{(s,\alpha,M,C) \in \Delta \\ s \in C,\ ■ \in M}} \varphi_\alpha \wedge \mathsf{X}\varphi(C \smallsetminus \{s\}) \Big) \wedge \bigwedge_{● \in \mathsf{Fin}(O)} \Big(\mathsf{FG} \bigvee_{\substack{(s,\alpha,M,C) \in \Delta \\ s \in C,\ ● \notin M}} \varphi_\alpha \wedge \mathsf{X}\varphi(C \smallsetminus \{s\}) \Big) \Bigg)$$

Intuitively, $\varphi_1(s)$ covers the case when a run leaves s after a finite number of looping transitions. Note that $\varphi_1(s)$ ignores acceptance marks on looping transitions of s as they play no role in acceptance of such runs. Further, $\varphi_2(s)$ describes the case when a run never leaves s. In this case, $\varphi_3(s)$ ensures that the infinite branch of the run staying in s forever is accepting. Indeed, $\varphi_3(s)$ says that the branch satisfies some prime implicant of the acceptance condition Φ as all Inf-marks of the prime implicant have to appear infinitely often on the branch and each Fin-mark of the prime implicant does not appear at all after a finite number of transitions.

Theorem 2. *Let $\mathcal{A}$ be an SLAA over an alphabet $\Sigma = 2^{AP'}$ and with initial state s_I. Further, let $\varphi(s_I)$ be the formula constructed as above. Then $L(\mathcal{A}) = L(\varphi(s_I))$.*

The statement can be proven straightforwardly by induction reflecting the inductive definition of the translation.

Theorems 1 and 2 imply that LTL and the class of SLAA with alphabets of the form $2^{AP'}$ have the same expressive power.

8 Implementation and Experimental Evaluation

We have implemented the presented translations in a tool called *LTL3TELA* which also offers dealternation algorithms for Inf-less SLAA and SLAA produced by the F, G-merging translation. Given an LTL formula, LTL3TELA first applies the formula optimization process implemented in SPOT. The processed formula is then translated to SLAA by one of the three presented translation, unreachable states are removed and transition reductions suggested by Babiak et al. [3] are applied.

Table 1. Reference table of tools used for the experimental evaluation.

Tool	Version	Homepage
LTL3BA	1.1.3	https://sourceforge.net/projects/ltl3ba
LTL3TELA	2.1.0	https://github.com/jurajmajor/ltl3tela
SPOT library	2.7.4	https://spot.lrde.epita.fr

8.1 Evaluation Settings

We compare the LTL to SLAA translation implemented in LTL3BA and the translations presented in this paper as implemented in LTL3TELA. Table 1 provides homepages and versions numbers of SPOT and both compared translators. All scripts and formulae used for the evaluation presented below are available in a Jupyter notebook that can be found at https://github.com/jurajmajor/ltl3tela/blob/master/Experiments/Evaluation_ICTAC19.ipynb.

The presented improvements can only be applied on certain kinds of formulae: formulae that contain at least one $\mathsf{F}\psi$ subformula where ψ contains some temporal subformula or $\mathsf{G} \bigwedge_i \psi_i$ subformula where at least one ψ_i is temporal. We call such formulae *mergeable*. We first evaluate how likely we can obtain a formula that is mergeable in Sect. 8.2 and then we present the impact of our merging technique on mergeable formulae in Sect. 8.3. We consider formulae that come from two sources.

(i) We use formulae collected from literature [10,12,15,21,26] that can be obtained using the tool `genltl` from SPOT [8]. For each such a formula, we added its negation, simplified all the formulae and removed duplicates and formulae equivalent to *tt* or *ff*. The resulting benchmark set contains 221 formulae.
(ii) We use the tool `randltl` from SPOT to generate random formulae. We generate formulae with up to five atomic propositions and with tree size equal to 15 (the default settings of `randltl`) before simplifications. We consider 4 different sets of random formulae. The generator allows the user to specify *priority* for each LTL operator which determines how likely the given operator appears in the generated formulae. By default, all operators (boolean and temporal) have priority 1 in `randltl`. For the sets *rand1*, *rand2*, and *rand4*, the number indicates the priority that was used for F and G. The last set called *randfg* uses priority 2 for F and G and sets 0 to all other temporal operators. For Sect. 8.2 we generated 1000 formulae for each priority setting, for Sect. 8.3 we generate for each priority setting 1000 formulae that are mergeable (and throw away the rest).

8.2 Mergeability

The set of formulae from literature contains mainly quite simple formulae. As a result, only 24 out of the 221 formulae are mergeable. For *rand1*, *rand2*, *rand4*, and *randfg* we have 302, 488, 697, and 802 mergeable formulae out of 1000, respectively. Consistently with intuition, the ratio of mergeable formulae increases considerably with F and G being more frequent.

Table 2. Comparison of LTL to SLAA translations on mergeable formulae.

	States					Acceptance marks				
	lit	rand1	rand2	rand4	randfg	lit	rand1	rand2	rand4	randfg
# of form.	24	1000	1000	1000	1000	24	1000	1000	1000	1000
LTL3BA	140	6253	6313	6412	5051	24	1000	1000	1000	1000
basic	140	6234	6287	6393	5051	24	997	1000	1000	1000
F-merging	110	5418	5296	5231	3926	46	1160	1244	1347	1343
F, G-merging	65	4595	4300	4015	2744	98	2971	3317	3677	2978

Table 3. Number of alternating automata that use only existential branching. The numbers of formulae are the same as in Table 2 for each set.

	Deterministic					Nonalternating				
	lit	rand1	rand2	rand4	randfg	lit	rand1	rand2	rand4	randfg
LTL3BA	0	5	2	0	0	6	148	114	89	144
basic	0	5	2	0	0	6	148	114	89	144
F-merging	2	64	53	40	73	6	171	146	119	192
F, G-merging	10	133	126	124	217	18	356	367	385	603

8.3 Comparison on Mergeable Formulae

Table 2 shows the cumulative numbers of states and acceptance marks of SLAA produced by LTL3BA and the translations presented in Sects. 3, 4, and 5 as implemented in LTL3TELA for mergeable formulae. The scatter plots in Fig. 6 offer more details on the improvement (in the number of states) of F, G-merging over basic translation. The missing plot for rand2 was omitted for space reasons as it is very similar to the one for rand1.

Merging of states has a positive impact also on the type of the produced automaton as it removes both some universal and nondeterministic branching from the automaton. Table 3 shows for each set of formulae and each translation the number of automata that have no universal branching (right part) and that are even deterministic (left part).

One can observe in the tables that the basic translation produces similar SLAA as LTL3BA. Further, the F-merging and F, G-merging translations bring gradual improvement both in the number of states and in the numbers of deterministic and nonalternating automata in comparison to the basic translation on all sets of formulae. The ratio of states that can be saved by merging grows with the increasing occurrence of F and G operators up to 45% (randfg) in the benchmarks that use randomly generated formulae.

The scatter plots reveal that most cases fit into the category *the* F, G*-merging translation saves up to 3 states.* But there are also cases where the F, G-merging translation reduces the resulting SLAA to 1 state only while the basic translation needs 8 or even more states. However, we still have one case where the basic translation produces smaller automaton than the F, G-merging (see Fig. 6, rand4) and F-merging translations.

Table 3 confirms by numbers that the F- and F, G-merging translations often build automata with fewer branching. The numbers of deterministic and nonalternating (without universal branching) automata are especially appealing for the F, G-merging translation on formulae from the set *randfg*.

On the downside, the presented translations produce SLAA with more acceptance marks than the translation of LTL3BA. This is the price we pay for small automata. Basic translation sometimes uses 0 acceptance marks if there is no F or U operator.

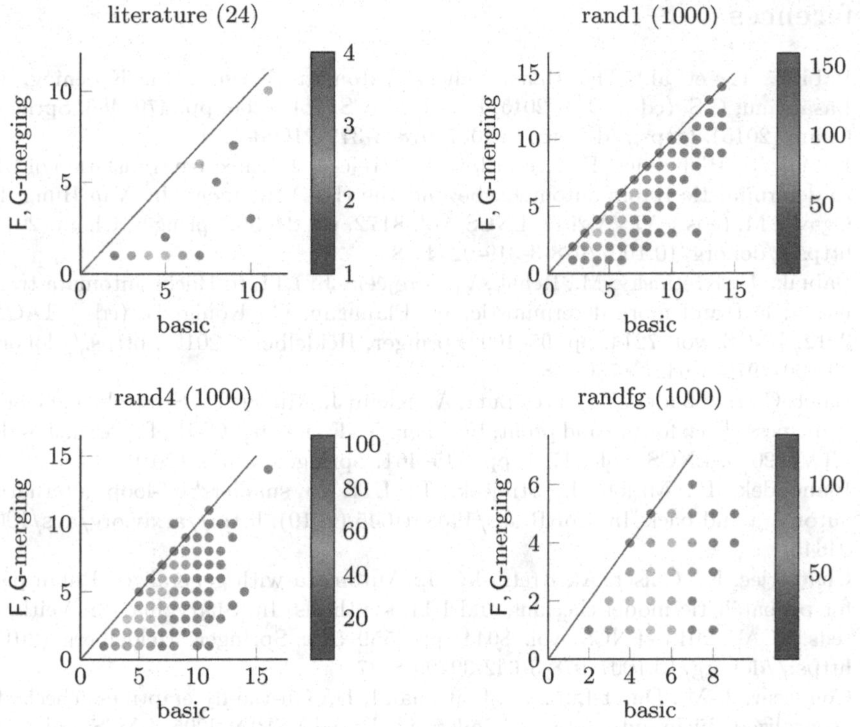

Fig. 6. Effect of F, G-merging on SLAA size for mergeable formulae. A dot represents the number of states of the SLAA produced by F, G-merging (y-axis) and by the basic translation (x-axis) for the same formula. The color of the dot reflects the number of dots at the position. (Color figure online)

9 Conclusion

We have presented a novel translation of LTL to self-loop alternating automata (SLAA) with Emerson-Lei acceptance condition. To our best knowledge, it is the first translation of LTL producing SLAA with other than Büchi or co-Büchi acceptance. Our experimental results demonstrated that the expressive acceptance condition allows to produce substantially smaller SLAA comparing to these produced by LTL3BA when F or G operators appear in the translated formula.

This work opens doors for research of algorithms processing SLAA with Emerson-Lei acceptance, in particular the algorithms transforming these SLAA to nonalternating automata with various degrees of determinism: nondeterministic, deterministic, or semi-deterministic (also known as limit-deterministic) automata. Our implementation can serve as a natural source of such automata and already helped to build a tool that can produce small deterministic and nondeterministic TELA.

References

1. Babiak, T., et al.: The Hanoi Omega-Automata Format. In: Kroening, D., Păsăreanu, C.S. (eds.) CAV 2015, Part I. LNCS, vol. 9206, pp. 479–486. Springer, Cham (2015). https://doi.org/10.1007/978-3-319-21690-4_31
2. Babiak, T., Blahoudek, F., Křetínský, M., Strejček, J.: Effective translation of LTL to deterministic Rabin automata: beyond the (F,G)-fragment. In: Van Hung, D., Ogawa, M. (eds.) ATVA 2013. LNCS, vol. 8172, pp. 24–39. Springer, Cham (2013). https://doi.org/10.1007/978-3-319-02444-8_4
3. Babiak, T., Křetínský, M., Řehák, V., Strejček, J.: LTL to Büchi automata translation: fast and more deterministic. In: Flanagan, C., König, B. (eds.) TACAS 2012. LNCS, vol. 7214, pp. 95–109. Springer, Heidelberg (2012). https://doi.org/10.1007/978-3-642-28756-5_8
4. Baier, C., Blahoudek, F., Duret-Lutz, A., Klein, J., Müller, D., Strejček, J.: Generic emptiness check for fun and profit. In: Chen, Y.-F., Cheng, C.-H., Esparza, J. (eds.) ATVA 2019. LNCS, vol. 11781, pp. 445–461. Springer, Cham (2019)
5. Blahoudek, F., Major, J., Strejček, J.: LTL to smaller self-loop alternating automata and back. In: CoRR abs/1908.04645 (2019). http://arxiv.org/abs/1908.04645
6. Chatterjee, K., Gaiser, A., Křetínský, J.: Automata with generalized Rabin pairs for probabilistic model checking and LTL synthesis. In: Sharygina, N., Veith, H. (eds.) CAV 2013. LNCS, vol. 8044, pp. 559–575. Springer, Heidelberg (2013). https://doi.org/10.1007/978-3-642-39799-8_37
7. Couvreur, J.-M., Duret-Lutz, A., Poitrenaud, D.: On-the-fly emptiness checks for generalized Büchi automata. In: Godefroid, P. (ed.) SPIN 2005. LNCS, vol. 3639, pp. 169–184. Springer, Heidelberg (2005). https://doi.org/10.1007/11537328_15
8. Duret-Lutz, A.: Manipulating LTL formulas using Spot 1.0. In: Van Hung, D., Ogawa, M. (eds.) ATVA 2013. LNCS, vol. 8172, pp. 442–445. Springer, Cham (2013). https://doi.org/10.1007/978-3-319-02444-8_31
9. Duret-Lutz, A., Lewkowicz, A., Fauchille, A., Michaud, T., Renault, É., Xu, L.: Spot 2.0 — a framework for LTL and ω-automata manipulation. In: Artho, C., Legay, A., Peled, D. (eds.) ATVA 2016. LNCS, vol. 9938, pp. 122–129. Springer, Cham (2016). https://doi.org/10.1007/978-3-319-46520-3_8
10. Dwyer, M.B., Avrunin, G.S., Corbett, J.C.: Property specification patterns for finite-state verification. In: Proceedings of FMSP 1998, pp. 7–15. ACM (1998)
11. Emerson, E.A., Lei, C.-L.: Modalities for model checking: branching time logic strikes back. Sci. Comput. Program. **8**(3), 275–306 (1987)
12. Etessami, K., Holzmann, G.J.: Optimizing Büchi automata. In: Palamidessi, C. (ed.) CONCUR 2000. LNCS, vol. 1877, pp. 153–168. Springer, Heidelberg (2000). https://doi.org/10.1007/3-540-44618-4_13
13. Gastin, P., Oddoux, D.: Fast LTL to Büchi automata translation. In: Berry, G., Comon, H., Finkel, A. (eds.) CAV 2001. LNCS, vol. 2102, pp. 53–65. Springer, Heidelberg (2001). https://doi.org/10.1007/3-540-44585-4_6
14. Hammer, M., Knapp, A., Merz, S.: Truly on-the-fly LTL model checking. In: Halbwachs, N., Zuck, L.D. (eds.) TACAS 2005. LNCS, vol. 3440, pp. 191–205. Springer, Heidelberg (2005). https://doi.org/10.1007/978-3-540-31980-1_13
15. Holeček, J., Kratochvíla, T., Řehák, V., Šafránek, D., Šimeček, P.: Verification results in Liberouter project. Technical report 03, 32 pp. CESNET, September 2004

16. Křetínský, J., Meggendorfer, T., Sickert, S., Ziegler, C.: Rabinizer 4: from LTL to your favourite deterministic automaton. In: Chockler, H., Weissenbacher, G. (eds.) CAV 2018. LNCS, vol. 10981, pp. 567–577. Springer, Cham (2018). https://doi.org/10.1007/978-3-319-96145-3_30
17. Loding, C., Thomas, W.: Alternating automata and logics over infinite words. In: van Leeuwen, J., Watanabe, O., Hagiya, M., Mosses, P.D., Ito, T. (eds.) TCS 2000. LNCS, vol. 1872, pp. 521–535. Springer, Heidelberg (2000). https://doi.org/10.1007/3-540-44929-9_36
18. Major, J., Blahoudek, F., Strejček, J., Sasaráková, M., Zbončáková, T.: `ltl3tela`: LTL to small deterministic or nondeterministic Emerson-Lei automata. In: Chen, Y.-F., Cheng, C.-H., Esparza, J. (eds.) ATVA 2019. LNCS, vol. 11781, pp. 357–365. Springer, Cham (2019)
19. Muller, D.E., Saoudi, A., Schupp, P.E.: Weak alternating automata give a simple explanation of why most temporal and dynamic logics are decidable in exponential time. In: Proceedings of LICS 1988, pp. 422–427. IEEE Computer Society (1988)
20. Müller, D., Sickert, S.: LTL to deterministic Emerson-Lei automata. In: Proceedings of GandALF 2017. EPTCS, vol. 256, pp. 180–194 (2017)
21. Pelánek, R.: BEEM: benchmarks for explicit model checkers. In: Bošnački, D., Edelkamp, S. (eds.) SPIN 2007. LNCS, vol. 4595, pp. 263–267. Springer, Heidelberg (2007). https://doi.org/10.1007/978-3-540-73370-6_17
22. Pelánek, R., Strejček, J.: Deeper connections between LTL and alternating automata. In: Farré, J., Litovsky, I., Schmitz, S. (eds.) CIAA 2005. LNCS, vol. 3845, pp. 238–249. Springer, Heidelberg (2006). https://doi.org/10.1007/11605157_20
23. Pnueli, A.: The temporal logic of programs. In: Proceedings of FOCS 1977, pp. 46–57. IEEE Computer Society (1977)
24. Renault, E., Duret-Lutz, A., Kordon, F., Poitrenaud, D.: Parallel explicit model checking for generalized Büchi automata. In: Baier, C., Tinelli, C. (eds.) TACAS 2015. LNCS, vol. 9035, pp. 613–627. Springer, Heidelberg (2015). https://doi.org/10.1007/978-3-662-46681-0_56
25. Rohde, G.S.: Alternating automata and the temporal logic of ordinals. Ph.D. thesis. University of Illinois at Urbana-Champaign (1997). ISBN 0-591-63604-2
26. Somenzi, F., Bloem, R.: Efficient Büchi automata from LTL formulae. In: Emerson, E.A., Sistla, A.P. (eds.) CAV 2000. LNCS, vol. 1855, pp. 248–263. Springer, Heidelberg (2000). https://doi.org/10.1007/10722167_21
27. Tauriainen, H.: Automata and linear temporal logic: translations with transition-based acceptance. Ph.D. thesis. Helsinki University of Technology, Laboratory for Theoretical Computer Science (2006). ISBN 951-22-8343-3
28. Vardi, M.Y.: Nontraditional applications of automata theory. In: Hagiya, M., Mitchell, J.C. (eds.) TACS 1994. LNCS, vol. 789, pp. 575–597. Springer, Heidelberg (1994). https://doi.org/10.1007/3-540-57887-0_116

Verification and Concurrency

Starvation-Free Monitors

Jafar Hamin(✉)

imec-DistriNet, Department of Computer Science, KU Leuven, Leuven, Belgium
jafar.hamin@cs.kuleuven.be

Abstract. Monitors are a synchronization construct which allows to keep a thread waiting until a specific resource for that thread is available. One potential problem with these constructs is starvation; a situation where a thread, competing for a resource, infinitely waits for that resource because other threads, that started competing for that resource later, get it earlier infinitely often. In this paper a modular approach to verify starvation-freedom of monitors is presented, ensuring that each time that a resource is released and its associated condition variable is notified each waiting thread approaches the front of the waiting queue; more specifically, the loop in which the wait command is executed (that checks the waiting condition) has a loop variant. To this end, we introduce notions of *publishable resources* and *publishable obligations*, which are published from the thread notifying a condition variable to all of the threads waiting for that condition variable. The publishable resources ensure the waiting threads that they are approaching the front of the waiting queue, by allowing to define an appropriate loop variant for the related loop. The publishable obligations ensure that for any thread waiting for a condition variable v there is another thread obliged to notify v, which only waits for waitable objects whose levels, some arbitrary numbers associated with each waitable object, are lower than the level of v (preventing circular dependencies). We encoded the proposed separation logic-based proof rules in the VeriFast program verifier and succeeded in verifying deadlock-freedom and starvation-freedom of two monitors, having no scheduling policy, which implement two common queue locking algorithms, namely *ticket lock* and *CLH lock*.

Keywords: Starvation-freedom · Monitors · Condition variables · Modular verification · Separation logic · Hoare logic

1 Introduction

Multithreaded programs use synchronizations such as locks and condition variables to control the execution of threads. Monitors are one of the popular constructs used to synchronize threads [32–34]. This construct, consisting of a lock

We thank three anonymous reviewers and also Prof. Bart Jacobs for their careful reading of our manuscript and their many insighful comments and suggestions. This research received funding from the European Union's Horizon 2020 research and innovation programme under grant agreement No. 731453 (project VESSEDIA).

R. M. Hierons and M. Mosbah (Eds.): ICTAC 2019, LNCS 11884, pp. 175–195, 2019.
https://doi.org/10.1007/978-3-030-32505-3_11

and some condition variables associated with that lock, provides some APIs for its clients, namely wait(v, l), causing the calling thread to release the lock l and wait for a notification on the condition variable v, and notify(v)/notifyAll(v), causing one/all thread(s) waiting for v to resume its/their execution. Each condition variable is associated with a lock; a thread must acquire the associated lock for waiting or notifying on a condition variable, and when a thread is notified it must reacquire the associated lock before it can continue execution.

Monitors allow to keep a thread waiting until a desired condition is met, e.g. an item is produced (in a producer-consumer program), a writer finishes its writing (in a readers-writer lock), or a specific number of threads finish a specific task (in a barrier). However, one potential problem with these constructs is starvation; a situation where a thread, competing for a resource, infinitely waits for that resource because other threads, that started competing for that resource later, get it earlier infinitely often. For example, a shared buffer allows starvation if it is possible that a thread starts a *take* operation but it never completes, while other threads' *take* operations complete infinitely often. As another example, a readers-writers lock allows starvation if it is possible that a thread attempts to enter the write lock but never succeeds, while other threads successfully acquire and release the write lock infinitely often.

Several approaches to verify termination, deadlock-freedom, liveness, and finite blocking of threads have been introduced. Some of these approaches only work with non-blocking algorithms [27], where the suspension of one thread cannot lead to the suspension of other threads. These approaches are not applicable for condition variables because suspension of a notifying thread causes other waiting threads to be blocked forever. Some other approaches are also presented to verify termination of programs using some blocking constructs such as channels [13,19,20] and semaphores [11]. These approaches are not general enough to cover condition variables because unlike channels and semaphores a notification on a condition variable is lost when there is no thread waiting for that condition variable. There are also some approaches [4,17,35] to verify correctness of programs in the presence of monitors. However, these approaches either only cover a very specific application of monitors, such as a buffer program with only one producer and one consumer, or are not modular and suffer from a long verification time when the size of the state space, such as the number of threads, is increased.

Recently, Hamin *et al.* [6,7] introduced a modular approach to verify absence of deadlock in monitors. This approach ensures that in any state of the execution there is a thread running, not waiting for a lock or condition variable, until the program terminates. The main idea behind this approach is to make sure that for any condition variable v for which a thread is waiting there exists a thread obliged to discharge an obligation for v that only waits for objects whose levels, some arbitrary numbers associated with each waitable object, are lower than the level of v. The notion of obligations in this approach, which is borrowed from the approach of [20], verifying deadlock-freedom of channels, and Kobayashi's type system [19], verifying deadlock-freedom of π-calculus processes, makes it

```
routine new_tvm()
{
 l := new_lock();
 v := new_cvar();
 tvm(l:=l, v:=v,
 active:=new_int(0),
 next:=new_int(0))
}

routine enter(tvm t)
{
 acquire(t.l);
 mine := t.next++;
 while(t.active<mine)
   wait(t.v, t.l);
 release(t.l)
}

routine leave(tvm t)
{
 acquire(t.l);
 t.active++;
 notifyAll(t.v);
 release(t.l)
}

routine main(){
 t:=new_tvm;
 fork(while(1){
  enter(t); // CS
  leave(t)
 });
 enter(t); // CS
 leave(t)}
```

Fig. 1. A monitor implementing a ticket algorithm.

possible that different modules (routines) of a program are individually verified. However, in this approach a thread which holds a resource and has an obligation to release that resource cannot transfer that obligation to the next thread, which acquires that resource. Later, this problem is solved in [9] by allowing obligations to be transferred from the notifying thread to the one notified. In this approach the transferred obligations are discharged from the notifying thread only if there is a waiting thread onto which the transferred obligations are loaded. However, in this approach obligations can be transferred using notify(v) but not using notifyAll(v), which wakes up all the threads waiting for the condition variable v.

This constraint does not allow this approach to verify a monitor implementing a ticket lock algorithm, shown in Fig. 1, for example, where the releasing thread notifies all the waiting threads. As shown in the routine new_tvm in this figure, a tvm structure consists of a variable *active*, storing the now serving ticket number, a variable *next*, storing the last ticket number, a lock l, protecting these variables from concurrent accesses, and a condition variable v, keeping a thread waiting until its ticket number equals the now serving ticket number. As shown in the routine enter, the running thread first acquires the lock l, then it assigns the last ticket number to its own ticket number, denoted by *mine*, and increases the last ticket number. While the now serving ticket number is lower than the ticket number of this thread it releases the lock l and waits for a notification on the condition variable v. When this thread is notified (and reacquires l) if its ticket number is equal (or lower than) the now serving ticket number it releases l and continues its execution. After this thread gets service, as shown in the routine leave, it acquires l, increases the now serving ticket number, notifies all threads waiting for v, and releases l. After this notification, the thread whose ticket number equals the now serving ticket number continues it execution and the rest of the threads wait for a notification again.

Ticket lock algorithms and CLH lock algorithms [2] are two common queue locking algorithms used to enforce fairness in concurrent constructs such as a concurrent queue and a readers-writers lock (see Fig. 2). To verify deadlock-freedom and starvation-freedom[1] of monitors which have no scheduling policy[2] and enforce fairness by implementing these algorithms, in addition to transferring obligations using notifyAll, it is also necessary to transfer some information from the notifying thread to the ones notified. This information must ensure the awoken threads that they are approaching the front of the waiting queue (for example, in Fig. 1 this information implies that *the now serving ticket number has been increased*). Although the approach presented in [6] allows permissions to be transferred through notifyAll, it considers permissions as linear resources, which cannot be in possession of more than one thread at the same time. Consequently, in that approach a notifying thread can transfer a permission p to n number of waiting threads only if it owns n instances of p.

```
routine new_rdwr(){
 l := new_lock;
 v_w := new_cvar; v_r := new_cvar;
 rdwr(l:=l, v_w:=v_w, v_r:=v_r, ar:=ar,
   active:=new_int(0), next:=new_int(0))}

routine acquire_read(rdwr b){
 acquire(b.l);
 while(b.active < b.next)
   wait(b.v_r, b.l);
 ar := ar + 1;
 release(b.l)}

routine release_read(rdwr b){
 acquire(b.l);
 ar := ar − 1;
 notifyAll(b.v_w);
 release(b.l)}

routine acquire_write(rdwr b)
{
 acquire(b.l);
 mine := b.next++;
 while(0 < b.ar ∨ b.active < mine)
   wait(b.v_w, b.l);
 release(b.l)
}

routine release_write(rdwr b)
{
 acquire(b.l);
 b.active++;
 notifyAll(b.v_w);
 if(b.active = b.next)
   notifyAll(b.v_r);
 release(b.l)
}
```

Fig. 2. A readers-writers lock which uses a ticket algorithm to make sure that no writer is starved, where *ar* denotes the number of the threads which are reading, the condition variable v_r keeps a reader waiting until all writers write, and the condition variable v_w keeps a writer waiting until its ticket number is equal the now serving ticket number and no other thread is reading.

In this paper we introduce a mechanism which allows obligations to be transferred from the notifying thread, executing notifyAll, to one of the notified threads. In this mechanism we make sure that a notified thread is successfully verified in the both cases where it receives the transferred obligations and

[1] Note that we assume that the lock associated with monitors is a fair lock.

[2] Note that in a monitor having no scheduling policy the blocked thread which is awoken by notify is not necessary chosen in priority order; this monitor do not keep the order in which threads start to wait.

where it does not receive the transferred obligations. Additionally, we introduce a notion of publishable resources which can be published from a thread to many other threads. This allows to transfer this information that *the waiting threads are approaching the front of the waiting queue* form a notifying thread to all of the threads notified. Having this information in a notified thread, it is possible to make a loop variant for the loop in which the wait command is executed (the loop that checks the waiting condition). Note that this loop variant ensures that the thread in the monitor which is waiting for a resource is not starved, i.e. each time that this resource is released and its associated condition variable is notified this thread approaches the front of the waiting queue. We encoded the proposed proof rules in the VeriFast program verifier [12,14,15] and verified that the program in Fig. 1 is deadlock-free and the loop in the routine enter is a finite loop, implying starvation-freedom of the related monitor[3].

This paper is structured as follows. Section 2 provides some background information on the existing approaches that verify deadlock-freedom of monitors. Section 3 introduces a mechanism to transfer obligations through broadcasts, which makes it possible to verify deadlock-freedom of monitors implementing some queue locking algorithms, which cannot be verified without such transfer. Section 4 introduces a mechanism to transfer some publishable resources through broadcasts, which makes it possible to verify starvation-freedom of monitors implementing some queue locking algorithms. Section 5 compares this approach with some related works and a conclusion is drawn in Sect. 6.

2 Deadlock-Free Monitors

Hamin *et al.* [6,7,9] introduced a modular approach for verifying absence of deadlock in programs synchronized using condition variables (CVs), where executing a command $\mathsf{wait}(v, l)$ on a CV v, which is associated with a lock l, releases l and suspends the running thread, and executing commands $\mathsf{notify}(v)$ or $\mathsf{notifyAll}(v)$ wake up one or all thread(s) waiting for CV v at which point they will try to reacquire l. This approach ensures absence of deadlock by making sure that for any CV v for which a thread is waiting there is a thread obliged to fulfill an obligation for v which only waits for waitable objects whose levels are lower than the level of v. In this approach when a thread acquires a lock l, the total number of waiting threads, and the total number of obligations of any CV v associated with l, denoted by $Wt(v)$ and $Ot(v)$ respectively, can be mentioned in the proof of that thread. Wt and Ot are actually two bags[4] which keep track of the total number of the waiting threads and the total number of the obligations of the condition variables associated with a lock, respectively. In order to ensure the mentioned constraint this approach makes sure that (1) if a command $\mathsf{wait}(v, l)$ is executed then $0 < Ot(v)$, i.e. there is an obligation for v in

[3] The proof of this program, verified by the VeriFast program verifier, and the proof of a CLH lock algorithm can be found in [8].

[4] We model bags of objects as functions from objects to natural numbers. We also use $\uplus$ indicating the union of two bags.

NewLock
$$\{\mathsf{true}\}\ \mathsf{newlock}\ \{\lambda l.\ \mathsf{ulock}(l, \{\!\!\}, \{\!\!\}) \land \mathsf{R}(l)=r\}$$

InitLock
$$\{\mathsf{ulock}(l, Wt, Ot) * inv(Wt, Ot) * \mathsf{obs}(O)\}\ \mathsf{g\_initl}(l)\ \{\lambda\_.\ \mathsf{lock}(l) * \mathsf{obs}(O) \land \mathsf{I}(l)=inv\}$$

Acquire
$$\{\mathsf{lock}(l) * \mathsf{obs}(O) \land l \prec O\}\ \mathsf{acquire}(l)$$
$$\{\lambda\_.\ \exists Wt, Ot.\ \mathsf{locked}(l, Wt, Ot) * \mathsf{I}(l)(Wt, Ot) * \mathsf{obs}(O \uplus \{\!l\!\})\}$$

Release
$$\{\mathsf{locked}(l, Wt, Ot) * \mathsf{I}(l)(Wt, Ot) * \mathsf{obs}(O \uplus \{\!l\!\})\}\ \mathsf{release}(l)\ \{\lambda\_.\ \mathsf{lock}(l) * \mathsf{obs}(O)\}$$

NewCV
$$\{\mathsf{true}\}\ \mathsf{new\_cvar}\ \{\lambda v.\ \mathsf{ucond}(v) \land \mathsf{R}(v)=r\}$$

InitCV
$$\{\mathsf{ucond}(v) * \mathsf{ulock}(l, Wt, Ot)\}\ \mathsf{g\_initc}(v)$$
$$\{\lambda\_.\ \mathsf{cond}(v, M, M') * \mathsf{ulock}(l, Wt, Ot) \land \mathsf{L}(v) = l\}$$

Wait
$$\{\mathsf{cond}(v, M, M') * \mathsf{locked}(l, Wt, Ot) * \mathsf{I}(l)(Wt \uplus \{\!v\!\}, Ot) * \mathsf{obs}(O \uplus \{\!l\!\}) \land 0 < Ot(v) \land$$
$$l = \mathsf{L}(v) \land v \prec O \land l \prec O \uplus M'\}\ \mathsf{wait}(v, l)\ \{\lambda\_.\ \mathsf{cond}(v, M, M') * \mathsf{obs}(O \uplus \{\!l\!\} \uplus M') *$$
$$\exists Wt', Ot'.\ \mathsf{locked}(l, Wt', Ot') * \mathsf{I}(l)(Wt', Ot') * M\}$$

Notify
$$\{\mathsf{obs}(O \uplus (0 < Wt(v)\ ?\ M' : \{\!\!\})) * \mathsf{cond}(v, M, M') * \mathsf{locked}(\mathsf{L}(v), Wt, Ot) *$$
$$(Wt(v)=0 \lor M)\}\ \mathsf{notify}(v)\ \{\lambda\_.\ \mathsf{obs}(O) * \mathsf{cond}(v, M, M') * \mathsf{locked}(\mathsf{L}(v), Wt - \{\!v\!\}, Ot)\}$$

NotifyAll
$$\{\mathsf{cond}(v, M, \{\!\!\}) * \mathsf{locked}(\mathsf{L}(v), Wt, Ot) * (\mathop{*}_{i:=1}^{Wt(v)} M)\}\ \mathsf{notifyAll}(v)$$
$$\{\lambda\_.\ \mathsf{cond}(v, M, \{\!\!\}) * \mathsf{locked}(\mathsf{L}(v), Wt[v{:=}0], Ot)\}$$

ChargeObligation
$$\{\mathsf{obs}(O) * \mathsf{locked}(\mathsf{L}(v), Wt, Ot)\}\ \mathsf{g\_charge}(v)$$
$$\{\lambda\_.\ \mathsf{obs}(O \uplus \{\!v\!\}) * \mathsf{locked}(\mathsf{L}(v), Wt, Ot \uplus \{\!v\!\})\}$$

DischargeObligation
$$\{\mathsf{obs}(O) * \mathsf{locked}(\mathsf{L}(v), Wt, Ot) \land (0 < Wt(v) \Rightarrow 1 < Ot(v))\}\ \mathsf{g\_discharge}(v)$$
$$\{\lambda\_.\ \mathsf{obs}(O - \{\!v\!\}) * \mathsf{locked}(\mathsf{L}(v), Wt, Ot - \{\!v\!\})\}$$

Fig. 3. Proof rules verifying deadlock-freedom of monitors, where $\mathsf{obs}(O)$ denote the bag of the obligations of the current thread; $\mathsf{R}(o)$ specifies the level of the waitable object o; $\mathsf{I}(l)$ specifies the invariant of the lock l; $Wt(v)$ and $Ot(v)$ in the permission $\mathsf{ulock}/\mathsf{locked}$ denote the total number of threads waiting for v and the total number of obligations for v, respectively; the parameters M and M' in the permission cond of a condition variable denote the permissions and the obligations which are transferred from the thread notifying that condition variable to the one notified, respectively; $\mathsf{L}(v)$ denotes the lock associated with the condition variable v; and $v \prec O \Leftrightarrow \forall o{\in}O.\ \mathsf{R}(v) < \mathsf{R}(o)$.

the system, (2) if an obligation for v is discharged then after this discharge the invariant $0 < Wt(v) \Rightarrow 0 < Ot(v)$ holds, i.e. if there is a thread waiting for v then after this discharge there are still some obligations for v in the system, and (3) a thread executes a command $\mathsf{wait}(v, l)$ only if the level of v is lower than the levels of the obligations of that thread.

A program in this approach can be successfully verified if each lock associated with some CVs has an appropriate invariant such that for any CV v associated with that lock this invariant implies $0 < Wt(v) \Rightarrow 0 < Ot(v)$. Accordingly, in this approach each lock invariant is parametrized over the bags Wt and Ot, which map all CVs associated with that lock to the number of their waiting threads and obligations, respectively.

The proof rules proposed in this approach are shown in Fig. 3. As shown in Rule NewLock, when a lock l is created an arbitrary level is assigned to that lock by the proof author, denoted by $\mathsf{R}(l)$, and an uninitialized lock permission $\mathsf{ulock}(l, \{\!\!\}, \{\!\!\})$ is produced. The second and the third parameters of this permission are two bags mapping the CVs associated with l to the total number of threads waiting for that CV and the total number of obligations of that CV in the system, respectively. As shown in Rule InitLock, using a ghost command[5] g_initl, this uninitialized lock permission can be converted to a duplicable lock permission $\mathsf{lock}(l)$ if the assertion resulting from applying the invariant of that lock, denoted by $\mathsf{I}(l)$, to the bags stored in the permission ulock is consumed (the permissions described by the invariant of this lock are transferred from the thread to the lock). As shown in Rule Acquire, when a thread acquires this lock the permissions described by the invariant of this lock are transferred from the lock to the thread. Additionally, a permission $\mathsf{locked}(l, Wt, Ot)$ is provided for the thread, where Wt and Ot are two bags mapping the CVs associated with l to their number of waiting threads and obligations, respectively, and are existentially quantified in the postcondition. Note that to prevent circular dependencies the precondition of this rule enforces that the level of l be lower than the levels of the obligations of the acquiring thread. Additionally, this lock is added to the bag of the obligations of this thread. As shown in rule Release, when this lock is released it is discharged from the bag of the obligations and the assertion resulting from applying the invariant of this lock to the bags stored in the permission locked is consumed. Additionally, the permission locked is converted to a permission lock.

As shown in Rule NewCV, when a CV is created an arbitrary level is assigned to it and an uninitialized permission ucond for that CV is produced. As shown in Rule InitCV, this permission can be converted to a duplicable permission cond if a lock is associated to this CV, denoted by $\mathsf{L}(v)$. Additionally, the transferred permissions and obligations of this CV, denoted by M and M', which are transferred from the notifying thread to the one notified, are also specified in this rule. These permissions and obligations are consumed when a command $\mathsf{notify}(v)$ is executed (if there is a thread waiting for v; see the precondition of Rule Notify), and are

[5] The ghost commands are inserted into the program for verification purposes and have no effect on the program's behavior.

produced when a command $\mathsf{wait}(v, l)$ is executed (see the postcondition of Rule WAIT). Note that $\mathsf{notifyAll}(v)$ transfers $Wt(v)$ instances of these permissions, denoted by $\mathop{*}\limits_{i:=1}^{Wt(v)} M$ (see the precondition of Rule NOTIFYALL). Also note that $\mathsf{notifyAll}(v)$ transfers no obligation. As shown in Rule WAIT, when a command $\mathsf{wait}(v, l)$ is executed, since l is going to be released and the number of the threads waiting for v is going to be increased, the result of applying the invariant of lock l to bags $Wt \uplus \{v\}$ and Ot must be consumed, where Wt and Ot are the bags stored in the permission locked of l. Additionally, the level of v must be lower than the levels of all obligations of the thread except for l. Note that the level of l must be lower than the levels of these obligations and those which are transfered from the notifying thread, denoted by M', since when the thread is woken up it tries to reacquire l. As previously mentioned, the precondition of this rule also makes sure that $0 < Ot(v)$. This precondition follows from $\mathsf{I}(l)(Wt \uplus \{v\}, Ot)$ provided that the invariant of l is properly defined such that for any CV v' associated with l this invariant implies that if there is a thread waiting for v' there is an obligation for v' in the system. Lastly the precondition of this command makes sure that v is associated with lock l, which is enforced by $\mathsf{L}(v) = l$. As shown in Rules NOTIFY/NOTIFYALL, when a CV v is notified, one/all instance(s) of v is/are removed from the bag Wt stored in the permission locked of the lock associated with v, if any. An obligation for a CV v is loaded/unloaded when that obligation is also loaded/unloaded onto/from the bag Ot stored in the permission locked of the lock associated with v, as shown in Rules CHARGEOBLIGATION and DISCHARGEOBLIGATION. However, an obligation for v is discharged only if after this discharge we have $0 < Wt(v) \Rightarrow 0 < Ot(v)$, which is enforced by the assertion $0 < Wt(v) \Rightarrow 1 < Ot(v)$ in the precondition of the rule DISCHARGEOBLIGATION.

3 Publishable Obligations

To verify deadlock-freedom of the program shown in Fig. 1 it is necessary to transfer an obligation of v from the thread running leave to the thread whose ticket number equals to the new value of *active*, if there exists such a thread. However, it is impossible to verify deadlock-freedom of this program using the proof system introduced by Hamin *et al.* [6,9] due to the following reasons: (1) this approach does not allow obligations to be transferred through broadcasts, because it is not clear which of the waiting threads should receive the transferred obligations, and (2) (even if such a transfer is possible) this approach allows obligations to be transferred through notifications only if there is a thread waiting for that condition variable ($0 < Wt(v)$), because it ensures that a thread will immediately receive the transferred obligations. Note that in this program there might be a situation where the thread whose ticket number equals the new value of *active* is not waiting for v but has been already awoken (by a broadcast issued beforehand) and is waiting to reacquire the lock associated with v.

INITCV

$$\{\mathsf{ucond}(v) * \mathsf{ulock}(l, Wt, Ot, It)\}\ \mathsf{g\_initc}(v)$$
$$\{\lambda\_.\ \mathsf{cond}(v, M, M') * \mathsf{ulock}(l, Wt, Ot, It) \wedge \mathsf{L}(v) = l\}$$

WAIT

$$\{\mathsf{cond}(v, M, M') * \mathsf{locked}(l, Wt, Ot, It) * \mathsf{I}(l)(Wt \uplus \{v\}, Ot, It \cup \{(v, id)\}) * \mathsf{obs}(O \uplus \{l\}) \wedge$$
$$l = \mathsf{L}(v) \wedge v \prec O \wedge l \prec O \uplus M' \wedge 0 < Ot(v)\}\ \mathsf{wait}(v, l)$$
$$\{\lambda\_.\ \mathsf{cond}(v, M, M') * M * \exists Wt', Ot', It'.\ \mathsf{I}(l)(Wt', Ot', It') *$$
$$(v, id) \in It'\ ?\ \mathsf{locked}(l, Wt', Ot', It' - \{(v, id)\}) * \mathsf{obs}(O \uplus \{l\})\ :$$
$$\mathsf{locked}(l, Wt', Ot', It') * \mathsf{obs}(O \uplus \{l\} \uplus M')\}$$

NOTIFYALL

$$\{\mathsf{obs}(O \uplus ((v, id) \in It\ ?\ M'\ :\ \{\})) * \mathsf{cond}(v, M, M') * \mathsf{locked}(\mathsf{L}(v), Wt, Ot, It) * (\mathop{*}_{i:=1}^{Wt(v)} M)\}$$
$$\mathsf{notifyAll}(v)\ \{\lambda\_.\ \mathsf{obs}(O) * \mathsf{cond}(v, M, M') * \mathsf{locked}(\mathsf{L}(v), Wt[v{:=}0], Ot, It - \{(v, id)\})\}$$

Fig. 4. New proof rules verifying deadlock-freedom of monitors allowing obligations to be transferred through broadcasts.

We address the problems above by extending the Hamin *et al.* approach such that, under some circumstances, we allow obligations to be transferred through broadcasts even if there is no thread waiting for the related condition variable. To this end, in addition to the bags Wt and Ot, we introduce a new set It consisting of condition variable-identifier pairs, such that for any condition variable v and identifier id if $(v, id) \in It$ then there exists a thread such that (1) this thread either (1–1) is waiting for v, or (1–2) is awoken by a notification on v but has not yet reacquired the lock associated with v and it also has not received any obligation through that notification, and (2) if (v, id) is removed from It then the transferred obligations, which are transferred through a notification on v, are loaded onto this thread. Having this set, we allow some obligations to be transferred through broadcasts on v from the notifying thread to a thread in It, if there exists such a thread.

The updated version of the proof rules is shown in Fig. 4, where, in addition to the bags Wt and Ot, the permissions locked/ulock and the lock invariants are parametrized over the set It (see Rules INITCV). As shown in Rule NOTIFYALL, the transferred obligations are discharged from the notifying thread only if the set It, stored in the associated locked permission, is not empty. After this notification a thread is removed from It, if any. As shown in Rule WAIT a thread having a permission $\mathsf{locked}(l, Wt, Ot, It)$ can wait for the condition variable associate with l only if there exists an identifier id such that the assertion resulting from applying the invariant of l to $Wt \uplus \{v\}, Ot$, and $It \cup s\{(v, id)\}$ is consumed, because this thread is going to wait for v. When this thread is awoken the result of applying the invariant of the associated lock to the two new bags Wt' and Ot' and the new set It' is produced for the thread. If (v, id) is still in It' it means that this thread has not yet been the target of an obligation transfer, hence the transferred obligations are not loaded onto this thread. Additionally, this thread is removed from the set It', stored in the related locked permission (because it is awoken and has acquired the associated lock). If (v, id) is not in It' anymore it means that this thread has been the target of an obligation transfer, hence the transferred obligations must be loaded onto this thread.

NEWCOUNTER
$\{\mathsf{true}\}\ \mathsf{g\_new\_ctr}\ \{\lambda g.\ \mathsf{ctr}(g, 0)\}$

INCCOUNTER
$\{\mathsf{ctr}(g, n)\}\ \mathsf{g\_inc}(g)\ \{\lambda\_.\ \mathsf{ctr}(g, n{+}1) * \mathsf{tic}(g)\}$

DECCOUNTER
$\{\mathsf{ctr}(g, n) * \mathsf{tic}(g)\}\ \mathsf{g\_dec}(g)\ \{\lambda\_.\ \mathsf{ctr}(g, n{-}1) \wedge 0{<}n\}$

NEWICOUNTER
$\{\mathsf{true}\}\ \mathsf{g\_new\_ictr}\ \{\lambda g.\ \mathsf{ictr}(g, 0)\}$

INCICOUNTER
$\{\mathsf{ictr}(g, n)\}\ \mathsf{g\_iinc}(g)\ \{\lambda\_.\ \mathsf{ictr}(g, n{+}1) * \mathsf{itic}(g)\}$

Fig. 5. Ghost counters and ghost monotonic counters.

It can be proved that any program verified by the mentioned proof rules, where the verification starts from an empty bag of obligations and also ends with such a bag, never deadlocks, i.e. it always has a running thread, not waiting for any waitable object such as a condition variable or a lock, until it terminates. We know that for any waitable object o all of these proof rules preserve the invariant $0 < Wt(o) \Rightarrow 0 < Ot(o)$, where $Wt(o)$ and $Ot(o)$ denote the total number of waiting threads and obligations for o in the system, respectively. Additionally, these rules ensure that a thread waits for a waitable object o only if the level of o is lower than the levels of the obligations of that thread. Note that these two invariants hold even when some obligations are transferred using notifyAll. Now consider a deadlocked state, where each thread of a verified program is waiting for an object. Among all of these objects take the one having a minimal wait level, namely o_{min}. By the invariants above there exists a thread having an obligation for o_{min} that is waiting for an object whose level is lower than the level of o_{min}, which contradicts minimality of the level of o_{min}.

Using the proposed proof rules it is possible to verify deadlock-freedom of the program shown in Fig. 1, as shown in Figs. 6 and 7. Note that to verify this program it is necessary to (1) keep track of the number of the elements in the set It, and (2) make sure that the variable *next* is not decreased at all. To meet the first condition we use the notion of *ghost counters* and corresponding *ghost counter tickets* [6], both of which are a particular kind of ghost resources[6]. Specifically, as shown in Fig. 5, there are three operations on a ghost counter: g_new_ctr, g_inc, and g_dec, where g_new_ctr allocates a new ghost counter whose *value* is zero and returns a *ghost counter identifier* g for it; g_inc(g) increments the value of the ghost counter with identifier g and produces a *ticket* for the counter; and g_dec(g) consumes a ticket for the ghost counter g and decrements the ghost counter's value. Since these are the only operations that manipulate

[6] Some logics for program verification, such as Iris [16], include general support for defining ghost resources such as our ghost counters. In particular, our ghost counters can be obtained in Iris as an instance of the *authoritative monoid* [16, p. 5].

$\mathsf{tvm}(\mathsf{tvm}\ t, \mathsf{gctr}\ gt, \mathsf{cond}\ v) ::= \mathsf{lock}(t.l) * \mathsf{cond}(v, \mathsf{true}, \{v\}) \wedge$
$\mathsf{R}(t.l) < \mathsf{R}(v) \wedge \mathsf{L}(v){=}t.l \wedge \mathsf{I}(t.l) = \mathsf{linv}(t) \wedge gt = t.gt \wedge v = t.v$

$\mathsf{linv}'(\mathsf{tvm}\ t) ::= \lambda Wt.\ \lambda Ot.\ \lambda It.$
$\exists active, next.\ t.active \overset{.5}{\longmapsto} active * m.next \overset{.5}{\longmapsto} next \wedge active \leqslant next * \mathsf{ictr}(t.gn, next) \wedge$
$(active < next \Rightarrow 0 < Ot(t.v)) \wedge$
$(next \leqslant active + 1 \Rightarrow Wt(t.v) = 0) \wedge$
$It = \{(v, id) \mid active{<}id{<}next\} *$
$\exists Tr.\ \mathsf{ctr}(t.gt, Tr) \wedge (active + 1 < next \Rightarrow next - active - 1 \leqslant Tr \wedge Tr \leqslant |It|)$

$\mathsf{linv}(\mathsf{tvm}\ t) ::= \lambda Wt.\ \lambda Ot.\ \lambda It.$
$\mathsf{linv}'(t)(Wt, Ot, It) * \exists ac, nx.\ t.active \overset{.5}{\longmapsto} ac * m.next \overset{.5}{\longmapsto} nx$

routine $\mathsf{new\_tvm}()\{$
req : $\{\mathsf{true}\}$
$l := \mathsf{new\_lock}();$
$\{\mathsf{ulock}(l, \{\!\}, \{\!\}, \{\}) \wedge \mathsf{R}(l){=}r - 1\}\ gt := \mathsf{g\_new\_ctr};\ gn := \mathsf{g\_new\_ictr};$
$\{\mathsf{ulock}(l, \{\!\}, \{\!\}, \{\}) * \mathsf{ctr}(gt, 0) * \mathsf{ictr}(gn, 0)\}$
$v := \mathsf{new\_cvar}();\ \mathsf{g\_initc}(v);$
$\{\mathsf{ulock}(l, \{\!\}, \{\!\}, \{\}) * \mathsf{ctr}(gt, 0) * \mathsf{ictr}(gn, 0) * \mathsf{cond}(v, \mathsf{true}, \{v\}) \wedge \mathsf{R}(v){=}r \wedge \mathsf{L}(v){=}l\}$
$t{:=}\mathsf{tvm}(l{:=}l, v{:=}v, active := \mathsf{new\_int}(0), next := \mathsf{new\_int}(0), gt{:=}gt, gn{:=}gn);$
$\mathsf{g\_initl}(l);\ t$
ens : $\{\lambda t.\ \mathsf{tvm}(t, gt, v) \wedge \mathsf{R}(v){=}r\}\}$

routine $\mathsf{enter}(\mathsf{tvm}\ t)\{$
req : $\{\mathsf{obs}(O) * \mathsf{tvm}(t, gt, v) \wedge v \prec O\}$
$\mathsf{acquire}(t.l);$
$\{\mathsf{obs}(O \uplus \{l\}) * \mathsf{tvm}(t, gt, v) * \exists Wt, Ot, It.\ \mathsf{locked}(l, Wt, Ot, It) * \mathsf{linv}(t)(Wt, Ot, It)\}$
$mine := t.next{+}{+};\ \mathsf{g\_ictr\_inc}(t.gn);\ \mathsf{g\_ctr\_inc}(gt);$
$\mathsf{if}(t.next = t.active{+}1)\ \mathsf{g\_charge}(v);$
inv : $\mathsf{tvm}(t, gt, v) * \mathsf{tic}(gt) * \exists active, next.\ t.active \overset{.5}{\longmapsto} active * m.next \overset{.5}{\longmapsto} next \wedge$
$mine < next * \exists Wt, Ot, It.\ \mathsf{locked}(l, Wt, Ot, It) *$
$mine \leqslant active\ ?\ \mathsf{obs}(O \uplus \{l, v\}) * \mathsf{linv}'(t)(Wt, Ot, It) :$
$\mathsf{obs}(O \uplus \{l\}) * \mathsf{linv}'(t)(Wt \uplus \{v\}, Ot, It \cup \{(v, mine)\})$
$\mathsf{while}(t.active < mine)$
$\mathsf{wait}(t.v, t.l);$
$\{\mathsf{obs}(O \uplus \{l, v\}) * \mathsf{tvm}(t, gt, v) *$
$\exists Wt, Ot, It.\ \mathsf{locked}(l, Wt, Ot, It) * \mathsf{linv}(t)(Wt, Ot, It) * \mathsf{tic}(gt)\}$
$\mathsf{release}(t.l)$
ens : $\{\mathsf{obs}(O \uplus \{v\}) * \mathsf{tvm}(t, gt, v) * \mathsf{tic}(gt)\}\}$

Fig. 6. The proof of deadlock-freedom of the program shown in Fig. 1 (part one).

ghost counters or ghost counter tickets, it follows that the value of a ghost counter g is always equal to the number of tickets for g in the system. Similarly, to meet the second condition we use a *ghost monotonic counter*, which is similar to a ghost counter but does not have an operation g_dec.

As shown in Figs. 6 and 7 the precondition and the postcondition of each routine are denoted by **req** and **ens**. Note that we use separation logic [26] to reason about the ownership of permissions. As shown in the routine new_tvm, after creating the lock l a permission ulock with two bags Wt and Ot, and a set It, which are initially empty, is produced for this lock. Additionally, we assign an arbitrary level, which is a level lower than the level of v, to this lock. Using the ghost commands g_new_ctr and g_new_ictr a ghost counter gt and a ghost monotonic counter gn is added to the resources of the current thread.

After creating and initializing the condition variable v a permission cond is produced for this condition variable where no permission but an obligation for v is transferred through any notification on this condition variable. We also assign r and l as the level and the associated lock of this condition variable, respectively. After creating the resulting tvm structure t, we initialize the lock l by consuming the assertion resulting from applying the invariant of l to the bags and the set in the permission ulock, i.e. $\mathsf{linv}(t)(\{\!\!\}, \{\!\!\}, \{\})$. Note that gt and gn are two ghost fields added to the structure tvm that hold the identifiers of the ghost counters gt and gn. Lastly, we encapsulate all the resulting permissions in a predicate tvm. This predicate is used in the precondition of the routine enter. Another precondition of this routine is that the level v must be lower that the levels of all obligations of the current thread ($v \prec O$), because this thread might wait for v and l (note that the level of l is lower than the level of v, which by $v \prec O$ implies $l \prec O$). After acquiring l, this lock is loaded onto the obligations of the current thread and the invariant of this lock is produced. After incrementing the ghost counters gt and gn (we leak a ticket $\mathsf{itic}(gn)$), and (conditionally) loading an obligation of v, the invariant of the loop, denoted by inv, holds. Note that this invariant ensures that if the body of the loop is executed ($t.active < mine$) there exists an obligation for v in the system ($0 < Ot(v)$), which is implied by $mine < next$. After execution of $\mathsf{wait}(v, l)$ there exists two bags Wt' and Ot' and a set It' such that the permissions described by $\mathsf{linv}(t)(Wt', Ot', It')$ is produced for this thread. If $(v, mine) \notin It'$, which means $mine \leqslant active$, then this thread receives the transferred obligations, which is $\{\!\!\{v\}\!\!\}$. Otherwise, if $(v, mine) \in It'$, which means $active < mine$, it does not receive these obligations and again waits for v. Lastly, after releasing l the post condition of this routine shows that an obligation of v is loaded onto the current thread and a permission $\mathsf{tic}(gt)$ is produced. As shown in the specification of the routine leave, this routine requires a permission $\mathsf{tic}(gt)$ and discharges an obligation for v. This permission ensures that if a thread is waiting to enter ($active + 1 < next$) the set It is not empty ($0 < |It|$), which makes it possible to transfer an obligation of v through the command notifyAll. If there is no thread waiting to enter we can safely discharge an obligation for v. Before releasing the lock l, we decrement the counter gt to be able to consume the assertion resulting from applying the invariant of l to the bags and the set in the related locked permission. The routine main is deadlock-free, because starting from an empty bag of obligations, the verification of this routine ends with such a bag too.

```
routine leave(tvm t){
req : {obs(O⊎{v}) ∗ tvm(t, gt, v) ∗ tic(gt) ∧ v ≺ O}
 acquire(t.l);
 {obs(O⊎{l, v}) ∗ tic(gt) ∗ tvm(t, gt, v) ∗
  ∃ Wt, Ot, It. locked(l, Wt, Ot, It) ∗ linv(t)(Wt, Ot, It)}
 t.active++; if(t.next = t.active) g_discharge(v);
 notifyAll(t.v); g_ctr_dec(gt);
 {obs(O⊎{l}) ∗ tvm(t, gt, v) ∗ ∃ Wt, Ot, It. locked(l, Wt, Ot, It) ∗ linv(t)(Wt, Ot, It)}
 release(t.l)
ens : {obs(O) ∗ tvm(t, gt, v)}}

routine main(){
req : {obs({})}
 t:=new_tvm;
 {obs({}) ∗ tvm(t, gt, v) ∧ R(v) = r}
 fork(
   inv : {obs({}) ∗ tvm(t, gt, v)}
   while(1){
     enter(t); // get service
     {obs({v}) ∗ tvm(t, gt, v) ∗ tic(gt)}
     leave(t)});
 {obs({}) ∗ tvm(t, gt, v)}
 enter(t); // get service
 {obs({v}) ∗ tvm(t, gt, v) ∗ tic(gt)}
 leave(t)
ens : {obs({})}}
```

Fig. 7. The proof of deadlock-freedom of the program shown in Fig. 1 (part two).

4 Publishable Resources

In Sect. 3 we proved that the program in Fig. 1 is deadlock-free, i.e. there is at least one thread in this program running. However, a stronger liveness property of this program is starvation-freedom, which implies that any thread calling enter eventually gets service (under the fairness assumption that every thread is scheduled to execute infinitely often). To achieve this goal we need to make sure that the loop in the routine enter eventually terminates, which means we need to provide a *variant* (ranking function) for this loop, e.g. an expression whose range is restricted to the non-negative integers and its value is decreased in each iteration of the loop.

Such a variant can be established for this loop if after each broadcast any thread waiting for v is ensured that the new value of the variable *active* is greater than the one before executing the wait command. This *information* must be transferred from the notifying thread to *all* of the waiting threads.

A first attempt to transfer this information is to create a ghost counter gp, keeping track of the value of *active*, and to transfer a ticket for this counter from the notifying threads to the ones notified. However, this attempt fails because

NEWPCOUNTER
$\{\mathsf{true}\}\ \mathsf{g\_new\_pctr}\ \{\lambda g.\ \mathsf{pctr}(g, 0)\}$

INCPCOUNTER
$\{\mathsf{pctr}(g, n)\}\ \mathsf{g\_pinc}(g)\ \{\lambda\_.\ \mathsf{pctr}(g, n{+}1) * \mathsf{ptic}(g, 1, 0)\}$

DECPCOUNTER
$\{\mathsf{pctr}(g, n) * \mathsf{ptic}(g, n', p) \wedge 0 < n'\}\ \mathsf{g\_pdec}(g)\ \{\lambda\_.\mathsf{ctr}(g, n{-}1) * \mathsf{ptic}(g, n'{-}1, p) \wedge n' \leqslant n\}$

MINIMUM
$\mathsf{pctr}(g, n) * \mathsf{ptic}(g, n', p) \Rightarrow \mathsf{pctr}(g, n) * \mathsf{ptic}(g, n', p) \wedge n' + p \leqslant n$

MERGE
$\mathsf{ptic}(g, n_1, p_1) * \mathsf{ptic}(g, n_2, p_2) \Leftrightarrow \mathsf{ptic}(g, n_1{+}n_2, p_1{+}p_2)$

$\mathsf{publishable}(\mathsf{ptic}(g, n, p)) = \mathsf{ptic}(g, 0, n) \qquad \mathsf{unpublishable}(\mathsf{ptic}(g, n, p)) = \mathsf{ptic}(g, 0, p{+}n)$

Fig. 8. Publishable ghost counters.

the notifying thread requires $Wt(v)$ instances of this ticket (while it can produce only one instance of this ticket).

To address this problem, instead of a ghost counter, we use a *publishable ghost counter*[7], shown in Fig. 8. Tickets of this counter consist of two values; the first value indicates the number of the tickets of the current thread which have not been published to any other thread, and the second value indicates the number of the tickets which have been already published to some other threads. A permission $\mathsf{ptic}(g, u, p)$, for example, indicates that the current thread has a total of $u{+}p$ tickets for g; u tickets unpublished and p tickets published. Accordingly, the value of the counter, shown in the predicate pctr indicates the maximum number of the tickets that a thread can have. Obviously, a thread having a ticket $\mathsf{ptic}(g, u, p)$ can publish u tickets for g ($\mathsf{ptic}(g, 0, u)$), and after publishing these tickets it has a ticket $\mathsf{ptic}(g, 0, p{+}u)$, which is not publishable anymore. Accordingly, we update the proof rule for notifyAll, as shown in Fig. 9, such that it transfers a publishable part of a resource from the notifying thread to the notified ones, and makes this resource unpublishable.

Having this rule, it is possible that the notifying thread in Fig. 1 having a ticket $\mathsf{ptic}(gp, 1, 0)$, publishes a ticket $\mathsf{ptic}(gp, 0, 1)$ to all of the waiting threads and informs them that the value of *active* has been increased, where gp is a publishable ghost counter keeping track of the value of *active*. Accordingly, the verification of this program is shown in Fig. 10 where the loop in the routine enter has an appropriate variant implying the termination of that loop[8].

[7] This conuter is an instance of a *max and plus united monoid* [22].

[8] Note that this loop terminates because (1) (we assume that) each thread is scheduled to execute infinitely often, (2) for any thread waiting for v there is an obligation for v which is discharged in a finite number of steps, and (3) each time that an obligations for v is discharged (and v is notified) the variant of this loop is decreased.

As shown in the routine new_tvm in this figure, a publishable ghost counter gp is created which keeps track of the value of $active$. As shown in the routine enter, this counter as well as a ticket $\mathsf{ptic}(gp, 0, activemin)$ is asserted in the invariant of the loop. Note that a ticket $\mathsf{ptic}(gp, 0, 0)$ can be produced by increasing and then decreasing the counter gp. Since whenever the thread is notified it receives a ticket $\mathsf{ptic}(gp, 0, 1)$ from the notifier thread, the value of $activemin$ in this ticket is increased (using the Rule MERGE in Fig. 8), which implies that the variant of the loop ($mine - active$) is decreased in each iteration. Additionally, by Rule MINIMUM we have $activemin \leqslant active$, and by the loop invariant we have $active \leqslant mine$, which implies that after each iteration the variant of the loop remains non-negative ($activemin \leqslant mine$). As shown in the routine leave, since the variable $active$ is increased, we can obtain a ticket $\mathsf{ptic}(gp, 1, 0)$ using the ghost command $\mathsf{g\_pctr\_inc}(t.gp)$. The publishable part of this ticket ($\mathsf{ptic}(gp, 0, 1)$) is transferred to the waiting threads through the command notifyAll, and the unpublishable part ($\mathsf{ptic}(gp, 0, 1)$) is leaked in this routine.

NOTIFYALL

$$\{\mathsf{obs}(O \uplus ((v, id) \in It \mathbin{?} M' : \{\!\!\}\!)) * \mathsf{cond}(v, M, M') * \mathsf{locked}(\mathsf{L}(v), Wt, Ot, It) * M_p \wedge \mathsf{publishable}(M_p) = M\}\ \mathsf{notifyAll}(v)$$
$$\{\lambda\_.\ \mathsf{obs}(O) * \mathsf{cond}(v, M, M') * \mathsf{locked}(\mathsf{L}(v), Wt[v{:=}0], Ot, It - \{(v, id)\}) * \mathsf{unpublishable}(M_p)\}$$

Fig. 9. Transferring publishable resources through broadcasts, where $\mathsf{publishable}(M_p)$ denotes the publishable part of the resource M_p and $\mathsf{unpublishable}(M_p)$ denotes the unpublishable part of M_p which remains after publishing M_p.

Publishable Points-to Resource. The notion of publishable resource can be also used to prove some other properties. For example, consider a publishable version of a points-to permission $x \xmapsto{(u,p)} v$, where u is the fraction of the ownership of the location x which has not been published, p is the fraction of the ownership of the location x which has been published, $\mathsf{publishable}(x \xmapsto{(u,p)} v) = x \xmapsto{(0,u)} v$, and $\mathsf{unpublishable}(x \xmapsto{(u,p)} v) = x \xmapsto{(0,p+u)} v$. Although a resource $x \xmapsto{(0,1)} v$ gives no information about the value stored in the location x and gives no permission to access x, a resource $x_1 \xmapsto{(0,1)} v_1 * x_2 \xmapsto{(0,1)} v_2$ ensures that $x_1 \neq x_2$. However, transferring or publishing these publishable and unpublishable resources through other synchronizations such as locks and channels requires future investigations.

$\mathsf{tvm}(\mathsf{tvm}\ t, \mathsf{gctr}\ gt, \mathsf{cond}\ v) ::= \mathsf{lock}(t.l) * \mathsf{cond}(v, \mathsf{ptic}(t.gp, 0, 1), \{v\}) \land$
$\mathsf{R}(t.l) < \mathsf{R}(v) \land \mathsf{L}(v){=}t.l \land \mathsf{I}(t.l) = \mathsf{linv}(t) \land gt = t.gt \land v = t.v$

$\mathsf{linv}'(\mathsf{tvm}\ t) ::= \lambda Wt.\ \lambda Ot.\ \lambda It.$
$\exists active, next.\ t.active \overset{.5}{\longmapsto} active * m.next \overset{.5}{\longmapsto} next \land active \leqslant next * \mathsf{ictr}(t.gn, next)$
$\land\ (active < next \Rightarrow 0 < Ot(t.v)) \land (next \leqslant active + 1 \Rightarrow Wt(t.v) = 0) \land$
$It = \{(v, id) \mid active{<}id{<}next\} *$
$\exists Tr.\ \mathsf{ctr}(t.gt, Tr) \land (active + 1 < next \Rightarrow next - active - 1 \leqslant Tr \land Tr \leqslant |It|)$

$\mathsf{linv}(\mathsf{tvm}\ t) ::= \lambda Wt.\ \lambda Ot.\ \lambda It.\ \mathsf{linv}'(t)(Wt, Ot, It) *$
$\exists active, next.\ t.active \overset{.5}{\longmapsto} active * m.next \overset{.5}{\longmapsto} next * \mathsf{pctr}(t.gp, active)$

routine $\mathsf{new\_tvm}()\{$ **req** : $\{\mathsf{true}\}$
$l := \mathsf{new\_lock}();$
$\{\mathsf{ulock}(l, \{\!\}, \{\!\}, \{\}) \land \mathsf{R}(l){=}r - 1\}\ gt := \mathsf{g\_new\_ctr};\ gn := \mathsf{g\_new\_ictr};\ gp := \mathsf{g\_new\_pctr};$
$\{\mathsf{ulock}(l, \{\!\}, \{\!\}, \{\}) * \mathsf{ctr}(gt, 0) * \mathsf{ictr}(gn, 0) * \mathsf{pctr}(gp, 0)\}$
$v := \mathsf{new\_cvar}();\ \mathsf{g\_initc}(v);$
$\{\mathsf{ulock}(l, \{\!\}, \{\!\}, \{\}) * \mathsf{ctr}(gt, 0) * \mathsf{ictr}(gn, 0) * \mathsf{pctr}(gp, 0) *$
$\mathsf{cond}(v, \mathsf{ptic}(gp, 0, 1), \{v\}) \land \mathsf{R}(v){=}r \land \mathsf{L}(v){=}l\}$
$t{:=}\mathsf{tvm}(l{:=}l, v{:=}v, active := \mathsf{new\_int}(0), next := \mathsf{new\_int}(0), gt{:=}gt, gn{:=}gn, gp{:=}gp);$
$\mathsf{g\_initl}(l);\ t$ **ens** : $\{\lambda t.\ \mathsf{tvm}(t, gt, v) \land \mathsf{R}(v){=}r\}\}$

routine $\mathsf{enter}(\mathsf{tvm}\ t)\{$ **req** : $\{\mathsf{obs}(O) * \mathsf{tvm}(t, gt, v) \land v \prec O\}$
$\mathsf{acquire}(t.l);$
$\{\mathsf{obs}(O \uplus \{l\}) * \mathsf{tvm}(t, gt, v) * \exists Wt, Ot, It.\ \mathsf{locked}(l, Wt, Ot, It) * \mathsf{linv}(t)(Wt, Ot, It)\}$
$mine := t.next{+}{+};\ \mathsf{g\_ictr\_inc}(t.gn);\ \mathsf{g\_ctr\_inc}(gt);$
$\mathsf{if}(t.next = t.active{+}1)\ \mathsf{g\_charge}(v);$
inv : $\mathsf{tvm}(t, gt, v) * \mathsf{tic}(gt) * \exists active, next.\ t.active \overset{.5}{\longmapsto} active * m.next \overset{.5}{\longmapsto} next *$
$\exists activemin.\ \mathsf{pctr}(t.gp, active) * \mathsf{ptic}(t.gp, 0, activemin) \land$
$mine < next * \exists Wt, Ot, It.\ \mathsf{locked}(l, Wt, Ot, It) *$
$mine \leqslant active\ ?\ \mathsf{obs}(O \uplus \{l, v\}) * \mathsf{linv}'(t)(Wt, Ot, It) :$
$\mathsf{obs}(O \uplus \{l\}) * \mathsf{linv}'(t)(Wt \uplus \{v\}, Ot, It \cup \{(v, mine)\})$
var : $mine - activemin$
$\mathsf{while}(t.active < mine)$
$\quad \mathsf{wait}(t.v, t.l);$
$\{\mathsf{obs}(O \uplus \{l, v\}) * \mathsf{tvm}(t, gt, v) *$
$\exists Wt, Ot, It.\ \mathsf{locked}(l, Wt, Ot, It) * \mathsf{linv}(t)(Wt, Ot, It) * \mathsf{tic}(gt)\}$
$\mathsf{release}(t.l)$
ens : $\{\mathsf{obs}(O \uplus \{v\}) * \mathsf{tvm}(t, gt, v) * \mathsf{tic}(gt)\}\}$

routine $\mathsf{leave}(\mathsf{tvm}\ t)\{$ **req** : $\{\mathsf{obs}(O \uplus \{v\}) * \mathsf{tvm}(t, gt, v) * \mathsf{tic}(gt) \land v \prec O\}$
$\mathsf{acquire}(t.l);$
$\{\mathsf{obs}(O \uplus \{l, v\}) * \mathsf{tic}(gt) * \mathsf{tvm}(t, gt, v) *$
$\exists Wt, Ot, It.\ \mathsf{locked}(l, Wt, Ot, It) * \mathsf{linv}(t)(Wt, Ot, It)\}$
$t.active{+}{+};\ \mathsf{g\_pctr\_inc}(t.gp);\ \mathsf{if}(t.next = t.active)\ \mathsf{g\_discharge}(v);$
$\mathsf{notifyAll}(t.v);\ \mathsf{g\_ctr\_dec}(gt);$
$\{\mathsf{obs}(O \uplus \{l\}) * \mathsf{tvm}(t, gt, v) * \exists Wt, Ot, It.\ \mathsf{locked}(l, Wt, Ot, It) * \mathsf{linv}(t)(Wt, Ot, It)\}$
$\mathsf{release}(t.l)$
ens : $\{\mathsf{obs}(O) * \mathsf{tvm}(t, gt, v)\}\}$

Fig. 10. The proof of termination of the monitor shown in Fig. 1.

5 Related Works

Separation Logic. Jung *et al.* [16] proposed a concurrent separation logic, namely *Iris*, for reasoning about safety of concurrent programs, as the logic in logical relations, to reason about type-systems and data-abstraction. In this logic user-defined protocols on shared state are expressed through *partial commutative monoids* and is enforced through *invariants*. However, the Iris program logic and many other logics such as [23,25] only prove per-thread safety (i.e. no thread ever crashes): their adequacy theorems state that the program does not reach a state where some thread cannot make a step. This works because these logics do not consider blocking constructs, where a thread may legitimately be stuck temporarily.

Liveness Properties. Several approaches to verify termination [5,21,28], total correctness [3], and lock freedom [10] of concurrent programs have been proposed. These approaches are only applicable to non-blocking algorithms, where the suspension of one thread cannot lead to the suspension of other threads. Consequently, they cannot be used to verify deadlock-freedom of programs using monitors and channels, where the suspension of a notifying/sending thread might lead a waiting thread to be infinitely blocked. In [24] a compositional approach to verify termination of multi-threaded programs is introduced, where *rely-guarantee reasoning* is used to reason about each thread individually while there are some assertions about other threads. In this approach a program is considered to be terminating if it does not have any infinite computations. As a consequence, it is not applicable to programs using monitors because a waiting thread that is never notified cannot be considered as a terminating thread.

Verification of Monitors Through Model Checking. There are also some other approaches addressing some common synchronization bugs of programs in the presence of condition variables. In [35], for example, an approach to identify some potential problems of concurrent programs consisting waits and notifies commands is presented. However, it does not take the order of execution of theses commands into account. In other words, it might accept an undesired execution trace where the waiting thread is scheduled after the notifying thread, that might lead the waiting thread to be infinitely suspended. [17] uses Petri nets to identify some common problems in multithreaded programs such as data races, lost signals, and deadlocks. However the model introduced for condition variables in this approach only covers the communication of two threads and it is not clear how it deals with programs having more than two threads communicating through condition variables. Recently, [1,4] have introduced an approach ensuring that every thread synchronizing under a set of condition variables eventually exits the synchronization block if that thread eventually reaches that block. This approach succeeds in verifying one of the applications of condition variables, namely the buffer. However, since this approach is not modular and relies on a Petri net analysis tool to solve the termination problem, it suffers from a long verification time when the size of the state space is increased, such that the verification of a buffer application having 20 producer and 18 consumer threads, for example, takes more than two minutes.

Modular Verification of Channels. Inspired by the notion of capabilities [18,19] and implicit dynamic frames [29–31], Leino *et al.* later integrated deadlock prevention into a verification system for an object-oriented and imperative programming language. In this approach each thread trying to receive a message from a channel must spend one credit for that channel, where a credit for a channel is obtained if a thread is obliged to discharge an obligation for that channel. A thread can discharge an obligation for a channel if it either sends a message on that channel or delegates that obligation to another thread. However this approach is not general enough to support condition variables, since, unlike channels, a notification on a condition variable is lost if there is no thread waiting for that condition variable.

Modular Verification of Monitors. Recently, Hamin *et al.* [6,7] introduced a modular approach to verify deadlock-freedom of monitors. Later, in [9] this approach is extended by allowing obligations to be transferred through notifications. However, in this approach obligations can be transferred using notify(v) but not using notifyAll(v). This constraint does not allow this approach to verify monitors implementing queue locking algorithms, for example, where the releasing thread notifies all the waiting threads. Additionally, in this approach it is impossible to transfer a permission, indicating that the awoken threads are approaching the front of the waiting queue, from the notifying thread to all of the notified threads. In this paper we introduce a mechanism which allows obligations to be transferred from the notifying thread, executing notifyAll, to one of the notified threads. Additionally, we introduce a notion of publishable resources which can be published from a thread to many other threads.

6 Conclusion

In this paper we introduce two mechanisms to transfer obligations and publishable resources through broadcasts. The first mechanism allows to transfer an obligation through notifyAll from the thread holding a resource to the next thread, which acquires that resource. The second mechanism allows to publish a resource from a notifying thread to many other notified threads. These mechanisms allow the modular approaches, verifying deadlock-freedom of monitors, to verify a wider range of interesting programs. Additionally, these mechanisms allows these approaches to verify termination of the loop in which the wait command is executed (the loop that checks the waiting condition), which ensures that a thread will not infinitely wait in a monitor. The notion of publishable resources introduced in this paper, used to prove termination of the loops, can be also used to prove some other properties. However, transferring or publishing these resources through other synchronizations such as locks and channels requires future investigations.

References

1. de Carvalho Gomes, P., Gurov, D., Huisman, M.: Specification and verification of synchronization with condition variables. In: Artho, C., Ölveczky, P.C. (eds.) FTSCS 2016. CCIS, vol. 694, pp. 3–19. Springer, Cham (2017). https://doi.org/10.1007/978-3-319-53946-1_1
2. Craig, T.: Building fifo and priorityqueuing spin locks from atomic swap. Technical Report TR 93-02-02, University of Washington, February 1993. (ftp tr/1993 ... (1993)
3. D'Osualdo, E., Farzan, A., Gardner, P., Sutherland, J.: Tada live: compositional reasoning for termination of fine-grained concurrent programs. arXiv preprint arXiv:1901.05750 (2019)
4. Gomes, P.d.C., Gurov, D., Huisman, M., Artho, C.: Specification and verification of synchronization with condition variables. Sci. Comput. Program. **163**, 174–189 (2018)
5. Hamin, J., Jacobs, B.: Modular verification of termination and execution time bounds using separation logic. In: 2016 IEEE 17th International Conference on Information Reuse and Integration (IRI), pp. 110–117. IEEE (2016)
6. Hamin, J., Jacobs, B.: Deadlock-free monitors. In: Ahmed, A. (ed.) ESOP 2018. LNCS, vol. 10801, pp. 415–441. Springer, Cham (2018). https://doi.org/10.1007/978-3-319-89884-1_15
7. Hamin, J., Jacobs, B.: Deadlock-free monitors: extended version. TR CW712, Department of Computer Science, KU Leuven, Belgium (2018). Full version http://www.cs.kuleuven.be/publicaties/rapporten/cw/CW712.abs.html
8. Hamin, J., Jacobs, B.: Deadlock-free monitors and channels, June 2019. https://doi.org/10.5281/zenodo.3363702
9. Hamin, J., Jacobs, B.: Transferring obligations through synchronizations. In: Donaldson, A.F. (ed.) 33rd European Conference on Object-Oriented Programming (ECOOP 2019). Leibniz International Proceedings in Informatics (LIPIcs), vol. 134, pp. 19:1–19:58. Schloss Dagstuhl-Leibniz-Zentrum fuer Informatik, Dagstuhl, Germany (2019). https://doi.org/10.4230/LIPIcs.ECOOP.2019.19, http://drops.dagstuhl.de/opus/volltexte/2019/10811
10. Hoffmann, J., Marmar, M., Shao, Z.: Quantitative reasoning for proving lock-freedom. In: 2013 28th Annual IEEE/ACM Symposium on Logic in Computer Science (LICS), pp. 124–133. IEEE (2013)
11. Jacobs, B.: Provably live exception handling. In: Proceedings of the 17th Workshop on Formal Techniques for Java-like Programs, p. 7. ACM (2015)
12. Jacobs, B.: Verifast 18.02. zenodo (2018). http://doi.org/10.5281/zenodo.1182724
13. Jacobs, B., Bosnacki, D., Kuiper, R.: Modular termination verification of single-threaded and multithreaded programs. ACM Trans. Program. Lang. Syst. (TOPLAS) **40**(3), 12 (2018)
14. Jacobs, B., Smans, J., Philippaerts, P., Vogels, F., Penninckx, W., Piessens, F.: Verifast: a powerful, sound, predictable, fast verifier for C and Java. NASA Form. Methods **6617**, 41–55 (2011)
15. Jacobs, B., Smans, J., Piessens, F.: A quick tour of the verifast program verifier. In: Ueda, K. (ed.) APLAS 2010. LNCS, vol. 6461, pp. 304–311. Springer, Heidelberg (2010). https://doi.org/10.1007/978-3-642-17164-2_21
16. Jung, R., et al.: Iris: monoids and invariants as an orthogonal basis for concurrent reasoning. ACM SIGPLAN Not. **50**(1), 637–650 (2015)

17. Kavi, K.M., Moshtaghi, A., Chen, D.J.: Modeling multithreaded applications using Petri nets. Int. J. Parallel Program. **30**(5), 353–371 (2002)
18. Kobayashi, N.: A type system for lock-free processes. Inf. Comput. **177**(2), 122–159 (2002)
19. Kobayashi, N.: A new type system for deadlock-free processes. In: Baier, C., Hermanns, H. (eds.) CONCUR 2006. LNCS, vol. 4137, pp. 233–247. Springer, Heidelberg (2006). https://doi.org/10.1007/11817949_16
20. Leino, K.R.M., Müller, P., Smans, J.: Deadlock-free channels and locks. In: Gordon, A.D. (ed.) ESOP 2010. LNCS, vol. 6012, pp. 407–426. Springer, Heidelberg (2010). https://doi.org/10.1007/978-3-642-11957-6_22
21. Liang, H., Feng, X., Shao, Z.: Compositional verification of termination-preserving refinement of concurrent programs. In: Proceedings of the Joint Meeting of the Twenty-Third EACSL Annual Conference on Computer Science Logic (CSL) and the Twenty-Ninth Annual ACM/IEEE Symposium on Logic in Computer Science (LICS), p. 65. ACM (2014)
22. Mokhov, A.: United monoids (2019). https://blogs.ncl.ac.uk/andreymokhov/united-monoids/
23. Nanevski, A., Ley-Wild, R., Sergey, I., Delbianco, G.A.: Communicating state transition systems for fine-grained concurrent resources. In: Shao, Z. (ed.) ESOP 2014. LNCS, vol. 8410, pp. 290–310. Springer, Heidelberg (2014). https://doi.org/10.1007/978-3-642-54833-8_16
24. Popeea, C., Rybalchenko, A.: Compositional termination proofs for multi-threaded programs. In: Flanagan, C., König, B. (eds.) TACAS 2012. LNCS, vol. 7214, pp. 237–251. Springer, Heidelberg (2012). https://doi.org/10.1007/978-3-642-28756-5_17
25. Raad, A., Villard, J., Gardner, P.: CoLoSL: concurrent local subjective logic. In: Vitek, J. (ed.) ESOP 2015. LNCS, vol. 9032, pp. 710–735. Springer, Heidelberg (2015). https://doi.org/10.1007/978-3-662-46669-8_29
26. Reynolds, J.C.: Separation logic: a logic for shared mutable data structures. In: Proceedings of the 17th Annual IEEE Symposium on Logic in Computer Science, 2002, pp. 55–74. IEEE (2002)
27. da Rocha Pinto, P., Dinsdale-Young, T., Gardner, P., Sutherland, J.: Modular termination verification for non-blocking concurrency. In: Thiemann, P. (ed.) ESOP 2016. LNCS, vol. 9632, pp. 176–201. Springer, Heidelberg (2016). https://doi.org/10.1007/978-3-662-49498-1_8
28. Rowe, R.N., Brotherston, J.: Automatic cyclic termination proofs for recursive procedures in separation logic. In: Proceedings of the 6th ACM SIGPLAN Conference on Certified Programs and Proofs, pp. 53–65. ACM (2017)
29. Smans, J., Jacobs, B., Piessens, F.: Implicit dynamic frames. In: Proceedings of the 10th ECOOP Workshop on Formal Techniques for Java-like Programs, pp. 1–12 (2008)
30. Smans, J., Jacobs, B., Piessens, F.: Implicit dynamic frames: combining dynamic frames and separation logic. In: Drossopoulou, S. (ed.) ECOOP 2009. LNCS, vol. 5653, pp. 148–172. Springer, Heidelberg (2009). https://doi.org/10.1007/978-3-642-03013-0_8
31. Smans, J., Jacobs, B., Piessens, F.: Implicit dynamic frames. ACM Trans. Program. Lang. Syst. (TOPLAS) **34**(1), 2 (2012)
32. Microsoft; API Specification: Sleepconditionvariablecs function (2019). https://docs.microsoft.com/en-us/windows/desktop/api/synchapi/nf-synchapi-sleepconditionvariablecs

33. Oracle; API Specification: Using condition variables (2019). https://docs.oracle.com/cd/E19120-01/open.solaris/816-5137/sync-21067/index.html
34. The Single UNIX; API Specification: Wait on a condition (2019). http://pubs.opengroup.org/onlinepubs/007908799/xsh/pthread_cond_wait.html
35. Wang, C., Hoang, K.: Precisely deciding control state reachability in concurrent traces with limited observability. In: McMillan, K.L., Rival, X. (eds.) VMCAI 2014. LNCS, vol. 8318, pp. 376–394. Springer, Heidelberg (2014). https://doi.org/10.1007/978-3-642-54013-4_21

Taming Concurrency for Verification Using Multiparty Session Types

Kirstin Peters[1,2(✉)], Christoph Wagner[1], and Uwe Nestmann[1]

[1] TU Berlin, Berlin, Germany
[2] TU Darmstadt, Darmstadt, Germany
kirstin.peters@cs.tu-darmstadt.de

Abstract. The additional complexity caused by concurrently communicating processes in distributed systems render the verification of such systems into a very hard problem. Multiparty session types were developed to govern communication and concurrency in distributed systems. As such, they provide an efficient verification method w. r. t. properties about communication and concurrency, like communication safety or progress. However, they do not support the analysis of properties that require the consideration of concrete runs or concrete values of variables. We sequentialise well-typed systems of processes guided by the structure of their global type to obtain interaction-free abstractions thereof. Without interaction, concurrency in the system is reduced to sequential and completely independent parallel compositions. In such abstractions, the verification of properties such as e. g. data-based termination that are not covered by multiparty session types, but rely on concrete runs or values of variables, becomes significantly more efficient.

Keywords: Concurrency · Verification · Multiparty session types

1 Introduction

Modern society is increasingly dependent on large-scale software systems that are distributed, collaborative, and communication-centred. One of the techniques developed to handle the additional complexity caused by distributed actors are *multiparty session types* (MPST) [19]. MPST allow to specify the desired behaviour of communication protocols as by-design correct types that are used to verify the communication structure of software products. The properties guaranteed by well-typed processes cover communication safety (all processes conform to globally agreed communication protocols) and liveness properties such as deadlock-freedom. Their main advantage is that their verification method is extremely efficient—in comparison to e. g. standard model checking.

MPST were developed to govern communication and concurrency in distributed systems. However, as it is typical for type systems, standard MPST variants (without dependable types) do not support the analysis of properties that require the consideration of concrete runs or concrete values of variables.

R. M. Hierons and M. Mosbah (Eds.): ICTAC 2019, LNCS 11884, pp. 196–215, 2019.
https://doi.org/10.1007/978-3-030-32505-3_12

The hardest part about the verification of distributed systems is the state space explosion that results from concurrent communication attempts, i. e., the exponential blow-up that results from computing all possible combinations of potential communication partners. The problem of concurrency mainly lies in the communication structure, which is already completely captured by MPST. We show that the knowledge of a program/system to be well-typed, allows us to sequentialise it following the structure of its global type and thereby to remove all communication. Accordingly, we show how we can benefit from the effort we spend on an MPST analysis of a system also for the verification of its properties that go beyond its communication structure.

We use the global type of a well-typed system to guide its sequentialisation. We refer to the result as *sequential global process* (SGP), although it might still contain parallel compositions, albeit only on completely independent parts. Since the structure of communication was already verified by the well-typedness proof, we can reduce communication to value updates. More precisely, we map well-typed systems that interact concurrently, to SGP-systems without any interaction mechanisms or name binders. Such SGP-systems consist of a vector of variables with values and a SGP-process that simulates the data flow of the original system. Therefore, we translate the reception of data in communication into updates of the vector in the SGP-system. By removing the communication we remove also the problem of state space explosion. Our translation is valid if the considered process is well-typed w. r. t. a (set of) global type(s). Thereby, we sequentialise communications that may happen concurrently in the original system but are sequential in global types. Note that such communications are always causally independent of each other, thus ordering them does not significantly influence the behaviour of the system, e. g. it does not influence what values are computed. Apart from such sequentialisations the original system and its abstraction into a SGP-system behave similarly.

Contributions. We provide an algorithm to remove communication from well-typed systems and thereby sequentialise them, while preserving the evolution of data of the original system. Deriving this algorithm was technically challenging but the result is a simple rewriting function and easy to automate.

Then we prove that, provided that the original system was well-typed, the algorithm produces a SGP-system that is closely related to the original system: the original system and its abstraction are related by a variant of operational correspondence [14] and are coupled similar [23]. With that, the derived SGP-system is a good abstraction of the original system that can be used instead of the original to verify properties on concrete data. Since the mapping into SGP-systems is usually linear and because SGP-systems do not contain any form of interaction or binders, properties can be checked more efficiently.

Finally, we provide a mapping—that is again a simple rewriting algorithm—from SGP-processes into Promela, the input language of the model checker Spin [16,17]. With that, the properties that are not already guaranteed by the MPST analysis but require the consideration of concrete runs or concrete data can be checked. Since the main challenge here is the sequentialisation of concurrent systems into interaction-free abstractions, the translation of SGP-systems into Promela is simple and can be used as a role model to obtain similar mappings for other model checkers.

Related Work. Intuitively, the technique that we present in this paper is a special case of partial order reduction (compare e. g. to [24]) as they can be found in model checkers. This technique tries to reduce the state space that has to be inspected in verification, by identifying different sequences of transitions that lead to similar states. Here, instead of searching for such similar states, we follow the structure of the global type, where well-typedness ensures that the generated abstraction captures the complete state space of the original system modulo coupled similarity.

The approach of [1] is very similar to this paper. Just as our algorithm, they rewrite a program (written in Haskell) by replacing communication with value updates, to obtain a sequential abstraction of the program on that verification—e. g. of termination based on values that are computed at runtime—can be done efficiently. The main difference is that [1] requires that the considered programs satisfy *symmetric non-determinism* whereas we require that programs are well-typed using MPST. Assuming asynchronous communication, symmetric non-determinism means that every receive in a given program location can receive only messages from either a single process or a set of symmetric processes, i. e., processes running the same code, at the same program location. MPST are more flexible, i. e., are not limited to systems that satisfy symmetric non-determinism. Hence, the method presented here can be applied to a larger class of programs.

Interestingly, we find a similar idea also in papers about the verification of distributed algorithms via invariants that use so-called *standard forms* (compare e. g. to [9,29]), where the global view gets constructed by gathering and combining all local processes. In case of [29] standard forms have their own TLA-like semantics that is 1-to-1 correspondent to the calculus semantics for proving properties on data. The main difference to these approaches is that we completely remove communication and present an algorithm to automatically derive this global view from a given well-typed system and its global type.

In [6] global types are translated into processes to mediate between multiparty and binary session types. These mediator processes capture the behaviour of global types—w. r. t. the communication structure and not values—to provide a disciplined communication exchange that allows to translate MPST into binary sessions. In contrast to this approach, we map processes onto processes and use global types to guide this mapping, where the communication structure is removed and our focus is on the evolution of data.

Choreographies [22] are global descriptions of distributed systems from which the distributed system is generated by endpoint projection. In contrast, we start with the distributed system and its global type. Note that global types describe solely the communication structure, i. e., interactions, of the system and do not contain any other implementation details of single peers. With that, MPST have an advantage in comparison to choreographies in industrial settings, where different parts are developed independently. Moreover, we also consider the case of interleaved MPST sessions. Our challenge is to derive a global description on how the data evolves in the system. This is related to the extraction of choreographies from distributed systems as discussed in [7,10]. However, without the global type as guide, the described algorithms to extract choreographies are exponential, whereas our algorithm is usually linear.

In [5] MPST are extended by assertions that allow to verify properties on data values provided that these properties are satisfied in all runs. In contrast, our approach allows to efficiently compute the exact values that are computed in concrete runs. Moreover, the language of assertions is limited to the language defined in [5] and an extension might require to redo some proofs, whereas here only the translation into Promela, i. e., the use of a concrete model checker, forces us to limit the languages of expressions and properties. The algorithm to sequentialise systems into SGP-systems does not rely on such limitations. A prominent example of a property that cannot be analysed statically is termination of a loop after computing some value. To prove such properties, a type system can use dependent types such as e. g. in [28]. In contrast to such extensions of MPST with dependable types, we do not add any complexity to the type system (or provide any new variant of MPST). Instead we provide a simple rewriting algorithm that transforms a well-typed system (after the type check) into an abstraction on that remaining properties on data can be verified with existing specialised tools.

Overview. In Sect. 2 we introduce multiparty session types very briefly. Section 3 describes how well-typed systems are sequentialised. Section 3.1 introduces a calculus for sequential global processes, Sect. 3.2 describes an algorithm to map into SGP-systems for the case of synchronous MPST and single sessions, and Sect. 3.3 discusses asynchronous variants of MPST and extends the algorithm to cover interleaved sessions. Section 4 shows operational correspondence and relates the original system and its abstraction by coupled similarity. Then, Sect. 5 illustrates how the sequentialisation can be used to verify properties of the original system. It discusses the limits of this method, i. e., what kind of properties cannot be analysed this way and presents a mapping from SGP-processes into Promela. We conclude in Sect. 6. Missing proofs and additional material can be found in [26].

2 Multiparty Session Types in a Nutshell

Our aim is to use the structure of a global type to remove communication—and with that the related concurrency—from the problem of verifying properties on the evolution of values. We conjecture that this procedure can be used for all kind of MPST variants but explain the method on a simple variant of synchronous MPST w. r. t. a single session. Later we extend our algorithm to asynchronous MPST with interleaved sessions. To explain the basic idea, we use a variant of multiparty synchronous session types as introduced in [2] with some alternations similar to variants as e. g. in [4,11,30].

MPST were developed to govern communication and concurrency in distributed systems. Therefore, systems are checked against a *global type*. Global types specify the desired communication structure from a global point of view. The specified communication structure of a global type describes a *session* and the participants of such a session are called *roles*. Here, they are given by

$$G ::= \mathrm{r}_1 \to \mathrm{r}_2 : \left\{ \mathsf{l}_i \left\langle \tilde{\mathsf{U}}_i \right\rangle . G_i \right\}_{i \in \mathrm{I}} \quad | \quad G_1, G_2 \quad | \quad (\mu \mathrm{t})\, G \quad | \quad \mathrm{t} \quad | \quad \mathsf{end}$$

The first construct specifies a communication from Role r_1 to r_2 that offers different branches for the receiver with respect to a label l_i that is transmitted by the sender, where $\tilde{\mathsf{U}}_i$ are the sorts (i. e., base types) of the transmitted values. If I is a singleton, we abbreviate communication with $\mathrm{r}_1 \to \mathrm{r}_2 : \mathsf{l}\left\langle \tilde{\mathsf{U}} \right\rangle . G$. The other constructs introduce parallel composition, recursion, and successful termination.

The systems, that we want to analyse, are modelled in a *session calculus*. As usual, we use an extension of the π-calculus [21] given by

$$\begin{aligned} P ::= {} & \bar{\mathsf{a}}[2..\mathrm{n}](\mathsf{s}).P \quad | \quad \mathsf{a}(\mathsf{s}[\mathrm{r}]).P \quad | \quad \mathsf{s}[\mathrm{r}_1, \mathrm{r}_2]!\mathsf{l}\langle \tilde{\mathsf{e}} \rangle .P \quad | \quad \mathsf{s}[\mathrm{r}_2, \mathrm{r}_1]?\, \{ \mathsf{l}_i(\tilde{x}_i).P_i \}_{i \in \mathrm{I}} \\ | \quad & \mathsf{if}\ \mathsf{c}\ \mathsf{then}\ P_1\ \mathsf{else}\ P_2 \quad | \quad P_1 \mid P_2 \quad | \quad \mathbf{0} \quad | \quad (\nu \mathsf{s})P \quad | \quad (\mu \mathrm{X})\, P \quad | \quad \mathrm{X} \end{aligned}$$

The first two constructs are used to initiate a session. The next two constructs model the sender and the receiver of a communication within a session, where the $\tilde{x}_i$ are *(input bounded) variables* that are instantiated as result of a communication by the received values. The remaining constructs introduce conditionals, parallel composition, termination, restriction, and recursion. Since we want to use the model checker Spin later, we restrict expressions e (the values that are transmitted in communication) and conditions c (used to guide conditionals) to functions that are known by Promela, the input language of Spin.

We use structural congruence ($\equiv$) to abstract from syntactically different but semantically similar processes, where $\equiv$ is the least congruence that satisfies alpha-conversion ($\equiv_\alpha$) and the rules:

$$\begin{gathered} P \mid \mathbf{0} \equiv P \qquad P_1 \mid P_2 \equiv P_2 \mid P_1 \qquad P_1 \mid (P_2 \mid P_3) \equiv (P_1 \mid P_2) \mid P_3 \\ (\mu \mathrm{X})\, P \equiv P \left\{ {}^{(\mu \mathrm{X})P} / \mathrm{X} \right\} \qquad (\nu \mathsf{s})(\nu \mathsf{s}')P \equiv (\nu \mathsf{s}')(\nu \mathsf{s})P \qquad (\nu \mathsf{s})\mathbf{0} \equiv \mathbf{0} \\ (\nu \mathsf{s})(P_1 \mid P_2) \equiv P_1 \mid (\nu \mathsf{s})P_2 \quad \mathsf{if}\ \mathsf{s} \notin \mathsf{fn}(P_1) \end{gathered}$$

The reduction semantics of the session calculus is given by the rules:

$$(\mathsf{Link})\ \frac{}{\bar{\mathsf{a}}[2..\mathrm{n}](\mathsf{s}).P_1 \mid \mathsf{a}(\mathsf{s}[2]).P_2 \mid \ldots \mid \mathsf{a}(\mathsf{s}[\mathrm{n}]).P_n \longmapsto (\nu \mathsf{s})(P_1 \mid P_2 \mid \ldots \mid P_n)}$$

$$(\mathsf{Com})\ \frac{j \in \mathrm{I}}{\mathsf{s}[\mathrm{r}_1, \mathrm{r}_2]!\mathsf{l}_j\langle \tilde{\mathsf{e}} \rangle .P \mid \mathsf{s}[\mathrm{r}_2, \mathrm{r}_1]?\, \{ \mathsf{l}_i(\tilde{x}_i).P_i \}_{i \in \mathrm{I}} \longmapsto P \mid \left(P_j \left\{ {}^{\tilde{\mathsf{e}}} / {}_{\tilde{x}_j} \right\} \right)}$$

$$(\mathsf{If\text{-}T})\ \frac{\mathsf{c}}{\mathsf{if}\ \mathsf{c}\ \mathsf{then}\ P_1\ \mathsf{else}\ P_2 \longmapsto P_1} \qquad (\mathsf{If\text{-}F})\ \frac{\neg \mathsf{c}}{\mathsf{if}\ \mathsf{c}\ \mathsf{then}\ P_1\ \mathsf{else}\ P_2 \longmapsto P_2}$$

$$(\mathsf{Par})\ \frac{P_1 \longmapsto P_1'}{P_1 \mid P_2 \longmapsto P_1' \mid P_2} \qquad (\mathsf{Res})\ \frac{P \longmapsto P'}{(\nu \mathsf{s})P \longmapsto (\nu \mathsf{s})P'}$$

$$(\mathsf{Struc})\ \frac{P_1 \equiv P_2 \quad P_2 \longmapsto P_2' \quad P_2' \equiv P_1'}{P_1 \longmapsto P_1'}$$

The Rule Link initialises a session s on the roles $1, \ldots, n$, where 1 requested the session on channel a and each i participates in the session as P_i. Communication within a session s is described by Rule Com, where in the case of matching roles and labels the continuations of sender and receiver are unguarded and the variables $\tilde{x}$ are replaced by the values $\tilde{e}$ in the receiver. The Rules If-T and If-F reduce conditionals as expected. The remaining rules allow for steps in various contexts and are standard.

Let $r(\cdot)$ return the roles used in a global type or a process. A process P has an *actor on* $c[r_1]$ if P has an unguarded subterm of the form $\bar{c}[2..n](s).P$ with $r_1 = 1$ or $c(s[r_1]).P$ (for session invitations) or an unguarded subterm of the form $c[r_1, r_2]!l\langle\tilde{e}\rangle.P$ or $c[r_1, r_2]?\{l_i(\tilde{x}_i).P_i\}_{i\in I}$ (for communication). Let $\mathsf{act}(P)$ be set of actors in P. If unambiguous, i. e., if there is only one session, we omit the session channel and abbreviate actors by their role.

The processes P are checked against their specification in *type judgements* $\Gamma \vdash P \triangleright \Delta$, where Γ, Δ are *type environments* that are built from the type information in the global type. A system that passes such a type check is denoted as *well-typed.* The design of MPST guarantees strong properties for the communication structure of well-typed systems.

Theorem 1. *Assume $\Gamma \vdash P \triangleright \Delta$, i. e., P is well-typed.*

Subject Reduction: *If $P \longmapsto P'$ then there is Δ' such that $\Gamma \vdash P' \triangleright \Delta'$.*
Linearity: *P has no two unguarded senders/receivers for the same actor.*
Progress: *If $P \longmapsto^* P'$ then either $P' \equiv \mathbf{0}$ or $P' \longmapsto P''$.*

To prove these properties, we have to reason about the *typing rules* that define under which circumstances a type judgement is valid. Due to space limitations, the typing rules as well as some other important aspects of MPST (e. g. projection and local types) and the proofs are postponed to [26]. Note that we do *not* introduce a new variant of MPST. Instead we rely on a standard MPST variant of that we introduced global types and the session calculus, because they are necessary to understand the remainder of this paper.

3 Sequentialising Well-Typed Systems

MPST are designed to analyse the communication structure of a system. Well-typed systems are guaranteed to satisfy properties like communication safety or progress. What remains, are safety and liveness properties that involve data.

We use the global type of a well-typed term to guide the sequentialisation of the implementation. The result is a kind of process that we call *sequential global process* (SGP), although it might still contain parallel compositions but only on completely independent parts. This abstraction of the implementation allows us to analyse properties on the values of data in the implementation without the problem of state space explosion that is caused by the concurrency of communication in the original system.

$$(\mathsf{Ass})\ \frac{}{\langle \mathcal{V}; \tilde{v} := \tilde{\mathsf{e}}.S\rangle \longmapsto \mathsf{eval}(\langle \mathcal{V}(\tilde{v}) := \tilde{\mathsf{e}}; S\rangle)} \qquad (\mathsf{Par})\ \frac{\langle \mathcal{V}; S_1\rangle \longmapsto \langle \mathcal{V}'; S_1'\rangle}{\langle \mathcal{V}; S_1 \parallel S_2\rangle \longmapsto \langle \mathcal{V}'; S_1' \parallel S_2\rangle}$$

$$(\mathsf{If\text{-}T})\ \frac{\mathsf{c}}{\langle \mathcal{V}; \mathsf{if}\ \mathsf{c}\ \mathsf{then}\ S_1\ \mathsf{else}\ S_2\rangle \longmapsto \langle \mathcal{V}; S_1\rangle} \quad (\mathsf{If\text{-}F})\ \frac{\neg\mathsf{c}}{\langle \mathcal{V}; \mathsf{if}\ \mathsf{c}\ \mathsf{then}\ S_1\ \mathsf{else}\ S_2\rangle \longmapsto \langle \mathcal{V}; S_2\rangle}$$

$$(\mathsf{Struc})\ \frac{\mathcal{S}_1 \equiv_{\mathcal{S}} \mathcal{S}_2 \quad \mathcal{S}_2 \longmapsto \mathcal{S}_2' \quad \mathcal{S}_2' \equiv_{\mathcal{S}} \mathcal{S}_1'}{\mathcal{S}_1 \longmapsto \mathcal{S}_1'}$$

Fig. 1. Reduction Semantics of SGP-Systems.

3.1 A Calculus for Sequential Global Processes

SGP-processes are simple processes consisting of assignments of values to variables, conditionals, parallelism, termination, and recursion.

Definition 1 (SGP-Processes). *SGP-processes are given by*

$$S ::= \tilde{v} := \tilde{\mathsf{e}}.S \quad | \quad \mathsf{if}\ \mathsf{c}\ \mathsf{then}\ S_1\ \mathsf{else}\ S_2 \quad | \quad S_1 \parallel S_2 \quad | \quad \mathbf{0} \quad | \quad (\mu \mathrm{X})\, S \quad | \quad \mathrm{X}$$

where $\tilde{\mathsf{e}}$ *are expressions to calculate a value,* c *are boolean conditions, and* X *process variables.*

SGP-processes introduce a new operator to assign values to variables in a vector. An assignment $\tilde{v} := \tilde{\mathsf{e}}.S$ describes a SGP-process that updates the variables $\tilde{v}$ by the values $\tilde{\mathsf{e}}$ and then continues as S, where $\tilde{v} := \tilde{\mathsf{e}}.S$ is short hand for $(v_1, \ldots, v_n) := (\mathsf{e}_1, \ldots, \mathsf{e}_n).S$. If $\tilde{v}$ (and accordingly also $\tilde{\mathsf{e}}$) is the empty sequence, then we abbreviate this empty assignment by $\tau.S$. Note that SGP-processes inherit the parallel operator not from processes but from global types. Thus, the parallel composition $S_1 \parallel S_2$ describes that S_1 and S_2 are independent, i. e., all variables that appear on both sides are used as read-only on both sides. The remaining operators for conditionals, successful termination, and recursion are inherited from processes. Note that SGP-processes do neither contain any interaction mechanisms nor name binders. But we still have branching via conditionals and recursion.

The SGP-processes are combined with a vector $\mathcal{V}$ of variables, that represents the current values of the local variables of all processes of the original distributed system. They consist of the input bounded variables of the implementation. A *SGP-system* $\langle \mathcal{V}; S\rangle$ then consists of a knowledge vector $\mathcal{V}$ and a SGP-process S.

Structural congruence on SGP-processes $\equiv_{\mathsf{S}}$ is the restriction of $\equiv$ on SGP-processes. Let $\equiv_{\mathcal{S}}$ be the least congruence that satisfies the rules $\mathcal{S} \equiv_{\mathcal{S}} \mathsf{eval}(\mathcal{S})$ and $\langle \mathcal{V}; S\rangle \equiv_{\mathcal{S}} \langle \mathcal{V}; S'\rangle$ if $S \equiv_{\mathsf{S}} S'$. We write $\mathcal{V}(\tilde{v}) := \tilde{\mathsf{e}}$ for the result of replacing, for all $v_i \in \tilde{v}$, the current value of the variable v_i in the vector $\mathcal{V}$ by the value that results from the evaluation of the expressions e_i. The semantics of SGP-systems is given in Fig. 1. We naturally extend substitution to SGP-systems, i. e., $\langle \mathcal{V}; S\rangle\, \sigma = \langle \mathcal{V}\sigma; S\sigma\rangle$. Let $\mathsf{eval}(\langle \mathcal{V}; S\rangle)$ be the result of replacing all variables

v in conditions and expressions in S that are not sequentially hidden after an assignment of v by the current value of v in $\mathcal{V}$, e. g. :

$$\begin{aligned}
&\mathsf{eval}(\langle(v = 5); \mathsf{if}\ v > 6\ \mathsf{then}\ \mathbf{0}\ \mathsf{else}\ v := v + 1.v := v + 1.\mathbf{0}\rangle) \\
&= \langle(v = 5); \mathsf{if}\ 5 > 6\ \mathsf{then}\ \mathbf{0}\ \mathsf{else}\ v := 5 + 1.v := v + 1.\mathbf{0}\rangle \\
&\longmapsto \langle(v = 5); v := 5 + 1.v := v + 1.\mathbf{0}\rangle \\
&\longmapsto \mathsf{eval}(\langle(v = 6); v := v + 1.\mathbf{0}\rangle) \quad = \langle(v = 6); v := 6 + 1.\mathbf{0}\rangle \\
&\longmapsto \mathsf{eval}(\langle(v = 7); \mathbf{0}\rangle) \quad = \langle(v = 7); \mathbf{0}\rangle
\end{aligned}$$

3.2 Mapping Well-Typed Systems onto SGP-Systems

We use the global type of a well-typed process P to sequentialise P into a SGP-process. Because of the parallel operator, SGP-processes are not completely sequential. However, since we remove communication and with it all forms of interaction from SGP-processes, parallel composition in SGP-processes is between independent parts only. More precisely, SGP-systems cover read and write operations on their vector of variables that simulate the evolution of knowledge in the original distributed system.

The main idea of the algorithm is simple. We fuse matching senders and receivers, i. e., the receiver that receives a message with the sender that transmitted this message, into a single SGP value assignment. The value assignment captures what the processes gain as new information from a communication. The problem is that finding the matching communication partners in the general π-calculusis NP-hard. In the π-calculusit is possible to have several matching receivers for a single sender or vice versa. Performing a communication step can unguard further senders and receivers. So different choices of matching pairs of communication partners and different orders in that communications are performed influence the further behaviour. To reduce the complexity of this problem, we use the type information that allows us to completely avoid the search for matching communication partners.

Firstly, well-typedness guarantees that there are no races at runtime, i. e., in no state there is more than one matching receiver for a sender and vice versa. This ensures, that for each well-typed system there is indeed a single SGP-abstraction that captures its overall behaviour, whereas without well-typedness (i. e., in the presence of races) several SGP-abstractions might be necessary to describe the behaviour of a single system. Secondly, well-typedness also ensures that there are no orphan communication partners, i. e., each sender will eventually meet a matching receiver and vice versa. Finally, the global type of a well-typed system tells us when and where communication takes place. Or, more precisely, the global type tells us one possible order of the communications and well-typedness ensures that all other possible orderings of communications of the system are similar (see Sect. 4). Accordingly, we do not search for matching communication partners but follow the structure of the global type. If the global type specifies that next there is a communication then we know that in the mentioned actors the respective send and receive action is indeed unguarded or guarded only by conditionals, that can be resolved without interactions.

Similarly, it is difficult in general π-calculussystems to identify at which point we have to introduce a global loop to translate the recursive behaviour of the single actors into a recursion of the global abstraction. Again we follow the structure of the global type and simply introduce a global loop if the global type loops, while ignoring the structure of recursion in the actors and only unfolding local recursion if necessary.

When we remove communication prefixes in order to obtain a SGP-process, we lose their respective scopes. To avoid ambiguities in SGP-systems and to clarify the owner of variables, we indicate input bounded variables, i. e., the variables a SGP-process may write on, by its corresponding actor. The variables indicated by an actor are the local knowledge of this actor. SGP-processes that are derived from well-typed processes will not have write access on variables of other actors but may read them to perform value updates.

The following mapping relies on the fact that the parallel composition $\prod_{i \in \mathrm{I}} P_i$ is well-typed w. r. t. the global type G. We prove in Theorem 2 below, that this mapping indeed produces a SGP-process in this case.

Definition 2. *The partial mapping* $\mathrm{SGP}\left(\{P_i\}_{i\in\mathrm{I}}, G\right)$ *is defined inductively as:*

1. $\mathbf{0}$, *if* $G = \mathsf{end}$
2. X_{t}, *else if* $G = \mathrm{t}$
3. $\mathrm{SGP}\left(\{P'_j\} \cup \{P_i\}_{i\in\mathrm{I}\setminus\{j\}}, G\right)$,
else if there is some $j \in \mathrm{I}$ *such that* $P_j = (\nu \mathsf{s})P'_j$
4. $\mathrm{SGP}\left(\{P_{j1}, P_{j2}\} \cup \{P_i\}_{i\in\mathrm{I}\setminus\{j\}}, G\right)$,
else if there is some $j \in \mathrm{I}$ *such that* $P_j = P_{j1} \mid P_{j2}$
5. $\mathrm{SGP}\left(\{P'_j\,\{^{(\mu\mathrm{X})P'_j}/_{\mathrm{X}}\}\} \cup \{P_i\}_{i\in\mathrm{I}\setminus\{j\}}, G\right)$,
else if there is $j \in \mathrm{I}$ *such that* $P_j = (\mu\mathrm{X})\, P'_j$
6. $\tilde{x}_m @\mathsf{s}[\mathrm{r}] := \tilde{\mathsf{e}}.\, \mathrm{SGP}\left(\{Q_m\,\{^{\tilde{x}_m @\mathsf{s}[\mathrm{r}]}/_{\tilde{x}_m}\}, Q\} \cup \{P_i\}_{i\in\mathrm{I}\setminus\{k,l\}}, G_m\right)$,
else if there are $k, l \in \mathrm{I}$, $m \in \mathrm{J} \subseteq \mathrm{J}'$ *such that*
$G = \mathrm{r}_1 \to \mathrm{r}_2 : \left\{\mathsf{l}_j\left\langle \tilde{\mathsf{U}}_j \right\rangle.G_j\right\}_{j\in\mathrm{J}}$, $P_k = \mathsf{s}[\mathrm{r}_1, \mathrm{r}_2]!\mathsf{l}_m\langle\tilde{\mathsf{e}}\rangle.Q$,
and $P_l = \mathsf{s}[\mathrm{r}_2, \mathrm{r}_1]?\,\{\mathsf{l}_j(\tilde{x}_j).Q_j\}_{j\in\mathrm{J}'}$
7. $\mathrm{SGP}\left(\{P_i\}_{i\in\mathrm{I}_1}, G_1\right) \parallel \mathrm{SGP}\left(\{P_j\}_{j\in\mathrm{I}_2}, G_2\right)$,
else if there are some $\mathrm{I}_1 \cup \mathrm{I}_2 = \mathrm{I}$ *such that* $G = G_1, G_2$,
$\bigcup_{i\in\mathrm{I}_1} \mathsf{r}(P_i) = \mathsf{r}(G_1)$, *and* $\bigcup_{j\in\mathrm{I}_2} \mathsf{r}(P_j) = \mathsf{r}(G_2)$
8. $(\mu\mathrm{X}_{\mathrm{t}})\, \mathrm{SGP}\left(\{P_i\}_{i\in\mathrm{I}}, G'\right)$, *else if* $G = (\mu\mathrm{t})\, G'$
9. $\tau.\, \mathrm{SGP}(\{P'_1, \ldots, P'_n\}, G)$,
else if $\{P_i\}_{i\in\mathrm{I}} = \{\bar{\mathsf{a}}[2..\mathrm{n}](\mathsf{s}).P'_1, \mathsf{a}(\mathsf{s}[2]).P'_2, \ldots, \mathsf{a}(\mathsf{s}[\mathrm{n}]).P'_n\}$
10. $\mathsf{if}\ \mathsf{c}\ \mathsf{then}\ \mathrm{SGP}\left(\{P_{j1}\} \cup \{P_i\}_{i\in\mathrm{I}\setminus\{j\}}, G\right)\ \mathsf{else}\ \mathrm{SGP}\left(\{P_{j2}\} \cup \{P_i\}_{i\in\mathrm{I}\setminus\{j\}}, G\right)$,
else if there is some $j \in \mathrm{I}$ *such that* $P_j = \mathsf{if}\ \mathsf{c}\ \mathsf{then}\ P_{j1}\ \mathsf{else}\ P_{j2}$

Note that the different cases of this definition are ordered. Thus, a conditional is not resolved (Case 10) unless none of the other cases can be applied. The first two cases provide the base cases for global types that are terminated (Case 1) or reduced to a type variable (Case 2). In these two cases the considered processes are ignored. The next three cases do not alter the type but prepare processes by removing restriction on session channels (Case 3)—which is safe because we will also remove all communication prefixes—splitting parallel compositions (Case 4), and unfolding recursion (Case 5). Because we require that process variables are guarded by a communication prefix, we cannot unfold the same recursion twice without applying another case in between.

The next three cases map processes that are well-typed w. r. t. a global type on communication (Case 6), parallel composition (Case 7), and recursion (Case 8), i. e., here we follow the structure of the global type to map the process. Case 6 unifies the sender and the receiver of a communication specified in the global type and maps it on corresponding value assignments. These assignments simulate the reception of the values $\tilde{\mathsf{e}}$ on the variables $\tilde{x}_m$@$\mathsf{s}[\mathsf{r}]$ of the receiver $\mathsf{s}[\mathsf{r}]$, where $\tilde{x}$@$\mathsf{s}[\mathsf{r}] = x_{1\mathsf{s}[\mathsf{r}]}, \ldots, x_{n\mathsf{s}[\mathsf{r}]}$ is the result of indicating the variables with the actor $\mathsf{s}[\mathsf{r}]$ of the receiving end. The substitution of $\tilde{x}_m$ into $\tilde{x}_m$@$\mathsf{s}[\mathsf{r}]$ ensures that names of different parallel branches are not confused.

This substitution does not remove all remaining name clashes but only the harmful clashes between parallel composed components of the considered system. Sequential composed input binders on the same variable and of the same actor are translated to the same name. If we apply this algorithm on processes that are not well-typed, we may still have parallel occurrences of syntactically the same name in parallel composed input binders. But well-typedness ensures that all such occurrences are linked to different actors and are thus distinguished. With that, we unify variables—that might have been denoted the same on purpose—and reduce the vector of variables in the SGP-system. Also, input bounded variables of different iterations of recursion are unified.

Case 7 maps a parallel composition of global types on a parallel composition of SGP-processes. Note that in both cases, parallel composition is between independent objects that have no means of interaction. To split the parallel components of the system accordingly, we rely on their roles. Well-typedness of the system ensures that it can be split as required.

Case 8 introduces recursion if the global type tells us to do so. This case does not alter the considered system or enforces any requirements on the structure of the system. Well-typedness ensures that the structure of the system w. r. t. recursion matches the recursion of the global type, but not necessarily that the system and the global type use recursion at the same time. For example $\overline{\mathsf{a}}[2](\mathsf{s}).\,(\mu \mathrm{X})\,\mathsf{s}[1,2]!\mathsf{l}\langle 5\rangle.\mathsf{s}[1,2]?\mathsf{l}'(x).\mathrm{X} \mid \mathsf{a}(\mathsf{s}[2]).\,(\mu \mathrm{X})\,\mathsf{s}[2,1]?\mathsf{l}(x).\mathsf{s}[2,1]!\mathsf{l}'\langle 6\rangle.\mathrm{X}$ is well-typed w. r. t. $1 \to 2 : \mathsf{l}\langle \mathbb{N}\rangle.\,(\mu \mathrm{t})\,2 \to 1 : \mathsf{l}'\langle \mathbb{N}\rangle.1 \to 2 : \mathsf{l}\langle \mathbb{N}\rangle.\mathrm{t}$, although the global type partially unfolds the recursion in comparison to the recursion of the process. Therefore, we rely on well-typedness and use only the global type to determine the correct place of recursion, where Case 5 allows to unfold recursion in processes. To ensure that the process variables of nested recursions are not confused, the Cases 2 and 8 use the type variable of the global type as index to distinguish process variables.

Case 9 unifies session invitations and the corresponding acceptances and maps them on an empty value assignment τ. The session invitation mechanism is already validated in the well-typedness proof and does not influence the data of the processes in the system. Thus, it is safe to ignore this step in SGP-processes.

Case 10 maps a conditional of one of the parallel components of the system to a conditional in the SGP-process. Global types do not consider conditionals of processes, but well-typedness ensures that both cases of the conditional have to follow the same global type. Because of that, both cases of the SGP-conditional inherit the same global type. To avoid unnecessary branching, we delay the mapping of conditionals until this is necessary e. g. to unguard a sender or receiver of a communication that is specified in the global type.

The mapping in Definition 2 is not deterministic, because it does not enforce an order in that several unguarded conditionals are mapped in Case 10 and this leads to different possible SGP-processes. Similarly, it is not specified in which order several restrictions in Case 3, several parallel compositions in Case 4, or several recursive processes in Case 5 are handled, though different orders in these cases will not lead to different SGP-processes. The other cases are guided by the global type or the fact that there is only one session. To obtain a deterministic version—and simplify the proofs—and to minimize the size of the computed SGP-process, we assume that Definition 2 gives precedence to parallel branches that implement (1) senders and (2) receivers that are unguarded in the global type and (3) smaller roles. However, different orders in that conditionals are handled lead to weakly bisimilar SGP-processes, because only unguarded conditionals are mapped, the translated subprocesses of the conditional are guarded by the resulting SGP-conditional, and an unguarded conditional will reduce to the same case in a single τ-step regardless of when we perform this step.

3.3 Asynchrony and Interleaved Sessions

The mapping in Definition 2 is designed for a synchronous variant of multiparty session types and only single sessions, because the syntax and semantics is simpler in these cases. However, the mapping in Definition 2 is *exactly the same* for the case of multiparty *asynchronous* session types as introduced in [19,20].

Note that the semantics of the session calculus defined in [19,20] use messages queues to reflect the asynchronous nature of communication. Sending and receiving are decoupled into two separate steps to transmit and then read from message queues. Nonetheless, when we remove communication in the mapping $\text{SGP}(\cdot, \cdot)$, we unify sending and receiving into value assignments as described in the Case 6 of Definition 2. This is because, SGP-processes are designed to track the evolution of data values of processes and therefore only the reception of values is relevant. Intuitively, value assignments of SGP-processes reflect the case that a participant of a session has learned new information by the reception of values and this information flow is covered by value assignments. To determine the correct point in the behaviour of the system in that a particular participant gains new information through the reception of values, we rely on the fact that for this communication to happen both communication partners, the sender and the receiver have to be unguarded. Well-typedness and the structure of the global type, guide us in the case of concurrently enabled communication prefixes.

The definition of well-typed processes for several interleaved sessions is more difficult. As described in [3], we have to ensure that actions of different sessions do not cause deadlocks by cyclic dependencies. Processes with only acyclic dependencies between interactions of different sessions are denoted as *globally progressing*. However, adapting SGP(·, ·) to allow for several interleaved sessions in processes that are globally progressing is straightforward. First we remove name clashes between session channels using alpha conversion. Then we adapt the mapping SGP(·, ·) of Definition 2 into SGP'(·, ·), where the latter expects a set $\{P_i\}_{i\in \mathrm{I}}$ and a set $\{(G_j, \mathsf{s}_j)\}_{j\in \mathrm{J}}$ of pairs of global types and session channels as input such that the parallel composition $\prod_{I\in \mathrm{I}} P_i$ is well-typed w.r.t. $\{(G_j, \mathsf{s}_j)\}_{j\in \mathrm{J}}$ and $\prod_{I\in \mathrm{I}} P_i$ does not contain name clashes between session channels.

Definition 3. $\mathrm{SGP'}\Big(\{P_i\}_{i\in \mathrm{I}}, \{(G_j, \mathsf{s}_j)\}_{j\in \mathrm{J}}\Big)$ *is defined inductively as:*

1. *(a)* $\mathbf{0}$, *if* $\mathrm{J} = \emptyset$
   (b) $\mathrm{SGP'}\Big(\{P_i\}_{i\in \mathrm{I}}, \{(G_j, \mathsf{s}_j)\}_{j\in \mathrm{J}\setminus\{k\}}\Big)$,
   else if there is some $k \in \mathrm{J}$ *such that* $G_k = \mathsf{end}$
2. $\mathrm{X}_{\mathcal{G}}$, *else if* $\mathcal{G} = \{G_j\}_{j\in \mathrm{J}} = \{\mathrm{t}_j\}_{j\in \mathrm{J}}$
3. $\mathrm{SGP'}\Big(\{P'_k\} \cup \{P_i\}_{i\in \mathrm{I}\setminus\{k\}}, \{(G_j, \mathsf{s}_j)\}_{j\in \mathrm{J}}\Big)$,
   else if there is $k \in \mathrm{I}$ *such that* $P_k = (\nu \mathsf{s})P'_k$
4. $\mathrm{SGP'}\Big(\{P_{k1}, P_{k2}\} \cup \{P_i\}_{i\in \mathrm{I}\setminus\{k\}}, \{(G_j, \mathsf{s}_j)\}_{j\in \mathrm{J}}\Big)$,
   else if there is some $k \in \mathrm{I}$ *such that* $P_k = P_{k1} \mid P_{k2}$
5. $\mathrm{SGP'}\Big(\{P'_k \{{}^{(\mu \mathrm{X}) P'_k}/_{\mathrm{X}}\}\} \cup \{P_i\}_{i\in \mathrm{I}\setminus\{k\}}, \{(G_j, \mathsf{s}_j)\}_{j\in \mathrm{J}}\Big)$,
   else if there is $k \in \mathrm{I}$ *such that* $P_k = (\mu \mathrm{X})\, P'_k$
6. $\tilde{x}_n@\mathsf{s}[\mathrm{r}] := \tilde{\mathsf{e}}.\,\mathrm{SGP'}(\mathcal{P}, \mathcal{G})$ *with* $\mathcal{P} = \{Q_n\{{}^{\tilde{x}_n@\mathsf{s}[\mathrm{r}]}/_{\tilde{x}_n}\}, Q\} \cup \{P_i\}_{i\in \mathrm{I}\setminus\{m,o\}}$ *and* $\mathcal{G} = \{(G_{l,n})\} \cup \{(G_j, \mathsf{s}_j)\}_{j\in \mathrm{j}\setminus\{L\}}$,
   else if there are $m, o \in \mathrm{I}$, $l \in \mathrm{J}$, $n \in \mathrm{K} \subseteq \mathrm{K}'$ *such that*
   $G_l = \mathrm{r}_1 \to \mathrm{r}_2 : \Big\{\mathsf{l}_k\Big\langle \tilde{\mathrm{U}}_k\Big\rangle.G_{l,k}\Big\}_{k\in \mathrm{K}}$, $P_m = \mathsf{s}[\mathrm{r}_1, \mathrm{r}_2]!\mathsf{l}_\mathsf{n}\langle \tilde{\mathsf{e}}\rangle.Q$,
   and $P_o = \mathsf{s}[\mathrm{r}_2, \mathrm{r}_1]?\{\mathsf{l}_\mathsf{k}(\tilde{x}_k).Q_k\}_{k\in \mathrm{K}'}$
7. *(a)* $\mathrm{SGP'}\Big(\{P_i\}_{i\in \mathrm{I}_1}, \{(G_j, \mathsf{s}_j)\}_{j\in \mathrm{J}_1}\Big) \parallel \mathrm{SGP'}\Big(\{P_i\}_{i\in \mathrm{I}_2}, \{(G_j, \mathsf{s}_j)\}_{j\in \mathrm{J}_2}\Big)$,
   else if there are some $\mathrm{I}_1 \cup \mathrm{I}_2 = \mathrm{I}$, $\mathrm{J}_1 \cup \mathrm{J}_2 = \mathrm{J}$ *such that* $\mathrm{J}_1 \cap \mathrm{J}_2 = \emptyset$ *and* $\bigcup_{i\in \mathrm{I}_k} \mathsf{act}(P_i) = \{\mathsf{s}_\mathsf{j}[\mathrm{r}] \mid j \in \mathrm{J}_k \wedge \mathrm{r} \in \mathsf{r}(G_j)\}$ *for* $k \in \{1, 2\}$
   (b) $\mathrm{SGP'}\Big(\{P_i\}_{i\in \mathrm{I}}, \{(G_{k1}, \mathsf{s}_k), (G_{k2}, \mathsf{s}_k)\} \cup \{(G_j, \mathsf{s}_j)\}_{j\in \mathrm{J}}\Big)$,
   else if there is $k \in \mathrm{J}$ *such that* $G_k = G_{k1}, G_{k2}$
8. $(\mu \mathrm{X}_{\mathcal{G}})\,\mathrm{SGP'}\Big(\{P_i\}_{i\in \mathrm{I}}, \Big\{(G'_j, \mathsf{s}_j)\Big\}_{j\in \mathrm{J}}\Big)$,
   else if $G_j = (\mu \mathrm{t}_j)\, G'_j$ *for all* $j \in \mathrm{J}$ *and* $\mathcal{G} = \{\mathrm{t}_j\}_{j\in \mathrm{J}}$
9. $\tau.\,\mathrm{SGP'}\Big(\{P'_1, \ldots, P'_n\} \cup \{P_i\}_{i\in \mathrm{I}\setminus\{k1,\ldots,kn\}}, \{(G_j, \mathsf{s}_j)\}_{j\in \mathrm{J}}\Big)$,
   else if there are $k1, \ldots, kn \in \mathrm{I}$ *such that* $P_{K1} = \bar{\mathsf{a}}[2..\mathrm{n}](\mathsf{s}).P'_1$, $P_{k2} = \mathsf{a}(\mathsf{s}[2]).P'_2$, …, $P_{kn} = \mathsf{a}(\mathsf{s}[\mathrm{n}]).P'_n$
10. $\mathsf{if}\ \mathsf{c}\ \mathsf{then}\ \mathrm{SGP'}(\{P_{k1}\} \cup \mathcal{P}, \mathcal{G})\ \mathsf{else}\ \mathrm{SGP'}(\{P_{k2}\} \cup \mathcal{P}, \mathcal{G})$
    with $\mathcal{P} = \{P_i\}_{i\in \mathrm{I}\setminus\{k\}}$ *and* $\mathcal{G} = \{(G_j, \mathsf{s}_j)\}_{j\in \mathrm{J}}$,
    else if there is $k \in \mathrm{I}$ *such that* $P_k = \mathsf{if}\ \mathsf{c}\ \mathsf{then}\ P_{k1}\ \mathsf{else}\ P_{k2}$

To deal with multiple sessiones (and their global types), we split Case 1 into a case to introduce the SGP-process $\mathbf{0}$ as soon as the set of considered global types is empty (Case 1a) and a case to remove terminated global types end from the set $\{G_j\}_{j\in J}$ (Case 1b). In a similar way, we split Case 7 into a case to introduce a parallel composition of SGP-processes if the considered sets of processes can be partitioned into two sets that implement the actors of different sessions (Case 7a) and a case to split parallel global types (Case 7b), i. e., to replace $\{G_j\}_{j\in J}$ by $\{G_{k1}, G_{k2}\} \cup \{G_j\}_{j\in J\setminus\{k\}}$ if there is a $k \in J$ such that $G_k = G_{k1}, G_{k2}$. The adaptation of the Cases 3, 4, 5, 6, 9, and 10 to multiple global types is straightforward. The Cases 2 and 8 for recursion, are replaced by variants that require all types of the considered set of global types to be reduced to a type variable or a recursive global type, respectively. With that we follow [3], that similarly requires that interleaved sessions can be joined in recursion. The remaining cases are straightforwardly adapted to sets of types.

Note that an implementation of this algorithm should exploit the acyclic dependency relation that is build according to [3] between interactions of different sessions. This relation tells us for the Cases 6 and 9, whether the required communication partners for a session are already unguarded or guarded by another session. In the latter case this communication will introduce a dependency from another session to this session and the respective case cannot be applied. Similarly, this relation tells us for Case 7a that it is possible to introduce a SGP parallel composition if and only if we can split the set of sessions into two disjoint sets such that there are no dependencies between the sessions in different sets.

We overload the definition of $\text{SGP}^*(\cdot, \cdot)$ for interleaved sessions. Let P be well-typed w. r. t. $\{(G_j, \mathsf{s}_j)\}_{j\in J}$ and $S = \text{SGP'}\left(\{P\}, \{(G_j, \mathsf{s}_j)\}_{j\in J}\right)$. Then the corresponding SGP-system is $\text{SGP}^*\left(\{P\}, \{(G_j, \mathsf{s}_j)\}_{j\in J}\right) = \langle \mathcal{V}; S\rangle$, where $\mathcal{V}$ is the vector of names in S.

Note that the results of Sect. 4 are proved in [26] for both variants: $\text{SGP}(\cdot, \cdot)$ and $\text{SGP'}(\cdot, \cdot)$. As we claim, we can extend this algorithm to all variants of MPST that satisfy linearity, i. e., all MPST variants we are aware of. This also includes variants with session delegation. Delegation can be handled similarly to session invitations using a substitution for the delegated session name.

4 Relating the Implementation and Its Sequentialisation

We show that for all processes P that are well-typed w. r. t. the global types $\{(G_j, \mathsf{s}_j)\}_{j\in J}$, the mapping $\text{SGP}\left(\{P\}, \{(G_j, \mathsf{s}_j)\}_{j\in J}\right)$ is defined and returns a SGP-process. Therefore, we show that all cases of Definition 3 except for Case 8 preserve well-typedness in their recursive calls. By an induction over J and the structure of the respective types, we show then that—after some preparation steps in the Cases 3, 4, 5, and 10 that do not alter the type—well-typedness ensures that the structure of the system is as required by the respective case to reduce the types. The case of a single session—if P is well-typed w. r. t. G then $\text{SGP}(\{P\}, G)$ is a SGP-process—is a special case of the following theorem.

Theorem 2. *If the process P is well-typed w. r. t. $\{(G_j, \mathsf{s}_j)\}_{j \in \mathrm{J}}$ then* $\mathrm{SGP'}\left(\{P\}, \{(G_j, \mathsf{s}_j)\}_{j \in \mathrm{J}}\right)$ *is defined and returns a SGP-process.*

Given a well-typed process, the computation of the mapping and the size of the constructed SGP-process are usually linear in the size of P combined with the sum of the sizes of its types. As discussed in [26], an exponential blow-up cannot be completely avoided but results only from not optimal conditionals, i. e., conditionals that are not only used to branch between alternative labels of an immediately following sender, or from actors that are not influenced by the choice of a branch of a communication in that the sender is preceded by such a conditional. By design, the algorithm in Definition 2 will even in these bad cases minimize the size of the generated SGP-process by delaying the mapping of conditionals as long as possible. In general the computation of the SGP-system is efficient, i. e., usually fast, and the construction does not suffer from the problem of state space explosion, i. e., the generated SGP-system is usually not considerably larger than the original system. Since the construction sequentialises the original system and thereby removes all forms of interaction and restriction, the verification of the SGP-abstraction is much easier than the verification of the original system.

It remains to show, that the SGP-abstraction of a well-typed system that is generated by $\mathrm{SGP}^*(\cdot, \cdot)$ and the original system are semantically similar enough, such that the analysis of the SGP-abstraction allows to verify properties of the original system. Intuitively, a well-typed system and its sequentionalisation into a SGP-system have the same steps, but SGP-systems may force an order on steps that are unordered in the original system. This happens for global types such as $1 \to 2 : \mathsf{l}\langle \mathbb{N} \rangle.3 \to 4 : \mathsf{l}\langle \mathbb{N} \rangle.\mathsf{end}$ that combine causally unrelated communications sequentially.

Example 1. Consider the global type $G = 1 \to 2 : \mathsf{l}\langle \mathbb{N} \rangle.3 \to 4 : \mathsf{l}\langle \mathbb{N} \rangle.\mathsf{end}$ that consists of two causally independent communications. The system

$$\begin{aligned} P = \bar{\mathsf{a}}[2..4](\mathsf{s}).\mathsf{s}[1,2]!!\mathsf{l}\langle 5 \rangle.\mathbf{0} \mid \mathsf{a}(\mathsf{s}[2]).\mathsf{s}[2,1]?\mathsf{l}(x).\mathbf{0} \\ \mid \mathsf{a}(\mathsf{s}[3]).\mathsf{s}[3,4]!!\mathsf{l}\langle 4 \rangle.\mathbf{0} \mid \mathsf{a}(\mathsf{s}[4]).\mathsf{s}[3,4]?\mathsf{l}(x).\mathbf{0} \end{aligned}$$

is a well-typed implementation of this global type. The algorithm of Definition 2 maps this process to the SGP-system $\mathrm{SGP}^*(P, G) = \langle (x_2, x_4) \,;\, S \rangle$, where $S = \tau.x_2{:=}5.x_4{:=}4.\mathbf{0}$. The process P has, modulo structural congruence, two maximal runs

$$P \longmapsto P' \begin{cases} \longmapsto (\nu \mathsf{s})(\mathsf{s}[3,4]!!\mathsf{l}\langle 4 \rangle.\mathbf{0} \mid \mathsf{s}[3,4]?\mathsf{l}(x).\mathbf{0}) \longmapsto \mathbf{0} \\ \longmapsto (\nu \mathsf{s})(\mathsf{s}[1,2]!!\mathsf{l}\langle 5 \rangle.\mathbf{0} \mid \mathsf{s}[2,1]?\mathsf{l}(x).\mathbf{0}) \longmapsto \mathbf{0} \end{cases}$$

where $P' = (\nu \mathsf{s})(\mathsf{s}[1,2]!!\mathsf{l}\langle 5 \rangle.\mathbf{0} \mid \mathsf{s}[2,1]?\mathsf{l}(x).\mathbf{0} \mid \mathsf{s}[3,4]!!\mathsf{l}\langle 4 \rangle.\mathbf{0} \mid \mathsf{s}[3,4]?\mathsf{l}(x).\mathbf{0})$. But the abstraction $\mathrm{SGP}^*(P, G)$ simulates only the sequence of steps at the top

$$\begin{aligned} &\langle (x_2 = 0, x_4 = 0) \,;\, S \rangle \longmapsto \langle (x_2 = 0, x_4 = 0) \,;\, x_2 := 5.x_4 := 4.\mathbf{0} \rangle \\ &\longmapsto \langle (x_2 = 5, x_4 = 0) \,;\, x_4 := 4.\mathbf{0} \rangle \longmapsto \langle (x_2 = 5, x_4 = 4) \,;\, \mathbf{0} \rangle \end{aligned}$$

in that first process 2 receives the value 5—and the SGP-process accordingly updates the variable x_2 of 2—and then 4 receives the value 4.

Except for such sequentialisations from the global type, the original system and its SGP-system are similar. In particular, this means that each step of the SGP-system can be simulated by the original system. Thus, SGP-systems do not introduce new behaviour.

Theorem 3 (Soundness). *If P is well-typed w. r. t. $\mathcal{G}$ then for all $\mathcal{S}'$ such that* $\mathrm{SGP}^*(\{P\}, \mathcal{G}) \longmapsto \mathcal{S}'$ *there exist some $P', \mathcal{G}'$ such that $P \longmapsto P'$, P' is well-typed w. r. t. $\mathcal{G}'$, and* $\mathrm{SGP}^*(\{P'\}, \mathcal{G}') \equiv_{\mathcal{S}} \mathcal{S}'$.

Although the SGP-system can perform intuitively the same steps as the original system, the order that is forced on some steps by the above discussed sequentialisations prevents us from obtaining the same result in the other direction. However, only the order of steps can differ. Because of that, whenever the original system does a step there is a sequence of steps bringing the original system towards a state that can be reached by the SGP-system. To prove this result, we rely on the observation that the behaviour of a SGP-process follows the global type it was constructed from and well-typedness forces processes to similarly follow the specification in their types. Since renamings of input binders change the vector of variables of a SGP-system, we assume that no alpha conversion is used to rename input binders in the following.

Theorem 4 (Completeness). *Let $\mathcal{G} = \{(G_j, \mathsf{s}_j)\}_{j \in \mathrm{J}}$. If the system P is well-typed w. r. t. $\mathcal{G}$ then for all $P \longmapsto P'$ there exist $P'', \mathcal{G}''$ such that $P' \longmapsto^* P''$, P'' is well-typed w. r. t. $\mathcal{G}''$, and* $\mathrm{SGP}^*(\{P\}, \mathcal{G}) \longmapsto^* \mathrm{SGP}^*(\{P''\}, \mathcal{G}'')$.

Interestingly, the combination of Theorem 3 and Theorem 4 is similar to *(weak) operational correspondence* as it is introduced in [14] as criterion for the quality of encodings. Using the results from [25], then the sequentialisation of a system is correspondence similar to the original system, where correspondence simulation $\precsim$ was introduced in [25].

Corollary 1. *If P is well-typed w. r. t. $\mathcal{G} = \{(G_j, \mathsf{s}_j)\}_{j \in \mathrm{J}}$ then* $\mathrm{SGP}^*(P, \mathcal{G}) \precsim P$.

Correspondence similarity is strictly weaker than bisimilarity, but it implies *coupled similarity.* Coupled similarity was introduced in [23] as a weaker variant of bisimilarity that allows to relate the distributed implementation to a global specification. Similarly, we relate the sequentialisation $\mathrm{SGP}^*(P, G)$ to the distributed implementation in P. As explained in [23], bisimilarity is in general too strict to relate a distributed implementation with a global specification. So, following the hierarchy in [12,13], coupled similarity (or the only slightly stronger correspondence simulation) is intuitively the strictest simulation relation that we could expect here.

5 Using SGP-Abstractions for Verification

To verify properties that are based on data values, we can use standard verification techniques such as model checking on the generated SGP-systems. The correspondence simulation $\mathrm{SGP}^*(P, G) \precsim P$ between the SGP-system and the original system P ensures that properties that are valid for the SGP-system are also valid for P, if these properties are preserved modulo $\precsim$. For the presented approach, this is the case for all properties on the values of data variables—that reflect the reception of values in the original system—that do not require to compare different such variables that are updated concurrently in the original system as explained in Example 1.

Accordingly, we cannot use this method to verify properties on the relation between variables that are updated concurrently in the original system. This is because, we use the structure of the global type to sequentialise. If—as in the case of G in Example 1—the global type combines independent communications sequentially, then the mapping $\mathrm{SGP}(P, \mathcal{G})$ forces an order on the corresponding value updates following the global types $\mathcal{G}$ and not the process P.

However, this problem occurs only w. r. t. communications that are causally unrelated, i. e., such properties are in general problematic in distributed systems. Since these values are altered independently in the original system, properties that relate their values will often not hold in all runs. The easiest way to avoid such false positives, is to compute the causal relation of the communications in $\mathcal{G}$. Remember that global types do not contain binders. Thus, computing the causality relation for a single session can be done in a similar way as in the π-calculus (compare e. g. to [27]), but does not have to bother about binders and scope extrusion. To obtain a causality relation for the case of globally interleaved sessions, we combine the relation that captures dependencies between the interactions of different sessions of [3] with the causality within a session. A property of the SGP-system then holds for the original system if it is invariant under different linearisations of this causal order.

To illustrate the verification of system properties, we use the model checker Spin [16,17] and translate SGP-systems into Promela, the input language of Spin. Therefore, we provide an algorithm to translate a SGP-process into Promela code. This algorithm serves as a role model to obtain similar mappings for other model checkers. We choose SPIN to illustrate how our algorithm for well-typed systems compares to the standard partial order reduction techniques that are implemented in SPIN and that work without the type information. Other implementations might prefer a model checker that is specialised on the analyses of data instead of concurrency issues such as NUXMV [8].

First we generate a preamble for the Promela program, i. e., declare variables and set their initial values. The variables are obtained from the vector of variables $\mathcal{V}$ in a SGP-system $\langle \mathcal{V}; S \rangle$. Sometimes the initial values are directly specified by the implementation or are given as parameters of the implementation. Otherwise, the developer has to pick suitable initial values respecting their respective sorts.

The translation into Promela consists of two layers. The outer layer $[\![S]\!]$ creates the proctype that is required by Promela and passes the term onto the inner layer $[\![S]\!]_i$. The inner layer is simple: Variable assignments are translated to variable assignments encapsulated in an atomic block if multiple assignments are done simultaneously. Recursion is implemented by introducing a label and jumping to that label when the recursion variable is called.

Definition 4 (Translation of SGP-Processes into Promela)

$$
\begin{aligned}
[\![S]\!] &= \text{active proctype ModelS() \{} \\
&\qquad [\![S]\!]_i \\
&\qquad \text{LEnd:} \\
&\quad \} \\
[\![\tilde{v} := \tilde{\mathsf{e}}.S]\!]_i &= \text{atomic } \{v_1 = e_1; \ldots ; v_n = e_n;\} \; [\![S]\!]_i \\
[\![\mathsf{if\ c\ then\ } S_1 \mathsf{\ else\ } S_2]\!]_i &= \text{if} \\
&\qquad :: \mathsf{c} \quad\; -> [\![S_1]\!]_i \\
&\qquad :: \text{else} -> [\![S_2]\!]_i \\
&\quad \text{fi} \\
[\![S_1 \parallel S_2]\!]_i &= \text{run}(S_1); \text{run}(S_2) \\
[\![\mathbf{0}]\!]_i &= \text{goto LEnd;} \\
[\![\tau.S]\!]_i &= \text{skip; } [\![S]\!]_i \\
[\![(\mu \mathrm{X})\, S]\!]_i &= \text{LX: } [\![S]\!]_i \\
[\![\mathrm{X}]\!]_i &= \text{goto LX;}
\end{aligned}
$$

where for each run(S_i) *a separate* proctype *is introduced with* $[\![S_i]\!]$.

Note that, if $\tilde{v}$ and $\tilde{e}$ are singletons in $\tilde{v} := \tilde{\mathsf{e}}.S$, the atomic block is omitted.

In [26] we present a toy example to illustrate our approach. It implements a simple auctioneer system consisting of an auctioneer and two bidders. After translating this example into a SGP-system and then into Promela, some properties given as LTL-formulae are verified in SPIN.

Moreover, to visualise the state-space explosion problem, we implemented the key-exchange part of the Needham-Schroeder protocol. We derived the sequentialised (s) version out of the distributed (d) version using our algorithm. The following table shows time and memory needed to check our Promela implementation of the Needham-Schroeder protocol. Spin crashed before it could compute the distributed versions for more than 6 participants.

participants	2(d)	2(s)	4(d)	4(s)	6(d)	6(s)	8(d)	8(s)	10(d)	10(s)
seconds	0.01	0	0.19	0	51.7	0.05	–	1.27	–	36.2
MB	128	128	137	129	1836	138	–	360	–	5809

6 Conclusions

We introduce a mapping from well-typed systems—that are distributed concurrent systems that interact by communication—into SGP-systems—that simulate the information flow of the original system from a global point of view and that do not have interactions. The algorithm to compute these SGP-systems is usually linear in the size of the original system. Without interactions and only finite vectors of variables, the verification of properties is significantly more efficient for SGP-systems than for the original system. The presented mapping ensures that properties that hold for the SGP-system are also valid for the original system modulo coupled similarity. Finally, we present a second mapping from SGP-systems into Promela; the input language of the Spin model checker.

To formalise the relation between the original system and its sequentialisation into a SGP-system, we relate them by correspondence simulation. Correspondence simulation was described in [25] to describe the relation that the criterion operational correspondence forces on processes and their encodings. As discussed in [14,15], operational correspondence is essential to reason about the quality of encodings between process calculi. In this sense, the presented mapping is a good encoding. Moreover, correspondence similarity implies coupled similarity. As discussed in [23], coupled similarity is a good way to relate central specifications—or in our case global sequentialisations—to their distributed implementations. Since bisimilarity is in general too strict to relate the original system and its sequentialisation, coupled similarity (or the only slightly stronger correspondence simulation) is intuitively the strictest simulation relation that we could expect here.

Multiparty session types are already a very efficient verification tool for all properties about the communication structure of systems. The presented sequentialisation allows us to benefit from their efficiency also in the verification of properties that are usually not in the range of type systems, because they require the consideration of concrete runs of the system or the values of variables.

Due to the interleaving of independent actions, the state space of a concurrent system is in the worst case exponentially larger than of its sequentialisation. As an example, we implemented the *Needham-Schroeder public key protocol* with 10 pairs of processes that interact with the same server (see [26]). Spin generated for the original system more than 35 million states (matching more than 154 million states while using more than 7,5GB memory) before crashing after 969 seconds. For the sequentialisation Spin computed the complete model in only 62 seconds generating 75 million states.

The Scribble project [18,31] provides a tool set that allows to specify and check multiparty session types. They also provide a tool to check a given implementation against a given type. The presented algorithms could support such tool sets by increasing the kinds of properties that can be analysed within such a tool set, while the efficiency of such tools is not negatively influenced. In fact, the derivation of SGP-abstractions can be directly integrated into the type check such that SGP-abstracts are produced as a by-product of type checking.

References

1. Bakst, A., Gleissenthall, K.V., Kıcı, R.G., Jhala, R.: Verifying distributed programs via canonical sequentialization. In: Proceedings of the ACM on Programming Languages 1(OOPSLA), 110:1–110:27 (2017). https://doi.org/10.1145/3133934
2. Bejleri, A., Yoshida, N.: Synchronous multiparty session types. Electron. Notes Theor. Comput. Sci. **241**, 3–33 (2009). https://doi.org/10.1016/j.entcs.2009.06.002
3. Bettini, L., Coppo, M., D'Antoni, L., De Luca, M., Dezani-Ciancaglini, M., Yoshida, N.: Global progress in dynamically interleaved multiparty sessions. In: van Breugel, F., Chechik, M. (eds.) CONCUR 2008. LNCS, vol. 5201, pp. 418–433. Springer, Heidelberg (2008). https://doi.org/10.1007/978-3-540-85361-9_33
4. Bocchi, L., Chen, T.-C., Demangeon, R., Honda, K., Yoshida, N.: Monitoring networks through multiparty session types. In: Beyer, D., Boreale, M. (eds.) FMOODS/FORTE -2013. LNCS, vol. 7892, pp. 50–65. Springer, Heidelberg (2013). https://doi.org/10.1007/978-3-642-38592-6_5
5. Bocchi, L., Honda, K., Tuosto, E., Yoshida, N.: A theory of design-by-contract for distributed multiparty interactions. In: Gastin, P., Laroussinie, F. (eds.) CONCUR 2010. LNCS, vol. 6269, pp. 162–176. Springer, Heidelberg (2010). https://doi.org/10.1007/978-3-642-15375-4_12
6. Caires, L., Pérez, J.A.: Multiparty session types within a canonical binary theory, and beyond. In: Albert, E., Lanese, I. (eds.) FORTE 2016. LNCS, vol. 9688, pp. 74–95. Springer, Cham (2016). https://doi.org/10.1007/978-3-319-39570-8_6
7. Carbone, M., Montesi, F., Schürmann, C.: Choreographies, logically. Distrib. Comput. **31**(1), 51–67 (2018). https://doi.org/10.1007/s00446-017-0295-1
8. Cavada, R., et al.: The nuXmv symbolic model checker. In: Biere, A., Bloem, R. (eds.) CAV 2014. LNCS, vol. 8559, pp. 334–342. Springer, Cham (2014). https://doi.org/10.1007/978-3-319-08867-9_22
9. Chandy, K.M., Lamport, L.: Distributed snapshots: determining global states of distributed systems. ACM Trans. Comput. Syst. (TOCS) **3**(1), 63–75 (1985). https://doi.org/10.1145/214451.214456
10. Cruz-Filipe, L., Larsen, K.S., Montesi, F.: The paths to choreography extraction. In: Esparza, J., Murawski, A.S. (eds.) FoSSaCS 2017. LNCS, vol. 10203, pp. 424–440. Springer, Heidelberg (2017). https://doi.org/10.1007/978-3-662-54458-7_25
11. Demangeon, R., Honda, K.: Nested protocols in session types. In: Koutny, M., Ulidowski, I. (eds.) CONCUR 2012. LNCS, vol. 7454, pp. 272–286. Springer, Heidelberg (2012). https://doi.org/10.1007/978-3-642-32940-1_20
12. Glabbeek, R.J.: The linear time — branching time spectrum II. In: Best, E. (ed.) CONCUR 1993. LNCS, vol. 715, pp. 66–81. Springer, Heidelberg (1993). https://doi.org/10.1007/3-540-57208-2_6
13. van Glabbeek, R.: The linear time - branching time spectrum i: the semantics of concrete, sequential processes. Handbook of Process Algebra, pp. 3–99 (2001)
14. Gorla, D.: Towards a unified approach to encodability and separation results for process calculi. Inf. Comput. **208**(9), 1031–1053 (2010). https://doi.org/10.1016/j.ic.2010.05.002
15. Gorla, D., Nestmann, U.: Full abstraction for expressiveness: history, myths and facts. Math. Struct. Comput. Sci. **26**(4), 639–654 (2014). https://doi.org/10.1017/S0960129514000279
16. Holzmann, G.J.: Design and Validation of Computer Protocols. Prentice Hall, Upper Saddle River (1991)

17. Holzmann, G.J.: The model checker SPIN. IEEE Trans. Software Eng. **23**(5), 279–295 (1997). https://doi.org/10.1109/32.588521
18. Honda, K., Mukhamedov, A., Brown, G., Chen, T.-C., Yoshida, N.: Scribbling interactions with a formal foundation. In: Natarajan, R., Ojo, A. (eds.) ICDCIT 2011. LNCS, vol. 6536, pp. 55–75. Springer, Heidelberg (2011). https://doi.org/10.1007/978-3-642-19056-8_4
19. Honda, K., Yoshida, N., Carbone, M.: Multiparty Asynchronous Session Types. In: Proceedings of POPL, vol. 43, pp. 273–284. ACM (2008). https://doi.org/10.1145/1328438.1328472
20. Honda, K., Yoshida, N., Carbone, M.: Multiparty asynchronous session types. J. ACM (JACM) 63(1) (2016). https://doi.org/10.1145/2827695
21. Milner, R., Parrow, J., Walker, D.: A calculus of mobile processes. Inf. Comput. **100**(1), 1–77 (1992). https://doi.org/10.1016/0890-5401(92)90008-4
22. Montesi, F.: Choreographic Programming. Ph.D. thesis, IT University of Copenhagen (2013). http://www.fabriziomontesi.com/files/choreographic_programming.pdf
23. Parrow, J., Sjödin, P.: Multiway synchronization verified with coupled simulation. In: Cleaveland, W.R. (ed.) CONCUR 1992. LNCS, vol. 630, pp. 518–533. Springer, Heidelberg (1992). https://doi.org/10.1007/BFb0084813
24. Peled, D.: Ten years of partial order reduction. In: Hu, A.J., Vardi, M.Y. (eds.) CAV 1998. LNCS, vol. 1427, pp. 17–28. Springer, Heidelberg (1998). https://doi.org/10.1007/BFb0028727
25. Peters, K., van Glabbeek, R.: Analysing and comparing encodability criteria. In: Proceedings of EXPRESS/SOS, EPTCS, vol. 190, pp. 46–60 (2015). https://doi.org/10.4204/EPTCS.190.4
26. Peters, K., Wagner, C., Nestmann, U.: taming concurrency for verification using multiparty session types (Technical Report). Technical report, TU Berlin/TU Darmstadt, https://arxiv.org (2019)
27. Priami, C.: Enhanced Operational Semantics for Concurrency. Ph.D. thesis, Universita' di Pisa-Genova-Udine, August 1996
28. Toninho, B., Yoshida, N.: Certifying data in multiparty session types. J. Logical Algebraic Methods Program. **90**, 61–83 (2017). https://doi.org/10.1016/j.jlamp.2016.11.005
29. Wagner, C., Nestmann, U.: States in process calculi. In: Proceedings of EXPRESS/SOS, EPTCS, vol. 160, pp. 48–62 (2014). https://doi.org/10.4204/EPTCS.160.6
30. Yoshida, N., Deniélou, P.-M., Bejleri, A., Hu, R.: Parameterised multiparty session types. In: Ong, L. (ed.) FoSSaCS 2010. LNCS, vol. 6014, pp. 128–145. Springer, Heidelberg (2010). https://doi.org/10.1007/978-3-642-12032-9_10
31. Yoshida, N., Hu, R., Neykova, R., Ng, N.: The scribble protocol language. In: Abadi, M., Lluch Lafuente, A. (eds.) TGC 2013. LNCS, vol. 8358, pp. 22–41. Springer, Cham (2014). https://doi.org/10.1007/978-3-319-05119-2_3

Towards a Call Behavior-Based Compositional Verification Framework for SysML Activity Diagrams

Samir Ouchani(✉)

LINEACT, Laboratoire d'Innovation Numérique École d'Ingénieur CESI, Aix-en-Provence, France
souchani@cesi.fr

Abstract. SysML activity diagram is a standard modeling language for complex systems. It supports systems' composition by providing the operator '*call behavior*'. In general, the verification of systems modeled with those diagram inherit the limitations of the developed built-in tools, especially the case of model checking. To address this shortcoming, we propose a compositional verification framework based on the *call behavior* operator to alleviate the state space explosion problem of model-checking. The framework decomposes a property into local sub-properties and verify them separately on the composed behavioral diagrams. Further, we propose to ignore the diagrams' artifacts that are useless with respect to the property under verification. We prove the soundness of the proposed approach by showing that the result deduced from the verification of the local properties is always preserved. The verification results are obtained by encoding SysML activity diagrams in the probabilistic model checker 'PRISM'. Finally, we demonstrate the effectiveness of our framework by verifying a set of properties on two use cases that require a large amount of memory and a considerable time processing.

Keywords: SysML · Activity diagrams · Model-checking · Compositional verification · Abstraction · PCTL · PRISM

1 Introduction

A major challenge in systems and software development process is to reduce as possible bugs by advancing the error detection at early stages of their life-cycles development. Experimentally, it has been shown that the cost of repairing a software flaw during maintenance is approximately 500 times higher than fixing it at early design phases [4]. Further, only 15% of flaws are detected in the initial design phase, whereas the cost of fixing them at this phase is extremely beneficial as compared to fixing them at the development and testing phases. Yet, a more ambitious challenge is to accelerate the verification process of a product based on its design artifacts. Here, we are interested on systems modeled by using modern and standard language like SysML [20]. The latter is a prominent object-oriented graphical language which today become defacto standard for software and systems modeling. Especially, SysML reuses a subset of UML packages [14] and extends others with specific systems' engineering features such as probability, time, and the rate. SysML covers mainly four perspectives of systems modeling: structure, behavior, requirement, and parametric diagrams. Particularly, SysML

R. M. Hierons and M. Mosbah (Eds.): ICTAC 2019, LNCS 11884, pp. 216–234, 2019.
https://doi.org/10.1007/978-3-030-32505-3_13

activity diagrams are behavioral diagrams used to model system behaviors at various levels of abstraction [15].

For the verification of SysML activity diagrams, model checking is the most popular used technique [23]. Model checking [5] is a formal and automatic verification technique that checks systems specifications expressed as temporal logic formula or automata-based formalism on finite state concurrent systems. Compared to qualitative model checking, quantitative verification techniques based on probabilistic model checkers [4,12] have recently gained popularity. Probabilistic verification offers the capability of measuring the satisfiability probability of a given property on systems that inherently exhibit probabilistic behavior. Despite its wide use, model checking in general is a resource-intensive process that requires a large amount of memory and time processing. This is due to the fact that the systems' state space may grow exponentially with the number of variables combined with the presence of concurrent behaviors. Consequently, it is of a major importance to reduce the verification process complexity.

To overcome this issue, various techniques have been explored [4,5] for qualitative model checking and then leveraged to the probabilistic case. Among these techniques, several solutions aim at optimizing the employed model checking algorithms by introducing symbolic data structures based on binary decision diagrams, while others target the analysis of the model itself. Besides, two classes of solutions are found in the literature: abstraction and compositional verification. The former provides a minimized representation of the global system under verification. Whereas, the latter avoids the construction of the considered global system. Abstraction techniques can be classified into four categories [5]: abstraction by state merging, on variables, by restriction, or by observer automata. Besides, the well-known compositional verification techniques [6] are: partitioned transition relation, lazy parallel composition, interface processes, and assume-guarantee.

In this paper, we are interested by the interface processes and the abstraction by restriction techniques that are consistent within the composition by call behaviors in SysML activity diagrams. The provided framework considers as input a system modeled with SysML activity diagrams and its requirements expressed in PCTL [21]. Then it decomposes a property into local sub-properties in order to verify them separately for each system's sub-component in parallel. Further, in order to accelerate more the verification process, it ignores the diagrams' artifacts that are useless with respect to the property and the local properties under verification. For verification, each system's component is transformed automatically into PRISM. Finally, the framework infers safely the verification result of the target property from the obtained results of the local properties. In a nutshell, the main contributions of this paper can be summarized as follows.

1. Proposing a complete formalization of the existing calculus dedicated to SysML activity diagrams.
2. Developing an efficient verification approach that reduces the verification costs overhead of probabilistic model checking.
3. Proving the soundness of the proposed approach.
4. Showing the effectiveness of the developed framework on two real use cases.

The next section compares our approach with the existing initiatives related to the verification of SysML activity diagrams. Then, the preliminaries needed for our work are presented in Sect. 3. Section 4 describes and formalizes SysML activity diagrams. Then, our compositional verification framework is detailed in Sect. 5 and Sect. 6 presents the experimental results. Finally, Sect. 7 concludes the paper and provides future directions.

2 Related Work

In this section, we survey the research initiatives dedicated mainly to the formalization and the verification of SysML diagrams and to the compositional verification of probabilistic systems.

Yuan et al. [16] construct a set of rules to transform UML state machines to Timed Automata (TA). They apply the query view transformation approach in order to produce TA encoded in UPPAAL input language. The properties to be verified against TA are expressed in LTL. Apvrille and Saqui-Sannes [3] apply structural analysis to SysML by using the TTool open-source toolkit. They translate a subset of SysML diagrams into a Petri net and solves an equation system built upon the incidence matrix of the net. Then, a push-button approach is applied to display verification results.

Ando et al. [1] express SysML state machine diagrams in CSP# processes that could be verified by the PAT model checker. This work includes only a sub-set of rules and experimenting the transformation on a toy case study. In addition, they did not detail the temporal logic that expresses the system requirements. Carrillo et al. [8] define SysML blocks in a refinement process. The structural architecture of a SysML block is given by the internal block diagram and the behavior of each sub-block is described by an interface automaton. Their main intention in a refinement process is to ensure the consistency and the compatibility between different blocks.

Ermeson et al. [7] verify the embedded realtime systems with energy constraints that are modeled using SysML State Machine diagram, and the MARTE UML Profile (Modeling and Analysis of Real-Time and Embedded systems) is used to specify ERTS's (Embedded Real-time Systems) constraints such as execution time and energy. They map only states and transitions into ETPN (Time Petri Net with Energy constraints). In their transformation, they don't give the transformation of actions in a given state even the semantics of the mutual exclusive and orthogonal states by taking just the internal states into consideration. Furthermore, they propose a similar methodology [2] that maps one SysML activity diagram to time Petri Net for requirement validation of embedded real-time systems with energy constraints. The computation model formalized as an ETPN is not well presented and it misses the representation of the energy consumption values. The authors do not provide a formal transformation for SysML elements even the values represented from MARTE profile. Also, they do not clarify why they represent each constraint in an action by a separate transition.

Ouchani et al. [24] introduce the abstraction by merging states to reduce the verification cost of a SysML activity diagram. In [22], the authors transform a diagram into an equivalent hierarchical form in order to help the abstraction developed in [24].

David et al. [18] introduced an extension of UML statecharts with randomly varying duration that allows probabilistic decision in state. The Input/Output (I/O) automata is used to provide a compositional semantics for statecharts. Also, probability distribution after a continuous or discrete time is introduced as an arbitrary operator. And in [17], they introduce means to specify system randomness within statecharts, and to verify probabilistic temporal properties. The model is represented as MDP, and the properties are expressed in PCTL.

Concerning the compositional verification for probabilistic systems, Feng et al. [11] discusses assume-guarantee technique for probabilistic system by focusing more on the learning algorithm to generate the minimal deterministic automata that represents a probabilistic safety property. And in [10], they propose the assume-guarantee approach where both the assumption and the guarantee properties are probabilistic safety properties such that assumptions are generated manually. Also in [9], they apply the assume-guarantee technique on synchronous systems modeled as DTMC, where assumptions are safety properties defined as probabilistic finite automata. To our knowledge, few probabilistic model checkers support abstraction and compositional verification techniques. As example, PRISM builds the symmetry reduction and LiQuor[1] implements the bi-simulation equivalence.

3 Preliminaries

In this section, we present the probabilistic automata as a modeling formalism and PCTL temporal logic as a specification language.

Probabilistic automata (PAs) [12] are a modeling formalism for systems that exhibit probabilistic and nondeterministic features. Definition 1 illustrates a PA where $Dist(S)$ denotes the set of convex distributions over S and $\mu = [\ldots, s_i \mapsto p_i, \ldots]$ is a distribution in $Dist(S)$ that assigns a probability $\mu(s_i) = p_i$ to the state s_i.

Definition 1 (Probabilistic Automaton). *A probabilistic automaton is a tuple* $M = (\bar{s}, \mathrm{S}, \mathrm{L}, \Sigma, \delta)$*, where:*

- $\bar{s}$ *is an initial state, such that* $\bar{s} \in \mathrm{S}$,
- S *is a finite set of states,*
- $\mathrm{L} : S \to 2^{AP}$ *is a labeling function that assigns to each state a set of atomic propositions taken from the set of atomic propositions (AP),*
- Σ *is a finite set of actions,*
- $\delta : \mathrm{S} \times \Sigma \to Dist(S)$ *is a probabilistic transition function assigning for each* $s \in \mathrm{S}$ *and* $\alpha \in \Sigma$ *a probabilistic distribution* $\mu \in Dist(S)$.

For PA's composition, this concept is modeled by the parallel composition as stipulated in Definition 2. During synchronization, each PA resolves its probabilistic choice independently. For transitions $s_1 \xrightarrow{\alpha} \mu_1$ and $s_2 \xrightarrow{\alpha} \mu_2$ that synchronize in α then the composed state (s_1', s_2') is reached from the state (s_1, s_2) with probability $\mu_1(s_1') \times \mu_2(s_2')$. In the no synchronization case, a PA takes a transition where the other remains in its current state with probability one.

Definition 2 (Parallel Composition of PAs). *The parallel composition of two PAs:* $M_1 = (\bar{s}_1, \mathrm{S}_1, \mathrm{L}_1, \Sigma_1, \delta_1)$ *and* $M_2 = (\bar{s}_2, \mathrm{S}_2, \mathrm{L}_2, \Sigma_2, \delta_2)$ *is a PA* $M = ((\bar{s}_1, \bar{s}_2), \mathrm{S}_1 \times \mathrm{S}_2, \mathrm{L}(s_1) \cup \mathrm{L}(s_2), \Sigma_1 \cup \Sigma_2, \delta)$*, where:* $\delta(\mathrm{S}_1 \times \mathrm{S}_2, \Sigma_1 \cup \Sigma_2)$ *is the set of transitions* $(s_1, s_2) \xrightarrow{\alpha} \mu_1 \times \mu_2$ *such that one of the following requirements is met.*

[1] http://www.i1.informatik.uni-bonn.de/baier/projectpages/LIQUOR/LiQuor.

1. $s_1 \xrightarrow{\alpha} \mu_1, s_2 \xrightarrow{\alpha} \mu_2$, *and* $\alpha \in \Sigma_1 \cap \Sigma_2$,
2. $s_1 \xrightarrow{\alpha} \mu_1, \mu_2 = [s_2 \mapsto 1]$, *and* $\alpha \in \Sigma_1 \backslash \Sigma_2$,
3. $\mu_1 = [s_1 \mapsto 1]$, $s_2 \xrightarrow{\alpha} \mu_2$, *and* $\alpha \in \Sigma_2 \backslash \Sigma_1$.

To verify a PA, we use PCTL to express its related specifications. The following grammar represents the PCTL syntax.

$$\phi ::= \top \mid ap \mid \phi \wedge \phi \mid \neg\phi \mid \mathrm{P}_{\bowtie p}[\psi]$$
$$\psi ::= \mathrm{X}\phi \mid \phi \mathrm{U}^{\leq k}\phi \mid \phi \mathrm{U}\phi$$

Where the term "⊤" means *true*, "*ap*" is an atomic proposition, $k \in \mathbb{N}$, $p \in [0,1]$, and $\bowtie \in \{<, \leq, >, \geq\}$. The operator "∧" represents the *conjunction* and "¬" is the *negation* operator, and P is the probabilistic operator. Also, "X", "$\mathrm{U}^{\leq k}$", and "U" are the *next*, the *bounded until*, and the *until* temporal logic operators, respectively.

To specify a satisfaction relation of a PCTL formula in a state "*s*", a class of adversaries has been defined to solve the nondeterminism in PAs. Hence, a PCTL formula should be satisfied under all adversaries. The satisfaction relation (⊨) of a PCTL formula is defined as follows, where "*s*" is a state and "π" is a path obtained by a memoryless adversary [12].

- $s \models \top$ is always satisfied.
- $s \models ap \Leftrightarrow ap \in \mathrm{L}(s)$ and L is a labeling function.
- $s \models \phi_1 \wedge \phi_2 \Leftrightarrow s \models \phi_1 \wedge s \models \phi_2$.
- $s \models \neg\phi \Leftrightarrow s \not\models \phi$.
- $s \models \mathrm{P}_{\bowtie p}[\psi] \Leftrightarrow \mathrm{P}(\{\pi \textit{ is a path starts from the state } s \mid \pi \models \psi\}) \bowtie p$.
- $\pi \models \mathrm{X}\phi \Leftrightarrow \pi(1) \models \phi$ where $\pi(1)$ is the second state of π.
- $\pi \models \phi_1 \mathrm{U}^{\leq k}\phi_2 \Leftrightarrow \exists i \leq k : \forall j < i,\ \pi(j) \models \phi_1 \wedge \pi(i) \models \phi_2$.
- $\pi \models \phi_1 \mathrm{U}\phi_2 \Leftrightarrow \exists\, k \geq 0 :\ \pi \models \phi_1 \mathrm{U}^{\leq k}\phi_2$.

4 SysML Activity Diagrams Formalization

In this section, we describe and formalize SysML activity diagrams by providing an adequate syntax and semantics.

As illustrated in Fig. 6, SysML activity diagrams are a graph-based representation where their main constructs (Fig. 1) can be decomposed into two categories: activity nodes and activity edges. The former contains three types: activity invocation, object and control nodes. Activity invocation includes receive and send signals, action, and call behavior. Activity control nodes are initial, flow final, activity final, decision, merge, fork, and join nodes. Activity edges are of two types: control flow and object flow. Control flow edges are used to show the execution path through the activity diagram and to connect activity nodes. Object flow edges are used to show the flow of data between activity nodes. Concurrency and synchronization are modeled using forks and joins, whereas, branching is modeled using decision and merge nodes. While a decision node specifies a choice between different possible paths based on the evaluation of a guard condition (and/or a probability distribution), a fork node indicates the beginning of multiple parallel control threads. Moreover, a merge node specifies a point from where different incoming control paths follow the same path, whereas a join node allows multiple parallel control threads to synchronize and rejoin. In addition, the call behavior action consumes its input tokens and invoke its specified behavior. The execution of the calling artifact is blocked until it receives a reply from the invoked behavior.

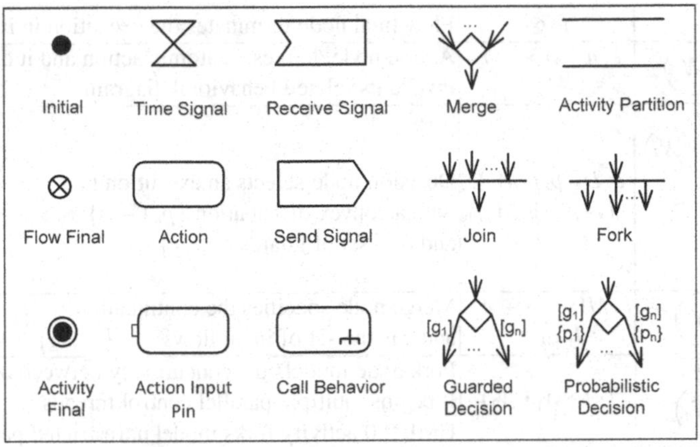

Fig. 1. SysML activity diagram constructs.

4.1 Syntax of SysML Activity Diagrams

The UML superstructure [14] specifies basic rules for the execution of the various nodes by explaining textually how tokens are passed from one node to another. For formalization, we present in Table 1 SysML activity diagrams constructs and their representation as NuAC terms. At the beginning, a first token starts flowing from the initial node and moves downstream from one node to another with respect to the foregoing set of control routing rules defined by the control nodes until reaching either an activity final or a flow final node.

However, activity diagram semantics as specified in the standard stay informal since it is explained textually. We present in Fig. 2 the Backus-Naur-Form of the new version of Activity Calculus (NuAC) that helps to formalize SysML activity diagrams. This version of NuAC calculus optimizes the syntax presented in [24] and allows for multiplicity in join, merge, fork, and decision constructs by exploiting their commutativity and associativity properties. We denote by $\mathscr{A}[\mathscr{N}]$ to specify $\mathscr{N}$ as a sub term of $\mathscr{A}$ and by $|\mathscr{A}|$ to denote a term $\mathscr{A}$ without tokens. For the call behavior case of $a \uparrow \mathscr{A}'$, we denote $\mathscr{A}[a \uparrow \mathscr{A}']$ by $\mathscr{A} \uparrow_a \mathscr{A}'$.

During the execution, the structure of the activity diagram is kept unmodified and the only changes is the tokens locus. The NuAC syntax was inspired by this idea so that a NuAC term presents a static structure while tokens are the only dynamic elements. We can distinguish two main syntactic terms: marked and unmarked. A marked NuAC term corresponds to an activity diagram with tokens. An unmarked NuAC term corresponds to the static structure of the diagram. A marked term is typically used to denote a reachable state that is characterized by the set of tokens locations in a given term.

Table 1. Rewriting activity diagram constructs in NuAC.

Activity Constructs	NuAC Terms	Description
	$l : \iota \rightarrowtail \mathcal{N}$	Initial node is activated when a diagram is invoked.
◉	$l : \odot$	Activity final node stops the execution of the diagram.
⊗	$l : \otimes$	Flow final node terminates the execution in its path.
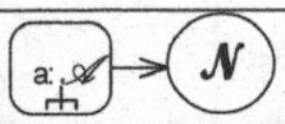	$l : a \uparrow \mathcal{A} \rightarrowtail \mathcal{N}$	Action node defines an atomic action and it can invoke its related behavioral diagram.
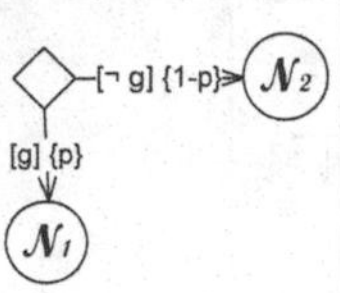	$l : D((p, g, \mathcal{N}),$ $(1-p, \neg g, \mathcal{N}))$	Decision node selects an execution path with a convex distribution $\{p, 1-p\}$ and/or a set of guards $\{g, \neg g\}$.
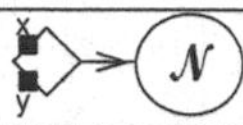	$l : M(x,y) \rightarrowtail \mathcal{N},$ l_x or l_y	Merge node specifies the continuation, and x is the set of input flows $x = \{x_1, x_2\}$.
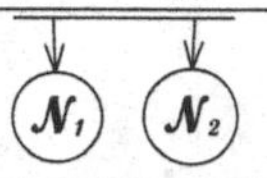	$l : F(\mathcal{N}_1, \mathcal{N}_2)$	Fork node models the concurrency between $\mathcal{N}_1$ and $\mathcal{N}_2$. It begins multiple parallel control threads. UML 2.0 activity forks model unrestricted parallelism.
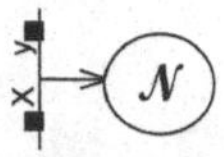	$l : J(x,y) \rightarrowtail \mathcal{N},$	Join node presents the synchronization and x is the set of input pins $x = \{x_1, x_2\}$.

$$
\begin{array}{lcl}
\mathcal{A} & ::= \varepsilon & \mid \overline{l : \iota}^{n} \rightarrowtail \mathcal{N} \\
\mathcal{N} & ::= \overline{\mathcal{N}}^{n} & \mid l : M(x,y) \rightarrowtail \mathcal{N} \mid l : J(x,y) \rightarrowtail \mathcal{N} \mid l : F(\mathcal{N}, \mathcal{N}) \mid \overline{l : a \uparrow \mathcal{A}}^{n} \rightarrowtail \mathcal{N} \\
 & & \mid l : D((p, g, \mathcal{N}), (1-p, \neg g, \mathcal{N})) \mid l : \otimes \mid l : \odot \mid l
\end{array}
$$

Fig. 2. Syntax of New Activity Calculus (NuAC).

To support multiple tokens, we augment the "overbar" operator with an integer n such that $\overline{\mathcal{N}}^{n}$ denotes a term marked with n tokens with the convention that $\overline{\mathcal{N}}^{1} = \overline{\mathcal{N}}$ and $\overline{\mathcal{N}}^{0} = \mathcal{N}$. Multiple tokens are needed when there are loops that encompass in their body a fork node. Furthermore, we use a prefix label for each node to reference it and uniquely use it in the case of a backward flow connection (case of merge or join). Particularly, labels are useful for connecting multiple incoming flows towards merge and join nodes. Let $\mathcal{L}$ be a collection of labels ranged over by $l_0, l_1, \cdots$ and $\mathcal{N}$ be any node (except initial) in the activity diagram. We write $l : \mathcal{N}$ to denote an l-labeled activity node $\mathcal{N}$. It is important to note that nodes with multi-inputs (e.g. join and merge) are visited as many times as they have incoming edges. Thus, as a syntactic convention, we use either the NuAC term (i.e. $l : M(x,y) \rightarrowtail \mathcal{N}$ for merge and $l : J(x,y) \rightarrowtail \mathcal{N}$ for join) if the current node is visited for the first time or its corresponding label (i.e. l_x or l_y) if the same node is encountered later during the traversal process. Also, we denote by $D((g,\mathscr{N}_1),(\neg g,\mathscr{N}_2))$ or $D((p,\mathscr{N}_1),(1-p,\mathscr{N}_2))$ to express a decision without probabilities or guards, respectively.

4.2 Semantics of SysML Activity Diagrams

The execution of SysML activity diagrams is based on token's flow. To give a meaning to this execution, we use structural operational semantics to formally describe how the computation steps of NuAC atomic terms take place. The operational semantics of NuAC is based on the informally specified tokens-passing rules defined in [14].

INIT-1	$\overline{l:\iota} \rightarrowtail \mathscr{N} \xrightarrow{l} l:\iota \rightarrowtail \overline{\mathscr{N}}$	
ACT-1	$\overline{l:a}^{m} \rightarrowtail \mathscr{N} \xrightarrow{l} \overline{l:a}^{m-1} \rightarrowtail \overline{\mathscr{N}}$	$\forall m>0$
ACT-2	$\overline{\overline{l:a}^{m} \rightarrowtail \mathscr{N}}^{n} \xrightarrow{l} \overline{\overline{l:a}^{m+1} \rightarrowtail \mathscr{N}}^{n-1}$	$\forall m \geq 0, n>0$
BH-1	$\dfrac{\mathscr{A} = l':\iota \rightarrowtail \mathscr{N}' \quad \forall n>0}{\overline{l:a\uparrow\mathscr{A}}^{n} \rightarrowtail \mathscr{N} \xrightarrow{l} \overline{l:a\uparrow \overline{l':\iota} \rightarrowtail \mathscr{N}'}^{n-1} \rightarrowtail \mathscr{N}}$	
BH-2	$\dfrac{\mathscr{A}[\overline{l':\odot}] \xrightarrow{l'} \lvert\mathscr{A}\rvert \quad \forall n>0}{\overline{l:a\uparrow\mathscr{A}}^{n} \rightarrowtail \mathscr{N} \xrightarrow{l'} \overline{l:a\uparrow\mathscr{A}}^{n} \rightarrowtail \overline{\mathscr{N}}}$	
FORK-1	$\overline{l:F(\mathscr{N}_1,\mathscr{N}_2)}^{m} \xrightarrow{l} \overline{l:F(\overline{\mathscr{N}_1},\overline{\mathscr{N}_2})}^{m-1}$	$\forall m>0$
PDEC-1	$\overline{l:D((p,g,\mathscr{N}_1),(1-p,\neg g,\mathscr{N}_2))}^{m} \xrightarrow{l}_{p} \overline{l:D((p,g,\overline{\mathscr{N}_1}),(1-p,\neg g,\mathscr{N}_2))}^{m-1}$	$\forall m>0$
MERG-1	$\mathscr{A}[\overline{l:M(x,y)\rightarrowtail\mathscr{N}}^{n},\overline{l_x}^{m},\overline{l_y}^{k}] \xrightarrow{l_x} \mathscr{A}[\overline{l:M(x,y)\rightarrowtail\overline{\mathscr{N}}}^{n},\overline{l_x}^{m-1},\overline{l_y}^{k}]$	$\forall m>0, k,n \geq 0$
MERG-2	$\mathscr{A}[\overline{l:M(x,y)\rightarrowtail\mathscr{N}}^{n},\overline{l_x}^{m},l_y] \xrightarrow{l_x} \mathscr{A}[\overline{l:M(x,y)\rightarrowtail\overline{\mathscr{N}}}^{n},\overline{l_x}^{m-1},l_y]$	$\forall m>0, n \geq 0$
JOIN-1	$\mathscr{A}[\overline{l:J(x,y)\rightarrowtail\mathscr{N}}^{n},\overline{l_x}^{m},\overline{l_y}^{k}] \xrightarrow{l_x} \mathscr{A}[\overline{l:J(x,y)\rightarrowtail\overline{\mathscr{N}}}^{n},\overline{l_x}^{m-1},\overline{l_y}^{k-1}]$	$\forall m,k>0, n \geq 0$
FLOWFINAL	$\mathscr{A}[\overline{l:\otimes}] \xrightarrow{l} \mathscr{A}[l:\otimes]$	
FINAL	$\mathscr{A}[\overline{l:\odot}] \xrightarrow{l} \lvert\mathscr{A}\rvert$	
ACTIVITY	$\dfrac{\mathscr{N} \xrightarrow{\alpha}_{p} \mathscr{N}'}{\mathscr{A}[\mathscr{N}] \xrightarrow{\alpha}_{p} \mathscr{A}[\mathscr{N}']}$	

Fig. 3. NuAC operational semantic rules.

We define Σ as the set of non-empty actions labeling the transitions (i.e. the alphabet of NuAC, to be distinguished from action nodes in activity diagrams). An element $\alpha \in \Sigma$ is the label of the executing active node. Let Σ^{o} be $\Sigma \cup \{o\}$ where o denotes the empty action. Let p be a probability value such that $p \in]0,1[$. The general form of a transition is $\mathscr{A} \xrightarrow{\alpha}_{p} \mathscr{A}'$ and $\mathscr{A} \xrightarrow{\alpha} \mathscr{A}'$ in the case of a Dirac (non probabilistic) transition. The probability value specifies the likelihood of a given transition to occur and it is denoted by $P(\mathscr{A},\alpha,\mathscr{A}')$. Figure 3 shows the operational semantic rules of NuAC. The semantics of SysML activity diagrams expressed using $\mathscr{A}$ as a result of the defined semantic rules can be described in terms of the PA stipulated in Definition 3. In addition, we propose in Table 2 the NuAC axioms that are proved by using NuAC semantic rules.

Definition 3 (NuAC-PA). *A probabilistic automata of a NuAC term $\mathcal{A}$ is the tuple $M_{\mathcal{A}} = (\bar{s}, \mathrm{L}, \mathrm{S}, \Sigma^o, \delta)$, where:*

- $\bar{s}$ *is an initial state, such that* $\mathrm{L}(\bar{s}) = \{\overline{l : \iota \rightarrowtail \mathcal{N}}\}$,
- $\mathrm{L} : S \rightarrow 2^{[\![\mathcal{L}]\!]}$ *is a labeling function where:* $[\![\mathcal{L}]\!] : \mathcal{L} \rightarrow \{\top, \bot\}$,
- S *is a finite set of states reachable from* $\bar{s}$, *such that,* $\mathrm{S} = \{s_{i:0 \leq i \leq n} : \mathrm{L}(s_i) \in \{\overline{\mathcal{N}}\}\}$,
- Σ^o *is a finite set of actions corresponding to labels in* $\mathcal{A}$,
- $\delta : \mathrm{S} \times \Sigma^o \rightarrow Dist(S)$ *is a partial probabilistic transition function such that, for each* $s \in S$ *and* $\alpha \in \Sigma^o$ *assigns a probabilistic distribution* μ, *where:*
  - *For* $S' \subseteq S$ *such that* $S' = \{s_{i:0 \leq i \leq n} : s \xrightarrow{\alpha}_{p_i} s_i\}$, *each transition* $s \xrightarrow{\alpha}_{p_i} s_i$ *satisfies one NuAC semantic rule and* $\mu(S') = \sum_{i=0}^{n} p_i = \sum_{i=0}^{n} \mu(s_i) = 1$.
  - *For each transition* $s \xrightarrow{\alpha}_{1} s''$ *satisfying a NuAC semantic rule,* μ *is defined such that* $\mu(s'') = 1$.

Table 2. Axioms for NuAC.

DA-1	$l : D((p, g, \mathcal{N}_1), (1-p, \neg g, \mathcal{N}_2)) = l : D((1-p, \neg g, \mathcal{N}_2), (p, g, \mathcal{N}_1))$
DA-2	$l : D((p, \mathcal{N}_1), (1-p, l' : D((p', \mathcal{N}_2), (1-p', \mathcal{N}_3)))) = l : D((p+p'-p \times p', l' : D((\frac{p}{p+p'-p\times p'}, \mathcal{N}_1), (\frac{p'-p\times p'}{p+p'-p\times p'}, \mathcal{N}_2))), (1-p-p'+p\times p', \mathcal{N}_3))$
DA-3	$l : D((p, g, \mathcal{N}_1), (1-p, \neg g, l' : D((p', g', \mathcal{N}_2), (1-p', \neg g', \mathcal{N}_3)))) = l : D((p, g, \mathcal{N}_1), (p'-p.p', \neg g \wedge g', \mathcal{N}_2), ((1-p)(1-p'), \neg g \wedge \neg g', \mathcal{N}_3))$
FA-1	$l : F(\mathcal{N}_1, \mathcal{N})_1 = \mathcal{N}_1$
FA-2	$l : F(\mathcal{N}_1, \mathcal{N}_2) = l : F(\mathcal{N}_2, \mathcal{N}_1)$
FA-3	$l : F(\mathcal{N}_1, l' : F(\mathcal{N}_2, \mathcal{N}_3)) = l : F(l' : F(\mathcal{N}_1, \mathcal{N}_2), \mathcal{N}_3) = l : F(\mathcal{N}_1, \mathcal{N}_2, \mathcal{N}_3)$
JA-1	$\mathcal{A}[l : J(x,y) \rightarrowtail \mathcal{N}', \mathcal{N} \rightarrowtail l_x, \mathcal{N} \rightarrowtail l_y] = \mathcal{A}[\mathcal{N} \rightarrowtail \mathcal{N}']$
JA-2	$l : J(x,y) \rightarrowtail \mathcal{N} = l : J(y,x) \rightarrowtail \mathcal{N}$
JA-3	$\mathcal{A}[l : J(x,x') \rightarrowtail \mathcal{N}, l' : J(y,z) \rightarrowtail l_{x'}] = \mathcal{A}[l : J(x,y,z) \rightarrowtail \mathcal{N}]$
MA-1	$\mathcal{A}[l : M(x,y) \rightarrowtail \mathcal{N}', \mathcal{N} \rightarrowtail l_x, \mathcal{N} \rightarrowtail l_y] = \mathcal{A}[\mathcal{N} \rightarrowtail \mathcal{N}']$
MA-2	$l : M(x,y) \rightarrowtail \mathcal{N} = l : M(y,x) \rightarrowtail \mathcal{N}$
MA-3	$\mathcal{A}[l : M(x,x') \rightarrowtail \mathcal{N}, l' : M(y,z) \rightarrowtail l_{x'}] = \mathcal{A}[l : M(x,y,z) \rightarrowtail \mathcal{N}]$
CA-1	$l : a \uparrow \varepsilon = a$
CA-2	$\mathcal{A}_1 \uparrow_{a_1} (\mathcal{A}_2 \uparrow_{a_2} \mathcal{A}_3) = (\mathcal{A}_1 \uparrow_{a_1} \mathcal{A}_2) \uparrow_{a_2} \mathcal{A}_3 = \mathcal{A}_1 \uparrow_{a_1} \mathcal{A}_2 \uparrow_{a_2} \mathcal{A}_3$

5 The Approach

Figure 4 depicts an overview of our compositional verification framework. It takes a set of SysML activity diagrams composed by the call behavior interface and a Probabilistic Computation Tree Logic (PCTL) [12] property as input. First, we develop an abstraction approach that restricts the verification of a PCTL property only on the influenced diagrams instead of the whole composition. Then, we propose a compositional verification approach by interface processes that distributes a PCTL property into local ones which helps to verify them separately for each diagram. For verification, we encode the diagrams into the PRISM input language [19]. Finally, we deduce the result of the main property from the results of the local properties that are verified separately for each called diagram.

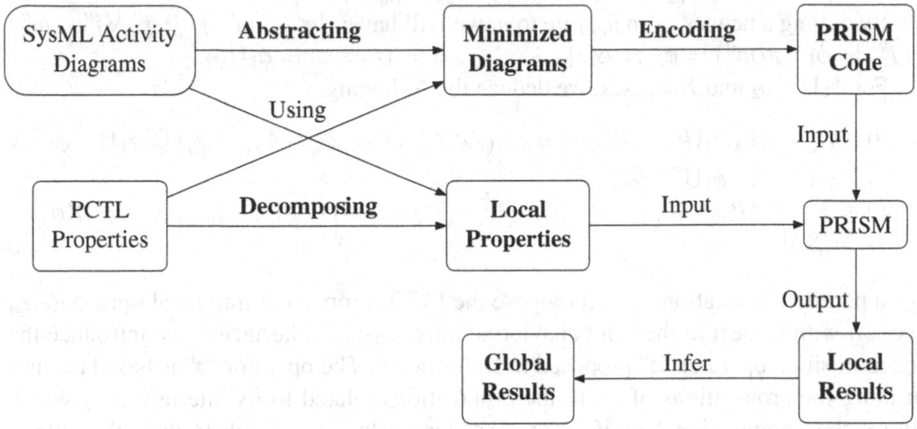

Fig. 4. A compositional verification framework.

5.1 The Compositional Verification

Let $\mathscr{A}$ be a SysML activity diagram with n call behaviors denoted by $\mathscr{A} = \mathscr{A}_0 \uparrow_{a_0} \mathscr{A}_1 \cdots \mathscr{A}_{i-1} \uparrow_{a_{i-1}} \mathscr{A}_i \cdots \mathscr{A}_{n-1} \uparrow_{a_{n-1}} \mathscr{A}_n$. In order to reduce the diagram $\mathscr{A}$, we apply NuAC axioms and introduce the reduction rule defined in Definition 4 to remove diagrams $\mathscr{A}_i$ that are not influenced by the property ϕ to be verified. The obtained diagram after applying the reduction rule is denoted by $\widehat{\mathscr{A}}$.

Definition 4. *Let $\mathscr{A}$ be a diagram that contains n call behaviors, AP_ϕ is the atomic propositions of the PCTL property ϕ, and $AP_{\mathscr{A}_i}$ is the atomic propositions of the behavioral diagram $\mathscr{A}_i$. Reducing $\mathscr{A}$ to the diagram $\widehat{\mathscr{A}}$ with respect to ϕ is obtained by applying the following rule.*

$$\frac{\forall 0 \le i \le n, AP_\phi \cap AP_{\mathscr{A}_i} = \emptyset}{\mathscr{A}_i = \varepsilon}$$

Below, Proposition 1 shows the satisfiability probability after reduction.

Proposition 1. *For a reduced diagram $\widehat{\mathscr{A}}$ of $\mathscr{A}$ with respect to ϕ, we have:*

$$[\widehat{\mathscr{A}} \models \phi] \Rightarrow [\mathscr{A} \models \phi].$$

Proof. The proof of this proposition follows an induction reasoning on the PCTL structure. First, we take the case of $\psi = \phi_1 \,\mathrm{U}\, \phi_2$.

By definition, for $0 \leq i \leq n$ where $AP_{\psi} \cap AP_{\mathscr{A}_i} = \emptyset$, then: $\mathscr{A}_i = \varepsilon$. The result is $\widehat{\mathscr{A}} = \mathscr{A}_0 \uparrow_{a_0} \mathscr{A}_1 \cdots \mathscr{A}_{k-1} \uparrow_{a_{k-1}} \mathscr{A}_k$ and $k \leq n$.

From the PCTL semantics, we have $[(\mathscr{A}_0 \uparrow_{a_0} \mathscr{A}_1 \cdots \mathscr{A}_{k-1} \uparrow_{a_{k-1}} \mathscr{A}_k) \models \psi] \Leftrightarrow \exists m, \forall j < m : \pi(j) \models \phi_1 \wedge \pi(m) \models \phi_2$ where $\pi(j)$ and $\pi(m)$ are the states i and j respectively in a path π of $\mathscr{A}$. And, by calling $\mathscr{A}_i$ in a_i using BH-1, the only changes in the path π are the propositions of $\mathscr{A}_i$ till executing BH-2, then: $\exists m' \geq m,\ j' \geq j,\ \forall j' < m' : \pi(j') \models \phi_1 \wedge \pi(m') \models \phi_2 \Leftrightarrow \mathscr{A}_0 \uparrow_{a_1} \cdots \uparrow_{a_k} \mathscr{A}_k \cdots \uparrow_{a_i} \mathscr{A}_i \models \psi$.

By calling a new $\mathscr{A}_{i+1}$ in a_{i+1} up to n, we will have: $\exists m'' \geq m',\ j'' \geq j',\ \forall j'' < m'' : \pi(j'') \models \phi_1 \wedge \pi(m'') \models \phi_2 \Leftrightarrow \mathscr{A}_0 \uparrow_{a_1} \cdots \uparrow_{a_n} \mathscr{A}_n \models \psi \Leftrightarrow \mathscr{A} \models \phi_1 \mathrm{U} \phi_2$.

For $\phi_1 \mathrm{U}^{\leq k} \phi_2$ and $\mathrm{X}\phi$ cases, we deduce the following.

- $\forall 0 \leq i \leq n, AP_{\phi} \cap AP_{\mathscr{A}_i} = \emptyset : [\mathscr{A}_i = \varepsilon \wedge (\mathscr{A}_0 \uparrow_{a_0} \mathscr{A}_1 \cdots \mathscr{A}_{n-1} \uparrow_{a_{n-1}} \mathscr{A}_n) \models \phi_1 \mathrm{U}^{\leq k} \phi_2] \Rightarrow [\exists k' \geq k : \mathscr{A} \models \phi_1 \mathrm{U}^{\leq k'} \phi_2]$.
- $\forall 0 \leq i \leq n, AP_{\phi} \cap AP_{\mathscr{A}_i} = \emptyset : [\mathscr{A}_i = \varepsilon \wedge (\mathscr{A}_0 \uparrow_{a_0} \mathscr{A}_1 \cdots \mathscr{A}_{n-1} \uparrow_{a_{n-1}} \mathscr{A}_n) \models \mathrm{X}\phi] \Rightarrow [\mathscr{A} \models \mathrm{X}\phi]$. □

For a parallel verification, we decompose the PCTL property ϕ into local ones $\phi_{i:0 \leq i \leq n}$ over $\mathscr{A}_i$ with respect to the call behavior actions $a_{i:0 \leq i \leq n}$ (interfaces), we introduce the decomposition operator "$\natural$" proposed in Definition 5. The operator "$\natural$" is based on substituting the propositions of $\mathscr{A}_i$ to the propositions related to its interface a_{i-1} which allows the compositional verification. We denote by $\phi[y/z]$ substituting the atomic proposition "z" in the PCTL property ϕ by the atomic proposition "y".

Definition 5 (PCTL Property Decomposition). *Let ϕ be a PCTL property to be verified on $\mathscr{A}_1 \uparrow_a \mathscr{A}_2$. The decomposition of ϕ into ϕ_1 and ϕ_2 is denoted by $\phi \equiv \phi_1 \natural_a \phi_2$ where $AP_{\mathscr{A}_i}$ are the atomic propositions of $\mathscr{A}_i$, then:*

1. $\phi_1 = \phi([l_a/AP_{\mathscr{A}_2}])$, *where l_a is the atomic proposition related to the action a in $\mathscr{A}_1$.*
2. $\phi_2 = \phi([\top/AP_{\mathscr{A}_1}])$.

The first rule is based on the fact that the only transition to reach a state in $\mathscr{A}_2$ from $\mathscr{A}_1$ is the transition of the action l_a (BH-1). The second rule ignores the existence of $\mathscr{A}_1$ while it kept unchanged till the execution of BH-2. To handle multiplicity for the operator "$\natural$", we have Property 1.

Property 1. The decomposition operator $\natural$ is associative for $\mathscr{A}_1 \uparrow_{a_1} \mathscr{A}_2 \uparrow_{a_2} \mathscr{A}_3$, i.e. :

$$\phi_1 \natural_{a_1} (\phi_2 \natural_{a_2} \phi_3) \equiv (\phi_1 \natural_{a_1} \phi_2) \natural_{a_2} \phi_3 \equiv \phi_1 \natural_{a_1} \phi_2 \natural_{a_2} \phi_3.$$

For the verification of ϕ on $\mathscr{A}_1 \uparrow_{a_1} \mathscr{A}_2$, Theorem 1 deduces the satisfiability of ϕ from the satisfiability of local properties ϕ_1 and ϕ_2 obtained by the operator $\natural$.

Theorem 1 (Compositional Verification). *The decomposition of the PCTL property ϕ by the decomposition operator $\natural$ for $\mathscr{A}_1 \uparrow_{a_1} \mathscr{A}_2$ is sound, i.e. :*

$$\frac{\mathscr{A}_1 \models \phi_1 \quad \mathscr{A}_2 \models \phi_2 \quad \phi = \phi_1 \natural_{a_1} \phi_2}{\mathscr{A}_1 \uparrow_{a_1} \mathscr{A}_2 \models \phi}$$

Proof. The proof of Theorem 1 follows a structural induction on the PCTL structure by using Definition 5. As an example, we take the until operator "U". Let $\phi = ap_1 \text{ U } ap_2$ where $ap_1 \in AP_{\mathscr{A}_1}$ and $ap_2 \in AP_{\mathscr{A}_2}$. By applying Definition 5, we have: $\phi_1 = ap_1 \text{ U } a_1$ and $\phi_2 = \top \text{ U } ap_2$. Let $\mathscr{A}_1 \models \phi_1 \Leftrightarrow \exists m_1, \forall j_1 < m_1 : \pi_1(j_1) \models ap_1 \wedge \pi_1(m_1) \models ap_1 \wedge a_1$ where π is a path in the NuAC PA of $\mathscr{A}$. For $\mathscr{A}_2 \models \phi_2 \Leftrightarrow \exists m_2, \forall j_2 < m_2 : \pi_2(j_2) \models \top \wedge \pi_2(m_2) \models ap_2$. To construct $\mathscr{A}_1 \uparrow_{a_1} \mathscr{A}_2$, BH-1 is the only transition to connect π_1 and π_2 which form: $\pi = \pi_1.\pi_2'$ such that $\pi_2'(i) = \pi_2(i) \cup \pi_1(m_1)$. Then: $\exists j \leq m$, $m = m_1 + m_2 : \pi(j) \models ap_1 \wedge \pi(m) \models ap_2 \Leftrightarrow \mathscr{A}_1 \uparrow_{a_1} \mathscr{A}_2 \models \phi$. □

Finally, Proposition 2 generalizes Theorem 1 to support the satisfiability of ϕ on an activity diagram $\mathscr{A}$ with n call behaviors.

Proposition 2 (CV-Generalization). *Let ϕ be a PCTL property to be verified on $\mathscr{A}$, such that: $\mathscr{A} = \mathscr{A}_0 \uparrow_{a_0} \cdots \uparrow_{a_{n-1}} \mathscr{A}_n$ and $\phi = \phi_0 \natural_{a_0} \cdots \natural_{a_{n-1}} \phi_n$, then:*

$$\frac{\mathscr{A}_0 \models \phi_0 \cdots \mathscr{A}_n \models \phi_n \qquad \phi = \phi_0 \natural_{a_0} \cdots \natural_{a_{n-1}} \phi_n}{\mathscr{A}_0 \uparrow_{a_0} \cdots \uparrow_{a_{n-1}} \mathscr{A}_n \models \phi}$$

Proof. We prove Proposition 2 by induction on n.

- The base step where "$n = 1$" is proved by Theorem 1.
- For the inductive step, first, we assume:

$$\frac{\mathscr{A}_0 \models \phi_0 \cdots \mathscr{A}_n \models \phi_n \qquad \phi = \phi_0 \natural_{a_0} \cdots \natural_{a_{n-1}} \phi_n}{\mathscr{A}_0 \uparrow_{a_0} \cdots \uparrow_{a_{n-1}} \mathscr{A}_n \models \phi}$$

  Let $\mathscr{A}' = \mathscr{A}_0 \uparrow_{a_0} \cdots \uparrow_{a_{n-1}} \mathscr{A}_n$ and $\phi' = \phi_0 \natural_{a_0} \cdots \natural_{a_{n-1}} \phi_n$. While $\natural$ and $\uparrow$ are associative operators, then: $\mathscr{A} = \mathscr{A}' \uparrow_{a_n} \mathscr{A}_{n+1}$ and $\phi = \phi' \natural_{a_n} \phi_{n+1}$. By assuming $\mathscr{A}_n \models \phi_n$ and applying Theorem 1, then:

$$\frac{\mathscr{A}' \models \phi' \quad \mathscr{A}_{n+1} \models \phi_{n+1} \qquad \mathscr{A} = \mathscr{A}' \uparrow_{a_n} \mathscr{A}_{n+1} \quad \phi = \phi' \natural_{a_n} \phi_{n+1}}{\mathscr{A} \models \phi}$$

5.2 The Encoding to PRISM

To encode a SysML activity diagram $\mathscr{A}$ into its equivalent PRISM code $\mathscr{P}$, we rely to the PRISM MDP formalism that refers to the PA[2] which coincides with the NuAC semantics. In PRISM, we define the NuAC transition $s \xrightarrow{l} \mu$ as a probabilistic command. Mainly, the probabilistic command takes the following form: $[l]$ $g \rightarrow p_1 : u_1 + \ldots + p_m : u_m$, which means, for the action "l" if the guard "g" is true, then, an update "u_i" is enabled with a probability "p_i". The guard "g" is a predicate of a conjunction form consisting to the evaluation of the atomic propositions related to the

[2] http://www.prismmodelchecker.org/doc/manual.pdf (The introduction section, line 10).

state s. The update u_i describes the evaluation of the atomic propositions related to the next state s_i of s such that $s \xrightarrow{l}_{p_i} s_i$ $(1 \leq i \leq m)$. For the Dirac case, the command is written simply by: $[l]\ g \longrightarrow u$.

The function Γ presented in Listing 1.1 produces the appropriate PRISM command for each NuAC term. The action label of a command is the label of its related term "l". The guard of this command depends on how the term is activated, therefore, a boolean proposition as a flag is assigned to define this activation. For simplicity, the flag related to a term labeled by l is denoted by a boolean proposition l that is initialized to false except for the initial node it is true which conforms to the premise of the NuAC rule "INIT-1". Concerning the command updates, they deactivate the propositions of a term $n \in \mathcal{A}$ and activate its successors. We define three useful functions: $L(n)$, $S(\mathcal{A}_i)$, and $E(\mathcal{A}_i)$ that return the label of a term n, the initial and the final terms of the diagram $\mathcal{A}_i$, respectively. For example, the call behavior action "$l: a \uparrow \mathcal{A}_i$" (line 32) produces two commands (line 34), and it calls the function Γ' (line 34). The first command in line 34 synchronizes with the first command in line 52 produced by the function Γ' in the action l from the diagram $\mathcal{A}$. Similarly, the second command in line 34 synchronizes with the command of line 56 in the action $L(E(\mathcal{A}_i))$ from the diagram $\mathcal{A}_i$. The first synchronization represents the NuAC rule BH-1 where the second represents the rule BH-2. The function Γ' is similar to the function Γ except for the initial and the final nodes as shown in lines 52 and 56, respectively. The generated PRISM fragment of each diagram $\mathcal{A}_i$ is bounded by two PRISM primitives: the module head "*Module* $\mathcal{A}_i$", and the module termination "*endmodule*".

1 $\Gamma:\ \mathcal{A} \rightarrow \mathcal{P}$
2 $\Gamma(\mathcal{A})$ = $\forall n \in \mathcal{A}$, $L(n=\iota)=\top$, $L(n \neq \iota)=\bot$, Case (n) **of**
3 $l:\iota \rightarrowtail \mathcal{N}$ $\Rightarrow$ **in** $\{[l]l \longrightarrow (l'=\bot)\&(L(\mathcal{N})'=\top);\} \cup \Gamma(\mathcal{N})$ **end**
4 $l:M(x,y) \rightarrowtail \mathcal{N}$ $\Rightarrow$ **in** $\{[l_x]l_x \longrightarrow (l'_x=\bot)\&(L(\mathcal{N})'=\top);\}$
5 $\cup\{[l_y]l_y \longrightarrow (l'_y=\bot)\&(L(\mathcal{N})'=\top);\} \cup \Gamma(\mathcal{N})$ **end**
6 $l:J(x,y) \rightarrowtail \mathcal{N}$ $\Rightarrow$ **in** $\{[l]l_x \wedge l_y \longrightarrow (l'_x=\bot)\&(l'_y=\bot)\&(L(\mathcal{N})'=\top);\} \cup \Gamma(\mathcal{N})$ **end**
7 $l:F(\mathcal{N}_1,\mathcal{N}_2) \Rightarrow$ **in** $\{[l]l \longrightarrow (l'=\bot)\&(L(\mathcal{N}_1)'=\top)\&(L(\mathcal{N}_2)'=\top);\} \cup \Gamma(\mathcal{N}_1) \cup \Gamma(\mathcal{N}_2)$ **end**
8 $l:D(\mathcal{A},p,g,\mathcal{N}_1,\mathcal{N}_2) \Rightarrow$
9 Case (p) **of** $]0,1[\ \Rightarrow$
10 **in** $\{[l]l \longrightarrow p:(l'=\bot)\&(l'_g=\top)+(1-p):(l'=\bot)\&(l'_{\neg g}=\top);\}$
11 $\cup\{[l_{\neg g}]l_g \wedge \neg g \longrightarrow (l'_{\neg g}=\bot)\&(L(\mathcal{N}_2)'=\top);\}$
12 $\cup\{[l_g]l_g \wedge g \longrightarrow (l'_g=\bot)\&(L(\mathcal{N}_1)'=\top);\} \cup \Gamma(\mathcal{N}_1) \cup \Gamma(\mathcal{N}_2)$**end**
13 Otherwise **in** $\{[l]l \longrightarrow (l'=\bot)\&(l'_g=\top);\} \cup \{[l]l \longrightarrow (l'=\bot)\&(l'_{\neg g}=\top);\}$
14 $\cup\{[l_g]l_g \wedge g \longrightarrow (l'_g=\bot)\&(L(\mathcal{N}_1)'=\top);\}$
15 $\cup\{[l_{\neg g}]l_g \wedge \neg g \longrightarrow (l'_{\neg g}=\bot)\&(L(\mathcal{N}_2)'=\top);\}$
16 $\cup \Gamma(\mathcal{N}_1) \cup \Gamma(\mathcal{N}_2)$**end**
17 $l:a\mathcal{B} \rightarrowtail \mathcal{N}$, Case $(\mathcal{B})$ **of**
18 $\uparrow \mathcal{A}_i \Rightarrow$
19 **in** $\{[l]l \rightarrow (l'=\bot);\}$
20 $\cup\{[L(E(\mathcal{A}_i))]L(E(\mathcal{A}_i)) \rightarrow (l'=\bot)\&(L(\mathcal{N})'=\top);\} \cup \Gamma'(\mathcal{A}_i)$; **end**
21 $\varepsilon \Rightarrow$ **in** $\{[l]l \longrightarrow (l'=\bot)\&(\mathcal{N}'=\top);\} \cup \Gamma(\mathcal{N}')$ **end**
22 $l:\otimes \Rightarrow$ **in** $[l]l \longrightarrow (l'=\bot)$; **end**
23 $l:\odot \Rightarrow$ **in** $[l]l \longrightarrow \&_{l \in \mathcal{L}}\ (l'=\bot)$;**end**
24 //Defining the function $\Gamma'(a \uparrow \mathcal{A}_i)$
25 $\Gamma':\ \mathcal{A} \rightarrow \mathcal{P}$
26 $\Gamma'(\mathcal{A}_i)$ = $\forall m \in \mathcal{A}_i : L(m)=\bot$, Case (m) **of**
27 $l:\iota \rightarrowtail \mathcal{N}$ $\Rightarrow$//The action l **and** the guard l are from the line 40.
28 **in** $\{[l]l \rightarrow (L(S(\mathcal{A}_i))'=\top)$;
29 $[L(S(\mathcal{A}_i))]L(S(\mathcal{A}_i)) \rightarrow (L(S(\mathcal{A}_i))'=\bot)\&(L(\mathcal{N})'=\top);\} \cup \Gamma(\mathcal{N})$ **end**
30 $l:\odot \Rightarrow$ **in** $[L(E(\mathcal{A}_i))]L(E(\mathcal{A}_i)) \rightarrow (L(E(\mathcal{A}_i))'=\bot)$; **end**
31 Otherwise $\Gamma(\mathcal{A})$;

Listing 1.1. Generating PRISM commands function.

6 Implementation and Experimental Results

For the purpose of providing experimental results demonstrating the efficiency and the validity of our framework, we verify a set of PCTL properties on the online shopping system [13] and the automated teller machine [13]. To this end, we compare the verification results "β", the verification cost in terms of the model size[3] "γ", and the verification time "δ" (sec) with and without applying our approach.

6.1 Online Shopping System

The online shopping system aims at providing services for purchasing online items. Figure 5a illustrates the corresponding SysML activity diagram. It contains four call-behavior actions[4], which are: "Browse Catalogue", "Make Order", "Process Order", and "Shipment" denoted by *a*, *b*, *c* and *d*, respectively. For simplicity, we take this order to denote their called diagrams by $\mathscr{A}_1$ to $\mathscr{A}_4$, respectively, where $\mathscr{A}_0$ denotes the main diagram. As an example, Fig. 5b expands the diagram related to the call behavior action "Process Order" and it is denoted by $\mathscr{A}_3$. The whole diagram is written by: $\mathscr{A} = \mathscr{A}_0 \uparrow_a \mathscr{A}_1 \uparrow_b \mathscr{A}_2 \uparrow_c \mathscr{A}_3 \uparrow_d \mathscr{A}_4$. Here, we propose to verify the properties Φ_1 and Φ_2 that are expressed in PCTL.

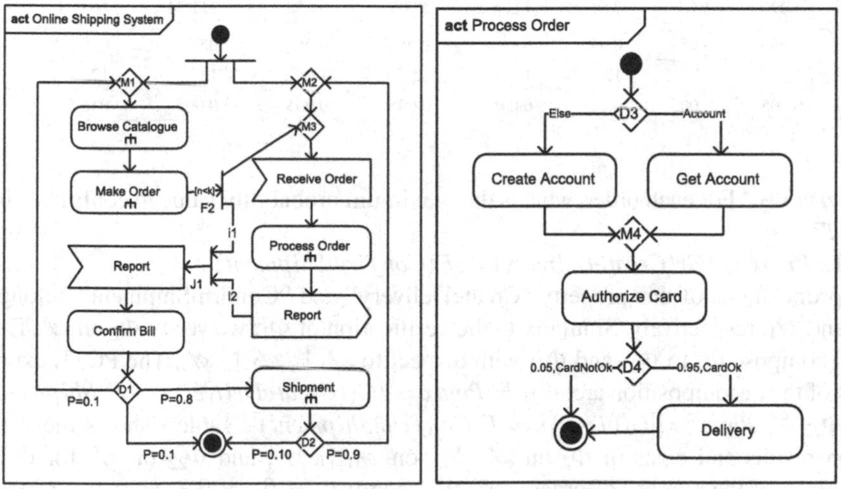

(a) Online Shopping System. (b) Process Order.

Fig. 5. SysML activity diagrams.

***Property Φ_1*.** "For each order, what is the minimum probability value to make a delivery after browsing the catalogue?"

[3] The model size is the number of transitions (edges).

[4] Each call-behavior action is represented by its proper diagram.

PCTL: $Pmin =?[(Browse\ Catalogue) \Rightarrow (F(Delivery))]$.

In this expression, the "Browse Catalogue" proposition is part of $\mathcal{A}_0$ and "*Delivery*" is a proposition of $\mathcal{A}_3$. For comparison, we verify first Φ_1 on $\mathcal{A}$. Then, by using Proposition 1, we reduce the verification of Φ_1 from $\mathcal{A}$ to $\mathcal{A}_0 \uparrow_c \mathcal{A}_3$. And, by using the decomposition rules of Definition 5, Φ_1 is decomposed into two properties: Φ_{11} and Φ_{12} such that: $\Phi_{11} \triangleq Pmin =?[(Browse\ Catalogue) \Rightarrow (F(Process\ Order))]$, and $\Phi_{12} \triangleq Pmin =?[(True) \Rightarrow (F(Delivery))]$. After the verification of Φ_1 on $\mathcal{A}$, Φ_{11} on $\mathcal{A}_0$ and Φ_{12} on $\mathcal{A}_3$, Table 3 summarizes the verification results and costs for different values of the number of orders "*n*". From the obtained results, we observe that the probability values are preserved where $\beta_1 = \beta_{11} \times \beta_{12}$. In addition, the size of the diagrams is minimized $\gamma_{11} + \gamma_{12} < \gamma_1$. Consequently, the verification time is reduced significantly $\delta_{11} + \delta_{12} \ll \delta_1$.

Table 3. The verification cost for properties Φ_1 Φ_{11}, and Φ_{12}.

n	3	4	5	6	7	8	9	10
β_1	0.76	0.76	0.76	0.76	0.76	0.76	0.76	0.76
γ_1	2,213,880	4,823,290	8,434,700	13,048,110	51,145,160	202,489,260	454,033,360	805,777,460
δ_1	10.764	24.364	44.098	72.173	358.558	1818.247	6297.234	17761.636
β_{11}	0.8	0.8	0.8	0.8	0.8	0.8	0.8	0.8
γ_{11}	5,486	7,266	9,046	10,826	12,606	14,386	16,166	17,946
δ_{11}	1.09	3.12	7.511	12.86	27.03	54.38	111.74	163.89
β_{12}	0.95	0.95	0.95	0.95	0.95	0.95	0.95	0.95
γ_{12}	12	12	12	12	12	12	12	12
δ_{12}	0.005	0.005	0.005	0.005	0.005	0.005	0.005	0.005

Property Φ_2**.** "For each order, what is the maximum probability value to confirm a shipment?"

PCTL: $Pmax =?[G((CreateDelivery) \Rightarrow F(ConfirmShipment)]$.

The propositions of this property "CreateDelivery" and "ConfirmShipment" belong to $\mathcal{A}_2$, and $\mathcal{A}_4$, respectively. Similarly to the verification of Φ_1, we verify Φ_2 on $\mathcal{A}$. Then, we decompose Φ_2 to Φ_{21} and Φ_{22} with respect to $\mathcal{A}_0 \uparrow_b \mathcal{A}_2 \uparrow_d \mathcal{A}_4$. The PCTL expressions of the decomposition are: $\Phi_{21} \triangleq Pmax =?[G((CreateDelivery) \Rightarrow F(Shipment)]$, and $\Phi_{22} \triangleq Pmax =?[G((True) \Rightarrow F(ConfirmShipment)]$. Table 4 shows the verification results and costs of Φ_2 on $\mathcal{A}$, Φ_{21} on $\mathcal{A}_0 \uparrow_b \mathcal{A}_2$, and Φ_{22} on $\mathcal{A}_4$ for different values of the number of orders "*n*". We found: $\beta_2 = \beta_{21} \times \beta_{22}$, $\gamma_{21} + \gamma_{22} < \gamma_2$ and $\delta_{21} + \delta_{22} \ll \delta_2$.

6.2 Automated Teller Machine

The Automated Teller Machine (ATM) is a system that interacts with a potential customer via a specific interface and communicates with the bank over an appropriate communication protocol. Figure 6 represents the ATM SysML activity diagram ($\mathcal{A}'$) composed of the main diagram ($\mathcal{A}'_0$) "Fig. 6-(a)" and three called diagrams: (a') Check Card ($\mathcal{A}'_1$)[5], (b') Authorize ($\mathcal{A}'_2$), and (c') Transaction ($\mathcal{A}'_3$) that is showed in Fig. 6-(b).

[5] The call behavior action "Check Card" is denoted by a' and calls the diagram $\mathcal{A}'_1$.

Table 4. The verification cost for properties Φ_2 Φ_{21}, and Φ_{22}.

n	3	4	5	6	7	8	9	10
β_2	0.9377	0.9377	0.9377	0.9377	0.9377	0.9377	0.9377	0.9377
γ_2	2,213,880	4,823,290	8,434,700	13,048,110	51,145,160	202,489,260	454033360	805,777,460
δ_2	33.394	78.746	168.649	354.211	2280.252	17588.755	34290.635	63097.014
β_{21}	0.9377	0.9377	0.9377	0.9377	0.9377	0.9377	0.9377	0.9377
γ_{21}	9614	12017	14420	16823	19226	21629	24032	26435
δ_{21}	4.775	12.301	32.852	83.337	274.9	450.81	586.43	652.76
β_{22}	1	1	1	1	1	1	1	1
γ_{22}	9	9	9	9	9	9	9	9
δ_{22}	0.003	0.003	0.003	0.003	0.003	0.003	0.003	0.003

Our goal is to measure the satisfiability probability of the PCTL properties Φ_3 and Φ_4 on $\mathscr{A}' = \mathscr{A}_0' \uparrow_{a'} \mathscr{A}_1' \uparrow_{b'} \mathscr{A}_2' \uparrow_{c'} \mathscr{A}_3'$.

Property Φ_3**.** "What is the minimum probability of authorizing a transaction after inserting a card". PCTL: $Pmin = ?[G(InstertCard \Rightarrow F(DebitAccount))]$.
After verifying Φ_3 on $\mathscr{A}'$, we verify Φ_{31} on $\mathscr{A}_0'$ and Φ_{32} on $\mathscr{A}_3'$ such that : $\Phi_{31} \triangleq Pmin = ?[G(InstertCard) \Rightarrow (F(Transaction))]$ and : $\Phi_{32} \triangleq Pmin = ?[G((True) \Rightarrow F(DebitAccount))]$. As a result we found the following: $\beta_3 = 0.8421$, $\gamma_3 = 606470$, $\delta_3 = 3.12$, $\beta_{31} = 0.8421$, $\gamma_{31} = 3706$, and $\delta_{31} = 0.64$, $\beta_{32} = 1$, $\gamma_{32} = 15$, and $\delta_{32} = 0.007$. From the obtained results, we found that the satisfiability probability is maintained $\beta_3 = \beta_{31} \times \beta_{32}$, with a considerable verification costs $\gamma_{31} + \gamma_{32} < \gamma_3$ and $\delta_{31} + \delta_{32} \ll \delta_3$.

Property Φ_4**.** "What is the maximum probability of inserting a card when it is not valid." PCTL: $Pmax = ?[(CardNotValid) \Rightarrow (F(InsertCard))]$.

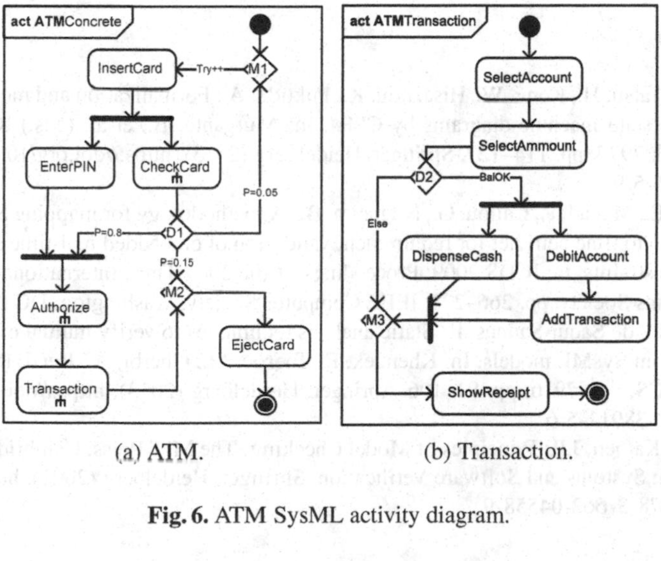

(a) ATM. (b) Transaction.

Fig. 6. ATM SysML activity diagram.

Similarly to the verification of Φ_3, instead of verifying Φ_4 on $\mathscr{A}'$ we verify Φ_{41} on $\mathscr{A}_1'$ and Φ_{42} on $\mathscr{A}_0'$ such that :
$\Phi_{41} \triangleq Pmax = ?[(CardNotValid) \Rightarrow (F(EndCheckCard))]$, and
$\Phi_{42} \triangleq Pmax = ?[(CheckCard) \Rightarrow (F(InsertCard))]$.

After verification, we found the following: $\beta_4 = 0.05$, $\gamma_4 = 606470$, $\delta_4 = 11.458$, $\beta_{41} = 1$, $\gamma_{41} = 11$, and $\delta_{41} = 0.004$, $\beta_{42} = 0.05$, $\gamma_{42} = 7211$, and $\delta_{42} = 1.584$. From these results, we have: $\beta_4 = \beta_{41} \times \beta_{42}$, $\gamma_{41} + \gamma_{42} < \gamma_4$ and $\delta_{41} + \delta_{42} \ll \delta_4$.

7 Conclusion

In this paper, we presented a compositional verification framework to improve the efficiency of probabilistic model-checking. More specifically, our target was verifying systems modeled using SysML activity diagrams composed by the call behavior interfaces. We improved their verification cost by introducing a probabilistic compositional verification approach based on decomposing a global PCTL property into local ones with respect to interfaces between diagrams. Moreover, the presented framework can ignore the called diagrams that are irrelevant to a given PCTL property. For verification, we proposed an algorithm to encode the composed diagrams into PRISM input language. Furthermore, we proposed a semantic for SysML activity diagrams that helps on proofs and to encode easily the diagrams in PRISM. We proved the soundness of the proposed framework by showing the satisfiability preservation of PCTL properties. In addition, we demonstrated the effectiveness of our framework by verifying real systems that are not symmetric, which mean, we can not benefit from the symmetry reduction built within the PRISM model checker. In future, we would like to extend our work by investigating several directions. First, we plan to extend our framework to handle more compositional verification techniques like assume-guaranty and integrate them within the PRISM implementation. Then, we explore more system features such as time and object. Finally, we intend to apply our framework on a large systems' applications.

References

1. Ando, T., Yatsu, H., Kong, W., Hisazumi, K., Fukuda, A.: Formalization and model checking of SysML state machine diagrams by CSP#. In: Murgante, B., et al. (eds.) ICCSA 2013. LNCS, vol. 7973, pp. 114–127. Springer, Heidelberg (2013). https://doi.org/10.1007/978-3-642-39646-5_9
2. Andrade, E., Maciel, P., Callou, G., Nogueira, B.: A methodology for mapping SysML activity diagram to time petri net for requirement validation of embedded real-time systems with energy constraints. In: ICDS 2009: Proceedings of the 2009 Third International Conference on Diagrams Society, pp. 266–271. IEEE Computer Society, Washington, DC (2009)
3. Apvrille, L., de Saqui-Sannes, P.: Static analysis techniques to verify mutual exclusion situations within SysML models. In: Khendek, F., Toeroe, M., Gherbi, A., Reed, R. (eds.) SDL 2013. LNCS, vol. 7916, pp. 91–106. Springer, Heidelberg (2013). https://doi.org/10.1007/978-3-642-38911-5_6
4. Baier, C., Katoen, J.P.: Principles of Model Checking. The MIT Press, Cambridge (2008)
5. Bérard, B.: Systems and Software Verification. Springer, Heidelberg (2001). https://doi.org/10.1007/978-3-662-04558-9

6. Berezin, S., Campos, S., Clarke, E.M.: Compositional reasoning in model checking. In: de Roever, W.-P., Langmaack, H., Pnueli, A. (eds.) COMPOS 1997. LNCS, vol. 1536, pp. 81–102. Springer, Heidelberg (1998). https://doi.org/10.1007/3-540-49213-5_4
7. Carneiro, E., Maciel, P., Callou, G., Tavares, E., Nogueira, B.: Mapping SysML state machine diagram to time petri net for analysis and verification of embedded real-time systems with energy constraints. In: ENICS 2008: Proceedings of the 2008 International Conference on Advances in Electronics and Micro-electronics, pp. 1–6. IEEE Computer Society, Washington, DC (2008)
8. Carrillo, O., Chouali, S., Mountassir, H.: Formalizing and verifying compatibility and consistency of SYSML blocks. SIGSOFT Softw. Eng. Notes **37**(4), 1–8 (2012)
9. Feng, L., Han, T., Kwiatkowska, M., Parker, D.: Learning-based compositional verification for synchronous probabilistic systems. In: Bultan, T., Hsiung, P.-A. (eds.) ATVA 2011. LNCS, vol. 6996, pp. 511–521. Springer, Heidelberg (2011). https://doi.org/10.1007/978-3-642-24372-1_40
10. Feng, L., Kwiatkowska, M., Parker, D.: Compositional verification of probabilistic systems using learning. In: Proceedings of the 2010 Seventh International Conference on the Quantitative Evaluation of Systems, QEST 2010, pp. 133–142. IEEE Computer Society (2010)
11. Feng, L., Kwiatkowska, M., Parker, D.: Automated learning of probabilistic assumptions for compositional reasoning. In: Giannakopoulou, D., Orejas, F. (eds.) FASE 2011. LNCS, vol. 6603, pp. 2–17. Springer, Heidelberg (2011). https://doi.org/10.1007/978-3-642-19811-3_2
12. Forejt, V., Kwiatkowska, M., Norman, G., Parker, D.: Automated verification techniques for probabilistic systems. In: Bernardo, M., Issarny, V. (eds.) SFM 2011. LNCS, vol. 6659, pp. 53–113. Springer, Heidelberg (2011). https://doi.org/10.1007/978-3-642-21455-4_3
13. Gomaa, H.: Software Modeling and Design: UML, Use Cases, Patterns, and Software Architectures. Cambridge University Press, Cambridge (2011)
14. Object Management Group. OMG Unified Modeling Language: Superstructure 2.1.2, November 2007
15. Holt, J., Perry, J.: SysML for Systems Engineering. Institution of Engineering and Technology Press, January 2007
16. Huang, X., Sun, Q., Li, J., Zhang, T.: MDE-based verification of SysML state machine diagram by UPPAAL. In: Yuyu Yuan, X.W., Yueming, L. (eds.) Trustworthy Computing and Services. CCIS, vol. 320, pp. 490–497. Springer, Berlin Heidelberg (2013)
17. Jansen, D.N., Hermanns, H., Katoen, J.-P.: A probabilistic extension of UML statecharts. In: Damm, W., Olderog, E.-R. (eds.) FTRTFT 2002. LNCS, vol. 2469, pp. 355–374. Springer, Heidelberg (2002). https://doi.org/10.1007/3-540-45739-9_21
18. Jansen, D.N., Hermanns, H., Katoen, J.-P.: A QoS-oriented extension of UML statecharts. In: Stevens, P., Whittle, J., Booch, G. (eds.) UML 2003. LNCS, vol. 2863, pp. 76–91. Springer, Heidelberg (2003). https://doi.org/10.1007/978-3-540-45221-8_7
19. Kwiatkowska, M., Norman, G., Parker, D.: PRISM 4.0: verification of probabilistic real-time systems. In: Gopalakrishnan, G., Qadeer, S. (eds.) CAV 2011. LNCS, vol. 6806, pp. 585–591. Springer, Heidelberg (2011). https://doi.org/10.1007/978-3-642-22110-1_47
20. Object Management Group. OMG Systems Modeling Language Specification, September 2007
21. Ouchani, S., Mohamed, O.A., Debbabi, M.: A security risk assessment framework for SYSML activity diagrams. In: 2013 IEEE 7th International Conference on Software Security and Reliability, pp. 227–236, June 2013
22. Ouchani, S.: Towards a fractionation-based verification: application on SYSML activity diagrams. In: Proceedings of the 34th ACM/SIGAPP Symposium on Applied Computing, SAC 2019, pp. 2032–2039. ACM (2019)

23. Ouchani, S., Debbabi, M.: Specification, verification, and quantification of security in model-based systems. Computing **97**(7), 691–711 (2015)
24. Ouchani, S., Ait Mohamed, O., Debbabi, M.: Efficient probabilistic abstraction for SysML activity diagrams. In: Eleftherakis, G., Hinchey, M., Holcombe, M. (eds.) SEFM 2012. LNCS, vol. 7504, pp. 263–277. Springer, Heidelberg (2012). https://doi.org/10.1007/978-3-642-33826-7_18

Context-Free Grammars for Deterministic Regular Expressions with Interleaving

Xiaoying Mou[1,2], Haiming Chen[1](✉), and Yeting Li[1,2]

[1] State Key Laboratory of Computer Science, Institute of Software, Chinese Academy of Sciences, Beijing, China
{mouxy,chm,liyt}@ios.ac.cn

[2] University of Chinese Academy of Sciences, Beijing, China

Abstract. In this paper, we study deterministic regular expressions with interleaving (IDREs). Regular expressions extended with interleaving are good at describing sequential and parallel data patterns. The interleaving makes these expressions exponentially more succinct than standard regular expressions. And when they obey the determinism constraint, they are more efficient for matching and parsing processes than general ones. However, the interleaving also makes the structure of expressions more complex, and the semantic definition of determinism is hard to follow, which prevent users from writing and developing IDREs. We address this problem by proposing a determinism checking method and a syntactic description for IDREs. Our checking method can locate the source of nondeterminism more efficiently and accurately than the existing method when the nondeterminism is caused by unary operators. We prove that the grammars of IDREs are context-free and suggest effective optimization rules to simplify the grammars.

Keywords: Context-free grammars · Determinism checking · Deterministic regular expressions · Interleaving · Syntax

1 Introduction

Regular expressions take an important role in computer science. They have been widely used in many applications, including programming languages [37], data query and process [8,10,11,27], XML schema languages [5,15,16], etc. Modern applications have various restrictions or extensions on regular expressions. One of the restrictions is the usage of deterministic regular expressions, which represent a subclass of regular languages. They are good at efficient matching of an input word against a regular expression. These expressions have been applied to the document markup language SGML in 1980s [24]. Nowadays, deterministic expressions are still important in XML schema languages, e.g., Document Type Definitions (DTD) [5] and XML Schema Definitions (XSD) [16], which are

Work supported by the National Natural Science Foundation of China under Grant Nos. 61872339 and 61472405.

R. M. Hierons and M. Mosbah (Eds.): ICTAC 2019, LNCS 11884, pp. 235–252, 2019.
https://doi.org/10.1007/978-3-030-32505-3_14

recommended by W3C (World Wide Web Consortium), and require the content models to be deterministic. Deterministic expressions are also used in other applications (e.g., [23,29]).

Roughly speaking, determinism means that when matching a word against an expression, a symbol in the word can be matched to only one position in the expression without looking ahead. For instance, an expression $E = (a + a \cdot b)$ is not deterministic, where $+$ denotes union (or disjunction) and $\cdot$ denotes concatenation, since it accepts the language $\{a, ab\}$, and then when a symbol a is input, it is impossible to decide which a in E to match without knowing whether a symbol b will follow or not. In other word, it needs to look ahead one position for making decision. Deterministic regular expressions show more efficient matching in $O(m + n \log\log m)$ time for an input word of length n and a deterministic expression of length m [20], while matching requires non-linear time for general regular expressions. Certainly, deterministic expressions also have other advantages, e.g., they behave better than general ones on several decision problems [30,31]. Deterministic regular expressions have been studied in the literature, also under the name of *one-unambiguous* regular expressions [7,9,13,20,25,26,33].

The interleaving operator is an extension of standard regular expressions and is also called shuffle [6,19,36]. Regular expressions with interleaving are double-exponentially more succinct than standard regular expressions [18]. For instance, given an expression $E = a\&b\&c$ (& represents interleaving), the language accepted by E is $\{abc, acb, bac, bca, cab, cba\}$. We can define a standard expression $E' = (abc + acb + bac + bca + cab + cba)$ accepting the same language as E, while E' is tedious and nondeterministic but E is succinct and deterministic. In practice, expressions with interleaving are used in a variety of applications, e.g., XML schema language Relax NG [15], RDF schema language ShEx [35], path queries [2], complex event semantics [28], concurrency and distributed systems [17,21], etc.

In this paper, we study deterministic regular expressions with interleaving (IDREs). The addition of interleaving makes the determinism checking more complex than that of the standard ones. For instance, a standard regular expression is deterministic if its corresponding Glushkov automaton is deterministic, and the Glushkov automaton is quadratic in the size of the expression in optimal worst case [7], but for an expression with interleaving, the number of states of the Glushkov automaton are exponential [6]. Besides, determinism is only defined in an arcane and semantic manner [8]. These facts make the studies on IDREs harder than the standard ones. And since the lack of theoretical studies on IDREs, many applications are limited. For instance, in XSD and ShEx, they only support two subclasses of IDREs. Hence, we focus on the determinism checking and syntactic description problems of IDREs. Such descriptions make it possible to automatically construct IDREs used in applications, such as benchmarking a validator for Relax NG, or incrementally constructing IDREs. Meanwhile, we hope our results can help users to better understand, use and develop these kinds of expressions.

Lots of work [7,13,25,26] focused on checking determinism of standard regular expressions and regular expressions with counting. Among these methods,

Groz et al. [20] gave the most efficient one, which supports deciding these two kinds of expressions in $\mathcal{O}(|E|)$ time by decomposing the syntax tree of the expression E. But for regular expressions with interleaving, only the work in [33] gave an $O(|\Sigma||E|)$ time algorithm to decide determinism that works on marked expressions with interleaving. Brüggemann-Klein [8] did a related study but checked determinism of a restricted case of interleaving.

There are several related works on describing and constructing deterministic expressions for users. Some works use Glushkov automata to describe deterministic expressions, and try to construct deterministic expressions from nondeterministic ones when possible [1,3,9] with the help of the automata. In [9], an exponential time algorithm was provided to decide whether a regular language, given by an arbitrary regular expression, is deterministic. An algorithm was further given there to construct an equivalent deterministic expression for a nondeterministic expression when possible, while the resulting expression could be double-exponentially large. Bex et al. [3] and Ahonen [1] constructed deterministic ones for nondeterministic expressions which may accept approximated languages. Some works try to learn deterministic expressions from a set of examples. Many restricted subclasses of standard deterministic expressions or IDREs (e.g., SORE and CHARE [4], DME [14], SIRE [32]) were defined, where each symbol occurs at most once. They also proposed corresponding algorithms to learn these subclasses of deterministic expressions.

The above works try to give alternative ways of constructing deterministic expressions, but they are only limited on approximate or restricted deterministic expressions. Chen et al. [12] proposed grammars of standard deterministic regular expressions and designed a generator, which inspired our work. We consider the addition of the interleaving and develop grammars to describe the whole class of IDREs. Then we can generate IDREs randomly based on the grammars.

The main contributions of this paper are listed as follows:

- We find a characterization of unmarked deterministic regular expressions with interleaving. Based on this, we propose an $O(|\Sigma||E|)$ time algorithm to check determinism of IDREs. Our algorithm needs no marking pre-processing and can locate the source of nondeterminism more efficiently and accurately than the method in [33], when the nondeterminism is caused by unary operators.
- We translate the above characterization into a derivation system, which shows how an IDRE can be constructed incrementally. Then we propose context-free grammars to describe IDREs using a 5-tuple nonterminal representation. Moreover, we suggest optimization rules to simplify the grammars effectively.
- We apply the method [12] to generate deterministic regular expressions with interleaving from the grammars.

The rest of this paper is organized as follows. Section 2 contains basic definitions. In Sect. 3 we study the characterization of IDREs and give an algorithm to check determinism based unmarked expressions. In Sect. 4 we develop the context-free grammars for IDREs and optimization rules. Experimental results are shown in Sect. 5. Section 6 concludes.

2 Preliminaries

2.1 Regular Expression with Interleaving

Let Σ be an alphabet, namely a finite non-empty set of symbols. The set of all finite words over Σ is denoted as Σ^*. The empty word is denoted as ε.

The interleaving of two words in Σ^* is a finite set of words defined inductively as follows: for $u, v \in \Sigma^*$ and $a, b \in \Sigma$, $u\&\varepsilon = \varepsilon\&u = \{u\}$; $au\&bv = \{aw \mid w \in u\&bv\} \cup \{bw \mid w \in au\&v\}$. Naturally, the interleaving of two languages L_1 and L_2 is: $L_1\&L_2 = \{w \in u\&v \mid u \in L_1, v \in L_2\}$ [36]. For instance, $L(a\&(bc)) = \{abc, bac, bca\}$.

Regular expressions with interleaving over Σ are defined by the following grammar:

$$\tau \to \emptyset \mid \varepsilon \mid a \mid (\tau \cdot \tau) \mid (\tau + \tau) \mid (\tau\&\tau) \mid (\tau^*) \quad (a \in \Sigma)$$

where $\emptyset$ represents the empty set, $\cdot$ represents concatenation, $+$ union (or disjunction), & interleaving and $*$ the Kleene star. The Kleene star has the highest precedence, followed by the concatenation, and the union and the interleaving have the lowest precedence. In practice, we usually omit parentheses as well when there is no ambiguity. We also use $E?$ as an abbreviation of $E + \varepsilon$ or $\varepsilon + E$. For an expression E, the language specified by E is denoted by $L(E)$. The languages of $E_1\&E_2$ is defined as $L(E_1\&E_2) = L(E_1)\&L(E_2)$; other cases of the languages are standard and readers can refer to [22] for the definitions. The size of an expression E, denoted by $|E|$, is the number of symbols and operators occurring in E. Without additional explanations, E refers to a regular expression with interleaving.

Expressions can be reduced by rules: $E + \emptyset = \emptyset + E = E$, $E \cdot \emptyset = \emptyset \cdot E = \emptyset$, $E\&\emptyset = \emptyset\&E = \emptyset$, $\emptyset^* = \varepsilon$, $E \cdot \varepsilon = \varepsilon \cdot E = E, E\&\varepsilon = \varepsilon\&E = E, \varepsilon^* = \varepsilon$. In the following, we only consider reduced regular expressions. Besides, for a reduced expression, it either does not contain $\emptyset$ or is $\emptyset$. So we will not consider the expression $\emptyset$ in the following.

2.2 Definitions for Analyzing Determinism

For an expression E, we can mark symbols with subscripts such that in the marked expression each marked symbol occurs only once. The marking of E is denoted by $\overline{E}$. The same notation will also be used for dropping subscripts from the marked form, namely $\overline{\overline{E}} = E$. The set of symbols that occur in E is denoted by $sym(E)$. A concise definition of determinism of an expression E can be given with the help of its marked form.

Definition 1 ([8]). *Let $\overline{E}$ be a marking of the expression E, $\overline{E}$ is deterministic iff for all words $uxv \in L(\overline{E}), uyw \in L(\overline{E})$ where $x, y \in sym(\overline{E})$ and $u, v, w \in sym(\overline{E})^*$, if $\overline{x} = \overline{y}$ then $x = y$. An expression E is deterministic iff $\overline{E}$ is deterministic.*

Example 1. The expression $E_1 = a^* \cdot a$ is not deterministic since one of its marked form $\overline{E}_1 = a_1^* \cdot a_2$, is not deterministic. $\overline{E}_1$ is nondeterministic, since for two words $a_1a_2, a_2 \in L(\overline{E}_1)$ there is $\overline{a}_1 = \overline{a}_2 = a$ but $a_1 \neq a_2$. For an expression $E_2 = a \cdot a^*$ with its marked form $\overline{E}_2 = a_1 \cdot a_2^*$, E_2 and $\overline{E}_2$ are deterministic.

These definitions are needed for analyzing the determinism of an expression.

$$\begin{aligned} \mathit{first}(E) &= \{a \mid au \in L(E), a \in \mathit{sym}(E), u \in \mathit{sym}(E)^*\} \\ \mathit{followlast}(E) &= \{a \mid uav \in L(E), u \in L(E), u \neq \varepsilon, a \in \mathit{sym}(E), v \in \mathit{sym}(E)^*\} \end{aligned}$$

The computing rules of the *first* set can be found in [7,33]. For the *followlast*, we give computing rules of the marked expressions.

Proposition 1 ([33]). *For a marked expression $\overline{E}$, computing rules of followlast are as follows: (fl represents followlast)*

$$\overline{E} = \varepsilon \text{ or } x: \ \mathit{fl}(\overline{E}) = \emptyset$$

$$\overline{E} = \overline{F} + \overline{G}: \ \mathit{fl}(\overline{E}) = \mathit{fl}(\overline{F}) \cup \mathit{fl}(\overline{G})$$

$$\overline{E} = \overline{F} \cdot \overline{G}: \ \mathit{fl}(\overline{E}) = \begin{cases} \mathit{fl}(\overline{G}) & \text{if } \varepsilon \notin L(\overline{G}) \\ \mathit{fl}(\overline{F}) \cup \mathit{fl}(\overline{G}) \cup \mathit{first}(\overline{G}) & \text{if } \varepsilon \in L(\overline{G}) \end{cases}$$

$$\overline{E} = \overline{F}\&\overline{G}: \mathit{fl}(\overline{E}) = \begin{cases} \mathit{fl}(\overline{F}) \cup \mathit{fl}(\overline{G}) & \text{if } \varepsilon \notin L(\overline{F}), \varepsilon \notin L(\overline{G}) \\ \mathit{fl}(\overline{F}) \cup \mathit{fl}(\overline{G}) \cup \mathit{first}(\overline{G}) & \text{if } \varepsilon \notin L(\overline{F}), \varepsilon \in L(\overline{G}) \\ \mathit{fl}(\overline{F}) \cup \mathit{fl}(\overline{G}) \cup \mathit{first}(\overline{F}) & \text{if } \varepsilon \in L(\overline{F}), \varepsilon \notin L(\overline{G}) \\ \mathit{fl}(\overline{F}) \cup \mathit{fl}(\overline{G}) \cup \mathit{first}(\overline{F}) \cup \mathit{first}(\overline{G}) & \text{if } \varepsilon \in L(\overline{F}), \varepsilon \in L(\overline{G}) \end{cases}$$

$$\overline{E} = \overline{F}? \ : \mathit{fl}(\overline{E}) = \mathit{fl}(\overline{F})$$

$$\overline{E} = \overline{F}^* \ : \mathit{fl}(\overline{E}) = \mathit{fl}(\overline{F}) \cup \mathit{first}(\overline{F})$$

We define a Boolean function λ to check whether an expression E can accept the empty word. $\lambda(E) = \mathit{true}$ if $\varepsilon \in L(E)$, otherwise $\lambda(E) = \mathit{false}$.

$$\lambda(\varepsilon) = \mathit{true} \quad \lambda(a) = \mathit{false}\ (a \in \Sigma) \quad \lambda(E_1 + E_2) = \lambda(E_1) \vee \lambda(E_2)$$
$$\lambda(E_1^*) = \mathit{true} \quad \lambda(E_1 \cdot E_2) = \lambda(E_1) \wedge \lambda(E_2) \quad \lambda(E_1\&E_2) = \lambda(E_1) \wedge \lambda(E_2)$$

3 Checking Determinism Based on Unmarked Expressions

As we can see in Definition 1, determinism is defined in an arcane and semantic manner based on marked expressions. Besides, the work in [33] derives from Definition 1 and also works on marked IDREs. When we directly use the characterization in [33] to describe IDREs, suppose that the alphabet is $\{a\}$, after the marking process, it includes $\{a_1, a_2, \ldots\}$. We are not sure the upper limit of subscript numbers, so the number of the alphabet is infinite. Thus the work on marked expressions cannot be applied here. In this section, we study a characterization of IDREs based on unmarked forms, and then to check determinism based on unmarked expressions.

3.1 A Characterization of Unmarked IDREs

In [33], they have proposed a characterization of marked IDREs, which lists all possible conditions causing nondeterminism. We rewrite the characterization based on the structure of an expression E, which lists conditions making sure E deterministic, as follows.

Lemma 1. *Let $\overline{E}$ be a marked expression.*
(1) $\overline{E} = \varepsilon$ or x $(x \in \overline{\Sigma})$: $\overline{E}$ is deterministic.
(2) $\overline{E} = \overline{E}_1 + \overline{E}_2$: $\overline{E}$ is deterministic iff $\overline{E}_1, \overline{E}_2$ are deterministic and $\overline{first(\overline{E}_1)} \cap \overline{first(\overline{E}_2)} = \emptyset$.
(3) $\overline{E} = \overline{E}_1 \cdot \overline{E}_2$: $\overline{E}$ is deterministic iff $\overline{E}_1, \overline{E}_2$ are deterministic, and

- *if $\lambda(\overline{E}_1)$, then $\overline{first(\overline{E}_1)} \cap \overline{first(\overline{E}_2)} = \emptyset$ and $\overline{followlast(\overline{E}_1)} \cap \overline{first(\overline{E}_2)} = \emptyset$;*
- *if $\neg\lambda(\overline{E}_1)$, then $\overline{followlast(\overline{E}_1) \cap first(\overline{E}_2)} = \emptyset$.*

(4) $\overline{E} = \overline{E}_1 \& \overline{E}_2$: $\overline{E}$ is deterministic iff $\overline{E}_1, \overline{E}_2$ are deterministic, and $\overline{sym(\overline{E}_1)} \cap \overline{sym(\overline{E}_2)} = \emptyset$.
(5) $\overline{E} = \overline{E}_1?$: $\overline{E}$ is deterministic iff $\overline{E}_1$ is deterministic.
(6) $\overline{E} = \overline{E}_1^*$: $\overline{E}$ is deterministic iff $\overline{E}_1$ is deterministic, and for all $x \in followlast(\overline{E}_1)$, for all $y \in first(\overline{E}_1)$, if $\overline{x} = \overline{y}$, then $x = y$.

To drop the marking operators, we refer to properties in [13], which has been proven for regular expressions with counting.

Lemma 2 ([13]). *For an expression E, there are:*

a. $first(E) = \overline{first(\overline{E})}$.
b. When E is deterministic, $followlast(E) = \overline{followlast(\overline{E})}$.

We can apply Lemma 2 to regular expressions with interleaving, since for the case $E_1 \& E_2$, computing rules of *first* and *followlast* we defined also follow their definitions. Hence, the proof is similar in [13]. Meanwhile, intuitively $sym(E) = \overline{sym(\overline{E})}$. Then we can drop the marking operators for claims (1)–(5) in Lemma 1. Next, we concentrate on the claim (6). To drop its marking operators, we define the function $\mathcal{W}$.

Definition 2. *The Boolean function $\mathcal{W}(E)$ is defined as:*

$$\begin{aligned}
&\mathcal{W}(\varepsilon) = \mathcal{W}(a) = true\\
&\mathcal{W}(E_1 + E_2) = \mathcal{W}(E_1) \wedge \mathcal{W}(E_2) \wedge \big(followlast(E_1) \cap first(E_2) = \emptyset\big)\\
&\qquad\qquad \wedge \big(followlast(E_2) \cap first(E_1) = \emptyset\big)\\
&\mathcal{W}(E_1 \cdot E_2) = \big(followlast(E_2) \cap first(E_1) = \emptyset\big) \wedge \Big(\big(\neg\lambda(E_1) \wedge \neg\lambda(E_2)\big)\\
&\qquad \vee \big(\lambda(E_1) \wedge \neg\lambda(E_2) \wedge \mathcal{W}(E_2)\big) \vee \big(\lambda(E_1) \wedge \lambda(E_2) \wedge \mathcal{W}(E_1) \wedge \mathcal{W}(E_2)\big)\\
&\qquad \vee \big(\neg\lambda(E_1) \wedge \lambda(E_2) \wedge \mathcal{W}(E_1) \wedge \big(first(E_1) \cap first(E_2) = \emptyset\big)\big)\Big)\\
&\mathcal{W}(E_1 \& E_2) = \mathcal{W}(E_1) \wedge \mathcal{W}(E_2)\\
&\mathcal{W}(E_1?) = \mathcal{W}(E_1^*) = \mathcal{W}(E_1)
\end{aligned}$$

The importance of the function $\mathcal{W}(E)$ can be seen from Lemma 3.

Lemma 3. *For an expression E, if E is deterministic, then two statements are equivalent:*
(1) for all $x \in followlast(\overline{E})$, for all $y \in first(\overline{E})$, if $\overline{x} = \overline{y}$, then $x = y$.
(2) $\mathcal{W}(E) = true$.

Proof. This lemma has been proved for regular expressions with counting [13]. Here, we only prove the case $E = E_1 \& E_2$. Since E is deterministic, E_1 and E_2 are deterministic by Definition 1 and Lemma 1.

(1) $\Rightarrow$ (2) : Based on Definition 2, $first(E) = first(E_1) \cup first(E_2)$, $followlast(E) \supseteq followlast(E_1) \cup followlast(E_2)$. Since (1) holds for E, we get that $\forall x \in followlast(\overline{E}_1)$ and $\forall y \in first(\overline{E}_1)$, or $\forall x \in followlast(\overline{E}_2)$ and $\forall y \in first(\overline{E}_2)$, if $\overline{x} = \overline{y}$, then $x = y$. By the inductive hypothesis, we get $\mathcal{W}(E_1) = true$ and $\mathcal{W}(E_2) = true$. Hence, $\mathcal{W}(E) = \mathcal{W}(E_1) \wedge \mathcal{W}(E_2) = true$.

(2) $\Rightarrow$ (1) : Since $\mathcal{W}(E) = true$, $\mathcal{W}(E_1) = \mathcal{W}(E_2) = true$. Let $x \in followlast(\overline{E}), y \in first(\overline{E})$, and $\overline{x} = \overline{y}$. When $\varepsilon \notin L(E_1)$ and $\varepsilon \notin L(E_2)$, $first(E) = first(E_1) \cup first(E_2)$, $followlast(E) = followlast(E_1) \cup followlast(E_2)$. Suppose that $x \in followlast(E_1)$ (*resp.* $x \in followlast(E_2)$), $y \in first(E_1)$ (*resp.* $x \in first(E_2)$). Since $\mathcal{W}(E_1) = true$ (*resp.* $\mathcal{W}(E_2) = true$) and E_1 (*resp.* E_2) is deterministic, there is $x = y$ by the inductive hypothesis. Suppose that $x \in followlast(E_1)$ (*resp.* $x \in followlast(E_2)$), $y \in first(E_2)$ (*resp.* $x \in first(E_1)$). Since E is deterministic, $sym(E_1) \cap sym(E_2) = \emptyset$ by Lemmas 1 and 2. Thus there is $x = y$. We can prove the other cases in a similar way.

Corollary 1. *For an expression $E = E_1^*$, E is deterministic iff E_1 is deterministic and $\mathcal{W}(E_1) = true$.*

Proof. It follows from Lemmas 1 and 3.

Then we obtain a characterization of unmarked IDREs.

Theorem 1. *Let E be an expression.*
(1) $E = \varepsilon$ or a : E is deterministic.
(2) $E = E_1 + E_2$: E is deterministic iff E_1, E_2 are deterministic and $first(E_1) \cap first(E_2) = \emptyset$.
(3) $E = E_1 \cdot E_2$: E is deterministic iff E_1, E_2 are deterministic, and

- *if $\lambda(\overline{E}_1)$, $first(E_1) \cap first(E_2) = \emptyset$ and $followlast(E_1) \cap first(E_2) = \emptyset$;*
- *if $\neg\lambda(\overline{E}_1)$, $followlast(E_1) \cap first(E_2) = \emptyset$.*

(4) $E = E_1 \& E_2$: E is deterministic iff E_1, E_2 are deterministic, and $sym(E_1) \cap sym(E_2) = \emptyset$.
(5) $E = E_1?$: E is deterministic iff E_1 is deterministic.
(6) $E = E_1^*$: E is deterministic iff E_1 is deterministic, and $\mathcal{W}(E_1) = true$.

Proof. Based on Lemma 2, we can drop the marking operators for claims (1)-(5) in Lemma 1. We replace the claim (6) with Corollary 1.

3.2 The Checking Algorithm *Determ_unmarked*

We complete an algorithm to decide determinism based on Theorem 1. By observation, not every subexpression should call the function $\mathcal{W}$. For an expression E, only a subexpression F is inside an iterative subexpression of E, then we run $\mathcal{W}(F)$. Referring to [13], this represents that F has the *continuing* property, denoted as $ct(F) = true$.

Definition 3 ([13]). *For a subexpression F of E, F is continuing (i.e., $ct(F) = true$), if for any word $w_1, w_2 \in L(F)$, there are $u, v \in (sym(E) \cup \flat)^*$ such that $u\flat w_1 \flat\flat w_2 \flat v \in L(E_{F_\flat})$, where $E_{F_\flat}$ is a new version of E by replacing F with $\flat F \flat$ ($\flat \notin sym(E)$).*

The function *markRable* is designed to compute the *continuing* property, shown in Fig. 1, where parameter E is the input expression, and the boolean parameter N is the value assigned to the current expression. The computation of *markRable* is actually a top-down traversal on the syntax tree of E. In this downward propagation, the value for N is inherited from upper level subexpressions to lower level subexpressions. Only in two cases the value will change, one is that when $E = E_1^*$, $ct(E_1) = true$; the other is that when $E = E_1 \cdot E_2$ and $ct(E) = true$, $ct(E_1) = \lambda(E_2)$ and $ct(E_2) = \lambda(E_1)$.

Hence, we can decompose the function $\mathcal{W}$ into each case in Theorem 1, and only when the function *ct* is true, the case runs conditions in $\mathcal{W}$. We use the following function $\mathcal{D}$ to check determinism, which is equivalent to Theorem 1 but is easier for implementation.

Definition 4. *The Boolean function $\mathcal{D}(E)$ is defined as follows:*

$$\mathcal{D}(\varepsilon) = \mathcal{D}(a) = true$$

$$\mathcal{D}(E_1 + E_2) = \mathcal{D}(E_1) \wedge \mathcal{D}(E_2) \wedge (first(E_1) \cap first(E_2) = \emptyset) \wedge \Big(\neg ct(E_1 + E_2) \vee \\ \big((followlast(E_1) \cap first(E_2) = \emptyset) \wedge (followlast(E_2) \cap first(E_1) = \emptyset)\big)\Big)$$

$$\mathcal{D}(E_1 \cdot E_2) = \mathcal{D}(E_1) \wedge \mathcal{D}(E_2) \wedge (followlast(E_1) \cap first(E_2) = \emptyset) \wedge \Big(\neg\lambda(E_1) \vee \\ (first(E_1) \cap first(E_2) = \emptyset)\Big) \wedge \Big(\neg ct(E_1 \cdot E_2) \vee \Big((followlast(E_2) \cap first(E_1) \\ = \emptyset) \wedge (\lambda(E_1) \vee \neg\lambda(E_2) \vee first(E_1) \cap first(E_2) = \emptyset)\Big)\Big)$$

$$\mathcal{D}(E_1 \& E_2) = \mathcal{D}(E_1) \wedge \mathcal{D}(E_2) \wedge (sym(E_1) \cap sym(E_2) = \emptyset)$$
$$\mathcal{D}(E_1?) = \mathcal{D}(E_1^*) = \mathcal{D}(E_1)$$

Theorem 2. *For an expression E, E is deterministic iff $\mathcal{D}(E) = true$.*

Proof. It follows from Theorem 1 and the definitions of $\mathcal{W}$ and *ct*.

The process for Theorems 2 is formalized in Algorithm 1. In the pseudocode, the computations for *first*, *followlast* and *sym* sets are omitted, which can easily be done by their computing rules.

Function *markRable*(*E: Expression, N: Boolean*):
  case $E = a$ or ε: $ct(E) = false$;
  case $E = E_1 + E_2$ or $E = E_1 \& E_2$:
    $ct(E) = N$;
    markRable(E_1, N); markRable(E_2, N);
  case $E = E_1 \cdot E_2$:
    $ct(E) = N$;
    if $ct(E)$, **then** {markRable(E_1, $\lambda(E_2)$); markRable(E_2, $\lambda(E_1)$,)}
    else { markRable(E_1, *false*); markRable(E_2, *false*);}
  case $E = E_1?$: $ct(E) = N$; markRable(E_1, N);
  case $E = E_1^*$: $ct(E) = N$; markRable(E_1, *true*);

Fig. 1. A procedure to recognize the continuing property by $markRable(E, false)$

Algorithm 1: Determ_unmarked (E: Expression)
input : a regular expression with interleaving E
output: true if the expression E is deterministic or false otherwise
if $E = a$ *or* $E = \varepsilon$ **then return** true
if $E = E_1 + E_2$ **then**
  if Determ_unmarked(E_1) and Determ_unmarked(E_2) **then**
    if $first(E_1) \cap first(E_2) \neq \emptyset$ **then**
      return false
    if $ct(E)$ and $(followlast(E_1) \cap first(E_2) \neq \emptyset$ or $followlast(E_2) \cap first(E_1) \neq \emptyset)$ **then return** false
    else return true
  else return false
if $E = E_1 \cdot E_2$ **then**
  if Determ_unmarked(E_1) and Determ_unmarked(E_2) **then**
    if $followlast(E_1) \cap first(E_2) \neq \emptyset$ or $(\lambda(E_1)$ and $first(E_1) \cap first(E_2) \neq \emptyset)$ **then return** false
    if $ct(E)$ and $\Big(followlast(E_2) \cap first(E_1) \neq \emptyset$ or $(first(E_1) \cap first(E_2) \neq \emptyset$ and $\neg\lambda(E_1)$and $\lambda(E_2))\Big)$ **then return** false
    else return true
  else return false
if $E = E_1 \& E_2$ **then**
  if Determ_unmarked(E_1) and Determ_unmarked(E_2) **then**
    return $sym(E_1) \cap sym(E_2) = \emptyset$
  else return false
if $E = E_1?$ *or* $E = E_1^*$ **then return** Determ_unmarked(E_1)

Theorem 3. *Determ_unmarked*(E) *runs in* $O(|\Sigma||E|)$ *time.*

Proof. First we construct the syntax tree for the expression E, and call the function $\lambda(E)$ and $markRable(E, false)$, which costs linear time. The checking process can be done only in one bottom-up and incremental traversal on the tree, at the same time, *first*, *followlast* and *sym* sets are computed in $O(|\Sigma||E|)$ time. Since the maximum number of symbols in *first*, *sym* or *followlast* set is $|\Sigma|$, emptiness test of $first(E_1) \cap first(E_2)$, $sym(E_1) \cap sym(E_2)$ or $first(E_1) \cap followlast(E_2)$ for subexpressions E_1 and E_2 can be computed in $O(|\Sigma|)$ time by means of a hash table.

Since there are at most $|E|$ nodes in the syntax tree, and the operations on each node can be completed in $O(|\Sigma|)$ time, the time complexity of $Determ\_unmarked(E)$ is $O(|\Sigma||E|)$. When the alphabet is fixed, the algorithm runs in linear time.

3.3 The Local Nondeterminism-Locating Feature

For some kind of nondeterministic expressions, $Determ\_unmarked(E)$ can locate the source of nondeterminism more efficiently than the method in [33] (equal to Lemma 1). We call it the local nondeterminism-locating feature, which can locally locate a nondeterministic subexpression without checking the whole expression, and thus is more advantageous for diagnosing purpose. Take $E = ((a \cdot a?\&b) \cdot (c + d)?)^*$ as an example, we run $Determ\_unmarked(E)$ and show the syntax tree of E in Fig. 2, where some computing results are also shown. For comparison, we also show the computation results by Lemma 1.

As we can see, checking by either $Determ\_unmarked(E)$ or Lemma 1, we get that E is not deterministic. For $Determ\_unmarked(E)$, the sequence of nodes in the checking process is : $1 \to 2 \to 3 \to 4$. At the node 4, it meets the 12th line in Algorithm 1 and returns *false*. However, by Lemma 1, the checking process will return $false$ in node 12 by checking nodes $1 \to 2 \to \cdots \to 11 \to 12$ step by step. $Determ\_unmarked(E)$ can locate on the subexpression $a \cdot a?$ causing nondeterminism, but for Lemma 1, it returns $((a \cdot a?\&b) \cdot (c + d)?)^*$.

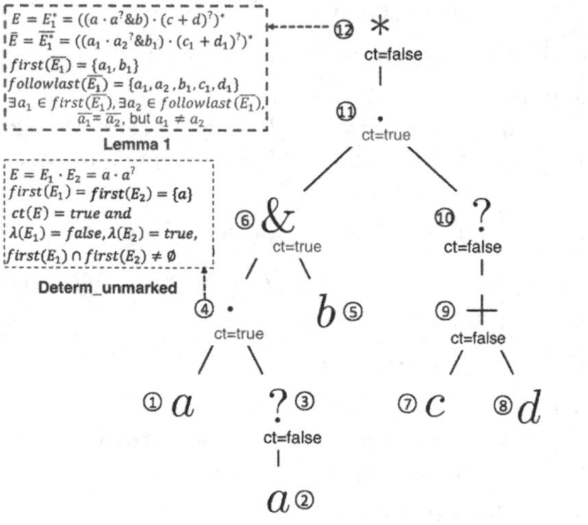

Fig. 2. The syntax tree of $((a \cdot a?\&b) \cdot (c + d)?)^*$

This example shows that for a nondeterministic expression E, when the source of nondeterminism is one of its subexpression E_1 and E_1 has the *continuing* property, our algorithm has the local nondeterminism-locating feature.

4 Syntactic Description for IDREs

4.1 The Derivation System

In this section, we give the grammars to define IDREs. We translate the characterization in Theorem 1 into a derivation system $\mathcal{T}$ for IDREs. The following notations are needed to define $\mathcal{T}$. Let $\vdash r$ mean that the regular expression r satisfies determinism. A derivation rule is of the form $\frac{\vdash r_1 \ldots \vdash r_n \quad c_1 \ldots c_m}{\vdash r}$, which means that if $\vdash r_1 \ldots \vdash r_n$ satisfy determinism and conditions $c_1 \ldots c_m$ hold, then r will also satisfy determinism. We say r is *derivable* iff there is derivation tree such that r is the root of the tree [34].

The system $\mathcal{T}$ consists of the following derivation rules, where each rule corresponds one claim in Theorem 1.

$$\textbf{Base: } \frac{}{\vdash \varepsilon \ \vdash a \quad a \in \Sigma} \qquad \textbf{Union: } \frac{\vdash r \ \vdash s \quad \mathit{first}(r) \cap \mathit{first}(s) = \emptyset}{\vdash r + s}$$

$$\textbf{Inter: } \frac{\vdash r \ \vdash s \quad \mathit{sym}(r) \cap \mathit{sym}(s) = \emptyset}{\vdash r \& s}$$

$$\textbf{Concat: } \frac{\vdash r \ \vdash s \quad \mathit{followlast}(r) \cap \mathit{first}(s) = \emptyset \quad \neg\lambda(r) \vee \mathit{first}(r) \cap \mathit{first}(s) = \emptyset}{\vdash r \cdot s}$$

$$\textbf{Opt: } \frac{\vdash r}{\vdash r?} \quad \textbf{Star: } \frac{\vdash r \quad \mathcal{W}(r)}{\vdash r^*}$$

Theorem 4. *A regular expression r is deterministic iff r is derivable from $\mathcal{T}$.*

Proof. It is proved based on Theorem 1 and derivation rules of $\mathcal{T}$.

4.2 The Context-Free Grammars for IDREs

As we can see from $\mathcal{T}$, to construct an IDRE, we must guarantee that all of its subexpressions are deterministic and they meet conditions on *first*, *followlast*, *sym* sets, and the functions λ and $\mathcal{W}$. Take the **Union** case $r + s$ as an example, we can give any deterministic expression r' with the satisfied condition that $\mathit{first}(r') \cap \mathit{first}(s) = \emptyset$ to replace r, and $r' + s$ is still deterministic.

Definition 5. *A context-free grammar is a 4-tuple $G = (N, T, P, S_0)$ with N and T disjoint sets of nonterminals and terminals respectively, P a set of productions of the form $A \to \alpha$ where $A \in N$, $\alpha \in (N \cup T)^*$, and $S_0 \in N$ the start symbol.*

In a grammar G, a nonterminal N_0 in N is *useful* if there is a terminal word w such that w is derived from N_0 denoted by $N_0 \overset{*}{\Rightarrow} w$. If all the nonterminals of a production are *useful*, then the production is *valid*. The language defined by G, denoted as $L(G)$, is the set of strings $L(G) = \{w \in T^* \mid S_0 \overset{*}{\Rightarrow} w\}$ and w is called a *sentence* of G if $w \in L(G)$.

We design the context-free grammars $G_{idre} = (N, T, P, S_0)$ to describe IDREs. Let $\Sigma = \{a_1, ..., a_n\}$, then $T = \{+, \cdot, \&, ?, *, (,), \emptyset, \varepsilon\} \cup \Sigma$. $N = \mathbb{R} \cup R$, where $\mathbb{R}$ is a finite set of nonterminal symbols. Each nonterminal in $\mathbb{R}$ is of the form $R^{F,L,S,\alpha,\beta}$, where $F, L, S \subseteq \Sigma^*$, $\alpha, \beta \subset [0, 1]$ (1 represents *true* and 0 represents *false*). $R^{F,L,S,\alpha,\beta}$ is intended to describe a set of expressions in IDREs and each expression r satisfies $F = \mathit{first}(r), L = \mathit{followlast}(r), S = \mathit{sym}(r), \alpha = \lambda(r)$ and $\beta = \mathcal{W}(r)$. We assume that all the nonterminal symbols are start symbols.

To construct P, we refer to derivation rules in $\mathcal{T}$ since each kind of rules corresponds to a class of productions. Meanwhile we add the computations of *first*, *followlast*, *sym*, λ and $\mathcal{W}$ in each class of productions. The productions P are shown as follows, where specific conditions are listed in Table 1.

$$\begin{array}{lll}
\textbf{Base}: & R^{\{a_i\},\emptyset,\{a_i\},0,1} \rightarrow a_i, i \in \{1,\cdots,n\} & R \rightarrow \emptyset \mid \varepsilon \\
\textbf{Union}: & R^{F,L,S,\alpha,\beta} \rightarrow \bigcup_{con1} (R^{F_1,L_1,S_1,\alpha_1,\beta_1} + R^{F_2,L_2,S_2,\alpha_2,\beta_2}) & \\
\textbf{Concat}: & R^{F,L,S,\alpha,\beta} \rightarrow \bigcup_{con2} (R^{F_1,L_1,S_1,\alpha_1,\beta_1} \cdot R^{F_2,L_2,S_2,\alpha_2,\beta_2}) & \\
\textbf{Inter}: & R^{F,L,S,\alpha,\beta} \rightarrow \bigcup_{con3} (R^{F_1,L_1,S_1,\alpha_1,\beta_1} \,\&\, R^{F_2,L_2,S_2,\alpha_2,\beta_2}) & \\
\textbf{Opt}: & R^{F,L,S,1,\beta} \rightarrow \bigcup_{\alpha_1 \in \{0,1\}} R^{F,L,S,\alpha_1,\beta}? & \\
\textbf{Star}: & R^{F,L,S,1,1} \rightarrow \bigcup_{L=F_1 \cup L_1, \alpha_1 \in \{0,1\}} R^{F_1,L_1,S,\alpha_1,1*} &
\end{array}$$

Table 1. The specific conditions for the productions in P

Condition	Details
*con*1	$F = F_1 \cup F_2,\ L = L_1 \cup L_2,\ S = S_1 \cup S_2,\ \alpha = \alpha_1 \vee \alpha_2,$ $\beta = (\beta_1 \wedge \beta_2 \wedge (F_1 \cap L_2 = \emptyset) \wedge (L_1 \cap F_2 = \emptyset)),\ F_1 \cap F_2 = \emptyset$
*con*2	$(\alpha_1 \wedge (F = F_1 \cup F_2 \wedge F_1 \cap F_2 = \emptyset)) \vee (\neg\alpha_1 \wedge (F = F_1)),$ $(\alpha_2 \wedge (L = L_1 \cup L_2 \cup F_2)) \vee (\neg\alpha_2 \wedge (L = L_2)),\ S = S_1 \cup S_2,\ \alpha = \alpha_1 \wedge \alpha_2,$ $\beta = (F_1 \cap L_2 = \emptyset) \wedge ((\neg\alpha_1 \wedge \neg\alpha_2) \vee (\alpha_1 \wedge \neg\alpha_2 \wedge \beta_2) \vee (\alpha_1 \wedge \alpha_2 \wedge \beta_1 \wedge \beta_2)$ $\vee(\neg\alpha_1 \wedge \alpha_2 \wedge \beta_1 \wedge F_1 \cap F_2 = \emptyset)),\quad L_1 \cap F_2 = \emptyset$
*con*3	$F = F_1 \cup F_2,\ S = S_1 \cup S_2,\ \alpha = \alpha_1 \wedge \alpha_2,\ \beta = \beta_1 \wedge \beta_2,\ S_1 \cap S_2 = \emptyset$ $(\neg\alpha_1 \wedge \neg\alpha_2 \wedge (L = L_1 \cup L_2)) \vee (\neg\alpha_1 \wedge \alpha_2 \wedge (L = L_1 \cup L_2 \cup F_2))$ $\vee(\alpha_1 \wedge \neg\alpha_2 \wedge (L = L_1 \cup L_2 \cup F_1)) \vee (\alpha_1 \wedge \alpha_2 \wedge (L = L_1 \cup L_2 \cup F_1 \cup F_2))$

We use the union ($\bigcup$) to denote a set of rules with the same left hand nonterminal. For a fixed alphabet, T, N and P are finite. Clearly, G_{idre} are context-free grammars. Compared with the grammars of standard deterministic expressions [12], we extend the set of parameters of nonterminals with S and add the **Inter** class of productions, to deal with interleaving. To be mentioned, G_{idre} only supports generating IDREs of reduced forms, excluding expressions like $\varepsilon \cdot r$, $r \cdot \emptyset$, $\varepsilon \& r$, etc.

Theorem 5. *IDREs can be defined by the context-free grammars* G_{idre}.

Proof. ($\Leftarrow$) We will prove that sentences generated by G_{idre} are IDREs. That is, in G_{idre}, if there exists a nonterminal $R^{F,L,S,\alpha,\beta} \stackrel{*}{\Rightarrow} r$, then r is deterministic satisfying $first(r) = F, followlast(r) = L, sym(r) = S, \lambda(r) = \alpha, \mathcal{W}(r) = \beta$. We prove it by induction on the length of the derivation $R^{F,L,S,\alpha,\beta} \stackrel{*}{\Rightarrow} r$.

Base case: If the derivation is only one-step, according to G_{idre}, then $r = a$ with $a \in \Sigma$. r is deterministic.

Inductive step: Suppose that the derivation has $k{+}1$ steps ($k \geq 1$) and derivations with no more k-steps satisfy the claim, i.e., there exists a corresponding nonterminal in G_{idre} generating a deterministic expression. If the $(k+1)$-steps derivation follows the **Inter** class of productions, i.e., $R^{F,L,S,\alpha,\beta} \Rightarrow R_1\&R_2 \overset{*}{\Rightarrow} r$. Since the derivation of R_1 or R_2 takes no more than k steps, there are $R_1 \overset{*}{\Rightarrow} r_1$ and $R_2 \overset{*}{\Rightarrow} r_2$ with deterministic expressions r_1, r_2. Then $r = r_1\&r_2$. According to computing rules of the **Inter**, r, r_1 and r_2 should meet all conditions in $con3$. The important one condition is $sym(r_1) \cap sym(r_2) = \emptyset$. Because r_1, r_2 are deterministic and $sym(r_1) \cap sym(r_2) = \emptyset$, we get r is deterministic by Theorem 1. Form other conditions, we know the value of F, L, S, α, β, such as $F = first(r_1) \cup first(r_2)$, $\alpha = \lambda(r_1) \wedge \lambda(r_2) \cdots$. Hence $F = first(r), L = followlast(r), S = sym(r), \alpha = \lambda(r), \beta = \mathcal{W}(r)$ from the computations of $first, followlast, sym, \lambda$ and $\mathcal{W}$. If the $(k+1)$-steps derivation follows others class of productions, the proof is in a similar way.

($\Rightarrow$) We will prove that all IDREs can be generated by G_{idre} by induction on the structure of an expression. That is, if an expression $r \in$ IDREs with $F = first(r), L = followlast(r), S = sym(r), \alpha = \lambda(r), \beta = \mathcal{W}(r)$, then there exists a useful nonterminal $R^{F,L,S,\alpha,\beta}$ in G_{idre}, such that $R^{F,L,S,\alpha,\beta} \overset{*}{\Rightarrow} r$.

For the case $r = a (a \in \Sigma)$, refer to **Base** in G_{idre}, there exists $R^{\{a\},\emptyset,\{a\},0,1}$.

For the case $r = r_1\&r_2$, since r is deterministic, then r_1, r_2 are deterministic by Theorem 1. By the inductive hypothesis, there exist nonterminals R_1 and R_2 in G_{idre} satisfying that $R_1^{F_1,L_1,S_1,\alpha_1,\beta_1} \overset{*}{\Rightarrow} r_1$ and $R_2^{F_2,L_2,S_2,\alpha_2,\beta_2} \overset{*}{\Rightarrow} r_2$, where $F_1 = first(r_1), L_1 = followlast(r_1), S_1 = sym(r_1), \alpha_1 = \lambda(r_1), \beta_1 = \mathcal{W}(r_1)$ and $F_2 = first(r_2), L_2 = followlast(r_2), S_2 = sym(r_2), \alpha_2 = \lambda(r_2), \beta_2 = \mathcal{W}(r_2)$. Referring to the computations of $first, followlast, sym, \lambda$ and $\mathcal{W}$, we get $first(r) = first(r_1) \cup first(r_2)$, $(\neg\lambda(r_1) \wedge \neg\lambda(r_2) \wedge followlast(r) = followlast(r_1) \cup followlast(r_2)) \vee (\neg\lambda(r_1) \wedge \lambda(r_2) \wedge followlast(r) = followlast(r_1) \cup followlast(r_2) \cup first(r_2)) \vee (\lambda(r_1) \wedge \neg\lambda(r_2) \wedge followlast(r) = followlast(r_1) \cup followlast(r_2) \cup first(r_1)) \vee (\lambda(r_1) \wedge \lambda(r_2) \wedge followlast(r) = followlast(r_1) \cup followlast(r_2) \cup first(r_1) \cup first(r_2))$, $sym(r) = sym(r_1) \cup sym(r_2)$ and $\lambda(r) = \lambda(r_1) \wedge \lambda(r_2)$, $\beta(r) = \beta(r_1) \wedge \beta(r_2)$. This meets the **Inter** in G_{idre}, hence there exists $R^{F,L,S,\alpha,\beta} \Rightarrow R_1\&R_2 \overset{*}{\Rightarrow} r_1\&r_2$, i.e., $R^{F,L,S,\alpha,\beta} \overset{*}{\Rightarrow} r$.

The other cases (i.e., $r = r_1 + r_2, r = r_1 \cdot r_2$, $r = r_1?$, $r = r_1^*$) can be proved similarly.

Theorem 6. *IDREs cannot be defined by regular grammars.*

Proof. It has been proved that standard deterministic regular expressions cannot be defined by regular grammars in [12]. Because standard deterministic regular expressions are not closed under homomorphisms as regular languages do. Hence, IDREs as a superset class of standard deterministic regular expressions, cannot be defined by regular grammars.

4.3 Optimization Rules for Valid Productions

For the grammars G_{idre}, since there are $2^{|\Sigma|}$ different *first*, *followlast* and *sym* sets, the number of nonterminal symbols are $2^{3|\Sigma|+2}$. Each production uses at most three nonterminals, then the number of possible productions of G_{idre} is $\mathcal{O}(2^{9|\Sigma|})$. Clearly, it is really a heavy task to recognize *valid* productions for G_{idre}. We give the following rules to help recognize *useful* nonterminals. These optimization rules can accelerate the process of simplifying G_{idre}, since the production constructed all by *useful* nonterminals must be *valid.*

Lemma 4. *The nonterminal symbol $R^{F,L,S,\alpha,\beta}$ in G_{idre} is useful iff it meets four conditions: (1) $S \cap F \neq \emptyset$; (2)$F \cup L \subseteq S$; (3) $(F \cap L = \emptyset) \rightarrow (\beta = 1)$; (4) $(\alpha = 0$ and $\beta = 1) \rightarrow (F \nsubseteq L)$.*

Proof. ($\Rightarrow$) The necessity of this lemma is easy to prove. (1) S is the *sym* set including all occurring symbols and F stores the first symbol. Since we have ignored the ε, S and F are not empty. So $S \cap F \neq \emptyset$. (2) F and L are subsets of S, so $F \cup L \subseteq S$. (3) When $F \cap L = \emptyset$, there exists no pair of symbols $a = b$ such that $a \in F, b \in L$. This case meets the statement (1) in Lemma 3, so $\mathcal{W} = true$ ($\beta = 1$). (4) When $\alpha = 0$ and $\beta = 1$, suppose that $F \subseteq L$, there at least exists a symbol $s \in F \cap L$. Since $\alpha = 0$, $R^{F,L,S,\alpha,\beta}$ can derive from rules Base, Union, Concat, or Inter. If it derives from Opt, Star, then $\alpha = 1$ contradicts the precondition. For case Base, we get $L = \emptyset$, which contradicts $s \in L$. For case Union $R^{F,L,S,\alpha,\beta} \rightarrow R^{F_1,L_1,S_1,\alpha_1,\beta_1} + R^{F_2,L_2,S_2,\alpha_2,\beta_2}$, referring to Table 1, we get $F = F_1 \cup F_2, L = L_1 \cup L_2$. Then s may exist in $F_1 \cap L_1$, $F_1 \cap L_2$, $F_2 \cap L_1$ or $F_2 \cap L_2$, but all these possible situations will cause $\mathcal{W} = false$ ($\beta = 0$) by Definition 2, contradicts the precondition $\beta = 1$. For cases Concat and Inter, they can be proved in a similar way.

($\Leftarrow$) The sufficiency of this lemma proves that when these four conditions hold, there exists an expression $r \in$ IDREs such that *first*$(r) = F$, *followlast*$(r) = L$, *sym*$(r) = S$, $\lambda(r) = \alpha$, $\mathcal{W}(r) = \beta$. That is $R^{F,L,S,\alpha,\beta}$ is useful.

When $F \cap L = \emptyset$, referring to conditions, we suppose $F = \{a_1, \ldots, a_m\}$ and $L = \{b_0, \ldots, b_p\}$ with a_i $(i \in [1, m]), b_j$ $(j \in [0, p])$ are distinct symbols, then $S = \{a_1, \ldots, a_m, b_0, \ldots, b_p\}$. By condition (3), there is $\beta = 1$. Thus we only need consider the following cases:

- Case 1: When $\alpha = 1$, we can construct $r = ((a_1 + \ldots + a_m) \cdot (b_0^* + \ldots + b_p^*))?$, which satisfies *first*$(r) = F$, *followlast*$(r) = L$, *sym*$(r) = S$, $\alpha = 1$, $\beta = 1$.
- Case 2: When $\alpha = 0$, since we have guaranteed $F \nsubseteq L$, the expression can be $r = (a_1 + \cdots + a_m) \ldots (b_0^* + \ldots + b_p^*)$, which satisfies *first*$(r) = F$, *followlast*$(r) = L$, *sym*$(r) = S$, $\alpha = 0$, $\beta = 1$.

When $F \cap L \neq \emptyset$, referring to conditions, we suppose $F = \{a_0, \ldots, a_m, c_1, \ldots, c_n\}$ and $L = \{b_0, \ldots, b_p, c_1, \ldots, c_n\}$, then $S = \{a_0, \ldots, a_m, b_0, \ldots, b_p, c_1, \ldots, c_n\}$. The cases include:

- Case 3: When $\alpha = 0$, $\beta = 0$, we can construct $r = (a_0 + \ldots + a_m + c_1 \ldots + c_n) \cdot (b_0^* + \ldots + b_p^* + c_1^* \ldots + c_n^*)$, which satisfies *first*$(r) = F$, *followlast*$(r) = L$, *sym*$(r) = S$, $\alpha = 0$, $\beta = 0$.

- Case 4: When $\alpha = 1$, $\beta = 0$, we can construct $r = ((a_0 + \ldots + a_m + c_1 \ldots + c_n) \cdot (b_0^* + \ldots + b_p^* + c_1^* \ldots + c_n^*))?$, which satisfies $\mathit{first}(r) = F$, $\mathit{followlast}(r) = L$, $\mathit{sym}(r) = S$, $\alpha = 1$, $\beta = 0$.
- Case 5: When $\alpha = 1$, $\beta = 1$, if $m \neq 0$, we can construct $r = (((a_1 + \ldots + a_m)\&(c_1^* + \ldots + c_n^*)) \cdot (b_0^* + \ldots + b_p^*))?$; if $m = 0$, $r = ((c_1\& \ldots \&c_n) \cdot (b_0^* + \ldots + b_p^*))^*$. Then r satisfies $\mathit{first}(r) = F$, $\mathit{followlast}(r) = L$, $\mathit{sym}(r) = S$, $\alpha = 1$, $\beta = 1$.
- Case 6: When $\alpha = 0$, $\beta = 1$, by condition (4), we get $F \not\subseteq L$. Hence there must exist a symbol $a \in F$ but $a \notin L$, then we change the assumption on F to $F = \{a_1, \ldots, a_m, c_1, \ldots, c_n\}$. The expression can be $r = ((a_1 + \ldots + a_m)\&(c_1^* + \ldots + c_n^*)) \cdot (b_0^* + \ldots + b_p^*)$, which satisfies $\mathit{first}(r) = F$, $\mathit{followlast}(r) = L$, $\mathit{sym}(r) = S$, $\alpha = 0$, $\beta = 1$.

5 Experiments

In this section, we implement the grammars over small alphabets to show the effectiveness of optimization rules in Lemma 4 by showing the sizes of nonterminals and productions. Then we generate sentences from the simplified grammars by the method in [12].

Grammars over Small Alphabets. We simplify the grammars G_{idre} by Lemma 4, and denote the simplified grammars as s-G_{idre}, which only consist of *useful* nonterminals and *valid* productions. Experimental results are shown in Table 2 over small alphabets. $|N|$ and $|P|$ represent the number of nonterminals and productions, respectively, and $|\Sigma|$ is the size of the alphabet.

Table 2. The comparison between grammars G_{idre} and s-G_{idre}

$\lvert\Sigma\rvert$	G_{idre}		s-G_{idre}		$\frac{\lvert P\rvert \text{ in } s\text{-}G_{idre}}{\lvert P\rvert \text{ in } G_{idre}}$ (100%)
	$\lvert N\rvert$	$\lvert P\rvert$	$\lvert N\rvert$	$\lvert P\rvert$	
1	33	2,097	6	15	0.668%
2	257	103,810	44	1,116	1.074%
3	2,049	4,926,467	282	46,865	0.951%
4	16,385	-	1,652	1,495,482	-

From Table 2, we observe that the optimization rules are effective, which are quite useful to reduce the number of productions making it reduced by about two orders of magnitude. When $n = 3$, about 99.05% of productions in G_{idre} are not *valid*. When $n = 4$, the executing time of generating productions of G_{idre} is too long, so we cannot record, which shows the necessity of our optimization rules.

For comparison, we also complete the simplified grammars for standard deterministic regular expressions in [12], denoted as s-G_{dre}. Results are shown in Fig. 3. With the increase of $|\Sigma|$, the numbers of nonterminals and productions are larger. s-G_{idre} increases more sharply than s-G_{dre}, especially for the $|P|$.

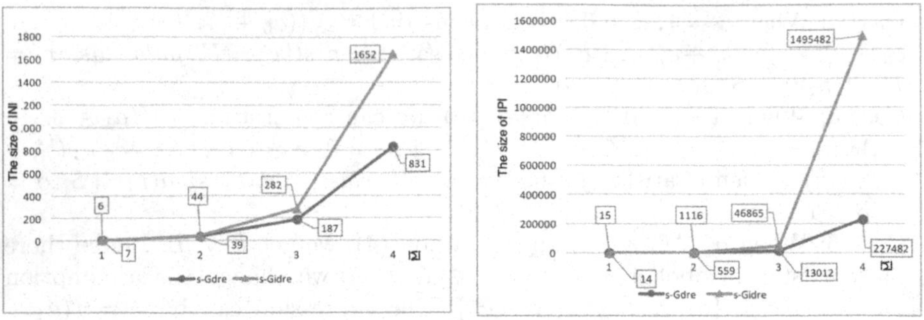

Fig. 3. The comparison between grammars $s\text{-}G_{idre}$ and $s\text{-}G_{dre}$

Sentences Generation by the Method in [12]**.** The *generator* in [12] is designed for efficient and random generation. Since it does not need to preprocess the whole grammar, this algorithm is good at efficient generation from large-scale grammars. As we can see from Fig. 3, even when $|\Sigma| = 4$, the productions size of $s\text{-}G_{idre}$ is more than 1 million. The grammars are too large to handle with the increasing of $|\Sigma|$. For Purdom's method, it works in the whole grammar, so we only record the generation results when $|\Sigma| \leq 4$, since the rest will cost too much time and storage. Since we have optimization rules to recognize all useful nonterminals in $s\text{-}G_{idre}$, the *generator* can be applied to $s\text{-}G_{idre}$ to generate IDREs without the limit of $|\Sigma|$. We only need to input the size of the alphabet $|\Sigma|$ and the length of sentence l, then we call the *generator* and obtain a random expression in IDREs with length no more than l.

We set $l = 30$ and use *generator* to successfully generate an expression in IDREs, when $|\Sigma|$ values from 1 to 7. The time cost is recorded in Table 3.

Table 3. Time cost for generating an expression in IDREs by *generator*

$\|\Sigma\|$	1	2	3	4	5	6	7
Time(s)	0.01	0.12	0.2	3	13	114	892

6 Conclusion

In this paper, we obtained a characterization of unmarked IDREs, and then developed an $\mathcal{O}(|\Sigma||E|)$ time algorithm to decide determinism worked on unmarked expressions. Besides, using the characterization, we developed the context-free grammars to describe the whole class of IDREs and gave optimization rules. Experimental results showed that our optimization rules are good at reducing the number of productions. We also apply the method [12] to generate IDREs from the grammars.

One future work is to use the grammars to help users to write IDREs interactively. We plan to improve the grammars and develop a tool to help users to write IDREs efficiently.

References

1. Ahonen, H.: Disambiguation of SGML content models. In: International Workshop on Principles of Document Processing, pp. 27–37 (1996)
2. Bárcenas, E., Genevès, P., Layaïda, N., Schmitt, A.: Query reasoning on trees with types, interleaving, and counting. In: Proceedings of the 22nd International Joint Conference on Artificial Intelligence, pp. 718–723 (2011)
3. Bex, G.J., Gelade, W., Martens, W., Neven, F.: Simplifying XML schema: effortless handling of nondeterministic regular expressions. In: Proceedings of the ACM SIGMOD International Conference on Management of Data, pp. 731–744 (2009)
4. Bex, G.J., Neven, F., Schwentick, T., Vansummeren, S.: Inference of concise regular expressions and DTDs. ACM Trans. Database Syst. **35**(2), 1–47 (2010)
5. Bray, T., Paoli, J., Sperberg-McQueen, C.M., Maler, E., Yergeau, F.: Extensible markup language (XML) 1.0, 5th edn. W3C Recommendation (2008). http://www.w3.org/TR/REC-xml/
6. Broda, S., Machiavelo, A., Moreira, N., Reis, R.: Automata for regular expressions with shuffle. Inf. Comput. **259**(2), 162–173 (2017)
7. Brüggemann-Klein, A.: Regular expressions into finite automata. Theoret. Comput. Sci. **120**(2), 197–213 (1993)
8. Brüggemann-Klein, A.: Unambiguity of extended regular expressions in SGML document grammars. In: European Symposium on Algorithms, pp. 73–84 (1993)
9. Brüggemann-Klein, A., Wood, D.: One-unambiguous regular languages. Inf. Comput. **140**(2), 229–253 (1998)
10. Buil-Aranda, C., Arenas, M., Corcho, O., Polleres, A.: Federating queries in SPARQL 11: Syntax, semantics and evaluation. Web Semant. Sci. Serv. Agents World Wide Web **18**(1), 1–17 (2013)
11. Che, D., Aberer, K., Özsu, T.: Query optimization in XML structured-document databases. J. Int. Conf. Very Large Databases **15**(3), 263–289 (2006)
12. Chen, H., Xu, Z., Lu, P.: Towards an effective syntax for deterministic regular expressions. Technical report, State Key Laboratory of Computer Science (2018). http://lcs.ios.ac.cn/~chm/papers.htm
13. Chen, H., Lu, P.: Checking determinism of regular expressions with counting. Inf. Comput. **241**, 302–320 (2015)
14. Ciucanu, R., Staworko, S.: Learning schemas for unordered XML. In: Proceedings of the 14th International Symposium on Database Programming Languages (2013)
15. Clark, J., Murata, M.: Relax NG specification. Organization for the Advancement of Structured Information Standards (OASIS) (2001)
16. Gao, S., Sperberg-McQueen, C.M., Thompson, H.S.: XML Schema Definition Language (XSD) 1.1 Part 1: Structures. W3C Recommendation (2012). http://www.w3.org/TR/2012/REC-xmlschema11-1-20120405/
17. Garg, V.K., Ragunath, M.T.: Concurrent regular expressions and their relationship to petri nets. Theoret. Comput. Sci. **96**(2), 285–304 (1992)
18. Gelade, W.: Succinctness of regular expressions with interleaving, intersection and counting. Theoret. Comput. Sci. **411**(31–33), 2987–2998 (2010)

19. Gischer, J.: Shuffle languages, petri nets, and context-sensitive grammars. Commun. ACM **24**(9), 597–605 (1981)
20. Groz, B., Maneth, S.: Efficient testing and matching of deterministic regular expressions. J. Comput. Syst. Sci. **89**, 372–399 (2017)
21. Hartmann, J., Vieira, M., Foster, H., Ruder, A.: A uml-based approach to system testing. ISSE **1**(1), 12–24 (2005)
22. Hopcroft, J.E., Ullman, J.D.: Introduction To Automata Theory, Languages And Computation. Addison-Wesley, Boston (2001)
23. Huang, X., Bao, Z., Davidson, S.B., Milo, T., Yuan, X.: Answering regular path queries on workflow provenance. In: International Conference on Data Engineering, pp. 375–386 (2015)
24. ISO, 8879: Information processing text and office systems Standard Generalized Markup Language (SGML) (1986). https://www.iso.org/standard/16387.html
25. Kilpeläinen, P.: Checking determinism of XML schema content models in optimal time. Inf. Syst. **36**(3), 596–617 (2011)
26. Kilpeläinen, P., Tuhkanen, R.: One-unambiguity of regular expressions with numeric occurrence indicators. Inf. Comput. **205**(6), 890–916 (2007)
27. Koch, C., Scherzinger, S., Schweikardt, N., Stegmaier, B.: Schema-based scheduling of event processors and buffer minimization for queries on structured data streams. In: Proceedings of the International Conference on Very Large Databases, pp. 228–239 (2004)
28. Li, Z., Ge, T.: PIE: approximate interleaving event matching over sequences. In: 31st IEEE International Conference on Data Engineering, ICDE 2015, Seoul, South Korea, 13–17 April 2015, pp. 747–758 (2015)
29. Losemann, K., Martens, W.: The complexity of evaluating path expressions in SPARQL. In: Proceedings of the 31st ACM SIGMOD-SIGACT-SIGART Symposium on Principles of Database Systems, pp. 101–112 (2012)
30. Losemann, K., Martens, W., Niewerth, M.: Closure properties and descriptional complexity of deterministic regular expressions. Theoret. Comput. Sci. **627**, 54–70 (2016)
31. Martens, W., Neven, F., Schwentick, T.: Complexity of decision problems for XML schemas and chain regular expressions. SIAM J. Comput. **39**(4), 1486–1530 (2009)
32. Peng, F., Chen, H.: Discovering restricted regular expressions with interleaving. In: Asia-Pacific Web Conference, pp. 104–115 (2015)
33. Peng, F., Chen, H., Mou, X.: Deterministic regular expressions with interleaving. In: Leucker, M., Rueda, C., Valencia, F.D. (eds.) ICTAC 2015. LNCS, vol. 9399, pp. 203–220. Springer, Cham (2015). https://doi.org/10.1007/978-3-319-25150-9_13
34. Pierce, B.C.: Types and Programming Languages. MIT Press, Cambridge (2002)
35. Staworko, S., Boneva, I., Gayo, J.E.L., Hym, S., Prud'hommeaux, E.G., Solbrig, H.R.: Complexity and expressiveness of ShEx for RDF. In: Proceedings of International Conference on Database Theory, pp. 195–211 (2015)
36. Ter Beek, M.H., Kleijn, J.: Infinite unfair shuffles and associativity. Theoret. Comput. Sci. **380**(3), 401–410 (2007)
37. Wall, L., Christiansen, T., Schwartz, R.L.: Programming Perl. O'Reilly & Associates Inc, Sebastopol (1999)

Privacy and Security

Completeness of Abstract Domains for String Analysis of JavaScript Programs

Vincenzo Arceri[1(✉)], Martina Olliaro[2,3(✉)], Agostino Cortesi[2(✉)], and Isabella Mastroeni[1(✉)]

[1] University of Verona, Verona, Italy
{vincenzo.arceri,isabella.mastroeni}@univr.it
[2] Ca' Foscari University of Venice, Venice, Italy
{martina.olliaro,cortesi}@unive.it
[3] Masaryk University of Brno, Brno, Czech Republic

Abstract. Completeness in abstract interpretation is a well-known property, which ensures that the abstract framework does not lose information during the abstraction process, with respect to the property of interest. Completeness has been never taken into account for existing string abstract domains, due to the fact that it is difficult to prove it formally. However, the effort is fully justified when dealing with string analysis, which is a key issue to guarantee security properties in many software systems, in particular for JavaScript programs where poorly managed string manipulating code often leads to significant security flaws. In this paper, we address completeness for the main JavaScript-specific string abstract domains, we provide suitable refinements of them, and we discuss the benefits of guaranteeing completeness in the context of abstract-interpretation based string analysis of dynamic languages.

Keywords: String abstract domains · Abstract interpretation completeness · String analysis

1 Introduction

Despite the growth of support for string manipulation in programming languages, string manipulation errors still lead to code vulnerabilities that can be exploited by malicious agents, causing potential catastrophic damages. This is even more true in the context of web applications, where common programming languages used for the web-based software development (e.g., JavaScript), offer a wide range of dynamic features that make string manipulation challenging.

String analysis is a static program analysis technique that computes, for each execution trace of the program given as input, the set of the possible string values that may reach a certain program point. String analysis, as others non-trivial analyses in the programming languages field, is an undecidable task. Thus, a certain degree of approximation is necessary in order to find evidence of bugs

R. M. Hierons and M. Mosbah (Eds.): ICTAC 2019, LNCS 11884, pp. 255–272, 2019.
https://doi.org/10.1007/978-3-030-32505-3_15

and vulnerabilities in string manipulating code. In the recent literature, different approximation techniques for string analysis have been developed, such as [7]: automata-based [6,9,37,38], abstraction-based [2–4,11,12,39], constraint-based [1,25,32,34], and grammar-based [28,36], and used, inter alia, with the purpose of detecting web application vulnerabilities [36–38].

In this paper we focus on string analysis by means of the abstract interpretation theory [13,14]. Abstract interpretation has been proposed by P. Cousot and R. Cousot in the 70s as a theory of sound abstraction (or approximation) of the semantics of computer programs, and nowadays it is widely integrated in software verification tools and used to rigorous mathematical proofs of approximations correctness. Since the introduction of the abstract interpretation theory, many abstract domains that represent properties of interest about numerical domains values have been designed [8,10,13,15,18,19,29,30,33]. On the other hand, just in the last few years, scientific community has taken an interest in the development of abstract domains for string analysis [2,4,11,12,22,26,31], some of them language specific, such as those defined as part of the JavaScript static analysers: TAJS [20], SAFE [24], and JSAI [21].

Desirable features of abstract interpretation are *soundness* and *completeness* [14]. If soundness (or correctness), as a basic requirement, actually is often guaranteed by static analysis tools, completeness is frequently not met. If completeness is satisfied, it means that the abstract computations do not lose information, during the abstraction process, with respect to a property of interest, and so the abstract interpretation can be considered optimal. In [17], authors highlighted the fact that completeness is an abstract domain property, and they presented a methodology to obtain complete abstract domains with respect to operations, by minimally extending or restricting the underlying domains.

1.1 Paper Contribution

Due to the important role played by JavaScript in the current landscape, its extensive use of strings, and the difficulties in statically analyse it, we believe that an improvement in the accuracy of JavaScript-specific string abstract domains can lead to a preciser reasoning about strings.

Thus, in this paper, we study the completeness property, with respect to some string operations of interest, of two JavaScript-specific string abstract domains, i.e., those defined as part of SAFE [24] and TAJS [20] static analysers. Finally, we define their complete versions, and we discuss the benefits of guaranteeing completeness in the context of abstract interpretation based string analysis of dynamic languages.

1.2 Paper Structure

Section 2 gives basics in mathematics and abstract interpretation. Section 3 presents important concepts related to the completeness property in abstract interpretation [17], that we will use through the whole paper. Moreover, a motivating example is given to show the importance to guarantee completeness in an abstract interpretation-based analysis with respect to strings. Section 4 defines our core language. Section 5 presents the completion of the string abstract domain integrated into SAFE [24] and TAJS [20] static analysers with respect to two operations of interest. Section 6 highlights the strengths and usefulness of the completeness approach to abstract-based static analysis of JavaScript string manipulating programs. Section 7 concludes and points out interesting aspects for future works.

2 Background

Mathematical Notation. Given a set S, we denote by S^* the set of all the finite sequences of elements of S and by S^n the set of all finite sequences of S of length n. If $s = s_0 \dots s_n \in S^*$, we denote by s_i the i-th element of s, and by $|s| = n+1$ its length. We denote by $s[x/i]$ the sequence obtained replacing s_i in s with x. Given two sets S and T, we denote with $\wp(S)$ the powerset of S, with $S \backslash T$ the set difference, with $S \subset T$ the strict inclusion relation, and with $S \subseteq T$ the inclusion relation between S and T. A set L with ordering relation $\leq$ is a poset and it is denoted by $\langle L, \leq \rangle$. A poset $\langle L, \leq \rangle$ is a lattice if $\forall x, y \in L$ we have that $x \vee y$ and $x \wedge y$ belong to L, and we say that it is also complete when for each $X \subseteq L$ we have that $\bigvee X, \bigwedge X \in L$. Given a poset $\langle L, \leq \rangle$ and $S \subseteq L$, we denote by $\max(S) = \{x \in S \mid \forall y \in S.\, x \leq y \Rightarrow x = y\}$ the set of the maximal elements of S in L. As usual, a complete lattice L, with ordering $\leq$, least upper bound (lub) $\vee$, greatest lower bound (glb) $\wedge$, greatest element (top) $\top$, and least element (bottom) $\bot$ is denoted by $\langle L, \leq, \vee, \wedge, \top, \bot \rangle$. An *upper closure operator* on a poset $\langle L, \leq \rangle$ is an operator $\rho : L \rightarrow L$ which is monotone, idempotent, and extensive (i.e., $x \leq \rho(x)$) and it can be uniquely identified by the set of its fix-points. The set of all closure operators on a poset L is denoted by $uco(L)$. Given $f : S \rightarrow T$ and $g : T \rightarrow Q$ we denote with $g \circ f : S \rightarrow Q$ their composition, i.e., $g \circ f = \lambda x. g(f(x))$. Given $f : S^n \rightarrow T$, $s \in S^n$ and $i \in [0, n)$, we denote by $f_s^i = \lambda z. f(s[z/i]) : S \rightarrow T$ a generic i-th unary restriction of f.

Abstract Interpretation. Abstract interpretation [13,14] is a theoretical framework for sound reasoning about program semantic properties of interest, and can be equivalently formalized either as Galois connections or closure operators on a given concrete domain, which is a complete lattice C [14]. Let C and A be complete lattices, a pair of monotone functions $\alpha : C \rightarrow A$ and $\gamma : A \rightarrow C$ forms a *Galois Connection* (GC) between C and A if for every $x \in C$ and for every $y \in A$ we have $\alpha(x) \leq_A y \Leftrightarrow x \leq_C \gamma(y)$. The function α (resp. γ) is the *left-adjoint* (resp. *right-adjoint*) to γ (resp. α), and it is additive (resp. co-additive). If $\langle \alpha, \gamma \rangle$ is a GC between C and A then $\gamma \circ \alpha \in uco(C)$. If C is a complete lattice, then $\langle uco(C), \sqsubseteq, \sqcup, \sqcap, \lambda x.C, \mathtt{id} \rangle$ forms a complete lattice [35], which is the set of all possible abstractions of C, where the bottom element is $\mathtt{id} = \lambda x.x$, and for every $\rho, \eta \in uco(C)$, ρ is *more concrete than* η iff $\rho \sqsubseteq \eta$ iff $\forall y \in C.\, \rho(y) \leq \eta(y)$ iff $\eta(C) \subseteq \rho(C)$, $(\sqcap_{i \in I} \rho_i)(x) = \wedge_{i \in I} \rho_i(x)$; $(\sqcup_{i \in I} \rho_i)(x) = x$ iff $\forall i \in I.\, \rho_i(x) = x$. The operator $\rho \in uco(C)$ is disjunctive when $\rho(C)$ is a join-sublattice of C which holds iff ρ is additive [14]. Let L be a complete lattice, then $X \subseteq L$ is a Moore family of L if $X = \mathcal{M}(X) = \{\wedge S \mid S \subseteq X\}$, where $\wedge\varnothing = \top$. The condition that any concrete element of C has the best abstraction in the abstract domain A, implies that A is a Moore family of C. We denote by $\mathcal{M}(X)$ the Moore closure of $X \subseteq C$, that is the least subset of C, which is a Moore family of C, and contains X. If $\langle\alpha, \gamma\rangle$ is a GC between C and A and $f : C \to C$ a concrete function, then $f^\sharp = \alpha \circ f \circ \gamma : A \to A$ is the best correct approximation of f in A. Let $\langle\alpha, \gamma\rangle$ be a GC between C and A, $f : C \to C$ be a concrete function and $f^\sharp : A \to A$ be an abstract function. The function $f^\sharp$ is a *sound approximation* of f if $\forall c \in C.\ \alpha(f(c)) \leq_A f^\sharp(\alpha(c))$. In abstract interpretation, there exist two notions of completeness. *Backward completeness* property focuses on complete abstractions of the inputs, while *forward completeness* focuses on complete abstractions of the outputs, both w.r.t. an operation of interest. In this paper, we focus on the more typical notion of completeness, i.e., backward completeness. Hence, when we will talk about completeness, we mean backward completeness. Given a GC $\langle\alpha, \gamma\rangle$ between C and A, a concrete function $f : C \to C$, and an abstract function $f^\sharp : A \to A$, then $f^\sharp$ is *backward complete* for f (for short complete) if $\forall c \in C.\ \alpha(f(c)) = f^\sharp(\alpha(c))$. If the backward completeness property is guaranteed, no loss of information arises during the input abstraction process, w.r.t. an operation of interest.

3 Making Abstract Interpretations Complete

In this section, we give the notions and methodologies that we will use through the whole paper (and proposed in [17]), in order to constructively build, from an initial abstract domain, a novel abstract domain that is complete w.r.t. an operation of interest. Finally, a motivating example showing the usefulness of completion of abstract domains for string analysis is given.

As reported in [17], it is worth noting that completeness is a property related to the underlying abstract domain. Starting from this fact, in [17], authors proposed a constructive method to manipulate the underlying incomplete abstract domain in order to get a complete abstract domain w.r.t. a certain operation. In particular, given two abstract domains A and B and an operator $f : A \to B$, the authors gave two different notions of completion of abstract domains w.r.t. f: the one that *adds* the minimal number of abstract points to the input abstract domain A or the other that *removes* the minimal number of abstract points from the output abstract domain B. The first approach captures the notion of *complete shell of* A, while the latter defines the *complete core of* B, both w.r.t. an operator f.

Complete Shell vs Complete Core. We will focus on the construction of complete shells of string abstract domains, rather than complete cores. This choice is guided by the fact that a complete core for an operation f removes abstract points from a starting abstract domain, and so, even if it is complete for f, the complete core could worsen the precision of other operations.

On the other hand, complete shells augment the starting abstract domains (adding abstract points), and consequently it can not compromise the precision of other operations.

Below, we provide two important theorems proved in [17] that give a constructive method to compute abstract domain complete shells, defined in terms of an upper closure operator ρ. Precisely, the latter theorems present two notions of complete shells: *i. complete shells of* ρ *relative to* η (where η is an upper closure operator), meaning that they are complete shells of operations defined on ρ that return results in η, and *ii. absolute complete shells of* ρ, meaning that they are complete shells of operations that are defined on ρ and return results in ρ.

Theorem 1 (*Complete shell of* ρ *relative to* η). *Let* C *and* D *be two posets and* $f : C^n \to D$ *be a continuous function. Given* $\rho \in uco(C)$, *then* $\mathcal{S}_f^\rho : uco(D) \to uco(C)$ *is the following domain transformer:*

$$\mathcal{S}_f^\rho(\eta) = \mathcal{M}(\rho \cup (\bigcup_{\substack{i \in [0,n) \\ x \in C^n, y \in \eta}} \max(\{z \in C \mid (f_x^i)(z) \leq_D y\})))$$

and it computes the complete shell of ρ *relative to* η.

As already mentioned above, the idea under the complete shell of ρ (input abstraction) relative to η (output abstraction) is to refine ρ adding the minimum number of abstract points to make ρ complete w.r.t. an operator f. From Theorem 1, this is obtained adding to ρ the maximal elements in C, whose f image is dominated by elements in η, at least in one dimension i. Clearly, the so-obtained abstraction may be not an upper closure operator for C. Hence, Moore closure operator is applied. On the other hand, absolute complete shells are involved in the case in which the operator f of interest has same input and output abstract domain, i.e., $f : C^n \to C$. In this case, given $\rho \in uco(C)$, absolute complete shells of ρ can be obtained as the greatest fix-point (*gfp*) of the domain transformer $\mathcal{S}_f^\rho$ (see Theorem 1), as stated by the following theorem.

Theorem 2 (*Absolute complete shell of* ρ). *Let* C *be a poset and* $f : C^n \to C$ *be a continuous function. Given* $\rho \in uco(C)$, *then* $\overline{\mathcal{S}}_f^\rho : uco(C) \to uco(C)$ *is the following domain transformer:*

$$\overline{\mathcal{S}}_f^\rho = gfp(\lambda\eta.\mathcal{S}_f^\rho(\eta))$$

and it computes the absolute complete shell of ρ.

The completeness property for the sign abstract domain, which approximates numerical values, has been discussed in [17]. The sign abstract domain is complete for the product operation, but it is not complete w.r.t. the sum. Indeed, the sign of $\mathsf{e}_1 + \mathsf{e}_2$ cannot be defined by simply knowing the sign of e_1 and e_2. In [17], authors computed the absolute complete shell of the sign domain w.r.t. the sum operation, and they showed it corresponds to the interval abstract domain [13].

3.1 Motivating Example

A common feature of dynamic languages, such as PHP or JavaScript, is to be not typed. Hence, in those languages, it is allowed to change the variable type through the program execution. For example, in PHP, it is completely legal to write fragments such as `$x=1;$x=true;`, where the type of the variable `x` changes from integer to boolean. The first attempt to static reasoning about variable types was to track the latter adopting the so-called *coalesced sum* abstract domain [5,23], in order to detect whether a certain variable has constant type through the whole program execution. In Fig. 1a, we report the coalesced sum abstract domain for an intra-procedural version of PHP [5], that tracks null, boolean, integer, float and string types[1]. Consider the formal semantics of the sum operation in PHP [16]. When one of the operand is a string, since the sum operation is feasible only between numbers, *implicit type conversion* occurs and converts the operand string to a number. In particular, if the prefix of the string is a number, it is converted to the maximum prefix of the string corresponding to a number, otherwise it is converted to 0. For example, the expression $e =$ `"2.4hello" + "4"` returns `4.4`. Let $+^\sharp$ be the abstract sum operation on the coalesced sum abstract domain. The type of the expression e is given by:

$$\alpha(\{\texttt{"2.4hello"}\}) +^\sharp \alpha(\{\texttt{"4"}\}) = \mathsf{String} +^\sharp \mathsf{String} = \top$$

The static type analysis based on the coalesced sum abstract domain returns $\top$ (i.e., any possible value), since the sum between two strings may return either an integer or a float value. Precisely, the coalesced sum abstract domain is not complete w.r.t. the PHP sum operation, since for any string σ and σ', it does not meet the completeness condition: $\alpha(\sigma + \sigma') = \alpha(\sigma) +^\sharp \alpha(\sigma')$, e.g., $\alpha(\sigma + \sigma') = \mathsf{Float} \neq \alpha(\sigma) +^\sharp \alpha(\sigma') = \top$. Intuitively, the coalesced sum abstract domain is not complete w.r.t. the sum operation due to the loss of precision that occurs during the abstraction process of the inputs, since the domain is not precise enough to distinguish between strings that may be implicitly converted to integers or floats.

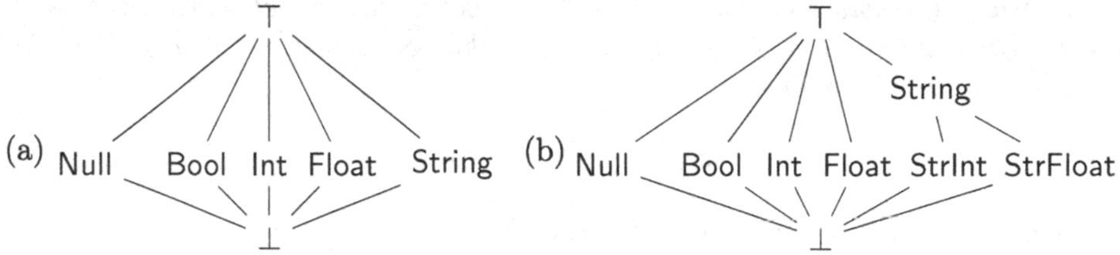

Fig. 1. (a) *Coalesced sum* abstract domain for PHP. (b) Complete shell of *coalesced sum* abstract domain w.r.t. the sum operation.

[1] Closing the coalesced sum abstract domain by the powerset operation, a more precise abstract domain is obtained, called *union type* abstract domain [23], that tracks the set of types of a certain variable during program execution.

```
a ::= n ∈ Int ∪ Float | a + a |
| a - a | a * a | a / a
| toNum(s)
b ::= true | false | b && b
| b || b | ! b
s ::= "s"
| concat(s1,s2)
e ::= x | a | b | s
bl ::= { } | { S }
S ::= x = e; | ; | bl
| if (b) bl1 else bl2
| while (b) bl
| S1 S2
```

where $x \in \textsc{Id}$, $c \in \Sigma$, $s \in \Sigma^*$

Fig. 2. μDyn syntax.

$$[\![x = \mathtt{e};]\!]\xi = \xi[x \leftarrow [\![\mathtt{e}]\!]\xi]$$

$$[\![\mathtt{if\ (b)\ bl_1\ else\ bl_2}]\!]\xi = \begin{cases} [\![\mathsf{bl}_1]\!]\xi & [\![\mathsf{b}]\!]\xi = \mathtt{true} \\ [\![\mathsf{bl}_2]\!]\xi & [\![\mathsf{b}]\!]\xi = \mathtt{false} \end{cases}$$

$$[\![\mathtt{while\ (b)bl}]\!]\xi = [\![\mathtt{if\ (b)\ \{bl\ while\ (b)bl\}\ else\ \{\}}]\!]\xi$$

$$[\![\{\}]\!]\xi = [\![;]\!]\xi = \xi \qquad [\![\{\mathsf{S}\}]\!]\xi = [\![\mathsf{S}]\!]\xi$$

$$[\![\mathsf{S}_1\mathsf{S}_2]\!]\xi = [\![\mathsf{S}_2]\!]([\![\mathsf{S}_1]\!]\xi)$$

$$[\![\mathtt{concat(s,s')}]\!]\xi = [\![\mathsf{s}]\!]\xi \cdot [\![\mathsf{s}']\!]\xi$$

$$[\![\mathtt{toNum(s)}]\!]\xi = \begin{cases} \mathcal{N}([\![\mathsf{s}]\!]\xi) & [\![\mathsf{s}]\!]\xi \in \Sigma^*_{\mathsf{Num}} \\ 0 & \text{otherwise} \end{cases}$$

Fig. 3. μDyn semantics.

Figure 1b shows the complete shell of the coalesced sum abstract domain w.r.t. the sum. The latter adds two abstract values to the original domain, namely StrFloat and StrInt, that correspond to the abstractions of the strings that may be implicitly converted to floats and to integers, respectively. Notice that, the type analysis on the novel abstract domain is now complete w.r.t. the sum operation. Indeed, the completeness condition also holds for the expression e, as shown below.

$$\begin{aligned} \alpha(\{\mathtt{"2.4hello"\ +\ "4"}\}) &= \mathsf{Float} \\ &= \alpha(\{\mathtt{"2.4hello"}\}) +^\sharp \alpha(\{\mathtt{"4"}\}) \\ &= \mathsf{StrFloat} +^\sharp \mathsf{StrInt} \\ &= \mathsf{Float} \end{aligned}$$

As pointed out above, guaranteeing completeness in abstract interpretation is a precious and desirable property that an abstract domain should aim to, since it ensures that no loss of precision occurs during the input abstraction process of the operation of interest. It is worth noting that *guessing* a complete abstract domain for a certain operation becomes particularly hard when the operation has a tricky semantics, such as in our example or, more in general, in dynamic languages operations. For this reason, complete shells become important since they are able to mathematically guarantee completeness for a certain operation, starting from an abstract domain of interest.

4 Core Language

We define μDyn, an imperative toy language expressive enough to handle some interesting behaviors related to strings in dynamic languages, e.g., implicit type conversion, and inspired by the JavaScript programming language [27].

μDyn syntax is reported in Fig. 2. The μDyn basic values are represented by the set $\text{VAL} = \text{INT} \cup \text{FLOAT} \cup \text{BOOL} \cup \text{STR}$, such that:

- $\text{INT} = \mathbb{Z}$ denotes the set of signed integers
- FLOAT denotes the set of signed decimal numbers[2]
- $\text{BOOL} = \{\texttt{true}, \texttt{false}\}$ denotes the set of booleans
- $\text{STR} = \Sigma^*$ denotes the set of strings over an alphabet Σ

  We consider Σ^* composed of two sets, namely $\Sigma^* = \Sigma^*_{\mathsf{Num}} \cup \Sigma^*_{\mathsf{NotNum}}$, where:
  - Σ^*_{Num} is the set of numeric strings (e.g., `"42"`, `"-7.2"`)
  - $\Sigma^*_{\mathsf{NotNum}}$ is the set of non numeric strings (e.g., `"foo"`, `"-2a"`)

  Moreover, we consider Σ^*_{Num} additionally composed of four sets:

$$\Sigma^*_{\mathsf{Num}} = \Sigma^*_{\mathsf{UInt}} \cup \Sigma^*_{\mathsf{UFloat}} \cup \Sigma^*_{\mathsf{SInt}} \cup \Sigma^*_{\mathsf{SFloat}}$$

  which correspond to the set of unsigned integer strings, unsigned float strings, signed integer strings and signed float strings, respectively.

μDyn programs are elements generated by S syntax rules. Program states $\text{STATE} : \text{ID} \rightarrow \text{VAL}$, ranging over ξ, are partial functions from identifiers to values. The concrete semantics of μDyn statements follows [5], and it is given by the function $[\![\cdot]\!]\cdot : \text{STMT} \times \text{STATE} \rightarrow \text{STATE}$, inductively defined on the structure of the statements, as reported in Fig. 3. We abuse notation in defining the concrete semantics of expressions: $[\![\cdot]\!]\cdot : \text{EXP} \times \text{STATE} \rightarrow \text{VAL}$. Figure 3 shows the formal semantics of two relevant expressions involving strings we focus on: `concat`, that concatenates two strings, and string-to-number operation, namely `toNum`, that takes a string as input and returns the number that it represents if the input string corresponds to a numerical strings, 0 otherwise. We denote by $\mathcal{N}(\sigma) \in \text{INT} \cup \text{FLOAT}$ the numeric value of a given string. For example, `toNum("4.2")` $= 4.2$ and `toNum("asd")` $= 0$.

5 Making JavaScript String Abstract Domains Complete

In this section, we study the completeness of two string abstract domains integrated into two state-of-the-art JavaScript static analysers based on abstract interpretation, that are SAFE [24] and TAJS [20]. Both the abstract domains track important information on JavaScript strings, e.g., SAFE tracks numeric strings, such as `"2.5"` or `"+5"`, and TAJS is able to infer when a string corresponds to an unsigned integer, that may be used as array index.

For the sake of readability, we recast the original string abstract domains for μDyn, following the notation adopted in [4]. Figure 4 depicts them. Notice that the original abstract domain part of SAFE analyser treats the string `"NaN"` as a numeric string. Since our core language does not provide the primitive value `NaN`, the corresponding string, i.e., `"NaN"`, has no particular meaning here, and it is treated as a non-numerical string.

For each string abstract domain D, we denote by $\alpha_D : \wp(\Sigma^*) \rightarrow D$ its abstraction function, by $\gamma_D : D \rightarrow \wp(\Sigma^*)$ its concretization function, and by $\rho_D : \wp(\Sigma^*) \rightarrow \wp(\Sigma^*) \in uco(D)$ the associated upper closure operator.

[2] Floats normally are represented in programming languages in the IEEE 754 double precision format. For the sake of simplicity, we use instead decimal numbers.

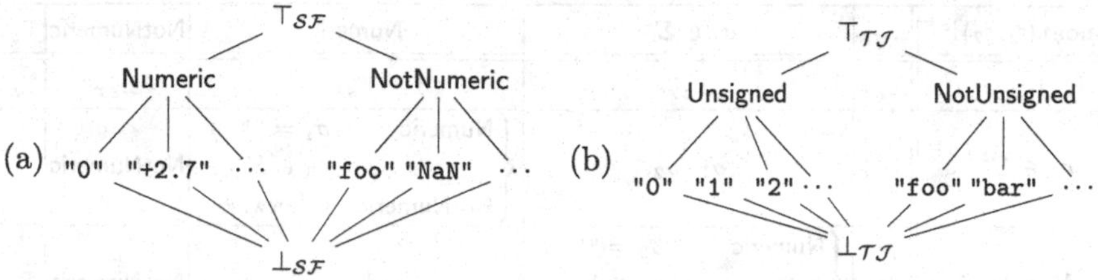

Fig. 4. (a) SAFE, (b) TAJS string abstract domains recasted for μDyn.

5.1 Completing SAFE String Abstract Domain

Figure 4a depicts the string abstract domain $\mathcal{SF}$, i.e., the recasted version of the domain involved into SAFE [24] static analyser. It splits strings into the abstract values: Numeric (i.e., numerical strings) and NotNumeric (i.e., all the other strings). Before reaching these abstract values, $\mathcal{SF}$ precisely tracks each string. For instance, $\alpha_{\mathcal{SF}}(\{\texttt{"+9.6"}, \texttt{"7"}\}) = \mathsf{Numeric}$, and $\alpha_{\mathcal{SF}}(\{\texttt{"+9.6"}, \texttt{"bar"}\}) = \top_{\mathcal{SF}}$.

We study the completeness of $\mathcal{SF}$ w.r.t. `concat` operation. Figure 5 presents the abstract semantics of the concatenation operation for $\mathcal{SF}$, that is:

$$[\![\texttt{concat}(\bullet, \bullet)]\!]^{\mathcal{SF}} : \mathcal{SF} \times \mathcal{SF} \to \mathcal{SF}$$

In particular, when both abstract values correspond to single strings, the standard string concatenation is applied (second row, second column). In the case in which one abstract value, involved in the concatenation, is a string and the other is Numeric (third row, second column and second row, third column) we distinguish two cases: if the string is empty or corresponds to an unsigned integer we can safely return Numeric, otherwise NotNumeric is returned. This happens because, when two float strings (hence numerical strings) are concatenated, a non-numerical string is returned (e.g., `concat("1.1","2.2")` = `"1.12.2"`). For the same reason, when both input abstract values are Numeric, the result is not guaranteed to be numerical, indeed, $[\![\texttt{concat}(\mathsf{Numeric}, \mathsf{Numeric})]\!]^{\mathcal{SF}} = \top_{\mathcal{SF}}$.

Lemma 1. *$\mathcal{SF}$ is not complete w.r.t.* `concat`*. In particular*[3], $\forall S_1, S_2 \in \wp(\Sigma^*)$ *we have that:*

$$\alpha_{\mathcal{SF}}([\![\texttt{concat}(S_1, S_2)]\!]) \subsetneq [\![\texttt{concat}(\alpha_{\mathcal{SF}}(S_1), \alpha_{\mathcal{SF}}(S_2))]\!]^{\mathcal{SF}}$$

Consider $S_1 = \{\texttt{"2.2"}, \texttt{"2.3"}\}$ and $S_2 = \{\texttt{"2"}, \texttt{"3"}\}$. The completeness property does not hold:

$$\alpha_{\mathcal{SF}}([\![\texttt{concat}(S_1, S_2)]\!]) = \mathsf{Numeric} \neq \top_{\mathcal{SF}} = [\![\texttt{concat}(\alpha_{\mathcal{SF}}(S_1), \alpha_{\mathcal{SF}}(S_2))]\!]^{\mathcal{SF}}$$

[3] We abuse notation denoting with $[\![\cdot]\!]$ the additive lift to set of basic values of the concrete semantics, i.e., the collecting semantics.

$[\![\texttt{concat}(\texttt{s}_1, \texttt{s}_2)]\!]^{\mathcal{SF}}$	$\bot_{\mathcal{SF}}$	$\sigma_2 \in \Sigma^*$	Numeric	NotNumeric	$\top_{\mathcal{SF}}$
$\bot_{\mathcal{SF}}$	$\bot_{\mathcal{SF}}$	$\bot_{\mathcal{SF}}$	$\bot_{\mathcal{SF}}$	$\bot_{\mathcal{SF}}$	$\bot_{\mathcal{SF}}$
$\sigma_1 \in \Sigma^*$	$\bot_{\mathcal{SF}}$	$\sigma_1 \cdot \sigma_2$	$\begin{cases} \text{Numeric} & \sigma_1 = \texttt{""} \text{ or } \\ & \sigma_1 \in \Sigma^*_{\text{UInt}} \\ \text{NotNumeric} & \text{otherwise} \end{cases}$	NotNumeric	$\top_{\mathcal{SF}}$
Numeric	$\bot_{\mathcal{SF}}$	$\begin{cases} \text{Numeric} & \sigma_2 = \texttt{""} \text{ or } \\ & \sigma_2 \in \Sigma^*_{\text{UInt}} \\ \text{NotNumeric} & \text{otherwise} \end{cases}$	$\top_{\mathcal{SF}}$	NotNumeric	$\top_{\mathcal{SF}}$
NotNumeric	$\bot_{\mathcal{SF}}$	NotNumeric	NotNumeric	NotNumeric	$\top_{\mathcal{SF}}$
$\top_{\mathcal{SF}}$	$\bot_{\mathcal{SF}}$	$\top_{\mathcal{SF}}$	$\top_{\mathcal{SF}}$	$\top_{\mathcal{SF}}$	$\top_{\mathcal{SF}}$

Fig. 5. SAFE concat abstract semantics.

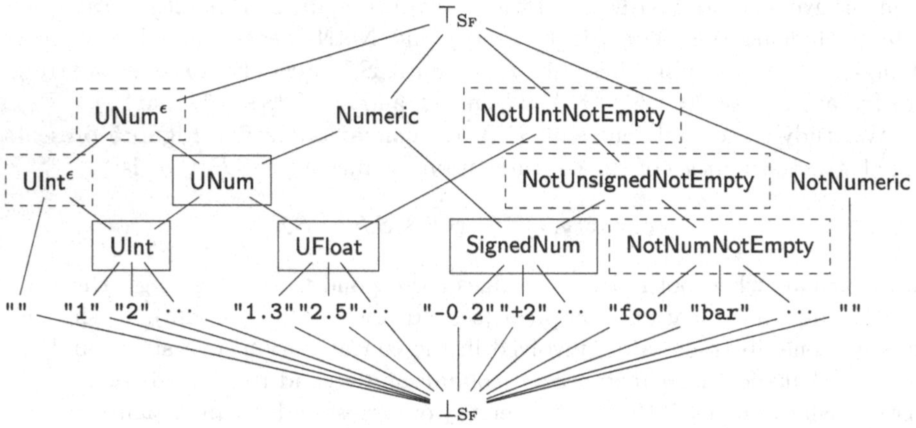

Fig. 6. Absolute complete shell of $\rho_{\mathcal{SF}}$ w.r.t. concat.

The $\mathcal{SF}$ abstract domain loses too much information during the abstraction process; information that can not be retrieved during the abstract concatenation. Intuitively, to gain completeness w.r.t. concat operation, $\mathcal{SF}$ should improve the precision of the numerical strings abstraction, e.g., discriminating between float and integer strings. Following Theorem 2, we can formally construct the absolute complete shell of $\rho_{\mathcal{SF}}$ w.r.t. concat operation $\overline{\mathcal{S}}^{\rho_{\mathcal{SF}}}_{\texttt{concat}}$, and we denote it by SF. This leads to a novel abstract domain, given in Fig. 6, that is complete for concat.

In particular, the points inside dashed boxes are the abstract values added during the iterative computations of SF, the points inside standard boxes are instead obtained by the Moore closure of the other points of the domain, while the remaining abstract values were already in $\mathcal{SF}$. The meaning of abstract values in SF is intuitive. In order to satisfy the completeness property, SF splits the Numeric abstract value, already taken into account in $\mathcal{SF}$, into all the strings corresponding to unsigned integer (UInt), unsigned floats (UFloat), and signed numbers (SignedNum). Moreover, particular importance is given to the empty string, since the novel abstract domain specifies whether each abstract value contains "". Indeed, the UInt^ϵ abstract value represents the strings corresponding to unsigned integer or to the empty string, and the UNum^ϵ abstract value represents the strings corresponding to unsigned numbers or to the empty string. An unexpected abstract value considered in SF is NotUnsignedNotEmpty, such that:

$$\gamma_{\mathrm{SF}}(\mathsf{NotUnsignedNotEmpty}) = \{\sigma \in \Sigma^* \mid \sigma \in \Sigma^*_{\mathsf{SInt}} \cup \Sigma^*_{\mathsf{SFloat}} \cup (\Sigma^*_{\mathsf{NotNum}} \setminus \{\texttt{""}\})\}$$

Namely, the abstract point whose concretization corresponds to the set of any non-numerical string, except the empty string, and any string corresponding to a signed number. This abstract point has been added to SF following the computation of the formula below:

$$\mathsf{NotUnsignedNotEmpty} \in \max(\{Z \in \wp(\Sigma^*) \mid [\![\mathtt{concat}(\mathsf{Numeric}, Z)]\!]\} \subseteq \gamma_{\mathcal{SF}}(\mathsf{NotNumeric}))$$

Informally speaking, we are wondering the following question: *which is the maximal set of strings s.t. concatenated to any possible numerical string will produce any possible non-numerical string?* Indeed, in order to be sure to obtain non-numerical strings, the maximal set doing so is exactly the set of any non-numerical non-empty string, and any string corresponding to a signed number, that is NotUnsignedNotEmpty.

Theorem 3. *ρ_{SF} is the absolute complete shell of $\rho_{\mathcal{SF}}$ w.r.t.* `concat` *operation and it is complete for it.*

For example, consider again $S_1 = \{\texttt{"2.2"}, \texttt{"2.3"}\}$ and $S_2 = \{\texttt{"2"}, \texttt{"3"}\}$. Given SF, the completeness condition holds:

$$\begin{aligned}\alpha_{\mathrm{SF}}([\![\mathtt{concat}(S_1, S_2)]\!]) = \mathsf{UFloat} &= [\![\mathtt{concat}(\alpha_{\mathrm{SF}}(S_1), \alpha_{\mathrm{SF}}(S_2))]\!]^{\mathrm{SF}} \\ &= [\![\mathtt{concat}(\mathsf{UFloat}, \mathsf{UInt})]\!]^{\mathrm{SF}}\end{aligned}$$

5.2 Completing TAJS String Abstract Domain

Figure 4b depicts the string abstract domain $\mathcal{TJ}$, the recasted version of the domain integrated into TAJS static analyser [20]. Differently from $\mathcal{SF}$, it splits the strings into Unsigned, that denotes the strings corresponding to unsigned numbers, and NotUnsigned, any other string. Hence, for example, $\alpha_{\mathcal{TJ}}(\{\texttt{"9"}, \texttt{"+9"}\}) = \top_{\mathcal{TJ}}$ and $\alpha_{\mathcal{TJ}}(\{\texttt{"9.2"}, \texttt{"foo"}\}) = \mathsf{NotUnsigned}$. As for $\mathcal{SF}$, before reaching these abstract values, $\mathcal{TJ}$ precisely tracks single string values.

In this section, we focus on the `toNum` (i.e., string-to-number) operation. Since this operation clearly involves numbers, in Fig. 7 we report the TAJS numerical abstract domain, denoted by $\mathcal{TJ}_{\mathsf{N}}$. The latter domain behaves similarly to $\mathcal{TJ}$, distinguishing between unsigned and not unsigned integers. Below we define the

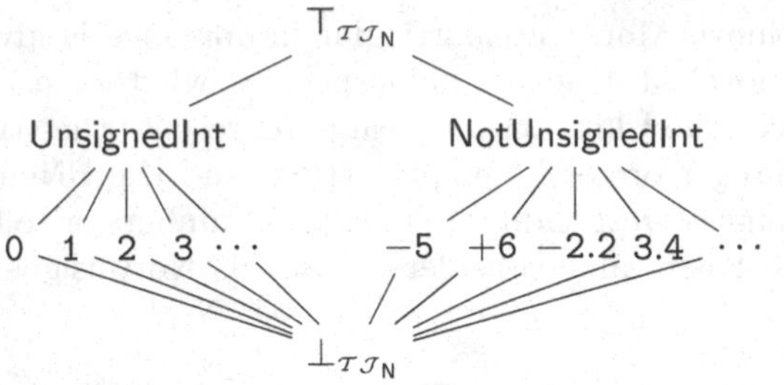

Fig. 7. TAJS numerical abstract domain.

abstract semantics of the string-to-number operation for $\mathcal{TJ}$. In particular, we define the function:

$$[\![\texttt{toNum}(\bullet)]\!]^{\mathcal{TJ}} : \mathcal{TJ} \rightarrow \mathcal{TJ}_{\mathsf{N}}$$

that takes as input a string abstract value in $\mathcal{TJ}$, and returns an integer abstract value in $\mathcal{TJ}_{\mathsf{N}}$.

$$[\![\texttt{toNum(s)}]\!]^{\mathcal{TJ}} = \begin{cases} \bot_{\mathcal{TJ}_{\mathsf{N}}} & [\![\mathsf{s}]\!]^{\mathcal{TJ}} = \bot_{\mathcal{TJ}} \\ [\![\texttt{toNum}(\sigma)]\!] & [\![\mathsf{s}]\!]^{\mathcal{TJ}} = \sigma \\ \mathsf{UnsignedInt} & [\![\mathsf{s}]\!]^{\mathcal{TJ}} = \mathsf{Unsigned} \\ \top_{\mathcal{TJ}_{\mathsf{N}}} & [\![\mathsf{s}]\!]^{\mathcal{TJ}} = \mathsf{NotUnsigned} \vee [\![\mathsf{s}]\!]^{\mathcal{TJ}} = \top_{\mathcal{TJ}} \end{cases}$$

When the input evaluates to $\bot_{\mathcal{TJ}}$, bottom is propagated and $\bot_{\mathcal{TJ}_{\mathsf{N}}}$ is returned (first row). While, if the input evaluates to a single string value, the abstract semantics relies on its concrete one (second row). When the input evaluates to the string abstract value Unsigned (third row), the integer abstract value UnsignedInt is returned. Finally, when the input evaluates to NotUnsigned or $\top_{\mathcal{JS}}$, the top integer abstract value is returned (fourth row).

Lemma 2. *$\mathcal{TJ}$ is not complete w.r.t.* `toNum`*. In particular, $\forall S \in \wp(\Sigma^*)$ we have that:*

$$\alpha_{\mathcal{TJ}}([\![\texttt{toNum}(S)]\!]) \subsetneq [\![\texttt{toNum}(\alpha_{\mathcal{TJ}}(S))]\!]^{\mathcal{TJ}}$$

Consider $S = \{\texttt{"2.3"}, \texttt{"3.4"}\}$. The completeness property does not hold:

$$\alpha_{\mathcal{TJ}}([\![\texttt{toNum}(S)]\!]) = \mathsf{NotUnsignedInt} \neq \top_{\mathcal{TJ}_{\mathsf{N}}} = [\![\texttt{toNum}(\alpha_{\mathcal{TJ}}(S))]\!]^{\mathcal{TJ}}$$

Again, the completeness condition does not hold because the $\mathcal{TJ}$ string abstract domain loses too much information during the abstraction process, and the latter information cannot be retrieved during the abstract `toNum` operation. In particular, when non-numeric strings and unsigned integer strings are converted to numbers by `toNum`, they are mapped to the same value, namely 0. Indeed, $\mathcal{TJ}$ does not differentiate between non-numeric and unsigned integer string values, and this is the principal cause of the $\mathcal{TJ}$ incompleteness w.r.t. `toNum`. Additionally, more precision can be obtained if we could differentiate numeric strings

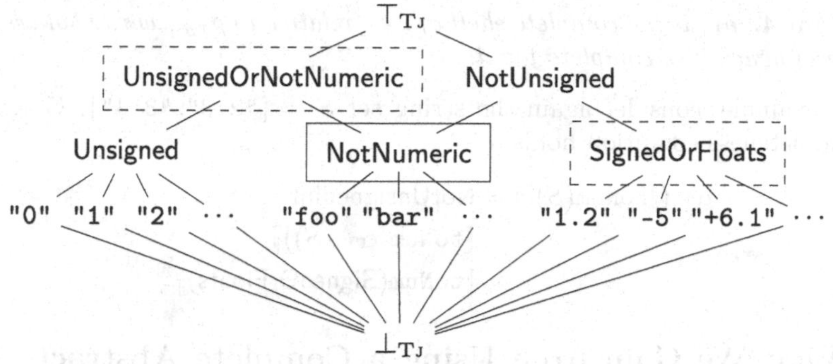

Fig. 8. Complete shell of $\rho_{\mathcal{TJ}}$ relative to $\rho_{\mathcal{TJ}_{\mathsf{N}}}$ w.r.t. toNum.

holding float numbers from those holding integer numbers. Thus, in order to make $\mathcal{TJ}$ complete w.r.t. toNum, we have to derive the complete shell of the $\mathcal{TJ}$ string abstract domain relative to the $\mathcal{TJ}_{\mathsf{N}}$ numerical abstract domain, applying Theorem 1. In particular, let $\rho_{\mathcal{TJ}}$ and $\rho_{\mathcal{TJ}_{\mathsf{N}}}$ be the upper closure operators related to $\mathcal{TJ}$ and $\mathcal{TJ}_{\mathsf{N}}$ abstract domains, respectively. By applying Theorem 1, we obtain $\mathcal{S}^{\rho_{\mathcal{TJ}}}_{\texttt{toNum}}(\mathcal{TJ}_{\mathsf{N}})$ (depicted in Fig. 8), i.e., the complete shell of $\rho_{\mathcal{TJ}}$ relative to $\rho_{\mathcal{TJ}_{\mathsf{N}}}$ w.r.t. toNum, and we denote it by TJ.

In particular, the abstract points inside dashed boxes are the abstract values added during the iterative computations of TJ, the points inside the standard boxes are instead obtained by the Moore closure of the other points of the domain, while the remaining abstract values were already in $\mathcal{TJ}$. A non-intuitive point added by TJ is SignedOrFloats, namely the abstract value s.t. its concretization contains any float string and the signed integers. This abstract point is added during the iterative computation of TJ, following the formula below:

$$\mathsf{SignedOrFloats} \in \max(\{Z \in \wp(\Sigma^*) \mid [\![\texttt{toNum}(Z)]\!] \subseteq \gamma_{\mathcal{TJ}}(\mathsf{NotUnsigned})\})$$

Informally speaking, we are wondering the following question: *which is the maximal set of strings Z s.t.* toNum(Z) *is dominated by* NotUnsigned*?* In order to obtain from toNum(Z) only values dominated by NotUnsigned, the maximal set doing so is exactly the set of the float strings and the signed strings. Other strings, such that: unsigned integer strings or not numerical strings are excluded, since they are both converted to unsigned integers, and they would violate the dominance relation.

Similarly, the abstract point UnsignedOrNotNumeric is added to the absolute complete shell TJ, when the following formula is computed:

$$\mathsf{UnsignedOrNotNumeric} = \max(\{Z \in \wp(\Sigma^*) \mid \texttt{toNum}(Z) \subseteq \gamma_{\mathcal{TJ}}(\mathsf{Unsigned})\})$$

In order to obtain from toNum(Z) only abstract values dominated by Unsigned, the maximal set doing so is exactly the set of the unsigned integer strings and the non-numerical strings, since the latter are converted to 0.

Theorem 4. *ρ_{TJ} is the complete shell of $\rho_{\mathcal{TJ}}$ relative to $\rho_{\mathcal{TJ}_{\mathrm{N}}}$ w.r.t. toNum operation and hence it is complete for it.*

For example, consider again the string set $S = \{\texttt{"2.3"}, \texttt{"3.4"}\}$. Given TJ, the completeness condition holds:

$$\begin{aligned}\alpha_{\mathrm{TJ}}(\llbracket \texttt{toNum}(S) \rrbracket) &= \mathsf{NotUnsignedInt} \\ &= \llbracket \texttt{toNum}(\alpha_{\mathrm{TJ}}(S)) \rrbracket^{\mathcal{TJ}} \\ &= \llbracket \texttt{toNum}(\mathsf{SignedOrFloats}) \rrbracket^{\mathcal{TJ}}\end{aligned}$$

6 What We Gain from Using a Complete Abstract Domain?

Now, we discuss and evaluate the benefits of adopting the complete shells reported in Sect. 5 and, more in general, complete domains, w.r.t. a certain operation. In particular, we compare the μDyn versions of the string abstract domains adopted by SAFE and TAJS with their corresponding complete shells, we discuss the complexity of the complete shells, and finally we argue how adopting complete abstract domains can be useful into static analysers.

Precision. In the previous section, we focused on the completeness of the string abstract domains integrated into SAFE and TAJS, for μDyn, w.r.t. two string operations, namely `concat` and `toNum`, respectively. While string concatenation is common in any programming language, `toNum` assumes critical importance in the dynamic language context, mostly where implicit type conversion is provided. Since type conversion is often hidden from the developer, aim to completeness of the analysis increases the precision of such operations. For instance, let `x` be a variable, at a certain program execution point. `x` may have concrete value in the set $S = \{\texttt{"foo"}, \texttt{"bar"}\}$. If S is abstracted into the starting TAJS string abstract domain, its abstraction will corresponds to Unsigned, losing the information about the fact that the concrete value of `x` surely does not contain numerical values. Hence, when the abstract value of S is used as input of `toNum`, the result will return $\top_{\mathcal{TJ}_{\mathrm{N}}}$, i.e., any possible concrete integer value. Conversely, abstracting S in TJ (the absolute complete shell of $\mathcal{TJ}$ relative to `toNum` discussed in Sect. 5.2) leads to a more precise abstraction, since TJ is able to differentiate between non-numerical and numerical strings. In particular, the abstract value of S in TJ is NotNumeric, and $\llbracket \texttt{toNum}(\mathsf{NotNumeric}) \rrbracket^{\mathcal{TJ}}$ will precisely return 0.

Adopting a complete shell w.r.t. a certain operation does not compromise the precision of the others. For example, consider again the original string abstract domain into TAJS static analyser and the following JavaScript fragment.

```
var obj = {
    "foo" : 1,
    "bar" : 2,
    "1.2" : 3,
    "2.2" : "hello"
}

y = obj[idx];
```

Suppose that the value of `idx` is the abstraction, in the starting TAJS string abstract domain, of the string set $S = \{$`"foo"`, `"bar"`$\}$, namely the abstract value NotUnsinged. The variable `idx` is used to access the property of the object `obj` at line 8 and, to guarantee soundness, it accesses *all* the properties of `obj`, included the fields `"1.2"` and `"2.2"`, introducing noise in the abstract computation, since `"1.2"` and `"2.2"` are false positives values introduced by the abstraction of the values of `idx`. If we analyse the same JavaScript fragment with the absolute complete shell (w.r.t. `toNum` operation) of the TAJS string abstract domain defined in Sect. 5.2, we obtain more precise results. Indeed, in this case, the value of `idx` corresponds to the abstract value NotNumeric, and when it is used to access the object `obj` at line 8, only `"foo"` and `"bar"` are accessed, since they are the only non-numerical string properties of `obj`.

Complexity of the Complete Shells. We evaluate the complexity of the complete shells we have provided in the previous section. As usual in static analysis by abstract interpretation, there exists a trade-off between precision and efficiency: choose a preciser abstract domain may compromise the efficiency of the abstract computations. A representative example is reported in [17]: the complete shell of the sign abstract domain w.r.t. addition is the interval abstract domain. Hence, starting from a finite height abstract domain (signs) we obtain an infinite height abstract domain (intervals). In particular, fix-point computations on signs converge, while on intervals may diverge. Indeed, after the completion, the interval abstract domain should be equipped also with a widening [13] in order to still guarantee termination. A worst-case scenario is when the complete shells w.r.t. a certain operation exactly corresponds to the collecting abstract domain, i.e., the concrete domain. Clearly, we cannot use the concrete domain due to undecidability reasons, but this suggest us to change the starting abstract domain, since it is not able to track any information related to the operation of interest. An example is the suffix abstract domain [12] with `substring` operation: since this abstract domain tracks only the common suffix of a strings set, it can not track the information about the indexes of the common suffix, and the complete shell of the suffix abstract domain w.r.t. `substring` would lead to the concrete domain. Hence, if the focus of the abstract interpreter is to improve the precision of the `substring` operation, we should change the abstract domain with a more precise one for `substring`, such as the finite state automata [6] abstract domain.

Consider now the complete shells reported in Sect. 5. The obtained complete shells still have finite height, hence termination is still guaranteed without the need to equip the complete shells with widening operators. Moreover, the complexity of the string operations of interest is preserved after completion. Indeed, in both TAJS and SAFE starting abstract domains, `concat` and `toNum` operations have constant complexity, respectively, and the same complexity is preserved in the corresponding complete shells. It is worth noting that also the complexity of the abstract domain-related operations, such as least upper bound, greatest lower bound and the ordering operator, is preserved in the complete shells. Hence, to conclude, as far as the complete shells we have reported for TAJS and SAFE are concerned, there is no worsening when we substitute the original string abstract domains with the corresponding complete shells, and this leads, as we have already mentioned before, to completeness during the input abstraction process w.r.t. the relative operations, namely `concat` for SAFE and `toNum` for TAJS.

False Positives Reduction. In static analysis, a certain degree of abstraction must be added in order to obtain decidable procedures to infer invariants on a generic program. Clearly, using less precise abstract domains lead to an increase of *false positive* values of the computed invariants. In particular, after a program is analysed, this burdens the phase of false positive detection: when a program is analysed, the phase after consists to detect which values of the invariants derived by the static analyser are spurious values, namely values that are not certainly computed by the concrete execution of the program of interest. In particular, using imprecise (i.e., not complete) abstract domains clearly augment the number of false positives in the abstract computation of the static analyser, burdening the next phase of detection of the spurious values. On the other hand, adopting (backward) complete abstract domains w.r.t. a certain operation reduce the numbers of false positives introduced during the abstract computations, at least in the input abstraction process. Clearly, in this way, the next phase of detection of false positives will be lighten since less noise has been introduced during the abstract computation of the invariants. Consider againt the JavaScript fragment reported in the previous paragraph. As we already discussed before, using the starting TAJS abstract domain to abstract the variable `idx` leads to a loss of precision, since the spurious value `"foo"` and `"bar"` are taken into account in its abstract value, namely Unsigned. Using the complete shell of TAJS w.r.t. `toNum` instead does not add noise when `idx` is used to access `obj`.

7 Conclusion

This paper addressed the problem of backward completeness in JavaScript-purpose string abstract domains, and provides, in particular, the complete shells of TAJS and SAFE string abstract domains w.r.t. `concat` and `toNum` operations. Our results can be easily applied also to JSAI string abstract domain [21], as it can be seen as an extension of the SAFE domain. The next issue we would like to investigate concerns forward completeness [17], meaning that no loss of precision occurs during the output abstraction process of a certain operation, and the integration of the completeness methodologies. As a final goal of our research, we aim to integrate the notion of complete shell into an industrial JavaScript static analyzer, so that, depending on the target program, an optimal string abstract domain is automatically selected from a set of domains and their complete shells, based on the specific string operations the program makes use of.

References

1. Abdulla, P.A., et al.: Norn: an SMT solver for string constraints. In: Kroening, D., Păsăreanu, C.S. (eds.) CAV 2015. LNCS, vol. 9206, pp. 462–469. Springer, Cham (2015). https://doi.org/10.1007/978-3-319-21690-4_29

2. Amadini, R., et al.: Reference abstract domains and applications to string analysis. Fundam. Inform. **158**(4), 297–326 (2018)
3. Amadini, R., Gange, G., Stuckey, P.J., Tack, G.: A novel approach to string constraint solving. In: Beck, J.C. (ed.) CP 2017. LNCS, vol. 10416, pp. 3–20. Springer, Cham (2017). https://doi.org/10.1007/978-3-319-66158-2_1
4. Amadini, R., et al.: Combining string abstract domains for Javascript analysis: an evaluation. In: Legay, A., Margaria, T. (eds.) TACAS 2017. LNCS, vol. 10205, pp. 41–57. Springer, Heidelberg (2017). https://doi.org/10.1007/978-3-662-54577-5_3
5. Arceri, V., Maffeis, S.: Abstract domains for type Juggling. Electr. Notes Theor. Comput. Sci. **331**, 41–55 (2017)
6. Arceri, V., Mastroeni, I.: Static Program Analysis for String Manipulation Languages. In: VPT 2019 (2019, to appear)
7. Bultan, T., Yu, F., Alkhalaf, M., Aydin, A.: String Analysis for Software Verification and Security. Springer, Cham (2017). https://doi.org/10.1007/978-3-319-68670-7
8. Chen, L., Miné, A., Cousot, P.: A sound floating-point polyhedra abstract domain. In: Ramalingam, G. (ed.) APLAS 2008. LNCS, vol. 5356, pp. 3–18. Springer, Heidelberg (2008). https://doi.org/10.1007/978-3-540-89330-1_2
9. Christensen, A.S., Møller, A., Schwartzbach, M.I.: Precise analysis of string expressions. In: Cousot, R. (ed.) SAS 2003. LNCS, vol. 2694, pp. 1–18. Springer, Heidelberg (2003). https://doi.org/10.1007/3-540-44898-5_1
10. Clarisó, R., Cortadella, J.: The octahedron abstract domain. Sci. Comput. Program. **64**(1), 115–139 (2007)
11. Cortesi, A., Olliaro, M.: M-string segmentation: a refined abstract domain for string analysis in C programs. In: TASE 2018, pp. 1–8 (2018)
12. Costantini, G., Ferrara, P., Cortesi, A.: A suite of abstract domains for static analysis of string values. Softw. Pract. Exper. **45**(2), 245–287 (2015)
13. Cousot, P., Cousot, R.: Abstract interpretation: a unified lattice model for static analysis of programs by construction or approximation of fixpoints. In: POPL 1977, pp. 238–252 (1977)
14. Cousot, P., Cousot, R.: Systematic design of program analysis frameworks. In: POPL 1979, pp. 269–282 (1979)
15. Cousot, P., Halbwachs, N.: Automatic discovery of linear restraints among variables of a program. In: POPL 1978, pp. 84–96 (1978)
16. Filaretti, D., Maffeis, S.: An executable formal semantics of PHP. In: ECOOP 2014 - Object-Oriented Programming - 28th European Conference, Uppsala, Sweden, July 28 - August 1, 2014. Proceedings, pp. 567–592 (2014)
17. Giacobazzi, R., Ranzato, F., Scozzari, F.: Making abstract interpretations complete. J. ACM **47**(2), 361–416 (2000)
18. Granger, P.: Static analysis of arithmetical congruences. Int. J. Comput. Math. - IJCM **30**, 165–190 (1989)
19. Granger, P.: Static analysis of linear congruence equalities among variables of a program. In: Abramsky, S., Maibaum, T.S.E. (eds.) CAAP 1991. LNCS, vol. 493, pp. 169–192. Springer, Heidelberg (1991). https://doi.org/10.1007/3-540-53982-4_10
20. Jensen, S.H., Møller, A., Thiemann, P.: Type analysis for JavaScript. In: Palsberg, J., Su, Z. (eds.) SAS 2009. LNCS, vol. 5673, pp. 238–255. Springer, Heidelberg (2009). https://doi.org/10.1007/978-3-642-03237-0_17
21. Kashyap, V., et al.: JSAI: a static analysis platform for JavaScript. In: FSE 2014, pp. 121–132 (2014)

22. Kim, S.-W., Chin, W., Park, J., Kim, J., Ryu, S.: Inferring grammatical summaries of string values. In: Garrigue, J. (ed.) APLAS 2014. LNCS, vol. 8858, pp. 372–391. Springer, Cham (2014). https://doi.org/10.1007/978-3-319-12736-1_20
23. Kneuss, E., Suter, P., Kuncak, V.: Phantm: PHP analyzer for type mismatch. In: FSE 2010, pp. 373–374 (2010)
24. Lee, H., Won, S., Jin, J., Cho, J., Ryu, S.: SAFE: formal specification and implementation of a scalable analysis framework for ECMAScript. In: FOOL 2012 (2012)
25. Liang, T., Reynolds, A., Tsiskaridze, N., Tinelli, C., Barrett, C., Deters, M.: An efficient SMT solver for string constraints. Formal Methods Syst. Des. **48**(3), 206–234 (2016)
26. Madsen, M., Andreasen, E.: String analysis for dynamic field access. In: CC 2014, pp. 197–217 (2014)
27. Maffeis, S., Mitchell, J.C., Taly, A.: An operational semantics for JavaScript. In: Ramalingam, G. (ed.) APLAS 2008. LNCS, vol. 5356, pp. 307–325. Springer, Heidelberg (2008). https://doi.org/10.1007/978-3-540-89330-1_22
28. Minamide, Y.: Static approximation of dynamically generated web pages. In: WWW 2005, pp. 432–441 (2005)
29. Miné, A.: The octagon abstract domain. Higher-Order Symbol. Comput. **19**(1), 31–100 (2006)
30. Oucheikh, R., Berrada, I., Hichami, O.E.: The 4-octahedron abstract domain. In: NETYS 2016, pp. 311–317 (2016)
31. Park, C., Im, H., Ryu, S.: Precise and scalable static analysis of jQuery using a regular expression domain. In: DLS 2016, pp. 25–36 (2016)
32. Saxena, P., Akhawe, D., Hanna, S., Mao, F., McCamant, S., Song, D.: A symbolic execution framework for JavaScript. In: S&P 2010, pp. 513–528 (2010)
33. Simon, A., King, A., Howe, J.M.: Two variables per linear inequality as an abstract domain. In: LOPSTR 2002, pp. 71–89 (2002)
34. Veanes, M., de Halleux, P., Tillmann, N.: Rex: symbolic regular expression explorer. In: ICST 2010, pp. 498–507 (2010)
35. Ward, M.: The closure operators of a lattice. Ann. Math. **43**(2), 191–196 (1942)
36. Wassermann, G., Su, Z.: Sound and precise analysis of web applications for injection vulnerabilities. In: PLDI 2007, pp. 32–41 (2007)
37. Yu, F., Alkhalaf, M., Bultan, T., Ibarra, O.H.: Automata-based symbolic string analysis for vulnerability detection. Formal Meth. Syst. Des. **44**(1), 44–70 (2014)
38. Yu, F., Bultan, T., Cova, M., Ibarra, O.H.: Symbolic string verification: an automata-based approach. In: Havelund, K., Majumdar, R., Palsberg, J. (eds.) SPIN 2008. LNCS, vol. 5156, pp. 306–324. Springer, Heidelberg (2008). https://doi.org/10.1007/978-3-540-85114-1_21
39. Yu, F., Bultan, T., Hardekopf, B.: String abstractions for string verification. In: Groce, A., Musuvathi, M. (eds.) SPIN 2011. LNCS, vol. 6823, pp. 20–37. Springer, Heidelberg (2011). https://doi.org/10.1007/978-3-642-22306-8_3

BCARET Model Checking for Malware Detection

Huu-Vu Nguyen(✉) and Tayssir Touili(✉)

LIPN, CNRS and University Paris 13, Villetaneuse, France
nguyen@lipn.univ-paris13.fr, touili@lipn.univ-paris13.fr

Abstract. The number of malware is growing fast recently. Traditional malware detectors based on signature matching and code emulation are easy to bypass. To overcome this problem, model-checking appears as an efficient approach that has been extensively applied for malware detection in recent years. Pushdown systems were proposed as a natural model for programs, as they allow to take into account the program's stack into the model. CARET and BCARET were proposed as formalisms for malicious behavior specification since they can specify properties that require matchings of calls and returns which is crucial for malware detection. In this paper, we propose to use BCARET for malicious behavior specification. Since BCARET formulas for malicious behaviors are huge, we propose to extend BCARET with variables, quantifiers and predicates over the stack. Our new logic is called SBPCARET. We reduce the malware detection problem to the model checking problem of PDSs against SBPCARET formulas, and we propose an efficient algorithm to model check SBPCARET formulas for PDSs.

1 Introduction

The number of malware is growing fast in recent years. Traditional approaches including signature matching and code emulation are not efficient enough to detect malwares. While attackers can use obfuscation techniques to hide their malware from the signature based malware detectors easily, the code emulation approaches can only track programs in certain execution paths due to the limited execution time. To overcome these limitations, model-checking appears as an efficient approach for malware detection, since model-checking allows to check the behaviors of a program in all its execution traces without executing it.

A lof of works have been investigated to apply model-checking for malware detection [2–4,7,10–12]. [4] proposed to use finite state graphs to model the program and use the temporal logic CTPL to specify malicious behaviours. However, finite graphs are not exact enough to model programs, as they don't allow to take into account the program's stack into the model. Indeed, the program's stack is usually used by malware writers for code obfuscation as explained in [5]. In addition, in binary codes and assembly programs, parameters are passed to

R. M. Hierons and M. Mosbah (Eds.): ICTAC 2019, LNCS 11884, pp. 273–291, 2019.
https://doi.org/10.1007/978-3-030-32505-3_16

functions by pushing them on the stack before the call is made. The values of these parameters are used to determine whether the program is malicious or not [6]. Therefore, being able to record the program's stack is critical for malware detection. To this aim, [10–13] proposed to use pushdown systems to model programs, and defined extensions of LTL and CTL (called SLTPL and SCTPL) to precisely and compactly describe malicious behaviors. However, these logics cannot specify properties that require matchings of calls and returns, which is crucial to describe malicious behaviours [8]. Let us consider the typical behaviour of a spyware to illustrate this. The typical behaviour of a spyware is seeking personal information (emails, bank account information,...) on local drives by searching files that match specific conditions. To do that, it has to search directories of the host to look for interesting files whose names match a certain condition. If a file is found, the spyware will invoke a payload to steal the information, then continue looking for the remaining matching files. If a folder is found, it will pass into the folder path and continue investigating the folder recursively. To obtain this behavior, the spyware first calls the API *FindFirstFileA* to search for the first matching file in a given folder path. After that, it has to check whether the call to the API function *FindFirstFileA* is successful or not. When the function call fails, the spyware will call the API *GetLastError*. Otherwise, when the function call succeeds, a search handle h will be returned by *FindFirstFileA*. There are two possibilities in this case. If the returned result is a folder, it will call the API function *FindFirstFileA* again to search for matching results in the found folder. If the returned result is a file, it will call the function *FindNextFileA* using h as first parameter to look for the remaining matching files. This behavior cannot be described by LTL or CTL since it requires to express that the return value of the API function *FindFirstFileA* should be used as input to the function *FindNextFileA*.

CARET was introduced to express linear-temporal properties that involve matchings of calls and returns [1] and CARET model-checking for PDSs was considered [6,7]. However, the above behaviour cannot be described by CARET since it is a branching-time property. To specify that behaviour naturally and intuitively, BCARET was introduced to express these branching-time properties that involve matchings of calls and returns [8]. Using BCARET, the above behavior can be expressed by the following formula:

$$\varphi_{sb} = \bigvee_{d \in D} EF^g \Bigg(call(FindFirstFileA) \wedge EX^a(eax = d) \wedge AF^a \\ \Big(call(GetLastError) \vee call(FindFirstFileA) \\ \vee \big(call(FindNextFileA) \wedge d\Gamma^* \big) \Big) \Bigg)$$

where the $\bigvee$ is taken over all possible memory addresses d that contain the values of search handles h in the program, EX^a is a BCARET operator saying that "next in some run, in the same procedural context"; EF^g is the standard CTL EF operator (eventually in some run), while AF^a is a BCARET operator stating that "eventually in all runs, in the same procedural context".

In binary codes and assembly programs, the return value of an API function is placed in the register eax. Therefore, the return value of $FindFirstFileA$ is the value of the register eax at the corresponding return-point of the call. Then, the subformula (call(FindFirstFileA) $\wedge$ $EX^a(eax = d)$) expresses that there is a call to the API function $FindFirstFileA$ whose return value is d (the abstract successor of a call is its corresponding return-point). A call to *FindNextFileA* requires a search handle h as parameter and h must be put on top of the program's stack (as parameters are passed through the stack in assembly programs). To express that d is on top of the program stack, we use the regular expression $d\Gamma^*$. Thus, the subformula [call(FindNextFileA) $\wedge$ $d\Gamma^*$] states that the API *FindNextFileA* is invoked with d as parameter (d stores the information of the search handle h). Therefore, φ_{sb} states that there is a call to the function $FindFirstFileA$ whose return value is d (the search handle), then, in all runs starting from that call, there will be either a call to the API $GetLastError$ or a call to the API function $FindFirstFileA$ or a call to the function $FindNextFileA$ in which d is used as a parameter.

However, it can be seen that this formula is huge, since it considers the disjunction (of different BCARET formulas) over all possible memory addresses d which contain the information of search handles h in the program. To represent it in a more compact fashion, we follow the idea of [4,6,10,12] and extend BCARET with variables, quantifiers, and predicates over the stack. We call our new logic SBPCARET. The above formula can be concisely described by a SBPCARET formula as follows:

$$\varphi'_{sb} = \exists x EF^g \Bigg(call(FindFirstFileA) \wedge EX^a(eax = x) \wedge AF^a \Bigg(call(GetLastError) \vee call(FindFirstFileA) \vee \Big(call(FindNextFileA) \wedge x\Gamma^* \Big) \Bigg) \Bigg)$$

Thus, we propose in this work to use pushdown systems (PDSs) to model programs, and SBPCARET formulas to specify malicious behaviors.We reduce the malware detection problem to the model checking problem of PDSs against SBPCARET formulas, and we propose an efficient algorithm to check whether a PDS satisfies a SBPCARET formula. Our algorithm is based on a reduction to the emptiness problem of Symbolic Alternating Büchi Pushdown Systems. This latter problem is already solved in [10].

The rest of paper is organized as follows. In Sect. 2, we recall the definitions of Pushdown Systems. Section 3 introduces our logic SBPCARET. Model checking SBPCARET for PDSs is presented in Sect. 4. Finally, we conclude in Sect. 5.

2 Pushdown Systems: A Model for Sequential Programs

Pushdown systems is a natural model that was extensively used to model sequential programs. Translations from sequential programs to PDSs can be found e.g.

in [9]. As will be discussed in the next section, to precisely describe malicious behaviors as well as context-related properties, we need to keep track of the call and return actions in each path. Thus, as done in [8], we adapt the PDS model in order to record whether a rule of a PDS corresponds to a *call*, a *return*, or another instruction. We call this model a *Labelled Pushdown System.* We also extend the notion of *run* in order to take into account matching returns of calls.

Definition 1. *A Labelled Pushdown System (PDS) $\mathcal{P}$ is a tuple $(P, \Gamma, \Delta, \sharp)$, where P is a finite set of control locations, Γ is a finite set of stack alphabet, $\sharp \notin \Gamma$ is a bottom stack symbol and Δ is a finite subset of $((P \times \Gamma) \times (P \times \Gamma^*) \times \{call, ret, int\})$. If $((p,\gamma),(q,\omega),t) \in \Delta$ $(t \in \{call, ret, int\})$, we also write $\langle p, \gamma\rangle \xrightarrow{t} \langle q, \omega\rangle \in \Delta$. Rules of Δ are of the following form, where $p \in P, q \in P, \gamma, \gamma_1, \gamma_2 \in \Gamma$, and $\omega \in \Gamma^*$:*

- *(r_1): $\langle p, \gamma\rangle \xrightarrow{call} \langle q, \gamma_1\gamma_2\rangle$*
- *(r_2): $\langle p, \gamma\rangle \xrightarrow{ret} \langle q, \epsilon\rangle$*
- *(r_3): $\langle p, \gamma\rangle \xrightarrow{int} \langle q, \omega\rangle$*

Intuitively, a rule of the form $\langle p, \gamma\rangle \xrightarrow{call} \langle q, \gamma_1\gamma_2\rangle$ corresponds to a call statement. Such a rule usually models a statement of the form $\gamma \xrightarrow{call\ proc} \gamma_2$. In this rule, γ is the control point of the program where the function call is made, γ_1 is the entry point of the called procedure,and γ_2 is the return point of the call. A rule r_2 models a return, whereas a rule r_3 corresponds to a *simple* statement (neither a call nor a return). A configuration of $\mathcal{P}$ is a pair $\langle p, \omega\rangle$, where p is a control location and $\omega \in \Gamma^*$ is the stack content. For technical reasons, we suppose w.l.o.g. that the bottom stack symbol $\sharp$ is never popped from the stack, i.e., there is no rule in the form $\langle p, \sharp\rangle \xrightarrow{t} \langle q, \omega\rangle \in \Delta$ $(t \in \{call, ret, int\})$. $\mathcal{P}$ defines a transition relation $\Rightarrow_{\mathcal{P}}$ $(t \in \{call, ret, int\})$ as follows: If $\langle p, \gamma\rangle \xrightarrow{t} \langle q, \omega\rangle$, then for every $\omega' \in \Gamma^*$, $\langle p, \gamma\omega'\rangle \Rightarrow_{\mathcal{P}} \langle q, \omega\omega'\rangle$. In other words, $\langle q, \omega\omega'\rangle$ is an immediate successor of $\langle p, \gamma\omega'\rangle$. Let $\overset{*}{\Rightarrow}_{\mathcal{P}}$ be the reflexive and transitive closure of $\Rightarrow_{\mathcal{P}}$.

A run of $\mathcal{P}$ from $\langle p_0, \omega_0\rangle$ is a sequence $\langle p_0, \omega_0\rangle\langle p_1, \omega_1\rangle\langle p_2, \omega_2\rangle$... where $\langle p_i, \omega_i\rangle \in P \times \Gamma^*$ s.t. for every $i \geq 0$, $\langle p_i, \omega_i\rangle \Rightarrow_{\mathcal{P}} \langle p_{i+1}, \omega_{i+1}\rangle$. Given a configuration $\langle p, \omega\rangle$, let $Traces(\langle p, \omega\rangle)$ be the set of all possible runs starting from $\langle p, \omega\rangle$.

2.1 Global and Abstract Successors

Let $\pi = \langle p_0, \omega_0\rangle\langle p_1, \omega_1\rangle$... be a run starting from $\langle p_0, \omega_0\rangle$. Over π, two kinds of successors are defined for every position $\langle p_i, \omega_i\rangle$:

- *global-successor*: The global-successor of $\langle p_i, \omega_i\rangle$ is $\langle p_{i+1}, \omega_{i+1}\rangle$ where $\langle p_{i+1}, \omega_{i+1}\rangle$ is an immediate successor of $\langle p_i, \omega_i\rangle$.
- *abstract-successor*: The abstract-successor of $\langle p_i, \omega_i\rangle$ is determined as follows:
  - If $\langle p_i, \omega_i\rangle \Rightarrow_{\mathcal{P}} \langle p_{i+1}, \omega_{i+1}\rangle$ corresponds to a call statement, there are two cases: (1) if $\langle p_i, \omega_i\rangle$ has $\langle p_k, \omega_k\rangle$ as a corresponding return-point in π, then, the abstract successor of $\langle p_i, \omega_i\rangle$ is $\langle p_k, \omega_k\rangle$; (2) if $\langle p_i, \omega_i\rangle$ does not have any corresponding return-point in π, then, the abstract successor of $\langle p_i, \omega_i\rangle$ is $\perp$.

- If $\langle p_i, \omega_i \rangle \Rightarrow_{\mathcal{P}} \langle p_{i+1}, \omega_{i+1} \rangle$ corresponds to a *simple* statement, the abstract successor of $\langle p_i, \omega_i \rangle$ is $\langle p_{i+1}, \omega_{i+1} \rangle$.
- If $\langle p_i, \omega_i \rangle \Rightarrow_{\mathcal{P}} \langle p_{i+1}, \omega_{i+1} \rangle$ corresponds to a return statement, the abstract successor of $\langle p_i, \omega_i \rangle$ is defined as $\perp$.

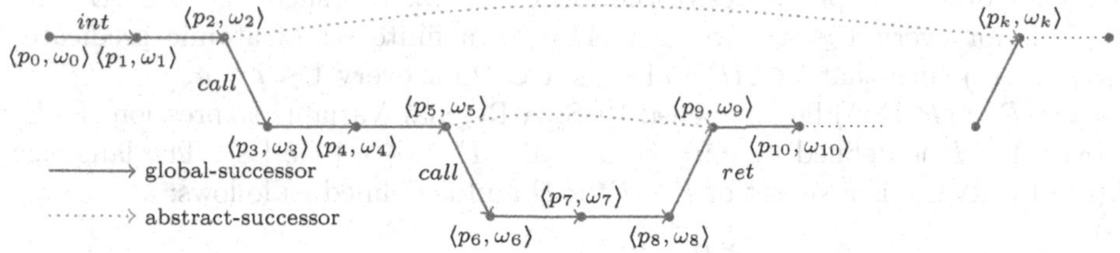

Fig. 1. Two kinds of successors on a run

For example, in Fig. 1:

- The global-successors of $\langle p_1, \omega_1 \rangle$ and $\langle p_2, \omega_2 \rangle$ are $\langle p_2, \omega_2 \rangle$ and $\langle p_3, \omega_3 \rangle$ respectively.
- The abstract-successors of $\langle p_2, \omega_2 \rangle$ and $\langle p_5, \omega_5 \rangle$ are $\langle p_k, \omega_k \rangle$ and $\langle p_9, \omega_9 \rangle$ respectively.

Let $\langle p, \omega \rangle$ be a configuration of a PDS $\mathcal{P}$. A configuration $\langle p', \omega' \rangle$ is defined as a global-successor of $\langle p, \omega \rangle$ iff $\langle p', \omega' \rangle$ is a global-successor of $\langle p, \omega \rangle$ over a run $\pi \in Traces(\langle p, \omega \rangle)$. Similarly, a configuration $\langle p', \omega' \rangle$ is defined as an abstract-successor of $\langle p, \omega \rangle$ iff $\langle p', \omega' \rangle$ is an abstract-successor of $\langle p, \omega \rangle$ over a run $\pi \in Traces(\langle p, \omega \rangle)$.

A *global-path* of $\mathcal{P}$ from $\langle p_0, \omega_0 \rangle$ is a sequence $\langle p_0, \omega_0 \rangle \langle p_1, \omega_1 \rangle \langle p_2, \omega_2 \rangle ...$ where $\langle p_i, \omega_i \rangle \in P \times \Gamma^*$ s.t. for every $i \geq 0$, $\langle p_{i+1}, \omega_{i+1} \rangle$ is a global-successor of $\langle p_i, \omega_i \rangle$. Similarly, an *abstract-path* of $\mathcal{P}$ from $\langle p_0, \omega_0 \rangle$ is a sequence $\langle p_0, \omega_0 \rangle \langle p_1, \omega_1 \rangle \langle p_2, \omega_2 \rangle ...$ where $\langle p_i, \omega_i \rangle \in P \times \Gamma^*$ s.t. for every $i \geq 0$, $\langle p_{i+1}, \omega_{i+1} \rangle$ is an abstract-successor of $\langle p_i, \omega_i \rangle$. For instance, in Fig. 1, $\langle p_0, \omega_0 \rangle \langle p_1, \omega_1 \rangle \langle p_2, \omega_2 \rangle \langle p_3, \omega_3 \rangle \langle p_4, \omega_4 \rangle \langle p_5, \omega_5 \rangle ...$ is a global-path, while $\langle p_0, \omega_0 \rangle \langle p_1, \omega_1 \rangle \langle p_2, \omega_2 \rangle \langle p_k, \omega_k \rangle ...$ is an abstract-path.

3 Malicious Behaviour Specification

In this section, we define the Stack Branching temporal Predicate logic of CAlls and RETurns (SBPCARET) as an extension of BCARET [8] with variables and regular predicates over the stack contents. The predicates contain variables that can be quantified existentially or universally. Regular predicates are expressed by regular variable expressions and are used to describe the stack content of PDSs.

3.1 Environments, Predicates and Regular Variable Expressions

Let $\mathcal{X} = \{x_1, ..., x_n\}$ be a finite set of variables over a finite domain $\mathcal{D}$. Let $B : \mathcal{X} \cup \mathcal{D} \to \mathcal{D}$ be an environment that associates each variable $x \in \mathcal{X}$ with a value $d \in \mathcal{D}$ s.t $B(d) = d$ for every $d \in \mathcal{D}$. Let $B[x \leftarrow d]$ be an environment obtained from B such that $B[x \leftarrow d](x) = d$ and $B[x \leftarrow d](y) = B(y)$ for every $y \neq x$. Let $Abs_x(B) = \{B' \in \mathcal{B} \mid \forall y \in \mathcal{X}, y \neq x, B(y) = B'(y)\}$ be the function that abstracts away the value of x. Let $\mathcal{B}$ be the set of all environments.

Let $AP = \{a, b, c, ...\}$ be a finite set of atomic propositions. Let $AP_{\mathcal{D}}$ be a finite set of atomic predicates of the form $b(\alpha_1, ..., \alpha_m)$ such that $b \in AP$ and $\alpha_i \in \mathcal{D}$ for every $1 \leq i \leq m$. Let $AP_{\mathcal{X}}$ be a finite set of atomic predicates $b(\alpha_1, ..., \alpha_n)$ such that $b \in AP$ and $\alpha_i \in \mathcal{X} \cup \mathcal{D}$ for every $1 \leq i \leq n$.

Let $\mathcal{P} = (P, \Gamma, \Delta)$ be a Labelled PDS. A Regular Variable Expression (RVE) e over $\mathcal{X} \cup \Gamma$ is defined by $e ::= \epsilon \mid a \in \mathcal{X} \cup \Gamma \mid e + e \mid e.e \mid e^*$. The language $L(e)$ of a RVE e is a subset of $P \times \Gamma^* \times \mathcal{B}$ and is defined as follows:

- $L(\epsilon) = \{(\langle p, \epsilon \rangle, B) \mid p \in P, B \in \mathcal{B}\}$
- for $x \in \mathcal{X}$, $L(x) = \{(\langle p, \gamma \rangle, B) \mid p \in P, \gamma \in \Gamma, B \in \mathcal{B} \text{ s.t } B(x) = \gamma\}$
- for $\gamma \in \Gamma$, $L(\gamma) = \{(\langle p, \gamma \rangle, B) \mid p \in P, B \in \mathcal{B}\}$
- $L(e_1.e_2) = \{(\langle p, \omega' \omega'' \rangle, B) \mid (\langle p, \omega' \rangle, B) \in L(e_1); (\langle p, \omega'' \rangle, B) \in L(e_2)\}$
- $L(e^*) = \{(\langle p, \omega \rangle, B) \mid \omega \in \{v \in \Gamma^* \mid (\langle p, v \rangle, B) \in L(e)\}^*\}$.

3.2 The Stack Branching Temporal Predicate Logic of CAlls and RETurns - SBPCARET

A SBPCARET formula is a BCARET formula where predicates and RVEs are used as atomic propositions and where quantifiers are applied to variables. For technical reasons, we assume w.l.o.g. that formulas are written in positive normal form, where negations are applied only to atomic predicates, and we use the *release operator* R as the dual of the until operator U. From now on, we fix a finite set of variables $\mathcal{X}$, a finite set of atomic propositions AP, a finite domain $\mathcal{D}$, and a finite set of RVEs $\mathcal{V}$. A SBPCARET formula is defined as follows, where $v \in \{g, a\}$, $x \in \mathcal{X}$, $e \in \mathcal{V}$, $b(\alpha_1, ..., \alpha_n) \in AP_{\mathcal{X}}$:

$$\begin{aligned}\varphi := & \, true \mid false \mid b(\alpha_1, ..., \alpha_n) \mid \neg b(\alpha_1, ..., \alpha_n) \mid e \mid \neg e \mid \varphi \vee \varphi \mid \varphi \wedge \varphi \mid \forall x \varphi \mid \\ & \exists x \varphi \mid EX^v \varphi \mid AX^v \varphi \mid E[\varphi U^v \varphi] \mid A[\varphi U^v \varphi] \mid E[\varphi R^v \varphi] \mid A[\varphi R^v \varphi]\end{aligned}$$

Let $\lambda : P \longrightarrow 2^{AP_{\mathcal{D}}}$ be a labelling function which associates each control location to a set of atomic predicates. Let φ be a SBPCARET formula over AP. Let $\langle p, \omega \rangle$ be a configuration of $\mathcal{P}$. Then we say that $\mathcal{P}$ satisfies φ at $\langle p, \omega \rangle$ (denoted by $\langle p, \omega \rangle \models_\lambda \varphi$) iff there exists an environment $B \in \mathcal{B}$ such that $\langle p, \omega \rangle$ satisfies φ under B (denoted by $\langle p, \omega \rangle \models_\lambda^B \varphi$). The satisfiability relation of a SBPCARET formula φ at a configuration $\langle p_0, \omega_0 \rangle$ under the environment B w.r.t. the labelling function λ, denoted by $\langle p_0, \omega_0 \rangle \models_\lambda^B \varphi$, is defined inductively as follows:

- $\langle p_0, \omega_0 \rangle \models_\lambda^B true$ for every $\langle p_0, \omega_0 \rangle$
- $\langle p_0, \omega_0 \rangle \not\models_\lambda^B false$ for every $\langle p_0, \omega_0 \rangle$
- $\langle p_0, \omega_0 \rangle \models_\lambda^B b(\alpha_1, ..., \alpha_n)$, iff $b(B(\alpha_1), ..., B(\alpha_n)) \in \lambda(p_0)$

- $\langle p_0, \omega_0\rangle \models_\lambda^B \neg b(\alpha_1, ..., \alpha_n)$, iff $b(B(\alpha_1), ..., B(\alpha_n)) \notin \lambda(p_0)$
- $\langle p_0, \omega_0\rangle \models_\lambda^B e$ iff $((\langle p_0, \omega_0\rangle, B) \in L(e)$
- $\langle p_0, \omega_0\rangle \models_\lambda^B \neg e$ iff $((\langle p_0, \omega_0\rangle, B) \notin L(e)$
- $\langle p_0, \omega_0\rangle \models_\lambda^B \varphi_1 \vee \varphi_2$ iff $((\langle p_0, \omega_0\rangle \models_\lambda^B \varphi_1$ or $\langle p_0, \omega_0\rangle \models_\lambda^B \varphi_2)$
- $\langle p_0, \omega_0\rangle \models_\lambda^B \varphi_1 \wedge \varphi_2$ iff $((\langle p_0, \omega_0\rangle \models_\lambda^B \varphi_1$ and $\langle p_0, \omega_0\rangle \models_\lambda^B \varphi_2)$
- $\langle p_0, \omega_0\rangle \models_\lambda^B \forall x\varphi$ iff for every $d \in \mathcal{D}$, $\langle p_0, \omega_0\rangle \models_\lambda^{B[x\leftarrow d]} \varphi$
- $\langle p_0, \omega_0\rangle \models_\lambda^B \exists x\varphi$ iff there exists $d \in \mathcal{D}$, $\langle p_0, \omega_0\rangle \models_\lambda^{B[x\leftarrow d]} \varphi$
- $\langle p_0, \omega_0\rangle \models_\lambda^B EX^g\varphi$ iff there exists a global-successor $\langle p', \omega'\rangle$ of $\langle p_0, \omega_0\rangle$ such that $\langle p', \omega'\rangle \models_\lambda^B \varphi$
- $\langle p_0, \omega_0\rangle \models_\lambda^B AX^g\varphi$ iff $\langle p', \omega'\rangle \models_\lambda^B \varphi$ for every global-successor $\langle p', \omega'\rangle$ of $\langle p_0, \omega_0\rangle$
- $\langle p_0, \omega_0\rangle \models_\lambda^B E[\varphi_1 U^g \varphi_2]$ iff there exists a global-path $\pi = \langle p_0, \omega_0\rangle\langle p_1, \omega_1\rangle\langle p_2, \omega_2\rangle...$ of $\mathcal{P}$ starting from $\langle p_0, \omega_0\rangle$ s.t. $\exists i \geq 0$, $\langle p_i, \omega_i\rangle \models_\lambda^B \varphi_2$ and for every $0 \leq j < i$, $\langle p_j, \omega_j\rangle \models_\lambda^B \varphi_1$
- $\langle p_0, \omega_0\rangle \models_\lambda^B A[\varphi_1 U^g \varphi_2]$ iff for every global-path $\pi = \langle p_0, \omega_0\rangle\langle p_1, \omega_1\rangle\langle p_2, \omega_2\rangle...$of $\mathcal{P}$ starting from $\langle p_0, \omega_0\rangle$, $\exists i \geq 0$, $\langle p_i, \omega_i\rangle \models_\lambda^B \varphi_2$ and for every $0 \leq j < i$, $\langle p_j, \omega_j\rangle \models_\lambda^B \varphi_1$
- $\langle p_0, \omega_0\rangle \models_\lambda^B E[\varphi_1 R^g \varphi_2]$ iff there exists a global-path $\pi = \langle p_0, \omega_0\rangle\langle p_1, \omega_1\rangle\langle p_2, \omega_2\rangle...$ of $\mathcal{P}$ starting from $\langle p_0, \omega_0\rangle$ s.t. for every $i \geq 0$, if $\langle p_i, \omega_i\rangle \not\models_\lambda^B \varphi_2$ then there exists $0 \leq j < i$ s.t. $\langle p_j, \omega_j\rangle \models_\lambda^B \varphi_1$
- $\langle p_0, \omega_0\rangle \models_\lambda^B A[\varphi_1 R^g \varphi_2]$ iff for every global-path $\pi = \langle p_0, \omega_0\rangle\langle p_1, \omega_1\rangle\langle p_2, \omega_2\rangle...$ of $\mathcal{P}$ starting from $\langle p_0, \omega_0\rangle$, for every $i \geq 0$, if $\langle p_i, \omega_i\rangle \not\models_\lambda^B \varphi_2$ then there exists $0 \leq j < i$ s.t. $\langle p_j, \omega_j\rangle \models_\lambda^B \varphi_1$
- $\langle p_0, \omega_0\rangle \models_\lambda^B EX^a\varphi$ iff there exists an abstract-successor $\langle p', \omega'\rangle$ of $\langle p_0, \omega_0\rangle$ such that $\langle p', \omega'\rangle \models_\lambda^B \varphi$
- $\langle p_0, \omega_0\rangle \models_\lambda^B AX^a\varphi$ iff $\langle p', \omega'\rangle \models_\lambda^B \varphi$ for every abstract-successor $\langle p', \omega'\rangle$ of $\langle p_0, \omega_0\rangle$
- $\langle p_0, \omega_0\rangle \models_\lambda^B E[\varphi_1 U^a \varphi_2]$ iff there exists an abstract-path $\pi = \langle p_0, \omega_0\rangle\langle p_1, \omega_1\rangle\langle p_2, \omega_2\rangle...$ of $\mathcal{P}$ starting from $\langle p_0, \omega_0\rangle$ s.t. $\exists i \geq 0$, $\langle p_i, \omega_i\rangle \models_\lambda^B \varphi_2$ and for every $0 \leq j < i$, $\langle p_j, \omega_j\rangle \models_\lambda^B \varphi_1$
- $\langle p_0, \omega_0\rangle \models_\lambda^B A[\varphi_1 U^a \varphi_2]$ iff for every abstract-path $\pi = \langle p_0, \omega_0\rangle\langle p_1, \omega_1\rangle\langle p_2, \omega_2\rangle...$ of $\mathcal{P}$, $\exists i \geq 0$, $\langle p_i, \omega_i\rangle \models_\lambda^B \varphi_2$ and for every $0 \leq j < i$, $\langle p_j, \omega_j\rangle \models_\lambda^B \varphi_1$
- $\langle p_0, \omega_0\rangle \models_\lambda^B E[\varphi_1 R^a \varphi_2]$ iff there exists an abstract-path $\pi = \langle p_0, \omega_0\rangle\langle p_1, \omega_1\rangle\langle p_2, \omega_2\rangle...$ of $\mathcal{P}$ starting from $\langle p_0, \omega_0\rangle$ s.t. for every $i \geq 0$, if $\langle p_i, \omega_i\rangle \not\models_\lambda^B \varphi_2$ then there exists $0 \leq j < i$ s.t. $\langle p_j, \omega_j\rangle \models_\lambda^B \varphi_1$
- $\langle p_0, \omega_0\rangle \models_\lambda^B A[\varphi_1 R^a \varphi_2]$ iff for every abstract-path $\pi = \langle p_0, \omega_0\rangle\langle p_1, \omega_1\rangle\langle p_2, \omega_2\rangle...$ of $\mathcal{P}$ starting from $\langle p_0, \omega_0\rangle$, for every $i \geq 0$, if $\langle p_i, \omega_i\rangle \not\models_\lambda^B \varphi_2$ then there exists $0 \leq j < i$ s.t. $\langle p_j, \omega_j\rangle \models_\lambda^B \varphi_1$

Other SBPCARET operators can be expressed by the above operators: $EF^g\varphi = E[true\ U^g\varphi]$, $EF^a\varphi = E[true\ U^a\varphi]$, $AF^g\varphi = A[true\ U^g\varphi]$, $AF^a\varphi = A[trueU^a\varphi]$,...

Closure. Given a SBPCARET formula φ, the closure $Cl(\varphi)$ is the set of all subformulae of φ, including φ. Let $AP^+(\varphi) = \{b(\alpha_1, ..., \alpha_n) \in AP_\mathcal{X} \mid b(\alpha_1, ..., \alpha_n) \in Cl(\varphi)\}$; $AP^-(\varphi) = \{b(\alpha_1, ..., \alpha_n) \in AP_\mathcal{X} \mid \neg b(\alpha_1, ..., \alpha_n) \in Cl(\varphi)\}$, $Reg^+(\varphi) = \{e \in \mathcal{V} \mid e \in Cl(\varphi)\}$, $Reg^-(\varphi) = \{e \in \mathcal{V} \mid \neg e \in Cl(\varphi)\}$.

4 SBPCARET Model-Checking for Pushdown Systems

In this section, we show how to do SBPCARET model-checking for PDSs. Let then $\mathcal{P}$ be a PDS, φ be a SBPCARET formula, and $\mathcal{V}$ be the set of RVEs occurring in φ. We follow the idea of [10] and use Variable Automata to represent RVEs.

4.1 Variable Automata

Given a PDS $\mathcal{P} = (P, \Gamma, \Delta)$ s.t. $\Gamma \subseteq \mathcal{D}$, a Variable Automaton (VA) [10] is a tuple $(Q, \Gamma, \delta, s, F)$, where Q is a finite set of states, Γ is the input alphabet, $s \in Q$ is an initial state; $F \subseteq Q$ is a finite set of accepting states; and δ is a finite set of transition rules of the form $p \xrightarrow{\alpha} \{q_1, ..., q_n\}$ where α can be x, $\neg x$, or γ, for any $x \in \mathcal{X}$ and $\gamma \in \Gamma$.

Let $B \in \mathcal{B}$. A run of VA on a word $\gamma_1, ..., \gamma_m$ under B is a tree of height m whose root is labelled by the initial state s, and each node at depth k labelled by a state q has h children labelled by $p_1, ..., p_h$ respectively, such that:

- either $q \xrightarrow{\gamma_k} \{p_1, ..., p_h\} \in \delta$ and $\gamma_k \in \Gamma$;
- or $q \xrightarrow{x} \{p_1, ..., p_h\} \in \delta$, $x \in \mathcal{X}$ and $B(x) = \gamma_k$;
- or $q \xrightarrow{\neg x} \{p_1, ..., p_h\} \in \delta$, $x \in \mathcal{X}$ and $B(x) \neq \gamma_k$.

A branch of the tree is accepting iff the leaf of the branch is an accepting state. A run is accepting iff all its branches are accepting. A word $\omega \in \Gamma^*$ is accepted by a VA under an environment $B \in \mathcal{B}$ iff the VA has an accepting run on the word ω under the environment B.

The language of a VA M, denoted by $L(M)$, is a subset of $(P \times \Gamma^*) \times \mathcal{B}$. $(\langle p, \omega\rangle, B) \in L(M)$ iff M accepts the word ω under the environment B.

Theorem 1. *[10] For every regular expression $e \in \mathcal{V}$, we can compute in polynomial time a Variable Automaton M s.t. $L(M) = L(e)$.*

Theorem 2. *[10] VAs are closed under boolean operations.*

4.2 Symbolic Alternating Büchi Pushdown Systems (SABPDSs)

Definition 2. *A Symbolic Alternating Büchi Pushdown System (SABPDS) is a tuple $\mathcal{BP} = (P, \Gamma, \Delta, F)$, where P is a set of control locations, $\Gamma \subseteq \mathcal{D}$ is stack alphabet, $F \subseteq P \times 2^{\mathcal{B}}$ is a set of accepting control locations and Δ is a finite set of transitions of the form $\langle p, \gamma\rangle \overset{\mathbb{R}}{\hookrightarrow} \{\langle p_1, \omega_1\rangle, ..., \langle p_n, \omega_n\rangle\}$ where $p \in P$, $\gamma \in \Gamma$, for every $1 \leq i \leq n$: $p_i \in P$, $\omega_i \in \Gamma^*$; and $\mathbb{R} : (\mathcal{B})^n \to 2^{\mathcal{B}}$ is a function that maps a tuple of environments $(B_1, ..., B_n)$ to a set of environments.*

A configuration of a SABPDS $\mathcal{BP}$ is a tuple $\langle [\![p, B]\!], \omega\rangle$, where $p \in P$ is the current control location, $B \in \mathcal{B}$ is an environment and $\omega \in \Gamma^*$ is the current stack content. Let $\langle p, \gamma\rangle \overset{\mathbb{R}}{\hookrightarrow} \{\langle p_1, \omega_1\rangle, ..., \langle p_n, \omega_n\rangle\}$ be a rule of Δ, then, for every $\omega \in$

Γ^*, $B, B_1, ..., B_n \in \mathcal{B}$, if $B \in \mathbb{R}(B_1, ..., B_n)$, then the configuration $\langle [\![p, B]\!], \gamma\omega\rangle$ (resp. $\{\langle [\![p_1, B_1]\!], \omega_1\omega\rangle, ..., \langle [\![p_n, B_n]\!], \omega_n\omega\rangle\}$) is an immediate predecessor (resp. successor) of $\{\langle [\![p_1, B_1]\!], \omega_1\omega\rangle, ..., \langle [\![p_n, B_n]\!], \omega_n\omega\rangle\}$ (resp. $\langle [\![p, B]\!], \gamma\omega\rangle$).

A run ρ of a SABPDS $\mathcal{BP}$ starting form an initial configuration $\langle [\![p_0, B_0]\!], \omega_0\rangle$ is a tree whose root is labelled by $\langle [\![p_0, B_0]\!], \omega_0\rangle$, and whose other nodes are labelled by elements in $P \times \mathcal{B} \times \Gamma^*$. If a node of ρ is labelled by a configuration $\langle [\![p, B]\!], \omega\rangle$ and has n children labelled by $\langle [\![p_1, B_1]\!], \omega_1\rangle, ..., \langle [\![p_n, B_n]\!], \omega_n\rangle$ respectively, then, $\langle [\![p, B]\!], \omega\rangle$ must be a predecessor of $\{\langle [\![p_1, B_1]\!], \omega_1\rangle, ..., \langle [\![p_n, B_n]\!], \omega_n\rangle\}$ in $\mathcal{BP}$. A path of a run ρ is an infinite sequence of configurations $c_0c_1c_2...$ s.t. c_0 is the root of ρ and c_{i+1} is one of the children of c_i for every $i \geq 0$. A path is accepting iff it visits infinitely often configurations with control locations in F. A run ρ is accepting iff every path of ρ is accepting. The language of $\mathcal{BP}$, $\mathcal{L}(\mathcal{BP})$, is the set of configurations c s.t. $\mathcal{BP}$ has an accepting run starting from c.

$\mathcal{BP}$ defines the reachability relation $\Rightarrow_{\mathcal{BP}}$: $2^{(P\times\mathcal{B})\times\Gamma^*} \rightarrow 2^{(P\times\mathcal{B})\times\Gamma^*}$ as follows: (1) $c \Rightarrow_{\mathcal{BP}} \{c\}$ for every $c \in P \times \mathcal{B} \times \Gamma^*$, (2) $c \Rightarrow_{\mathcal{BP}} C$ if C is an immediate successor of c; (3) if $c \Rightarrow_{\mathcal{BP}} \{c_1, c_2, ..., c_n\}$ and $c_i \Rightarrow_{\mathcal{BP}} C_i$ for every $1 \leq i \leq n$, then $c \Rightarrow_{\mathcal{BP}} \bigcup_{i=1}^{n} C_i$. Given $c_0 \Rightarrow_{\mathcal{BP}} C'$, then, $\mathcal{BP}$ has an accepting run from c_0 iff $\mathcal{BP}$ has an accepting run from c' for every $c' \in C'$.

Theorem 3. *[10] The membership problem of SABPDS can be solved effectively.*

Functions of $\mathbb{R}$. In what follows, we define several functions of $\mathbb{R}$ which will be used in the next sections. These functions were first defined in [10].

1. $id(B) = \{B\}$. This is the identity function.
2.
$$equal(B_1, ..., B_n) = \begin{cases} B_1 & \text{if } B_i = B_j \text{ for every } 1 \leq i, j \leq n; \\ \emptyset & \text{otherwise} \end{cases}$$
This function checks whether all the environments are equal and returns $\{B_1\}$ (which is also equal to B_i for every i). Otherwise, it returns the emptyset.
3.
$$meet^x_{\{c_1,...,c_n\}}(B_1, ..., B_n) = \begin{cases} Abs_x(B_1) & \text{if } B_i(x) = c_i \text{ for } 1 \leq i \leq n, \\ & \text{and } B_i(y) = B_j(y) \text{ for } y \neq x, 1 \leq i, j \leq n; \\ \emptyset & \text{otherwise} \end{cases}$$
This function checks whether (1) $B_i(x) = c_i$ for every $1 \leq i \leq n$ (2) for every $y \neq x$; every $1 \leq i, j \leq n$ $B_i(y) = B_j(y)$. If the conditions are satisfied, it returns $Abs_x(B_1)$[1], otherwise it returns the emptyset.

[1] $Abs_x(B_1)$ is as defined in Sect. 3.1.

4.

$$join_c^x(B_1, ..., B_n) = \begin{cases} B_1 & \text{if } B_i(x) = c \text{ for } 1 \leq i \leq n \\ & \text{and } B_i = B_j \text{ for } 1 \leq i,j \leq n; \\ \emptyset & \text{otherwise} \end{cases}$$

This function checks whether $B_i(x) = c$ for every i. If this condition is satisfied, $equal(B_1, ..., B_n)$ is returned, otherwise, the emptyset is returned.

5.

$$join_c^{\neg x}(B_1, ..., B_n) = \begin{cases} B_1 & \text{if } B_i(x) \neq c \text{ for } 1 \leq i \leq n \\ & \text{and } B_i = B_j \text{ for } 1 \leq i,j \leq n; \\ \emptyset & \text{otherwise} \end{cases}$$

This function checks whether $B_i(x) \neq c$ for every i. If this condition is satisfied, $equal(B_1, ..., B_n)$ is returned, otherwise, the emptyset is returned.

4.3 From SBPCARET Model Checking of PDSs to the Membership Problem in SABPDSs

Let $\mathcal{P} = (P, \Gamma, \Delta)$ be a PDS. We suppose w.l.o.g. that $\mathcal{P}$ has a bottom stack symbol $\sharp$ that is never popped from the stack. Let AP be a set of atomic propositions. Let φ be a SBPCARET formula over AP, $\lambda : P \longrightarrow 2^{AP_{\mathcal{D}}}$ be a labelling function. Given a configuration $\langle p_0, \omega_0 \rangle$, we propose in this section an algorithm to check whether $\langle p_0, \omega_0 \rangle \models_\lambda \varphi$, i.e., whether there exists an environment B s.t. $\langle p_0, \omega_0 \rangle \models_\lambda^B \varphi$. Intuitively, we compute an SABPDS $\mathcal{BP}_\varphi$ s.t. $\langle p, \omega \rangle \models_\lambda^B \varphi$ iff $\langle [\![(\!| p, \varphi |\!), B]\!], \omega \rangle \in \mathcal{L}(\mathcal{BP}_\varphi)$ for every $p \in P$, $\omega \in \Gamma^*$, $B \in \mathcal{B}$. Then, to check if $\langle p_0, \omega_0 \rangle \models_\lambda \varphi$, we will check whether there exists a $B \in \mathcal{B}$ s.t. $\langle [\![(\!| p_0, \varphi |\!), B]\!], \omega_0 \rangle \in \mathcal{L}(\mathcal{BP}_\varphi)$.

Let $Reg^+(\varphi) = \{e_1, ..., e_k\}$ and $Reg^-(\varphi) = \{e_{k+1}, ..., e_m\}$. Using Theorems 1 and 2; for every $1 \leq i \leq k$, we can compute a VA $M_{e_i} = (Q_{e_i}, \Gamma, \delta_{e_i}, s_{e_i}, F_{e_i})$ s.t. $L(M_{e_i}) = L(e_i)$. In addition, for every $k+1 \leq j \leq m$, we can compute a VA $M_{\neg e_j} = (Q_{\neg e_j}, \Gamma, \delta_{\neg e_j}, s_{\neg e_j}, F_{\neg e_j})$ s.t. $L(M_{\neg e_j}) = (P \times \Gamma^*) \times \mathcal{B} \setminus L(e_j)$. Let $\mathcal{M}$ be the union of all these automata, $\mathcal{S}$ and $\mathcal{F}$ be respectively the union of all states and final states of these automata.

Let $\mathcal{BP}_\varphi = (P', \Gamma', \Delta', F)$ be the SABPDS defined as follows:

- $P' = P \cup (P \times Cl(\varphi)) \cup \mathcal{S} \cup \{p_\bot\}$
- $\Gamma' = \Gamma \cup (\Gamma \times Cl(\varphi)) \cup \{\gamma_\bot\}$
- $F = F_1 \cup F_2 \cup F_3 \cup F_4$ where
  - $F_1 = \{[\![(\!| p, b(\alpha_1, ..., \alpha_n) |\!), \beta]\!] \mid b(\alpha_1, ..., \alpha_n) \in AP^+(\varphi)$, and $\beta = \{B \in \mathcal{B} \mid b(B(\alpha_1), ..., B(\alpha_n)) \in \lambda(p)\}$
  - $F_2 = \{[\![(\!| p, \neg b(\alpha_1, ..., \alpha_n) |\!), \beta]\!] \mid b(\alpha_1, ..., \alpha_n) \in AP^-(\varphi)$, and $\beta = \{B \in \mathcal{B} \mid b(B(\alpha_1), ..., B(\alpha_n)) \notin \lambda(p)\}$

- $F_3 = P \times Cl_R(\varphi) \times \mathcal{B}$ where $Cl_R(\varphi)$ is the set of formulas of $Cl(\varphi)$ in the form $E[\varphi_1 R^v \varphi_2]$ or $A[\varphi_1 R^v \varphi_2]$ ($v \in \{g, a\}$)
- $F_4 = \mathcal{F} \times \mathcal{B}$

The transition relation Δ' is the smallest set of transition rules defined as follows: For every $p \in P$, $\phi \in Cl(\varphi)$, $\gamma \in \Gamma$ and $t \in \{call, ret, int\}$:

($\hbar$1) If $\phi = b(\alpha_1, ..., \alpha_n)$, then, $\langle (\!|p, \phi|\!), \gamma \rangle \xhookrightarrow{id} \langle (\!|p, \phi|\!), \gamma \rangle \in \Delta'$
($\hbar$2) If $\phi = \neg b(\alpha_1, ..., \alpha_n)$, then, $\langle (\!|p, \phi|\!), \gamma \rangle \xhookrightarrow{id} \langle (\!|p, \phi|\!), \gamma \rangle \in \Delta'$
($\hbar$3) If $\phi = \phi_1 \wedge \phi_2$, then, $\langle (\!|p, \phi|\!), \gamma \rangle \xhookrightarrow{equal} [\langle (\!|p, \phi_1|\!), \gamma \rangle, \langle (\!|p, \phi_2|\!), \gamma \rangle] \in \Delta'$
($\hbar$4) If $\phi = \phi_1 \vee \phi_2$, then, $\langle (\!|p, \phi|\!), \gamma \rangle \xhookrightarrow{id} \langle (\!|p, \phi_1|\!), \gamma \rangle \in \Delta'$ and $\langle (\!|p, \phi|\!), \gamma \rangle \xhookrightarrow{id} \langle (\!|p, \phi_2|\!), \gamma \rangle \in \Delta'$
($\hbar$5) If $\phi = \exists x \phi_1$, then, $\langle (\!|p, \phi|\!), \gamma \rangle \xhookrightarrow{meet^x_{\{c\}}} \langle (\!|p, \phi_1|\!), \gamma \rangle \in \Delta'$ for every $c \in \mathcal{D}$
($\hbar$6) If $\phi = \forall x \phi_1$, then, $\langle (\!|p, \phi|\!), \gamma \rangle \xhookrightarrow{meet^x_{\mathcal{D}}} [\langle (\!|p, \phi_1|\!), \gamma \rangle, ..., \langle (\!|p, \phi_1|\!), \gamma \rangle] \in \Delta'$ where $\langle (\!|p, \phi_1|\!), \gamma \rangle$ is repeated m times in the right-hand side, where m is the number of elements in $\mathcal{D}$
($\hbar$7) If $\phi = EX^g \phi_1$, then
$\langle (\!|p, \phi|\!), \gamma \rangle \xhookrightarrow{id} \langle (\!|q, \phi_1|\!), \omega \rangle \in \Delta'$ for every $\langle p, \gamma \rangle \xrightarrow{t} \langle q, \omega \rangle \in \Delta$
($\hbar$8) If $\phi = AX^g \phi_1$, then,
$\langle (\!|p, \phi|\!), \gamma \rangle \xhookrightarrow{equal} [\langle (\!|q_1, \phi_1|\!), \omega_1 \rangle, ..., \langle (\!|q_n, \phi_1|\!), \omega_n \rangle] \in \Delta'$, where for every $1 \leq i \leq n$, $\langle p, \gamma \rangle \xrightarrow{t} \langle q_i, \omega_i \rangle \in \Delta$ and these transitions are all the transitions of Δ that are in the form $\langle p, \gamma \rangle \xrightarrow{t} \langle q, \omega \rangle$ that have $\langle p, \gamma \rangle$ on the left hand side.
($\hbar$9) If $\phi = EX^a \phi_1$, then,
(a) $\langle (\!|p, \phi|\!), \gamma \rangle \xhookrightarrow{id} \langle q, \gamma' (\!|\gamma'', \phi_1|\!) \rangle \in \Delta'$ for every $\langle p, \gamma \rangle \xrightarrow{call} \langle q, \gamma'\gamma'' \rangle \in \Delta$
(b) $\langle (\!|p, \phi|\!), \gamma \rangle \xhookrightarrow{id} \langle (\!|q, \phi_1|\!), \omega \rangle \in \Delta'$ for every $\langle p, \gamma \rangle \xrightarrow{int} \langle q, \omega \rangle \in \Delta$
(c) $\langle (\!|p, \phi|\!), \gamma \rangle \xhookrightarrow{id} \langle p_\perp, \gamma_\perp \rangle \in \Delta'$ for every $\langle p, \gamma \rangle \xrightarrow{ret} \langle q', \epsilon \rangle \in \Delta$
($\hbar$10) If $\phi = AX^a \phi_1$, then,
$\langle (\!|p, \phi|\!), \gamma \rangle \xhookrightarrow{equal} [\langle p_1, \gamma'_1 (\!|\gamma''_1, \phi_1|\!) \rangle, ..., \langle p_m, \gamma'_m (\!|\gamma''_m, \phi_1|\!) \rangle, \langle (\!|q_1, \phi_1|\!), \omega_1 \rangle, ..., \langle (\!|q_n, \phi_1|\!), \omega_n \rangle, \langle p_\perp, \gamma_\perp \rangle, ..., \langle p_\perp, \gamma_\perp \rangle] \in \Delta'$, where $\langle p_\perp, \gamma_\perp \rangle$ is repeated k times in the right-hand side s.t.:
(a) for every $1 \leq i \leq m$, $\langle p, \gamma \rangle \xrightarrow{call} \langle p_i, \gamma'_i \gamma''_i \rangle \in \Delta$ and these transitions are all the transitions of Δ that are in the form $\langle p, \gamma \rangle \xrightarrow{call} \langle q, \gamma'\gamma'' \rangle$ that have $\langle p, \gamma \rangle$ on the left hand side.
(b) for every $1 \leq i \leq n$, $\langle p, \gamma \rangle \xrightarrow{int} \langle q_i, \omega_i \rangle \in \Delta$ and these transitions are all the transitions of Δ that are in the form $\langle p, \gamma \rangle \xrightarrow{int} \langle q, \omega \rangle$ that have $\langle p, \gamma \rangle$ on the left hand side.
(c) for every $1 \leq i \leq k$, $\langle p, \gamma \rangle \xrightarrow{ret} \langle q'_i, \epsilon \rangle \in \Delta$ and these transitions are all the transitions of Δ that are in the form $\langle p, \gamma \rangle \xrightarrow{ret} \langle q', \epsilon \rangle$ that have $\langle p, \gamma \rangle$ on the left hand side.
($\hbar$11) If $\phi = E[\phi_1 U^g \phi_2]$, then,

(a) $\langle (\!|p,\phi|\!), \gamma\rangle \xhookrightarrow{id} \langle (\!|p,\phi_2|\!), \gamma\rangle \in \Delta'$
(b) $\langle (\!|p,\phi|\!), \gamma\rangle \xhookrightarrow{equal} [\langle (\!|p,\phi_1|\!), \gamma\rangle, \langle (\!|q,\phi|\!), \omega\rangle] \in \Delta'$ for every $\langle p,\gamma\rangle \xrightarrow{t} \langle q,\omega\rangle \in \Delta$

(ℏ12) If $\phi = E[\phi_1 U^a \phi_2]$, then,
(a) $\langle (\!|p,\phi|\!), \gamma\rangle \xhookrightarrow{id} \langle (\!|p,\phi_2|\!), \gamma\rangle \in \Delta'$
(b) $\langle (\!|p,\phi|\!), \gamma\rangle \xhookrightarrow{equal} [\langle (\!|p,\phi_1|\!), \gamma\rangle, \langle q, \gamma'(\!|\gamma'',\phi|\!)\rangle] \in \Delta'$ for every $\langle p,\gamma\rangle \xrightarrow{call} \langle q,\gamma'\gamma''\rangle \in \Delta$
(c) $\langle (\!|p,\phi|\!), \gamma\rangle \xhookrightarrow{equal} [\langle (\!|p,\phi_1|\!), \gamma\rangle, \langle (\!|q,\phi|\!), \omega\rangle] \in \Delta'$ for every $\langle p,\gamma\rangle \xrightarrow{int} \langle q,\omega\rangle \in \Delta$
(d) $\langle (\!|p,\phi|\!), \gamma\rangle \xhookrightarrow{id} \langle p_\bot,\gamma_\bot\rangle \in \Delta'$ for every $\langle p,\gamma\rangle \xrightarrow{ret} \langle q',\epsilon\rangle \in \Delta$

(ℏ13) If $\phi = A[\phi_1 U^g \phi_2]$, then,
(a) $\langle (\!|p,\phi|\!), \gamma\rangle \xhookrightarrow{id} \langle (\!|p,\phi_2|\!), \gamma\rangle \in \Delta'$
(b) $\langle (\!|p,\phi|\!), \gamma\rangle \xhookrightarrow{equal} [\langle (\!|p,\phi_1|\!), \gamma\rangle; \langle (\!|q_1,\phi|\!), \omega_1\rangle, ..., \langle (\!|q_n,\phi|\!), \omega_n\rangle] \in \Delta'$ where for every $1 \leq i \leq n$,$\langle p,\gamma\rangle \xrightarrow{t} \langle q_i,\omega_i\rangle \in \Delta$ and these transitions are all the transitions of Δ that are in the form $\langle p,\gamma\rangle \xrightarrow{t} \langle q,\omega\rangle$ that have $\langle p,\gamma\rangle$ on the left hand side.

(ℏ14) If $\phi = A[\phi_1 U^a \phi_2]$, then,
(a) $\langle (\!|p,\phi|\!), \gamma\rangle \xhookrightarrow{id} \langle (\!|p,\phi_2|\!), \gamma\rangle \in \Delta'$
(b) $\langle (\!|p,\phi|\!), \gamma\rangle \xhookrightarrow{equal} [\langle (\!|p,\phi_1|\!), \gamma\rangle; \langle p_1, \gamma_1'(\!|\gamma_1'',\phi|\!)\rangle, ..., \langle p_m, \gamma_m'(\!|\gamma_m'',\phi|\!)\rangle;$ $\langle (\!|q_1,\phi|\!), \omega_1\rangle, ..., \langle (\!|q_n,\phi|\!), \omega_n\rangle, \langle p_\bot,\gamma_\bot\rangle, ..., \langle p_\bot,\gamma_\bot\rangle] \in \Delta'$, where $\langle p_\bot,\gamma_\bot\rangle$ is repeated k times in the right-hand side s.t.:
  - for every $1 \leq i \leq m$, $\langle p,\gamma\rangle \xrightarrow{call} \langle p_i,\gamma_i'\gamma_i''\rangle \in \Delta$ and these transitions are all the transitions of Δ that are in the form $\langle p,\gamma\rangle \xrightarrow{call} \langle q,\gamma'\gamma''\rangle$ that have $\langle p,\gamma\rangle$ on the left hand side.
  - for every $1 \leq i \leq n$, $\langle p,\gamma\rangle \xrightarrow{int} \langle q_i,\omega_i\rangle \in \Delta$ and these transitions are all the transitions of Δ that are in the form $\langle p,\gamma\rangle \xrightarrow{int} \langle q,\omega\rangle$ that have $\langle p,\gamma\rangle$ on the left hand side.
  - for every $1 \leq i \leq k$, $\langle p,\gamma\rangle \xrightarrow{ret} \langle q_i',\epsilon\rangle \in \Delta$ and these transitions are all the transitions of Δ that are in the form $\langle p,\gamma\rangle \xrightarrow{ret} \langle q',\epsilon\rangle$ that have $\langle p,\gamma\rangle$ on the left hand side.

(ℏ15) If $\phi = E[\phi_1 R^g \phi_2]$, then, we add to Δ' the rule:
(a) $\langle (\!|p,\phi|\!), \gamma\rangle \xhookrightarrow{equal} [\langle (\!|p,\phi_2|\!), \gamma\rangle, \langle (\!|p,\phi_1|\!), \gamma\rangle] \in \Delta'$
(b) $\langle (\!|p,\phi|\!), \gamma\rangle \xhookrightarrow{equal} [\langle (\!|p,\phi_2|\!), \gamma\rangle, \langle (\!|q,\phi|\!), \omega\rangle] \in \Delta'$ for every $\langle p,\gamma\rangle \xrightarrow{t} \langle q,\omega\rangle \in \Delta$

(ℏ16) If $\phi = A[\phi_1 R^g \phi_2]$, then, we add to Δ' the rule:
(a) $\langle (\!|p,\phi|\!), \gamma\rangle \xhookrightarrow{equal} [\langle (\!|p,\phi_2|\!), \gamma\rangle, \langle (\!|p,\phi_1|\!), \gamma\rangle] \in \Delta'$
(b) $\langle (\!|p,\phi|\!), \gamma\rangle \xhookrightarrow{equal} [\langle (\!|p,\phi_2|\!), \gamma\rangle; \langle (\!|q_1,\phi|\!), \omega_1\rangle, ..., \langle (\!|q_n,\phi|\!), \omega_n\rangle] \in \Delta'$ where for every $1 \leq i \leq n$, $\langle p,\gamma\rangle \xrightarrow{t} \langle q_i,\omega_i\rangle \in \Delta$ and these transitions are all the transitions of Δ that are in the form $\langle p,\gamma\rangle \xrightarrow{t} \langle q,\omega\rangle$ that have $\langle p,\gamma\rangle$ on the left hand side.

(ℏ17) If $\phi = E[\phi_1 R^a \phi_2]$, then,

(a) $\langle (\!| p, \phi |\!), \gamma \rangle \xrightarrow{equal} [\langle (\!| p, \phi_2 |\!), \gamma \rangle, \langle (\!| p, \phi_1 |\!), \gamma \rangle] \in \Delta'$

(b) $\langle (\!| p, \phi |\!), \gamma \rangle \xrightarrow{equal} [\langle (\!| p, \phi_2 |\!), \gamma \rangle, \langle q, \gamma' (\!| \gamma'', \phi |\!) \rangle] \in \Delta'$ for every $\langle p, \gamma \rangle \xrightarrow{call} \langle q, \gamma' \gamma'' \rangle \in \Delta$

(c) $\langle (\!| p, \phi |\!), \gamma \rangle \xrightarrow{equal} [\langle (\!| p, \phi_2 |\!), \gamma \rangle, \langle (\!| q, \phi |\!), \omega \rangle] \in \Delta'$ for every $\langle p, \gamma \rangle \xrightarrow{int} \langle q, \omega \rangle \in \Delta$

(d) $\langle (\!| p, \phi |\!), \gamma \rangle \xrightarrow{id} \langle p_\perp, \gamma_\perp \rangle \in \Delta'$ for every $\langle p, \gamma \rangle \xrightarrow{ret} \langle q', \epsilon \rangle \in \Delta$

(ℏ18) If $\phi = A[\phi_1 R^a \phi_2]$, then,

(a) $\langle (\!| p, \phi |\!), \gamma \rangle \xrightarrow{equal} [\langle (\!| p, \phi_2 |\!), \gamma \rangle, \langle (\!| p, \phi_1 |\!), \gamma \rangle] \in \Delta'$

(b) $\langle (\!| p, \phi |\!), \gamma \rangle \xrightarrow{equal} [\langle (\!| p, \phi_2 |\!), \gamma \rangle; \langle p_1, \gamma_1' (\!| \gamma_1'', \phi |\!) \rangle, ..., \langle p_m, \gamma_m' (\!| \gamma_m'', \phi |\!) \rangle; \langle (\!| q_1, \phi |\!), \omega_1 \rangle, ..., \langle (\!| q_n, \phi |\!), \omega_n \rangle, \langle p_\perp, \gamma_\perp \rangle, ..., \langle p_\perp, \gamma_\perp \rangle] \in \Delta'$, where $\langle p_\perp, \gamma_\perp \rangle$ is repeated k times in the right-hand side s.t.:

- for every $1 \leq i \leq m$, $\langle p, \gamma \rangle \xrightarrow{call} \langle p_i, \gamma_i' \gamma_i'' \rangle \in \Delta$ and these transitions are all the transitions of Δ that are in the form $\langle p, \gamma \rangle \xrightarrow{call} \langle q, \gamma' \gamma'' \rangle$ that have $\langle p, \gamma \rangle$ on the left hand side.
- for every $1 \leq i \leq n$, $\langle p, \gamma \rangle \xrightarrow{int} \langle q_i, \omega_i \rangle \in \Delta$ and these transitions are all the transitions of Δ that are in the form $\langle p, \gamma \rangle \xrightarrow{int} \langle q, \omega \rangle$ that have $\langle p, \gamma \rangle$ on the left hand side.
- for every $1 \leq i \leq k$, $\langle p, \gamma \rangle \xrightarrow{ret} \langle q_i', \epsilon \rangle \in \Delta$ and these transitions are all the transitions of Δ that are in the form $\langle p, \gamma \rangle \xrightarrow{ret} \langle q', \epsilon \rangle$ that have $\langle p, \gamma \rangle$ on the left hand side.

(ℏ19) for every $\langle p, \gamma \rangle \xrightarrow{ret} \langle q, \epsilon \rangle \in \Delta$:

- $\langle q, (\!| \gamma'', \phi_1 |\!) \rangle \xrightarrow{id} \langle (\!| q, \phi_1 |\!), \gamma'' \rangle \in \Delta'$ for every $\gamma'' \in \Gamma$, $\phi_1 \in Cl(\varphi)$

(ℏ20) $\langle p_\perp, \gamma_\perp \rangle \xrightarrow{id} \langle p_\perp, \gamma_\perp \rangle \in \Delta'$

(ℏ21) for every $\langle p, \gamma \rangle \xrightarrow{t} \langle q, \omega \rangle \in \Delta$: $\langle p, \gamma \rangle \xrightarrow{id} \langle q, \omega \rangle \in \Delta'$

(ℏ22) If $\phi = e$, e is a regular expression, then, $\langle (\!| p, \phi |\!), \gamma \rangle \xrightarrow{id} \langle s_e, \gamma \rangle \in \Delta'$

(ℏ23) If $\phi = \neg e$, e is a regular expression, then, $\langle (\!| p, \phi |\!), \gamma \rangle \xrightarrow{id} \langle s_{\neg e}, \gamma \rangle \in \Delta'$

(ℏ24) for every transition $q \xrightarrow{\alpha} \{q_1, .., q_n\}$ in $\mathcal{M}$: $\langle q, \gamma \rangle \xrightarrow{\mathbb{R}} [\langle q_1, \epsilon \rangle, ..., \langle q_n, \epsilon \rangle] \in \Delta'$, where:

(a) $\mathbb{R} = equal$ iff $\alpha = \gamma$

(b) $\mathbb{R} = join_\gamma^x$ iff $\alpha = x \in \mathcal{X}$

(c) $\mathbb{R} = join_\gamma^{\neg x}$ iff $\alpha = \neg x$ and $x \in \mathcal{X}$

(ℏ25) for every $q \in \mathcal{F}$, $\langle q, \sharp \rangle \xrightarrow{id} \langle q, \sharp \rangle \in \Delta'$

Roughly speaking, the SABPDS $\mathcal{BP}_\varphi$ is a kind of product between $\mathcal{P}$ and the SBPCARET formula φ which ensures that $\mathcal{BP}_\varphi$ has an accepting run from $\langle [\![(\!| p, \varphi |\!), B]\!], \omega \rangle$ iff the configuration $\langle p, \omega \rangle$ satisfies φ under the environment B. The form of the control locations of $\mathcal{BP}_\varphi$ is $[\![(\!| p, \phi |\!), B]\!]$ where $\phi \in Cl(\varphi)$, $B \in \mathcal{B}$. Let us explain the intuition behind our construction:

- If $\phi = b(\alpha_1, ..., \alpha_n)$, then, for every $\omega \in \Gamma^*$, $\langle p, \omega \rangle \models_\lambda^B \phi$ iff $b(B(\alpha_1), ..., B(\alpha_n)) \in \lambda(p)$. Thus, for such a B, $\mathcal{BP}_\varphi$ should have an accepting run from $\langle [\![(\!| p, b(\alpha_1, ..., \alpha_n) |\!), B]\!], \omega \rangle$ iff $b(B(\alpha_1), ..., B(\alpha_n)) \in \lambda(p)$. This is ensured by the transition rules in **(ℏ1)** which add a loop at $\langle [\![(\!| p, b(\alpha_1, ..., \alpha_n) |\!), B]\!], \omega \rangle$ and the fact that $[\![(\!| p, b(\alpha_1, ..., \alpha_n) |\!), B]\!] \in F$ (because it is in F_1). The function id in **(ℏ1)** ensures that the environments before and after are the same.
- If $\phi = \neg b(\alpha_1, ..., \alpha_n)$, then, for every $\omega \in \Gamma^*$, $\langle p, \omega \rangle \models_\lambda^B \phi$ iff $b(B(\alpha_1), ..., B(\alpha_n)) \notin \lambda(p)$. Thus, for such a B, $\mathcal{BP}_\varphi$ should have an accepting run from $\langle [\![(\!| p, \neg b(\alpha_1, ..., \alpha_n) |\!), B]\!], \omega \rangle$ iff $b(B(\alpha_1), ..., B(\alpha_n)) \notin \lambda(p)$. This is ensured by the transition rules in **(ℏ2)** which add a loop at $\langle [\![(\!| p, \neg b(\alpha_1, ..., \alpha_n) |\!), B]\!], \omega \rangle$ and the fact that $[\![(\!| p, \neg b(\alpha_1, ..., \alpha_n) |\!), B]\!] \in F$ (because it is in F_2). The function id in **(ℏ2)** ensures that the environments before and after are the same.
- If $\phi = \phi_1 \wedge \phi_2$, then, for every $\omega \in \Gamma^*$, $\langle p, \omega \rangle \models_\lambda^B \phi$ iff $(\langle p, \omega \rangle \models_\lambda^B \phi_1$ and $\langle p, \omega \rangle \models_\lambda^B \phi_2)$. This is ensured by the transition rules in **(ℏ3)** stating that $\mathcal{BP}_\varphi$ has an accepting run from $\langle [\![(\!| p, \phi_1 \wedge \phi_2 |\!), B]\!], \omega \rangle$ iff $\mathcal{BP}_\varphi$ has an accepting run from both $\langle [\![(\!| p, \phi_1 |\!), B]\!], \omega \rangle$ and $\langle [\![(\!| p, \phi_2 |\!), B]\!], \omega \rangle$. **(ℏ4)** is similar to **(ℏ3)**.
- If $\phi = \exists x \phi_1$, then, for every $\omega \in \Gamma^*$, $\langle p, \omega \rangle \models_\lambda^B \phi$ iff there exists $c \in \mathcal{D}$ s.t. $\langle p, \omega \rangle \models_\lambda^{B[x \leftarrow c]} \phi_1$. This is ensured by the transition rules in **(ℏ5)** stating that $\mathcal{BP}_\varphi$ has an accepting run from $\langle [\![(\!| p, \exists x \phi_1 |\!), B]\!], \omega \rangle$ iff there exists $c \in \mathcal{D}$ s.t. $\mathcal{BP}_\varphi$ has an accepting run from $\langle [\![(\!| p, \phi_1 |\!), B[x \leftarrow c]]\!], \omega \rangle$ since $B \in meet^x_{\{c\}}(B[x \leftarrow c])$
- If $\phi = \forall x \phi_1$, then, for every $\omega \in \Gamma^*$, $\langle p, \omega \rangle \models_\lambda^B \phi$ iff for every $c \in \mathcal{D}$, $\langle p, \omega \rangle \models_\lambda^{B[x \leftarrow c]} \phi_1$. This is ensured by the transition rules in **(ℏ6)** stating that $\mathcal{BP}_\varphi$ has an accepting run from $\langle [\![(\!| p, \forall x \phi_1 |\!), B]\!], \omega \rangle$ iff for every $c \in \mathcal{D}$, $\mathcal{BP}_\varphi$ has an accepting run from $\langle [\![(\!| p, \phi_1 |\!), B[x \leftarrow c]]\!], \omega \rangle$ since if $\mathcal{D} = \{c_1, ..., c_m\}$, then, $B \in meet^x_{\mathcal{D}}(B[x \leftarrow c_1], ..., B[x \leftarrow c_m])$
- If $\phi = EX^g \phi_1$, then, for every $\omega \in \Gamma^*$, $\langle p, \omega \rangle \models_\lambda^B \phi$ iff there exists an immediate successor $\langle p', \omega' \rangle$ of $\langle p, \omega \rangle$ s.t. $\langle p', \omega' \rangle \models_\lambda^B \phi_1$. This is ensured by the transition rules in **(ℏ7)** stating that $\mathcal{BP}_\varphi$ has an accepting run from $\langle [\![(\!| p, EX^g \phi_1 |\!), B]\!], \omega \rangle$ iff there exists an immediate successor $\langle p', \omega' \rangle$ of $\langle p, \omega \rangle$ s.t. $\mathcal{BP}_\varphi$ has an accepting run from $\langle [\![(\!| p', \phi_1 |\!), B]\!], \omega' \rangle$. **(ℏ8)** is similar to **(ℏ7)**.
- If $\phi = E[\phi_1 U^g \phi_2]$, then, for every $\omega \in \Gamma^*$, $\langle p, \omega \rangle \models_\lambda^B \phi$ iff $\langle p, \omega \rangle \models_\lambda^B \phi_2$ or $(\langle p, \omega \rangle \models_\lambda^B \phi_1$ and there exists an immediate successor $\langle p', \omega' \rangle$ of $\langle p, \omega \rangle$ s.t. $\langle p', \omega' \rangle \models_\lambda^B \phi)$. This is ensured by the transition rules in **(ℏ11)** stating that $\mathcal{BP}_\varphi$ has an accepting run from $\langle [\![(\!| p, E[\phi_1 U^g \phi_2] |\!), B]\!], \omega \rangle$ iff $\mathcal{BP}_\varphi$ has an accepting run from $\langle [\![(\!| p, \phi_2 |\!), B]\!], \omega \rangle$ (by the rules in **(ℏ11)(a)** or ($\mathcal{BP}_\varphi$ has an accepting run from both $\langle [\![(\!| p, \phi_1 |\!), B]\!], \omega \rangle$ and $\langle [\![(\!| p', \phi |\!), B]\!], \omega' \rangle$ where $\langle p', \omega' \rangle$ is an immediate successor of $\langle p, \omega \rangle$) (by the rules in **(ℏ11)(b)**). **(ℏ13)** is similar to **(ℏ11)**.
- If $\phi = E[\phi_1 R^g \phi_2]$, then, for every $\omega \in \Gamma^*$, $\langle p, \omega \rangle \models_\lambda^B \phi$ iff $((\langle p, \omega \rangle \models_\lambda^B \phi_2$ and $\langle p, \omega \rangle \models_\lambda^B \phi_1)$ or $(\langle p, \omega \rangle \models_\lambda^B \phi_2$ and there exists an immediate successor $\langle p', \omega' \rangle$ of $\langle p, \omega \rangle$ s.t. $\langle p', \omega' \rangle \models_\lambda^B \phi)$. This is ensured by the transition rules in **(ℏ15)** stating that $\mathcal{BP}_\varphi$ has an accepting run from $\langle [\![(\!| p, E[\phi_1 R^g \phi_2] |\!), B]\!], \omega \rangle$ iff $\mathcal{BP}_\varphi$

has an accepting run from both $\langle [\![(\!| p, \phi_2 |\!), B]\!], \omega \rangle$ and $\langle [\![(\!| p, \phi_1 |\!), B]\!], \omega \rangle$ (by the rules in **(ħ15)(a)**); or $\mathcal{BP}_\varphi$ has an accepting run from both $\langle [\![(\!| p, \phi_2 |\!), B]\!], \omega \rangle$ and $[\![(\!| p', \phi |\!), B]\!], \omega' \rangle$ where $\langle p', \omega' \rangle$ is an immediate successor of $\langle p, \omega \rangle$ (by the rules in **(ħ15)(b)**). In addition, for R^g formulas, the *stop* condition is not required, i.e, for a formula $\phi_1 R^g \phi_2$ that is applied to a specific run, we don't require that ϕ_1 must eventually hold. To ensure that the runs on which ϕ_2 always holds are accepted, we add $[\![(\!| p, \phi |\!), B]\!]$ to the Büchi accepting condition F (via the subset F_3 of F). **(ħ16)** is similar to **(ħ15)**.

- If $\phi = EX^a \phi_1$, then, for every $\omega \in \Gamma^*$, $\langle p, \omega \rangle \vDash^B_\lambda \phi$ iff there exists an abstract-successor $\langle p_k, \omega_k \rangle$ of $\langle p, \omega \rangle$ s.t. $\langle p_k, \omega_k \rangle \vDash^B_\lambda \phi_1$ (A1) . Let $\pi \in Traces(\langle p, \omega \rangle)$ be a run starting from $\langle p, \omega \rangle$ on which $\langle p_k, \omega_k \rangle$ is the abstract-successor of $\langle p, \omega \rangle$. Over π, let $\langle p', \omega' \rangle$ be the immediate successor of $\langle p, \omega \rangle$. In what follows, we explain how we can ensure this.

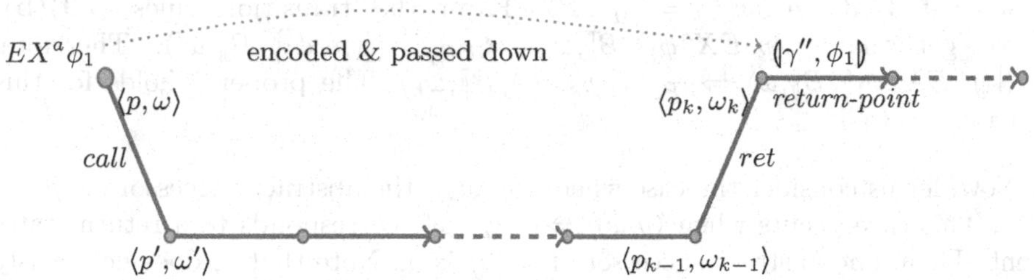

Fig. 2. $\langle p, \omega \rangle \Rightarrow_\mathcal{P} \langle p', \omega' \rangle$ corresponds to a call statement

1. Firstly, we show that for every abstract-successor $\langle p_k, \omega_k \rangle \neq \perp$ of $\langle p, \omega \rangle$, $\langle [\![(\!| p, EX^a \phi_1 |\!), B]\!], \omega \rangle \Rightarrow_{\mathcal{BP}_\varphi} \langle [\![(\!| p_k, \phi_1 |\!), B]\!], \omega_k \rangle$ where $B \in \mathcal{B}$. There are two possibilities:

- If $\langle p, \omega \rangle \Rightarrow_\mathcal{P} \langle p', \omega' \rangle$ corresponds to a call statement. Let us consider Fig. 2 to explain this case. $\langle [\![(\!| p, \phi |\!), B]\!], \omega \rangle \Rightarrow_{\mathcal{BP}_\varphi} \langle [\![(\!| p_k, \phi_1 |\!), B]\!], \omega_k \rangle$ is ensured by the rules in **(ħ9)(a)**, the rules in **(ħ21)** and the rules in **(ħ19)** as follows: rules in **(ħ9)(a)** allow to record ϕ_1 in the return point of the call, rules in **(ħ21)** allow to mimic the run of the PDS $\mathcal{P}$ and rules in **(ħ19)** allow to extract and put back ϕ_1 when the return-point is reached. In what follows, we show in more details how this works: Let $\langle p, \gamma \rangle \xrightarrow{call} \langle p', \gamma' \gamma'' \rangle$ be the rule associated with the transition $\langle p, \omega \rangle \Rightarrow_\mathcal{P} \langle p', \omega' \rangle$, then we have $\omega = \gamma \omega''$ and $\omega' = \gamma' \gamma'' \omega''$. Let $\langle p_{k-1}, \omega_{k-1} \rangle \Rightarrow_\mathcal{P} \langle p_k, \omega_k \rangle$ be the transition that corresponds to the *ret* statement of this call on π.Let then $\langle p_{k-1}, \beta \rangle \xrightarrow{ret} \langle p_k, \epsilon \rangle \in \Delta$ be the corresponding return rule. Then, we have necessarily $\omega_{k-1} = \beta \gamma'' \omega''$, since as explained in Sect. 2, γ'' is the return address of the call. After applying this rule, $\omega_k = \gamma'' \omega''$. In other words, γ'' will be the topmost stack symbol at the corresponding return point of the call. So, in order to ensure that $\langle [\![(\!| p, \phi |\!), B]\!], \omega \rangle \Rightarrow_{\mathcal{BP}_\varphi} \langle [\![(\!| p_k, \phi_1 |\!), B]\!], \omega_k \rangle$, we proceed as follows: At the call

$\langle p,\gamma\rangle \xrightarrow{call} \langle p',\gamma'\gamma''\rangle$, we encode the formula ϕ_1 into γ'' by the rule in **(ℏ9)(a)** stating that $\langle (\!| p, EX^a\phi_1 |\!), \gamma\rangle \overset{id}{\hookrightarrow} \langle p', \gamma'(\!|\gamma'', \phi_1|\!)\rangle \in \Delta'$. This allows to record ϕ_1 in the corresponding return point of the stack. After that, the rules in **(ℏ21)** allow $\mathcal{BP}_\varphi$ to mimic the run π of $\mathcal{P}$ from $\langle p', \omega'\rangle$ till the corresponding return-point of this call, where $(\!|\gamma'', \phi_1|\!)$ is the topmost stack symbol. More specifically, the following sequence of $\mathcal{P}$: $\langle p', \gamma'\gamma''\omega''\rangle \overset{*}{\Rightarrow}_{\mathcal{P}} \langle p_{k-1}, \beta\gamma''\omega''\rangle \overset{*}{\Rightarrow}_{\mathcal{P}} \langle p_k, \gamma''\omega''\rangle$ will be mimicked by the following sequence of $\mathcal{BP}_\varphi$: $\langle [\![p', B]\!], \gamma'(\!|\gamma'', \phi_1|\!)\omega''\rangle \Rightarrow_{\mathcal{BP}_\varphi} \langle [\![p_{k-1}, B]\!], \beta(\!|\gamma'', \phi_1|\!)\omega''\rangle \Rightarrow_{\mathcal{BP}_\varphi} \langle [\![p_k, B]\!], (\!|\gamma'', \phi_1|\!)\omega''\rangle$ using the rules of **(ℏ21)**. At the return-point, we extract ϕ_1 from the stack and encode it into p_k by adding the transition rules in **(ℏ19)** $\langle p_k, (\!|\gamma'', \phi_1|\!)\rangle \overset{id}{\hookrightarrow} \langle (\!|p_k, \phi_1|\!), \gamma''\rangle$. Therefore, we obtain that $\langle [\![(\!|p, \phi|\!), B]\!], \omega\rangle \Rightarrow_{\mathcal{BP}_\varphi} \langle [\![(\!|p_k, \phi_1|\!), B]\!], \omega_k\rangle$. The property holds for this case.

- If $\langle p, \omega\rangle \Rightarrow_{\mathcal{P}} \langle p', \omega'\rangle$ corresponds to a simple statement. Then, the abstract successor of $\langle p, \omega\rangle$ is its immediate successor $\langle p', \omega'\rangle$. Thus, we get that $\langle p_k, \omega_k\rangle = \langle p', \omega'\rangle$. From the transition rules **(ℏ9)(b)**, we get that $\langle [\![(\!|p, EX^a\phi_1|\!), B]\!], \omega\rangle \Rightarrow_{\mathcal{BP}_\varphi} \langle [\![(\!|p', \phi_1|\!), B]\!], \omega'\rangle$. Therefore, $\langle [\![(\!|p, EX^a\phi_1|\!), B]\!], \omega\rangle \Rightarrow_{\mathcal{BP}_\varphi} \langle [\![(\!|p_k, \phi_1|\!), B]\!], \omega_k\rangle$. The property holds for this case.

2. Now, let us consider the case where $\langle p_k, \omega_k\rangle$, the abstract successor of $\langle p, \omega\rangle$, is $\perp$. This case occurs when $\langle p, \omega\rangle \Rightarrow_{\mathcal{P}} \langle p', \omega'\rangle$ corresponds to a return statement. Then, one abstract successor of $\langle p, \omega\rangle$ is $\perp$. Note that $\perp$ does not satisfy any formula, i.e., $\perp$ does not satisfy ϕ_1. Therefore, from $\langle [\![(\!|p, EX^a\phi_1|\!), B]\!], \omega\rangle$, we need to ensure that the path of $\mathcal{BP}_\varphi$ reflecting the possibility in (A1) that $\langle p_k, \omega_k\rangle \models^B_\lambda \phi_1$ is not accepted. To do this, we exploit additional trap configurations. We use $p_\perp$ and $\gamma_\perp$ as trap control location and trap stack symbol to obtain these trap configurations. To be more specific, let $\langle p, \gamma\rangle \xrightarrow{ret} \langle p', \epsilon\rangle$ be the rule associated with the transition $\langle p, \omega\rangle \Rightarrow_{\mathcal{P}} \langle p', \omega'\rangle$, then we have $\omega = \gamma\omega''$ and $\omega' = \omega''$. We add the transition rule in **(ℏ9)(c)** to allow $\langle [\![(\!|p, EX^a\phi_1|\!), B]\!], \omega\rangle \Rightarrow_{\mathcal{BP}_\varphi} \langle [\![p_\perp, B]\!], \gamma_\perp\omega''\rangle$. Since a run of $\mathcal{BP}_\varphi$ includes only infinite paths, we equip these trap configurations with self-loops by the transition rules in **(ℏ20)**, i.e., $\langle [\![p_\perp, B]\!], \gamma_\perp\omega''\rangle \Rightarrow_{\mathcal{BP}_\varphi} \langle [\![p_\perp, B]\!], \gamma_\perp\omega''\rangle$. As a result, we obtain a corresponding path in $\mathcal{BP}_\varphi$: $\langle [\![(\!|p, EX^a\phi_1|\!), B]\!], \omega\rangle \Rightarrow_{\mathcal{BP}_\varphi} \langle [\![p_\perp, B]\!], \gamma_\perp\omega''\rangle \Rightarrow_{\mathcal{BP}_\varphi} \langle [\![p_\perp, B]\!], \gamma_\perp\omega''\rangle$. Note that this path is not accepted by $\mathcal{BP}_\varphi$ because $[\![p_\perp, B]\!] \notin F$.

In summary, for every abstract-successor $\langle p_k, \omega_k\rangle$ of $\langle p, \omega\rangle$, if $\langle p_k, \omega_k\rangle \neq \perp$, then, $\langle [\![(\!|p, EX^a\phi_1|\!), B]\!], \omega\rangle \Rightarrow_{\mathcal{BP}_\varphi} \langle [\![(\!|p_k, \phi_1|\!), B]\!], \omega_k\rangle$; otherwise $\langle [\![(\!|p, EX^a\phi_1|\!), B]\!], \omega\rangle \Rightarrow_{\mathcal{BP}_\varphi} \langle [\![p_\perp, B]\!], \gamma_\perp\omega''\rangle \Rightarrow_{\mathcal{BP}_\varphi} \langle [\![p_\perp, B]\!], \gamma_\perp\omega''\rangle$ which is not accepted by $\mathcal{BP}_\varphi$. Therefore, (A1) is ensured by the transition rules in **(ℏ9)** stating that $\mathcal{BP}_\varphi$ has an accepting run from $\langle [\![(\!|p, EX^a\phi_1|\!), B]\!], \omega\rangle$ iff there exists an abstract successor $\langle p_k, \omega_k\rangle$ of $\langle p, \omega\rangle$ s.t. $\mathcal{BP}_\varphi$ has an accepting run from $\langle [\![(\!|p_k, \phi_1|\!), B]\!], \omega_k\rangle$.

– If $\phi = AX^a\phi_1$: this case is ensured by the transition rules in **(ℏ10)** together with **(ℏ19)** and **(ℏ21)**. The intuition of **(ℏ10)** is similar to that of **(ℏ9)**.

– If $\phi = E[\phi_1 U^a \phi_2]$, then, for every $\omega \in \Gamma^*$, $\langle p, \omega\rangle \models_\lambda^B \phi$ iff $\langle p, \omega\rangle \models_\lambda^B \phi_2$ or $(\langle p, \omega\rangle \models_\lambda^B \phi_1$ and there exists an abstract successor $\langle p_k, \omega_k\rangle$ of $\langle p, \omega\rangle$ s.t. $\langle p_k, \omega_k\rangle \models_\lambda^B \phi)$ (A2) . Let $\pi \in Traces(\langle p, \omega\rangle)$ be a run starting from $\langle p, \omega\rangle$ on which $\langle p_k, \omega_k\rangle$ is the abstract-successor of $\langle p, \omega\rangle$. Over π, let $\langle p', \omega'\rangle$ be the immediate successor of $\langle p, \omega\rangle$.

1. Firstly, we show that for every abstract-successor $\langle p_k, \omega_k\rangle \neq \perp$ of $\langle p, \omega\rangle$, $\langle [\![(\!|p, \phi|\!), B]\!], \omega\rangle \Rightarrow_{\mathcal{BP}_\varphi} \{\langle [\![(\!|p, \phi_1|\!), B]\!], \omega\rangle, \langle [\![(\!|p_k, \phi|\!), B]\!], \omega_k\rangle\}$ where $B \in \mathcal{B}$. There are two possibilities:

- If $\langle p, \omega\rangle \Rightarrow_\mathcal{P} \langle p', \omega'\rangle$ corresponds to a call statement. From the rules in **($\hbar$12)(b)**, we get that $\langle [\![(\!|p, \phi|\!), B]\!], \omega\rangle \Rightarrow_{\mathcal{BP}_\varphi} \{\langle [\![(\!|p, \phi_1|\!), B]\!], \omega\rangle, \langle p', \omega'\rangle\}$ where $\langle p', \omega'\rangle$ is the immediate successor of $\langle p, \omega\rangle$. Thus, to ensure that $\langle [\![(\!|p, \phi|\!), B]\!], \omega\rangle \Rightarrow_{\mathcal{BP}_\varphi} \{\langle [\![(\!|p, \phi_1|\!), B]\!], \omega\rangle, \langle [\![(\!|p_k, \phi|\!), B]\!], \omega_k\rangle\}$, we only need to ensure that $\langle p', \omega'\rangle \Rightarrow_{\mathcal{BP}_\varphi} \langle [\![(\!|p_k, \phi|\!), B]\!], \omega_k\rangle$. As for the case $\phi = EX^a\phi_1$, $\langle p', \omega'\rangle \Rightarrow_{\mathcal{BP}_\varphi} \langle [\![(\!|p_k, \phi|\!), B]\!], \omega_k\rangle$ is ensured by the rules in **($\hbar$21)** and the rules in **($\hbar$19)**: rules in **($\hbar$21)** allow to mimic the run of the PDS $\mathcal{P}$ before the return and rules in **($\hbar$19)** allow to extract and put back ϕ_1 when the return-point is reached.
- If $\langle p, \omega\rangle \Rightarrow_\mathcal{P} \langle p', \omega'\rangle$ corresponds to a simple statement. Then, the abstract successor of $\langle p, \omega\rangle$ is its immediate successor $\langle p', \omega'\rangle$. Thus, we get that $\langle p_k, \omega_k\rangle = \langle p', \omega'\rangle$. From the transition rules **($\hbar$12)(c)**, we get that $\langle [\![(\!|p, E[\phi_1 U^a \phi_2]|\!), B]\!], \omega\rangle \Rightarrow_{\mathcal{BP}_\varphi} \{\langle [\![(\!|p, \phi_1|\!), B]\!], \omega\rangle, \langle [\![(\!|p', \phi|\!), B]\!], \omega'\rangle\}$. Therefore, $\langle [\![(\!|p, E[\phi_1 U^a \phi_2]|\!), B]\!], \omega\rangle \Rightarrow_{\mathcal{BP}_\varphi} \{\langle [\![(\!|p, \phi_1|\!), B]\!], \omega\rangle, \langle [\![(\!|p_k, \phi|\!), B]\!], \omega_k\rangle\}$. In other words, $\mathcal{BP}_\varphi$ has an accepting run from both $\langle [\![(\!|p, \phi_1|\!), B]\!], \omega\rangle$ and $\langle [\![(\!|p_k, \phi|\!), B]\!], \omega_k\rangle$ where $\langle p_k, \omega_k\rangle$ is an abstract successor of $\langle p, \omega\rangle$. The property holds for this case.

2. Now, let us consider the case where $\langle p_k, \omega_k\rangle = \perp$. As explained previously, this case occurs when $\langle p, \omega\rangle \Rightarrow_\mathcal{P} \langle p', \omega'\rangle$ corresponds to a return statement. Then, the abstract successor of $\langle p, \omega\rangle$ is $\perp$. Note that $\perp$ does not satisfy any formula, i.e., $\perp$ does not satisfy ϕ. Therefore, from $\langle [\![(\!|p, E[\phi_1 U^a \phi_2]|\!), B]\!], \omega\rangle$, we need to ensure that the path reflecting the possibility in (A2) that $(\langle p, \omega\rangle \models_\lambda^B \phi_1$ and $\langle p_k, \omega_k\rangle \models_\lambda^B \phi)$ is not accepted by $\mathcal{BP}_\varphi$. This is ensured as for the case $\phi = EX^a\phi_1$ by the transition rules in **($\hbar$12)(d)**.

In summary, for every abstract-successor $\langle p_k, \omega_k\rangle$ of $\langle p, \omega\rangle$, if $\langle p_k, \omega_k\rangle \neq \perp$, then, $\langle [\![(\!|p, E[\phi_1 U^a \phi_2]|\!), B]\!], \omega\rangle \Rightarrow_{\mathcal{BP}_\varphi} \{\langle [\![(\!|p, \phi_1|\!), B]\!], \omega\rangle, \langle [\![(\!|p_k, E[\phi_1 U^a \phi_2]|\!), B]\!], \omega_k\rangle\}$; otherwise $\langle [\![(\!|p, E[\phi_1 U^a \phi_2]|\!), B]\!], \omega\rangle \Rightarrow_{\mathcal{BP}_\varphi} \langle [\![p_\perp, B]\!], \gamma_\perp \omega''\rangle \Rightarrow_{\mathcal{BP}_\varphi} \langle [\![p_\perp, B]\!], \gamma_\perp \omega''\rangle$ which is not accepted by $\mathcal{BP}_\varphi$. Therefore, (A2) is ensured by the transition rules in **($\hbar$12)** stating that $\mathcal{BP}_\varphi$ has an accepting run from $\langle [\![(\!|p, E[\phi_1 U^a \phi_2]|\!), B]\!], \omega\rangle$ iff $\mathcal{BP}_\varphi$ has an accepting run from $\langle [\![(\!|p, \phi_2|\!), B]\!], \omega\rangle$; or $\mathcal{BP}_\varphi$ has an accepting run from both $\langle [\![(\!|p, \phi_1|\!), B]\!], \omega\rangle$ and $\langle [\![(\!|p_k, E[\phi_1 U^a \phi_2]|\!), B]\!], \omega_k\rangle$ where $\langle p_k, \omega_k\rangle$ is an abstract successor of $\langle p, \omega\rangle$.

– The intuition behind the rules corresponding to the cases $\phi = A[\phi_1 U^a \phi_2]$, $\phi = A[\phi_1 R^a \phi_2]$ are similar to the previous case.

– If $\phi = e(e \in \mathcal{V})$. Given $p \in P$, $e \in \mathcal{V}$, $\omega \in \Gamma^*$, we get that the SABPDS $\mathcal{BP}_\varphi$ should accept $\langle [\![(\!| p, e |\!), B]\!], \omega \rangle$ iff $(\langle p, \omega \rangle, B) \in L(M_e)$. To check whether $(\langle p, \omega \rangle, B) \in L(M_e)$, we let $\mathcal{BP}_\varphi$ go to state s_e, the initial state corresponding to p in M_e by adding rules in **(ℏ22)**; and then, from this state, we will check whether ω is accepted by M_e under B. This is ensured by the transition rules in **(ℏ24)** and **(ℏ25)**. **(ℏ24)** lets $\mathcal{BP}_\varphi$ mimic a run of M_e on ω under B, which includes three possibilities:

- if $\mathcal{BP}_\varphi$ is in a state $[\![q, B]\!]$ with γ on the top of the stack where $\gamma \in \Gamma$, and if $q \xrightarrow{\gamma} \{q_1, ..., q_n\}$ is a transition rule in M_e, then, $\mathcal{BP}_\varphi$ will move to states $[\![q_1, B]\!], ..., [\![q_n, B]\!]$ and pop γ from its stack. Note that popping γ allows us to check the rest of the word. This is ensured by the rules corresponding to **(ℏ24)(a)**. Then function *equal* ensures that all these environments are the same.
- if $\mathcal{BP}_\varphi$ is in a state $[\![q, B]\!]$ with γ on the top of the stack, and if $q \xrightarrow{x} \{q_1, ..., q_n\}$ is a transition rule in M_e where $x \in \mathcal{X}$, then, $\mathcal{BP}_\varphi$ can mimic a run of M_e under B iff $B(x) = \gamma$. If this condition is guaranteed, $\mathcal{BP}_\varphi$ will move to states $[\![q_1, B]\!], ..., [\![q_n, B]\!]$ and pop γ from its stack. Again, popping γ allows us to check the rest of the word. This is ensured by the rules corresponding to **(ℏ24)(b)**. Then function $join_\gamma^x$ ensures that all these environments are the same B and $B(x) = \gamma$.
- Similar to **(ℏ24)(b)**, **(ℏ24)(c)** deals with the cases where $q \xrightarrow{\neg x} \{q_1, ..., q_n\}$ is a transition rule in M_e where $x \in \mathcal{X}$.

In each VA M_e, a configuration is accepted if the run with the word ω reaches a final state in F_e; i.e., if $\mathcal{BP}_\varphi$ reaches a state $q \in F_e$ with an empty stack, i.e., with a stack containing the bottom stack symbol $\sharp$. Thus, we should add $F_e \times \mathcal{B}$ as a set of accepting control locations in $\mathcal{BP}_\varphi$. This is why F_4 is a set of accepting control locations. In addition, since $\mathcal{BP}_\varphi$ only recognizes infinite paths, **(ℏ25)** adds a loop on every configuration $\langle [\![q, B]\!], \sharp \rangle$ where $q \in F_e$.

– If $\phi = \neg e(e \in \mathcal{V})$. This case is ensured by the transition rules in **(ℏ23)**, **(ℏ24)** and **(ℏ25)**. The intuition behind this case is similar to the case $\phi = e$.

We can show that:

Theorem 4. *Given a PDS $\mathcal{P} = (P, \Gamma, \Delta)$, a set of atomic propositions AP, a labelling function $\lambda : AP_\mathcal{D} \to 2^P$ and a SBPCARET formula φ, we can compute an SABPDS $\mathcal{BP}_\varphi$ such that for every configuration $\langle p, \omega \rangle$, for every $B \in \mathcal{B}$, $\langle p, \omega \rangle \models_\lambda^B \varphi$ iff $\mathcal{BP}_\varphi$ has an accepting run from the configuration $\langle [\![(\!| p, \varphi |\!), B]\!], \omega \rangle$.*

5 Conclusion

In this paper, we present a new logic SBPCARET and show how it can precisely and succinctly specify malicious behaviors. We then propose an efficient algorithm for SBPCARET model-checking for PDSs. Our algorithm is based on reducing the model checking problem to the emptiness problem of Symbolic Alternating Büchi Pushdown Systems.

References

1. Alur, R., Etessami, K., Madhusudan, P.: A temporal logic of nested calls and returns. In: Jensen, K., Podelski, A. (eds.) TACAS 2004. LNCS, vol. 2988, pp. 467–481. Springer, Heidelberg (2004). https://doi.org/10.1007/978-3-540-24730-2_35
2. Bergeron, J., Debbabi, M., Desharnais, J., Erhioui, M.M., Lavoie, Y., Tawbi, N.: Static detection of malicious code in executable programs. Int. J. Req. Eng. **184–189**, 79 (2001)
3. Christodorescu, M., Jha, S.: Static analysis of executables to detect malicious patterns. In: Proceedings of the 12th Conference on USENIX Security Symposium - Volume 12, SSYM 2003, Berkeley, CA, USA, p. 12. USENIX Association (2003)
4. Kinder, J., Katzenbeisser, S., Schallhart, C., Veith, H.: Detecting malicious code by model checking. In: Julisch, K., Kruegel, C. (eds.) DIMVA 2005. LNCS, vol. 3548, pp. 174–187. Springer, Heidelberg (2005). https://doi.org/10.1007/11506881_11
5. Lakhotia, A., Kumar, E.U., Venable, M.: A method for detecting obfuscated calls in malicious binaries. IEEE Trans. Softw. Eng. **31**(11), 955–968 (2005)
6. Nguyen, H.-V., Touili, T.: CARET model checking for malware detection. In: SPIN 2017 (2017)
7. Nguyen, H.-V., Touili, T.: CARET model checking for pushdown systems. In: SAC 2017 (2017)
8. Nguyen, H.-V., Touili, T.: Branching temporal logic of calls and returns for pushdown systems. In: Furia, C.A., Winter, K. (eds.) IFM 2018. LNCS, vol. 11023, pp. 326–345. Springer, Cham (2018). https://doi.org/10.1007/978-3-319-98938-9_19
9. Schwoon, S.: Model-checking pushdown systems. Dissertation, Technische Universität München, München (2002)
10. Song, F., Touili, T.: Pushdown model checking for malware detection. In: Flanagan, C., König, B. (eds.) TACAS 2012. LNCS, vol. 7214, pp. 110–125. Springer, Heidelberg (2012). https://doi.org/10.1007/978-3-642-28756-5_9
11. Song, F., Touili, T.: Efficient malware detection using model-checking. In: Giannakopoulou, D., Méry, D. (eds.) FM 2012. LNCS, vol. 7436, pp. 418–433. Springer, Heidelberg (2012). https://doi.org/10.1007/978-3-642-32759-9_34
12. Song, F., Touili, T.: LTL model-checking for malware detection. In: Piterman, N., Smolka, S.A. (eds.) TACAS 2013. LNCS, vol. 7795, pp. 416–431. Springer, Heidelberg (2013). https://doi.org/10.1007/978-3-642-36742-7_29
13. Song, F., Touili, T.: PoMMaDe: pushdown model-checking for malware detection. In: SIGSOFT 2013 (2013)

DABSTERS: A Privacy Preserving e-Voting Protocol for Permissioned Blockchain

Marwa Chaieb[1](✉), Mirko Koscina[2], Souheib Yousfi[3], Pascal Lafourcade[4], and Riadh Robbana[3]

[1] LIPSIC, Faculty of Sciences of Tunis, University Tunis El-Manar, Tunis, Tunisia
chaiebmarwa.insat@gmail.com
[2] Département d'Informatique, École normale supérieure, Paris, France
[3] LIPSIC, INSAT, University of Carthage, Tunis, Tunisia
[4] LIMOS, University Clermont Auvergne, CNRS UMR6158, Aubiére, France

Abstract. With the immutability property and decentralized architecture, Blockchain technology is considered as a revolution for several topics. For electronic voting, it can be used to ensure voter privacy, the integrity of votes, and the verifiability of vote results. More precisely permissioned Blockchains could be the solution for many of the e-voting issues. In this paper, we start by evaluating some of the existing Blockchain-based e-voting systems and analyze their drawbacks. We then propose a fully-decentralized e-voting system based on permissioned Blockchain. Called DABSTERS, our protocol uses a blinded signature consensus algorithm to preserve voters privacy. This ensures several security properties and aims at achieving a balance between voter privacy and election transparency. Furthermore, we formally prove the security of our protocol by using the automated verification tool, ProVerif, with the Applied Pi-Calculus modeling language.

Keywords: Permissioned Blockchain · Electronic voting · Blind signature · Formal verification · Applied Pi-Calculus · ProVerif

1 Introduction

Voting is the cornerstone of a democratic country. The list of security properties that must respect a secure voting protocol includes the following features. **Eligibility:** only registered voters can vote and only one vote per voter is counted. If the voter is allowed to vote more than once, the most recent ballot will be tallied and all others must be discarded. **Individual verifiability:** the voter him/herself must be able to verify that his/her ballot was cast as intended and counted as cast. **Universal verifiability:** after the tallying process, the results are published and must be verifiable by everybody. **Vote-privacy:** the connection between a voter and his/her vote must not be reconstructable without his/her help. **Receipt-freeness:** a voter cannot prove to a potential coercer

R. M. Hierons and M. Mosbah (Eds.): ICTAC 2019, LNCS 11884, pp. 292–312, 2019.
https://doi.org/10.1007/978-3-030-32505-3_17

that he/she voted in a particular way. **Coercion resistance:** even when a voter interacts with a coercer during the voting process, the coercer will be not sure whether the voter obeyed their demand or not. **Integrity:** ballots are not altered or deleted during any step of the election. **Fairness:** no partial results are published before tallying has ended, otherwise voters may be influenced by these results and vote differently. **Robustness:** the system should be able to tolerate some faulty votes. **Vote-and-go:** a voter does not need to wait for the end of the voting phase or trigger the tallying phase. **Voting policy:** specify if a voter has the right to vote more than once or he/she has not the right to change his/her opinion once he/she casted a vote.

Traditionally, during an election, the voter goes to a polling station and makes his/her choice in an anonymous manner, without any external influence. To perform the tally, we need to trust a central authority. From this comes the risk of electoral fraud. The tallying authority has the possibility to falsify votes and thus to elect a candidate who should not be elected. It is also possible for the registration authority to allow ineligible voters to vote. Hence, voting becomes useless and we notice a decrease in voter turnout in elections. Decentralized systems can be a good alternative to traditional voting since we need a secure, verifiable and privacy preserving e-voting systems for our elections. Blockchain is a distributed ledger that operates without the need to a trusted party. Expanding e-voting into Blockchain technology could be the solution to alleviate the present issues in voting.

Due to the proliferation of Blockchain implementations, the European Blockchain Observatory and Forum has published a technical report [14] where it recommends the use of private or permissioned Blockchains for sensitive data storage, which is the architecture implemented in an e-voting system. In this Blockchain architecture, the user credentials are generated by a Certificate Authority (CA). Hence, the users must be enrolled into the system through the CA before joining the network. This model is suitable for an e-voting system because the user management can rely on the Blockchain platform, due to their formal enrolling process. The advantage of having a minimum level of trust through our knowing the participants is that we can achieve security for the Blockchain replication process by using Byzantine Agreement as a consensus mechanism. Although permissioned Blockchains have several features suitable for services that involve sensitive data, such as user personal information, they have drawbacks related to transactions and user linkability. This is due to the fact that each user credential, public key pair and certificates, are issued for specific users that were previously enrolled in the CA. In order to overcome this drawback, we use, in this paper, the Okamoto-Schnorr blind signature scheme to sign the transactions without linking the user to it. This model allows validating transactions without exposing the user's identification, and therefore maintaining the privacy of the votes.

Related Work: In the last few decades, a considerable number of Blockchain-based e-voting protocols have been proposed to address the security issues of traditional voting protocols. Due to the limitation on the number of pages, we give a brief overview of some of these systems and evaluate their security in Table 1, in which we use the following abbreviations[1].

- *Open Vote Network (OVN)* [15]: It is a self-tallying, boardroom scale e-voting protocol implemented as a smart contract in Ethereum. This protocol guarantees voter's privacy and removes the need to trust the tallying authorities whether to ensure the anonymity of voters or to guarantee the verifiability of elections. However, it suffers from several security issues. For example, it supports only elections with two options (yes or no) and with a maximum of 50 voters due to the mathematical tools that they used and to the gas limit for blocks imposed by Ethereum. Additionally, this protocol does not provide any mechanism to ensure coercion resistance and needs to trust the election administrator to ensure voter's eligibility. Open Vote Network is not resistant to the misbehavior of a dishonest miner who can invalidate the election by modifying voters' transactions before storing them on blocks. Dishonest voter can also invalidate the election by sending an invalid vote or by abstaining during the voting phase.
- *E-Voting with Blockchain: An E-Voting Protocol with Decentralization and Voter Privacy (EVPDVP)* [10]: Implemented on a private network that uses the Ethereum Blockchain API, this protocol uses the blind signature to ensure voters privacy. It needs a central authority (CA) as a trusted party to ensure voters eligibility and allow voters to change and update their votes. To ensure fairness, voters include in their ballots a digital commitment of their choices instead of the real identity of the chosen candidate. To tally ballots, voters must broadcast to the network a ballot opening message during the counting phase.
- *Verify-Your-Vote: A Verifiable Blockchain-based Online Voting Protocol (VYV)* [7]: It is an online e-voting protocol that uses Ethereum Blockchain as a bulletin board. It is based on a variety of cryptographic primitives, namely Elliptic Curve Cryptography [9], pairings [4,19] and Identity Based Encryption [5]. The combination of security properties in this protocol has numerous advantages. It ensures voter's privacy because the Blockchain is characterized by the anonymity of its transactions. It also ensures fairness, individual and universal verifiability because the ballot structure includes counter-values, which serve as receipts for voters, and homomorphism of pairings. However, the registration phase of this protocol is centralized. A unique authority, which is the registration agent, is responsible for verifying the eligibility of voters and registering them. A second problem is inherent in the use of Ethereum because each transaction sent by the protocol entities in the Blockchain passes through miners who validate it, put it in the current block and execute the consensus algorithm. Any dishonest miner in the election Blockchain can modify transactions before storing them on blocks. Additionally, this protocol is not coercion resistent.
- *TIVI* [21]: It is a commercial online voting solution based on biometric authentication, designed by the company Smartmatic. It checks the elec-

[1] TCA: Trusted Central Authority; SV: Single Vote; MV: Multiple Votes.

tor's identity via a selfie using facial recognition technology. TIVI ensures the secrecy of votes so long as the encryption remains uncompromised. It provides also voters' privacy thanks to its mixing phase and offers the possibility to follow votes by the mean of a QR code stored during voting phase and checked later via a smartphone application. However, this system does not provide any mechanism to protect voters from coercion or to ensure receipt-freeness. Additionally, TIVI uses the Ethereum Blockchain as a ballot box so it is not resistant to misbehaving miners that could invalidate the election by modifying votes before storing them on the election Blockchain.

- *Follow My Vote (FMV)* [8]: It is a commercial online voting protocol that uses the Ethereum Blockchain as a ballot box. A trusted authority authenticates eligible voters and provides them with pass-phrases needed in case of changing their votes in the future. Voters can watch the election progress in real time as votes are cast. It includes an authentication phase which ensures voters' eligibility. It allows voters to locate their votes, and check that they are both present and correct using their voters' IDs. Nevertheless, this voting system requires a trusted authority to ensure votes confidentiality and hide the correspondence between the voters' real identities and their voting keys. If this authority is corrupted, votes are no longer anonymous. Votes secrecy is not verified by this system because votes are cast without being encrypted. Moreover, the ability to change votes, coupled with the ability to observe the election in real time compromise fairness property. This system is not coercion resistance and is not universally verifiable because we have no way to verify that the votes present in the election final result are cast by eligible voters.
- *BitCongress* [11]: A commercial online voting platform based on a combination of three networks which are: Bitcoin, Counterparty (a decentralized asset creation system and decentralized asset exchange) and a Smart Contract Blockchain. It aims at preventing double voting by using the time stamp system of the Bitcoin Blockchain. This platform does not ensure voters eligibility because it allows any Bitcoin address to register for the election. It performs the tally using, by default, a modified version of Borda count and a Quota Borda system for large scale elections. It ensures individual and universal verifiability but it is not coercion resistent.
- *Platform-independent Secure Blockchain-based Voting System (PSBVS)* [22]: Implemented in the Hyperledger Fabric Blockchain [2], this protocol uses Paillier cryptosystem [18] to encrypt votes before being cast, proof of knowledge to ensure the correctness and consistence of votes, and Short Linkable Ring Signature (SLRS) [3] to guarantee voters privacy. In the other hand, this protocol does not include a registration phase in which we verify, physically or by using biometric techniques, the eligibility of the voter. A voter can register him/herself by simply providing his/her e-mail address, identity number or an invitation URL with a desired password. These mechanisms are not sufficient to verify the eligibility of a voter and information like e-mail address or identity number can be known by people other than the voter him/herself. Also, with reference to the definition of coercion resistance given by

Juels et al. [12], this protocol is not coercion resistant. In fact, if a voter gives his/her secret key to a coercer, the coercer can vote in the place of the voter who cannot modify this vote later. We mention here that the coerced voter cannot provide a fake secret key to the coercer because a vote with a fake secret key is rejected by the smart contract.

Table 1. Security evaluation of OVN, EVPDVP, VYV, TIVI, FMV, BitCongress, PSBVS and DABSTERS.

	OVN	EVPDVP	VYV	TIVI	FMV	BitCongress	PSBVS	DABSTERS
Eligibility	TCA	TCA	TCA	✓	TCA	X	X	✓
Individual verif	✓	✓	✓	✓	✓	✓	✓	✓
Universal verif	✓	✓	✓	X	X	✓	✓	✓
Vote-Privacy	✓	✓	✓	✓	TCA	✓	✓	✓
Receipt-freeness	X	✓	✓	X	X	X	✓	✓
Coercion resistance	X	X	X	X	X	X	X	X
Fairness	X	✓	✓	✓	X	X	✓	✓
Integrity	✓	✓	X	✓	X	✓	✓	✓
Robustness	X	✓	✓	✓	✓	✓	✓	✓
Vote-and-go	X	X	✓	✓	✓	✓	✓	✓
Voting policy	SV	MV	MV	SV	MV	MV	SV	MV

Contributions: In this paper, we aim at designing a secure online e-voting protocol that addresses the security issues mentioned in the related work section by using the Blockchain technology and a variety of cryptographic primitives. Called DABSTER, our protocol uses a new architecture based on permissioned Blockchain and blind signature. It satisfies the following security properties: eligibility, individual verifiability, universal verifiability, vote-privacy, receipt-freeness, fairness, integrity and robustness. Our contributions can be summarized as follows:

- A new architecture of trust for electronic voting systems. This architecture is based on permissioned Blockchain and on a blind consensus which provides voter's privacy and vote's integrity.
- A secure and fully distributed electronic voting protocol based on our propounded architecture.
- A detailed security evaluation of the protocol and a formal security proof using the Applied Pi-Calculus modeling language and the automated verification tool ProVerif.

Outline: In the next section, we give an overview of the Byzantine Fault Tolerante (BFT) with blind signature consensus algorithms. Then in Sect. 3, we describe our proposed e-voting protocol, DABSTERS, and give its different stakeholders and phases as well as the structure of each voter's ballot. In Sect. 4, we evaluate the security of our protocol using Proverif when it is possible. The conclusion is a summary of DABSTERS and a proposal for ongoing evaluation of its performance.

2 Background

We give a definition of the Okamoto-Schnorr blind signature, before using it in a Byzantine based consensus.

2.1 Blind Signature

Let p and q be two large primes with $q|p-1$. Let G be a cyclic group of prime order q, and g and h be generators of G. Let $H : \{0,1\}^* \rightarrow \mathbb{Z}_q$ be a cryptographic hash function.

Key Generation: Let $(r,s) \xleftarrow{r} \mathbb{Z}_q$ and $y = g^r h^s$ be the A's private and public key, respectively.

Blind signature protocol:

1. A chooses $(t,u) \xleftarrow{r} \mathbb{Z}_q$, computes $a = g^t h^u$, and sends a to the user.
2. The user chooses $(\beta, \gamma, \delta) \xleftarrow{r} \mathbb{Z}_q$ and computes the blind version of a as $\alpha = ag^{-\beta}h^{-\gamma}y^{\delta}$, and $\varepsilon = H(M, \alpha)$. Then calculates $e = \varepsilon - \delta \bmod q$, and sends e to the A.
3. A computes $S = u - es \bmod q$ and $R = t - er \bmod q$, sends (S, R) to the user.
4. The user calculates $\rho = R - \beta \bmod q$ and $\sigma = S - \gamma \bmod q$.

Verification: Given a message $M \in \{0,1\}^*$ and a signature $(\rho, \sigma, \varepsilon)$, we have $\alpha = g^{\rho} h^{\sigma} y^{\varepsilon} \bmod p$.

The Okamoto-Schnorr blind signature scheme is suitable with a private Blockchain architecture due to the blinding process that can be performed by the same authority responsible of the enrollment process (see Fig. 1, where the authority A blindly signs a message for the user). We use *BlindSign(M,*$(\beta, \sigma, \gamma), y$*)* and *VerifyBlindSign(M,* $(\rho, \delta, \varepsilon), y$*)* to blind sign and to verify the blinded signature, respectively using Okamoto-Schnorr, where M corresponds to the message to be signed, (β, σ, γ) to the secret values randomly chosen, $(\rho, \delta, \varepsilon)$ to the blinded signature; and y to the RA's public key. The result obtained from the function *BlindSign* corresponds to the blinded signature $(\rho, \sigma, \varepsilon)$. On the other hand, the function *VerifyBlindSign* returns a response *valid* or *invalid*.

2.2 BFT Based Consensus Algorithm

Now, considering a permissioned Byzantine Fault Tolerance (BFT) based consensus protocol like the one introduced in Hyperledger Fabric [2]. In this protocol, the digital signature is used as a user authentication method without protecting the user privacy. Hence, for a privacy preserving consensus protocol, we need to add the following properties to the BFT based consensus algorithm:

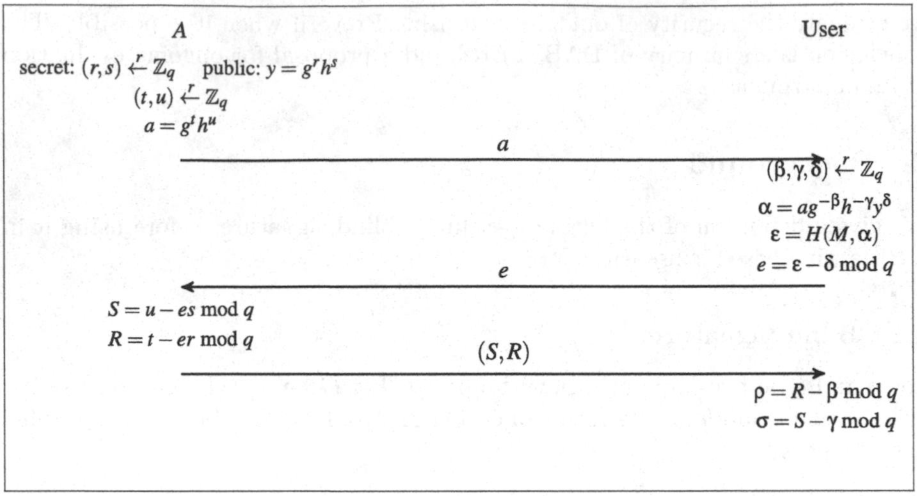

Fig. 1. Okamoto-Schnorr blind signature diagram, where $y \xleftarrow{r} \mathbb{Z}_q$ means that y is randomly chosen in $\mathbb{Z}_q$.

- Alice sends a newly signed transaction to the registration authority (RA) which is responsible for the enrollment of Alice.
- Alice's signature is validated only by the RA.
- The RA anonymises Alice's identity.
- The RA signs the transaction sent by Alice to the network.
- All the node of the transactions validation process can validate the RA's signature.
- The RA signature cannot be duplicated.

Now, to keep the privacy of the client and peers involved in the transactional process, we need to hide his ID and make his signature blind. However, we do not address the ID hiding process with any particular mechanism. Therefore, we consider that the ID is replaced by a value corresponding to the anonymised user ID, and this process can be performed by using different schemes. As presented in [13], to address the issue related to the digital signature, we replace the signing mechanism used in the original protocol by the Okamoto-Schnorr blind signature scheme [16]. In order to maintain the consistency and liveness that the protocol has, we keep the transactional flow. However, the steps are modified in order to accept the new blind signature scheme to authenticate the clients and peers.

The transactional process based on our BFT consensus algorithm with Blind Signature consists of the following steps:

1. **Initiating Transactions:** The client c_{bc} generates a message M to execute an operation o_{bc} in the network with a blinded signature by using *Blind-Sign*$((M,\beta,\sigma,\gamma),y)$.
2. **Transaction Proposal:** The submitting peer sp_{bc} receives the message M coming from the client c_{bc}, validates the client blinded signature by using *Ver-*

ifyBlindSign(M,$(\rho, \delta, \varepsilon), y$) and proposes a transaction with the client instruction o_{bc}.

3. **Transaction Endorsement:** The endorser peers ep_{bc} validate the client blinded signature using *VerifyBlindSign*(M,$(\rho, \delta, \varepsilon), y$) and verify if the transaction is valid by simulating the operation o_{bc} using his local version of the Blockchain. Then, the endorser peers generate signed transactions with the result of the validation process and send it to the submitting peer sp_{bc}.
4. **Broadcasting to Consensus:** The submitting peer sp_{bc} collects the endorsement coming from the endorsing peers connected to the network. Once sp_{bc} collects enough valid answers from the endorsing peers, it broadcasts the transaction proposal with the endorsements to the ordering service.
5. **Commitment:** All the transactions are ordered within a block, and are validated with their respective endorsement. Then, the new block is spread through the network to be committed by the peers.

3 Description of DABSTERS

Our protocol is implemented over a new architecture of trust. It is based on a BFT-based consensus protocol [2] and on a blinded signature consensus protocol, called *BlindCons* [13], presented in Sect. 2. It eliminates the risk of invalidating the election because of dishonest miners who modify the transactions before storing them on blocks. We also propose a distributed enrollment phase to reduce the need to trust election agents and impose the publication of the list of eligible registered voters at the end of the enrollment phase. This list is auditable and verifiable by all parties.

Our scheme unfolds in 5 stages. It starts with an enrollment phase in which registration authorities (RAs) verify the eligibility of voters by verifying the existence of their names and their identity card numbers in a list published beforehand and containing the names of all persons who have the right to vote. Then, all eligible voters are registered and provided with credentials. The enrollment phase is offline. At the end of this phase, RAs construct a list containing the names of all registered eligible voters and their ID card numbers. This list can be rejected or published on the election Blockchain during the validation phase. Once the list is validated, we move to the third stage which is voting phase. Each eligible voter (V_i) initiates a transaction in which he/she writes his/her encrypted vote, signs the transaction using his/her credential and sends it to the RAs to check his/her signature and blind it. Then, the voter sends the transaction with the blinded signature and his/her anonymous ID (his/her credential) to the consensus peers to be validated and stored in the election Blockchain anonymously. After validating and storing all votes in our Blockchain, tallying authorities (TAs) read these encrypted votes from the network, decrypt them, and proceed to the tally. The final stage is the verification phase. During this phase, voters make sure that their votes have been considered correctly and check the accuracy of the tally. The individual verifiability is ensured due to the structure of our ballots and the universal verification is ensured thanks to the homomorphism property of pairings. Except the enrollment phase, all the phases of our protocol are on-chain. Therefore, we call the BFT based consensus protocol with each transaction initiated by authorities and the BlindCons with each transaction initiated by eligible voters because we do not need to hide the identity of our authorities but we need to ensure voter's privacy. In the following, we give a detailed description of the role of our protocol stakeholders, the structure of our ballot, the different protocol phases and the two consensus.

3.1 Protocol Stakeholders

DABSTERS involves three main entities:

- *Registration authorities (RAs):* they verify the eligibility of every person wishing to register to the election and provide eligible voters by their credentials which are constructed by cooperation between all RAs.
- *Eligible voters (V):* every eligible voter (V_i) has the right to vote more than once before the end of the voting phase and only his/her last vote is counted. Voters have the possibility to verify that their votes are cast as intended and counted as cast during the verification phase. Also, they can check the accuracy of the election final result but they are not obliged to participate in the verification phase (they can vote and go).
- *Tallying authorities (TAs):* the protocol includes as many tallying authorities as candidates. Before the voting phase, they construct n ballots, where n is the number of registered voters. Thus, every voter has a unique ballot which is different from the other ballots. TAs encrypt ballots and send them to voters during the voting phase. They decrypt votes and calculate the election final result during the tallying phase and publish the different values that allow voters to check the accuracy of the count during the verification phase.

DABSTERS also involves observers and election organizers who have the right to host the Blockchain peers to ensure the correctness of the execution of the protocol.

3.2 Ballot Structure:

As illustrated in Fig. 2, each ballot is composed of a unique bulletin number BN calculated as follows: $BN = \{g, D\}_{PK_A}$, where g is a generator of an additive cyclic group G, D is a random number and PK_A is the administrator's public key. It contains also a set of candidates' names $name_j$ and candidates' pseudo IDs, denoted C_j, which are the positions of candidates in the ballot, calculated from an initial order and an offset value. In addition, each ballot includes a set of

Ballot number ***BN***			
Pseudo "ID C_j"	**Candidate's** "name $name_j$"	**Choice**	**Counter-value** "$CV_{BN,name_j,k}$"
0	Paul	☐	$CV_{BN,name_0,0}$
1	Nico	☐	$CV_{BN,name_1,1}$
2	Joel	☐	$CV_{BN,name_2,2}$

Fig. 2. Ballot structure [17]

counter-values $CV_{BN,name_j,k}$ that are receipts for each voter. They are calculated using the following formula: $CV_{BN,name_j,k} = e(Q_{namej}, S_k \cdot Q_{BN})$; where $e(.,.)$ is the pairing function, S_k is the secret key of the tallying authority TA_k, $Q_{namej} = H_1(name_j)$ and $Q_{BN} = H_1(BN)$ are two points of the elliptic curve E.

3.3 Protocol Stages

Our solution includes the following phases:

Enrollment Phase: Every person who has the right to vote and desires to do so, physically goes to the nearest registration station. He/she provides his/her national identity card to the RAs, who verify his/her eligibility by checking if his/her name and ID card number exists in a list, previously published, contains all persons that are able to participate in the election. If he/she is an eligible voter, the RAs save the number of his/her ID card and provide him with a credential that allows him to participate in the voting process. Voters' credentials are calculated using elliptic curve cryptography and have this form:

$Credential_{V_i} = S_M \cdot H_1(ID_{V_i})$ where:

- $S_M = S_1 \cdot S_2 \ldots, S_R$ is a secret master key calculated by cooperation between all RAs. Each registration authority participates with its secret fragment S_r; $r \in \{1 \ldots R\}$,
- H_1 is an hash function defined as follows: $H_1 : \{0,1\}^* \rightarrow G_1$; G_1 an additive cyclic group of order prime number q,
- ID_{V_i} is the number of the ID card of the voter V_i.

Validation Phase: After registering all eligible voters, RAs create a list containing the names and the identity card numbers of all registered voters. This list should be viewable and verifiable by voters, election organizers and observers. Thus, RAs send this list in a transaction on our election Blockchain. This transaction passes through the five steps of the BFT based consensus protocol to be validated if the list is correct or rejected if the list contains names of ineligible voters.

- **Step1: Transaction initiation.** RAs generate the list of eligible voters to be validated by the network. The list is sent to a submitter peer. In the case of an offline or misbehaving submitter peer, RAs send the transaction to the next submitter peer.
  This step is illustrated by Fig. 3.
  - ID_{RA} is the ID of the registration authorities,
  - $Write(List)$ is the operation invoked by the RAs to be executed by the network. It consists of writing the list of eligible voters and their ID card numbers in the Blockchain,
  - *List* is the payload of the submitted transaction, which is the list of registered voters to be published on the Blockchain,

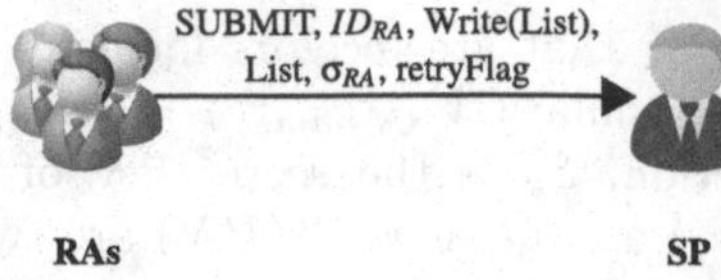

Fig. 3. Step1: Transaction initiation.

  - σ_{RA} is the signature of the registration authorities,
  - *retryFlag* is a boolean variable to identify whether to retry the submission of the transaction in case of the transaction fails.
- **Step2: Transaction proposal.** The submitter peer receives the transaction and verifies the RAs signature. Then prepares a transaction proposal to be sent to the endorsing peers. Endorsing peers are composed of some voters, election organizers and observers who desire to host the Blockchain peers. This step is described in Fig. 4.

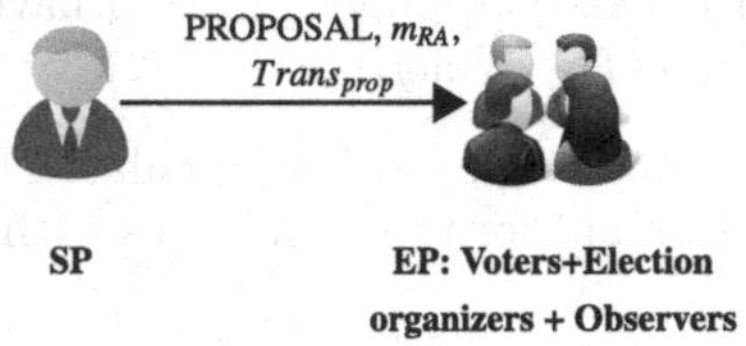

Fig. 4. Step2: Transaction proposal.

  - $m_{RA} = (ID_{RA}, Write(List), List, \sigma_{RA})$
  - $Trans_{prop} = (SP, Write(List), List, StateUpdate, VerDep)$:
    * *StateUpdate* corresponds to the state machine after simulate locally the operation coming in *Write(List)*.
    * *VerDep* is the version dependency associated to the variable to be created or modified. It is used to keep the consistency of the variables across the different machine state version.
- **Step3: Transaction endorsement.** Each endorser peer verifies the signature of the registration authorities σ_{RA} coming in m_{RA} and checks that the list of eligible voters in m_{RA} and $Trans_{prop}$ is the same. Then, each endorser verifies the eligibility of all names and ID card numbers included in the list. If they are all valid, the endorser peer generates *a transaction valid message* to be sent to the submitter peer (Fig. 5). But if the list includes names of ineligible voters, the endorser peer generates *a transaction invalid message* (Fig. 6).
  - Tx_{ID} is the transaction ID,
  - σ_{EP} is the signature of the endorser peer.
  - *Error*: can has the following values:

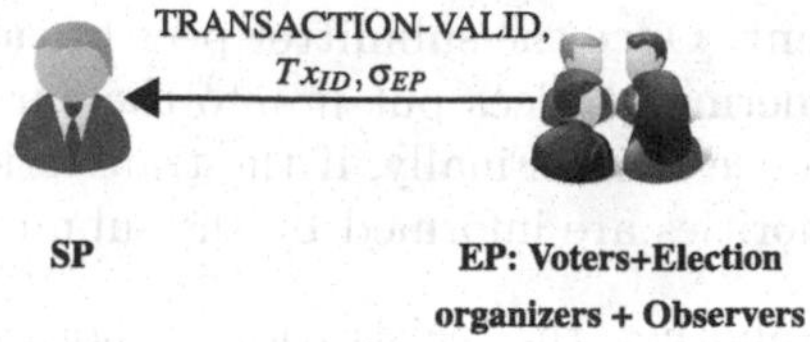

Fig. 5. Step3: Transaction endorsement: valid transaction.

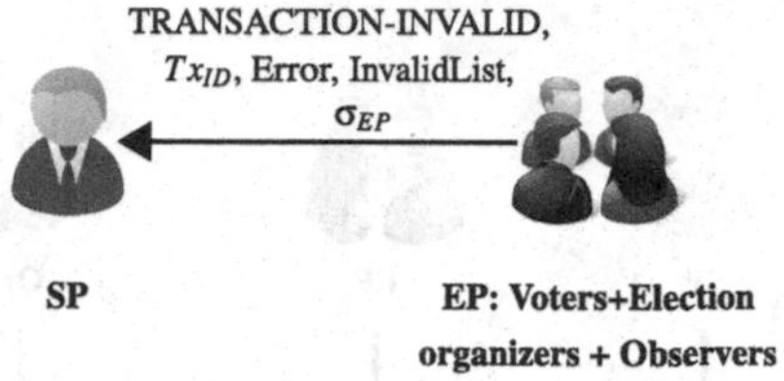

Fig. 6. Step3: Transaction endorsement: invalid transaction.

  * INCORRECT-STATE: when the endorser tries to validate the transaction with a different local version of the Blockchain than the one coming in the transaction proposal.
  * INCORRECT-VERSION: when the version of the variable where the list will be recorded differs from the one referred in the transaction proposal.
  * REJECTED: for any other reason.
- InvalidList: is the list of ineligible names that were included in the list sent by the RAs.

– **Step4: Broadcasting to Consensus.** The submitter peer waits for the response from the endorser peers. When it receives enough *Transaction Valid messages* adequately signed, the peer stores the endorsing signatures into packaged called endorsement. Once the transaction is considered endorsed, the peer invokes the consensus services by using *broadcast(blob)*, where *blob* = $(Trans_{prop}, endorsement)$ (Fig. 7).
The number of responses and endorsement to consider the transaction proposal as endorsed is equal to 50% + 1 of the total number of endorser peers. If the transaction has failed to collect enough endorsements, it abandons this transaction and notifies the RAs.

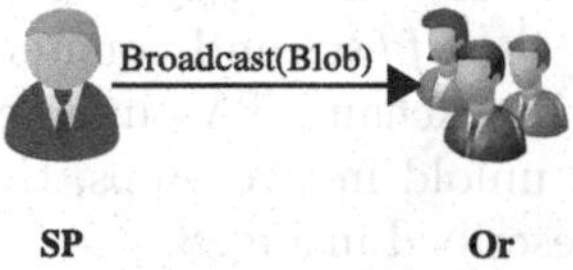

Fig. 7. Step4: Broadcasting to consensus

- **Step5: Commitment.** Once the submitter peer broadcasts the transaction to consensus, the ordering services put it into the current block, which will be sent to all peers once built. Finally, if the transaction was not validated, the registration authorities are informed by the submitter peer SP.

In the case of an invalid list, the registration authorities have to correct the list and restart the validation phase. We move to the next phase (which is the voting phase) only when we obtain a valid list of registered voters.

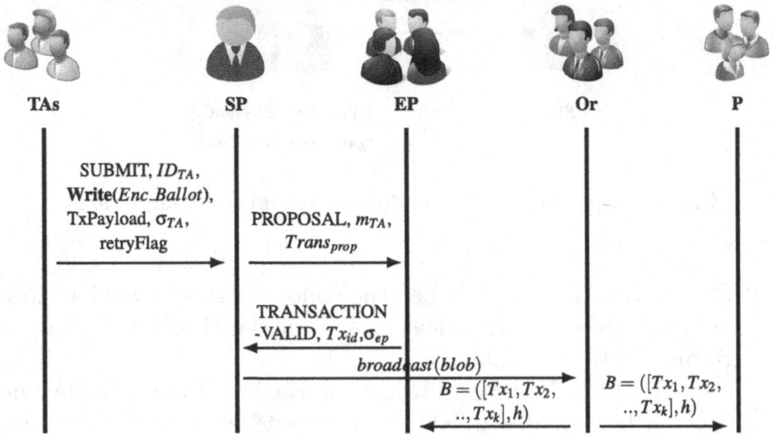

Fig. 8. Interaction between TAs and peers.

Voting Phase: Two entities participate during this phase:

- The tallying authorities who have constructed ballots before the beginning of the voting phase. To construct a ballot, TAs calculate, locally, the unique ballot number $BN = \{g, D\}_{PK_{TA}}$, the offset value $offset = H(g) \ mod \ m$ and the counter-values $CV_{BN,name_j,k} = e(Q_{namej}, S_k \cdot Q_{BN})$, where g is a generator of G an additive cyclic group of order a prime number, D is a random number, PK_{TA} is TAs' public key, m is the number of candidates, $e(.,.)$ is the pairing function, S_k is the secret key of the tallying authority TA_k, $Q_{namej} = H_1(name_j)$ and $Q_{BN} = H_1(BN)$ are two points of the elliptic curve E. Then, TAs choose, randomly, a blank ballot for each voter, encrypt it with the voter's public key and transmit it to the corresponding voter via the Blockchain. Ballots are sent encrypted because they contain secret information like the BN, the $offset$ and counter-values. To send encrypted ballots to voters via the Blockchain, TAs interact with the BFT consensus peers. These interactions unfold in five steps, the same steps as those presented in Sect. 3.3, and described in Fig. 8.
  1. **Transaction initiation.** TAs initiate a transaction and send it to a submitter peer SP. The transaction contains their ID (ID_{TA}), the list of encrypted ballots, the transaction payload, their signature (σ_{TA}) and the value of the variable retryFlag.

2. **Transaction proposal.** SP verifies the TAs signature and prepares a transaction proposal $Trans_{prop} = (SP, Write(Enc\_Ballot), Enc\_Ballot, stateUpdate, VerDep)$ to be sent to the endorsing peer with the TAs message $m_{TA} = (ID_{TA}, Write(Enc\_Ballot), Enc\_Ballot, \sigma_{TA})$.
3. **Transaction endorsement.** EP verifies the σ_{TA} coming in m_{TA}, simulates the transaction proposal and validates that the *stateUpdate* and *verDep* are correct. If the validation process is successful, the endorser peer generates a transaction valid message to be sent to the submitter peer.
4. **Broadcasting to consensus.** When the SP receives a number of *Transaction Valid message* equals to 50% + 1 of the total number of endorser peers, adequately signed, he stores the endorsing signatures into an endorsement package and invokes the consensus services by using $broadcast(blob)$; where $blob = (Trans_{prop}, endorsement)$.
5. **Commitment.** Ordering services (Or) add the transaction to a block. Once they collect enough endorsed transactions, they broadcast the new block to all other peers. A block has the following form: $B = ([tx_1, tx_2, \ldots, tx_k]; h)$ where h corresponds to the hash value of the previous block.

- Every eligible voter retrieves his/her ballot, decrypts it using his/her secret key and encrypts his/her vote by voting then sends it in a transaction through the Blockchain. To encrypt his/her vote, the voter uses the Identity Based Encryption [5] and encrypts his/her ballot number BN with $Q_{C_j} = H_1(C_j)$ where C_j is the pseudo ID of the chosen candidate. Thus, each encrypted vote has the following form: $Enc\_Vote = \{BN\}_{Q_{C_j}}$.

  To be read from the Blockchain or be written on it, voters' transactions pass through the blinded signature consensus. We model in Fig. 9 the steps through which a transaction of an eligible voter passes. We take the example of a transaction containing an encrypted vote. During the interactions between TAs and peers, we use the digital signature as user authentication method without protecting the TAs privacy because we do not need to hide the identity of our protocol authorities. However, when it comes to interactions between voters and peers, we need to preserve voters' privacy by blinding their signatures. The privacy preserving consensus adds two steps:

  (i) The signature of each eligible voter is blinded automatically after the vote is cast by the function $BlindSign(M, (\beta, \gamma, \delta), PK_{RA})$, where $M = (Credential_{V_i} || Write(Enc\_Vote) || Enc\_Vote || retryFlag)$ is the message to be signed, (β, γ, δ) are secret values randomly chosen by the voter and PK_{RA} is the public key of the RAs.

  (ii) RAs blind the signature of each eligible voter by providing him the tuple (R, S), allowing the voter to construct his/her blinded signature $(\rho, \sigma, \varepsilon)$ to be used during his/her interactions with the peers.

The other steps are the same as the BFT based consensus, but instead of sending their signatures, the voters send their blinded signatures provided by the RAs.

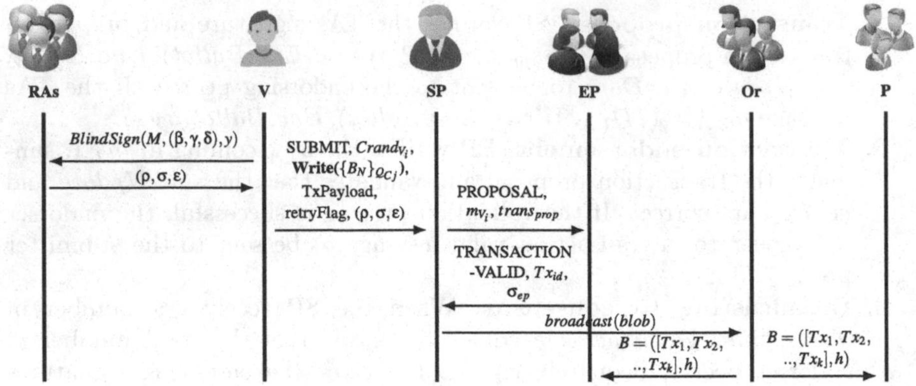

Fig. 9. Interactions between eligible voter and BlindCons peers.

1. **Initiating transaction:** $<$ SUBMIT, $Credential_{V_i}$, $Write(Enc\_Vote)$, Enc_vote, retryFlag, $(\rho, \sigma, \varepsilon) >$
2. **Transaction Proposal:** $<$ PROPOSAL, $m_{V_i}, trans_{prop} >$
3. **Transaction Endorsement:** $<$ TRANSACTION-VALID, $Tx_{id}, \sigma_{ep} >$
4. **Broadcasting to consensus:** broadcast(blob)
5. **Commitment:** $B = ([Tx_1, Tx_2, \ldots, Tx_k], h)$

The voters who intend to verify that their votes were properly counted must memorize the counter-values that correspond to their chosen candidates.

Tallying Phase: After all votes have been cast, TAs proceed to the tally. We have as many TAs as candidates. Each tallying authority TA_k is responsible for counting the number of votes for a specific pseudo ID C_j: for example the first tallying authority TA_1 decrypts, with its secret key $S_1 \cdot Q_{C_1}$, all bulletins that were encrypted with the public key Q_{C_1} (certainly these ballots contain votes for candidates with $C_j = 0$). TA_k starts by initiating a transaction to read encrypted votes from the Blockchain. This transaction passes through the five steps of the BFT based consensus. Then, it decrypts the votes with its secret key S_k that were encrypted with Q_{C_j} in order to reveal the bulletin number BN. Then, it reconstructs the ballot, identifies the chosen candidate, and added to the corresponding counter. At the end of this phase, TA_k publishes the count for each candidate using the following formula: $\sigma_{k,name_j} = \sum_{i=1}^{l_j} S_k \cdot Q_{BN_i}$; Where l_j is the number of votes received by the candidate j, S_k is the private key of the tallying authority k, $Q_{BN_i} = H_1(BN_i)$ and BN_i is the ballot number of the vote i that corresponds to the candidate with name $name_j$.

Verification Phase: This phase allows voters to check that their votes were counted as cast and that the election final result corresponds to the sum of all eligible votes. It includes two sub-phases. During the first one, TAs calculate the list of chosen counter-values from each ballot number and the name of the chosen candidate, and publish this list on the Blockchain. Each eligible voter can read this list and verify the existence of his/her counter-value to be sure that his/her vote was counted correctly. The second sub-phase uses the homomorphism of pairings to check the accuracy of the tally. Using the published counts and the reconstructed counter-values, we can verify that the announced result corresponds to the sum of all eligible votes, as follows:

$$\begin{aligned}\prod_{i=1}^{l} CV_{BN_i} &= \prod_{k=1}^{m}\prod_{j=1}^{m}\prod_{i=1}^{l_j} CV_{BN_{i,name_j},k} = \prod_{k=1}^{m}\prod_{j=1}^{m}\prod_{i=1}^{l_j} e(Q_{name_j}, S_k \cdot Q_{BN_i}) \\ &= \prod_{k=1}^{m}\prod_{j=1}^{m} e(Q_{name_j}, \sum_{i=1}^{l_j} S_k \cdot Q_{BN_i}) = \prod_{k=1}^{m}\prod_{j=1}^{m} e(Q_{name_j}, \sigma_{k,name_j}) \qquad (1)\end{aligned}$$

where $l = \sum_{j=1}^{m} l_j$ is the total number of votes. These equalities use the bilinear property of pairing: $\prod_{i=1}^{l_j} e(Q_{name_j}, S_k \cdot Q_{BN_i}) = e(Q_{name_j}, \sum_{i=1}^{l_j} S_k \cdot Q_{BN_i})$.

4 Security Evaluation of DABSTER

Thanks to the use of the BFT based consensus, the BlindCons and a variety of cryptographic primitives, our protocol ensures several security properties. We discuss the security properties ensured by our protocol and prove, formally, that our solution guarantees vote secrecy, vote privacy, and voter's authentication.

4.1 Informal Security Evaluation

We evaluate our protocol according to a list of security properties that must respect a secure and practical voting system.

- **Eligible voter:** The registration and the validation phases of our protocol ensure that only eligible voters participate in the voting process. During the registration phase, RAs verify the identity of each voter via a face to face meeting and only eligible voters are provided with credentials. During the validation phase, RAs send the list of registered voters to the consensus peers, which are composed of voters, election organizers and observers, in order to verify the eligibility of all registered voters and validate or reject this list.
- **Individual verifiability:** This property is ensured by our protocol because our ballot structure includes counter-values. These values serve as receipts for voters and enable them to verify that their votes have been cast as intended without disclosing who they voted for. In fact, counter-values are calculated using the following formula: $CV_{BN,name_j,k} = e(Q_{name_j}, S_k \cdot Q_{BN})$. Thus, we cannot get the name of the candidate from the value of $CV_{BN,name_j,k}$.

- **Universal verifiability:** From the parameters published by the TAs during the verification phase, everyone can verify the accuracy of the final result by checking the Eq. (1).
- **Vote-Privacy:** This property is ensured thanks to the BlindCons. Before interacting with the consensus peers, RAs blind the signature of all eligible voters to hide their real identities. Voters' transactions are signed by the blind signature issued by the RAs and not with the voter's signature. Thus voters' identities are kept private and no one can link a vote to a voter.
- **Receipt-freeness:** In our case, a voter cannot find his/her vote from the counter-value $CV_{BN,name_j,k}$ and the other public parameters. He/she cannot therefore prove that he/she voted for a given candidate.
- **Coercion resistance:** Our protocol is not resistant to coercion. A coercer can force a voter to vote for a certain candidate and check his/her submission later using the counter-value.
- **Integrity:** The BFT based consensus and the blind signature algorithm prevent votes from being altered while keeping the voter's secrecy. Each transaction is stored in the Blockchain after being validated by 50%+1 of the endorsing peers. This eliminates the risk of modifying transactions before storing them. We mention here that the BFT consensus is based on the assumption that 2/3 of the endorsing peers are honest.
- **Fairness:** During the voting phase, each eligible voter encrypts his/her ballot number BN with $Q_{C_j} = H_1(C_j)$ where C_j is the pseudo-ID of the desired candidate. Ballot numbers are secret and candidates' pseudo-IDs do not reflect the real identities of candidates thanks to the offset value, so nobody can identify the chosen candidate from the encrypted vote. Thus, we cannot get partial results before the official count.
- **Robustness:** Our scheme is resistant to the misbehavior of dishonest voters which cannot invalidate the election by casting an invalid vote or by refusing to cast a vote.
- **Vote-and-go:** Our protocol does not need the voter to trigger the tallying phase, they can cast their votes and quit before the voting ends.
- **Voting policy:** DABSTERS gives the possibility to eligible voters to vote more than once and only their last votes are counted. It means that we have a maximum of one vote per voter in the final tally. In fact, every eligible voter has a unique valid credential which is sent with his/her vote in the transaction.

4.2 Formal Security Evaluation

ProVerif is a fully automated and efficient tool to verify security protocols. It is capable of proving reachability properties, correspondence assertions, and observational equivalence. To perform an automated security analysis using this verification tool, we model our protocol in the Applied Pi-Calculus [1] which is a language for modeling and analyzing security protocols. It is a variant of the Pi-Calculus extended with equational theory over terms and functions and provides an intuitive syntax for studying concurrency and process interaction.

The Applied Pi-Calculus allows to describe several security goals and to determine whether the protocol meets these goals or not. To describe our protocol with the Applied Pi calculus, we need to define a set of names, a set of variables and a signature that consists of the function symbols which will be used in order to define terms. These function symbols have arities and types. To represent the encryption, decryption, signature, blind signature and hash operations, we use the following function symbols: `pk(skey)`, `aenc(x,pk(skey))`, `adec(x,skey)`, `spk(sskey)`, `sign(x,sskey)`, `checksign(x,spk(sskey))`, `BlindSign(x,smkey)`, `checkBlindSign(x, spk(smkey))`, `H1(x)`. Intuitively, the `pk` function generates the corresponding public key of a given secret key, `aenc` and `adec` stand, respectively, for asymmetric encryption and asymmetric decryption, `aenc` and `adec` follow this equation: `adec(aenc(x,y),pk(y))=x`. The `spk` function generates the corresponding public key of a given signature secret key, `sign` and `checksign` provide, respectively, the signature of a given message and the verification of the signature. They respect the following equation: `checksign(sign(x,y),spk(y))=x`. `BlindSign` and `checkBlind- Sign` stand, respectively, for blind sign and check blinded signature, `BlindSign` and `checkBlindSign` follow the equation `checkBlindSign(BlindSign(x,y),spk(y))=x`. We also assume the hash operation which is denoted with the function `H1`.

Because of the limitation on the number of pages, we put all ProVerif codes online[2] and give only the queries, the results of executing these codes, and the time it takes ProVerif to prove the properties in Table 2 (Execution times are expressed in seconds).

Table 2. ProVerif results and execution times.

Property to evaluate	Description	Result	Exec time
Vote secrecy	To capture the value of a given vote, an attacker has to intercept the values of two parameters: the ballot number BN and the pseudo ID of the chosen candidate Cj	Proved	0.012 s
Voter's authentication	We use correspondence assertion to prove this property	Proved	0.010 s
Vote privacy	To express vote privacy we prove the observational equivalence between two instances of our process that differ only in the choice of candidates	Proved	0.024 s

4.3 Blockchain Security Evaluation

DABSTERS Blockchain protocol has the following security properties.

[2] http://sancy.univ-bpclermont.fr/~lafourcade/DABSTERS_FormalVerif/.

Consistency: A Blockchain protocol achieves consistency if it is capable of ensuring that each valid transaction sent to the network will stay immutable in the Blockchain.

Definition 1 (Consistency). *A Blockchain protocol* $\mathbb{P}$ *is* $T-$ *consistent if a valid transaction* tx *is confirmed and stays immutable in the Blockchain after* $T-$ *round of new blocks.*

Theorem 1. *DABSTERS Blockchain protocol is 1-consistent.*

Proof. The consistency is achieved by agreeing on the validity of the transaction through a Byzantine Agreement process. Hence, the probability to not settling it in a new block is negligible if the transaction has at least 50% + 1 of valid endorsement and the network have at most $\lfloor \frac{n-1}{3} \rfloor$ out of total n malicious peers, as it has been shown in [6,14] under the terminology of safeness. The protocol achieves consistency after a new block is created (1-consistency) due to the chain is growing without forks.

Liveness: A consensus protocol ensures liveness if a honest client submits a valid transaction and a new block is generated with the transaction in it. Hence, the protocol must ensure that the Blockchain growths if valid clients generate valid transactions.

Definition 2 (Liveness). *A consensus protocol* $\mathbb{P}$ *ensures* liveness for a Blockchain C *if* $\mathbb{P}$ *ensures that after a period of time* t*, the new version of the Blockchain* C' *is* $C' > C$*, if a valid client* $c_{i_{bc}}$ *has broadcasted a valid transaction* tx_i *during the time* t*.*

Theorem 2. *DABSTERS Blockchain protocol achieves liveness.*

Proof. Our protocol is a BFT-based consensus. Thus, liveness is achieved if after the transaction validation process, the network agrees in new block B with the transactions broadcasted by the clients during a period of time t. Hence, for valid transactions tx_i, where $i \in \mathbb{N}_0$, issued by valid a client c_i during a period of time t, the probability that $C' = C$ is neglected if we have at most $\lfloor \frac{n-1}{3} \rfloor$ out of total n malicious peers [6].

Blindness: We use the definition of *blindness* defined by Schnorr in [20]. A signature is properly blinded if the signer cannot get any information about the signature if the receiver follows the protocol correctly.

Definition 3 (Blind signature). *A signing scheme is* blind *if the signature* $(m, \rho, \sigma, \varepsilon)$ *generated by following correctly the protocol, is statistically independent of the interaction* (a, e, R, S) *with that provides the view to the signer.*

Theorem 3. *Okamoto-Schnorr signature* $(m, \rho, \delta, \varepsilon)$ *is statistically independent to the interaction* (a, e, R, S) *between the authority* A *and the user.*

Proof. We recall how the protocol works. To generate a blind signature $(m, \rho, \sigma, \varepsilon)$ the user chooses randomly $(\beta, \gamma, \delta) \in \mathbb{Z}_q$ to respond to the commitment a generated by A with the challenge $e = H(m, ag^{\beta}h^{\gamma}y^{\delta}) - \delta \bmod q$. The authority A then sends $(R, S) = (t - er, u - es)$ to finally obtain the signature by calculating $(\rho, \sigma) = (R - \beta, S - \gamma)$. Hence, for the constant interaction (a, e, R, S) and a unique set (β, γ, δ) randomly chosen per signature, we generate a signature $(m, \rho, \delta, \varepsilon) = (m, R - \beta, S - \gamma, e + \gamma)$ that is uniformly distributed over all the signatures of the message m due to the random $(\beta, \gamma, \delta) \leftarrow \mathbb{Z}_q$ [20].

5 Conclusion

We proposed a fully decentralized electronic voting system that combines several security properties. This protocol, called DABSTERS, uses a new architecture that allows enhancement of the security of e-voting systems and guarantees the trustworthiness required by voters and election organizers. DABSTERS is designed to be implemented on private Blockchains and uses a new blinded signature consensus algorithm to guarantee vote integrity and voter's privacy due to the unlinkability property that the blinded signature has. Future work will be dedicated to evaluating the performance and the scalability of DABSTERS.

Acknowledgments. This work has received funding from the European Union's Horizon 2020 research and innovation programme under the Grant Agreement No. 826404, Project CUREX.

References

1. Abadi, M., Blanchet, B., Fournet, C.: The applied pi calculus: mobile values, new names, and secure communication. J. ACM **65**(1), 1:1–1:41 (2018)
2. Androulaki, E., et al.: Hyperledger fabric: a distributed operating system for permissioned blockchains. In: Proceedings of the Thirteenth EuroSys Conference, EuroSys 2018, Porto, Portugal, 23–26 April 2018, pp. 30:1–30:15. ACM (2018)
3. Au, M.H., Chow, S.S.M., Susilo, W., Tsang, P.P.: Short linkable ring signatures revisited. In: Atzeni, A.S., Lioy, A. (eds.) EuroPKI 2006. LNCS, vol. 4043, pp. 101–115. Springer, Heidelberg (2006). https://doi.org/10.1007/11774716_9
4. Boneh, D.: Pairing-based cryptography: past, present, and future. In: Wang, X., Sako, K. (eds.) ASIACRYPT 2012. LNCS, vol. 7658, p. 1. Springer, Heidelberg (2012). https://doi.org/10.1007/978-3-642-34961-4_1
5. Boneh, D., Franklin, M.K.: Identity-based encryption from the weil pairing. SIAM J. Comput. **32**(3), 586–615 (2003)
6. Castro, M., Liskov, B.: Practical byzantine fault tolerance. In: Proceedings of the Third USENIX 1999 (1999)
7. Chaieb, M., Yousfi, S., Lafourcade, P., Robbana, R.: Verify-your-vote: a verifiable blockchain-based online voting protocol. In: Themistocleous, M., Rupino da Cunha, P. (eds.) EMCIS 2018. LNBIP, vol. 341, pp. 16–30. Springer, Cham (2019). https://doi.org/10.1007/978-3-030-11395-7_2
8. Followmyvote. Follow my vote (2012). https://followmyvote.com/

9. Hankerson, D., Menezes, A.: Elliptic curve cryptography. In: van Tilborg, H.C.A., Jajodia, S. (eds.) Encyclopedia of Cryptography and Security. Springer, New York (2005)
10. Hardwick, F.S., Akram, R.N., Markantonakis, K.: E-voting with blockchain: an e-voting protocol with decentralisation and voter privacy. CoRR, abs/1805.10258 (2018)
11. BIT Congress Inc., Bitcongress. http://cryptochainuni.com/wp-content/uploads/BitCongress-Whitepaper.pdf
12. Juels, A., Catalano, D., Jakobsson, M.: Coercion-resistant electronic elections. In: Chaum, D., et al. (eds.) Towards Trustworthy Elections. LNCS, vol. 6000, pp. 37–63. Springer, Heidelberg (2010). https://doi.org/10.1007/978-3-642-12980-3_2
13. Koscina, M., Lafourcade, P., Manset, D., Naccache, D.: Blindcons: a consensus algorithm for privacy preserving private blockchains. Technical report, LIMOS (2018). http://sancy.univ-bpclermont.fr/~lafourcade/BlindCons.pdf
14. Li, J., Maziéres, D.: Beyond one-third faulty replicas in byzantine fault tolerant systems. In: NSDI (2007)
15. McCorry, P., Shahandashti, S.F., Hao, F.: A smart contract for boardroom voting with maximum voter privacy. In: Kiayias, A. (ed.) FC 2017. LNCS, vol. 10322, pp. 357–375. Springer, Cham (2017). https://doi.org/10.1007/978-3-319-70972-7_20
16. Okamoto, T.: Provably secure and practical identification schemes and corresponding signature schemes. In: Brickell, E.F. (ed.) CRYPTO 1992. LNCS, vol. 740, pp. 31–53. Springer, Heidelberg (1993). https://doi.org/10.1007/3-540-48071-4_3
17. Narayan, S., Udaya, P., Teague, V.: A secure electronic voting scheme using identity based public key cryptography. In: Proceedings of SAR-SSI 2007, Annecy, France, 12–16 June 2007 (2007)
18. Paillier, P.: Paillier encryption and signature schemes. In: van Tilborg, H.C.A., Jajodia, S. (eds.) Encyclopedia of Cryptography and Security, 2nd edn, pp. 902–903. Springer, New York (2011)
19. Rossi, F., Schmid, G.: Identity-based secure group communications using pairings. Comput. Netw. **89**, 32–43 (2015)
20. Schnorr, C.P.: Security of blind discrete log signatures against interactive attacks. In: Qing, S., Okamoto, T., Zhou, J. (eds.) ICICS 2001. LNCS, vol. 2229, pp. 1–12. Springer, Heidelberg (2001). https://doi.org/10.1007/3-540-45600-7_1
21. Smartmatic. Tivi (2016). http://www.smartmatic.com/voting/online-voting-tivi/
22. Yu, B., et al.: Platform-independent secure blockchain-based voting system. In: Chen, L., Manulis, M., Schneider, S. (eds.) ISC 2018. LNCS, vol. 11060, pp. 369–386. Springer, Cham (2018). https://doi.org/10.1007/978-3-319-99136-8_20

Enhanced Models for Privacy and Utility in Continuous-Time Diffusion Networks

Daniele Gorla[1], Federica Granese[1,2(✉)], and Catuscia Palamidessi[2]

[1] Department of Computer Science, Sapienza University of Rome, Rome, Italy
granese.1615552@studenti.uniroma1.it
[2] INRIA Saclay and LIX, Palaiseau, France
federica.granese@inria.fr

Abstract. Controlling the propagation of information in social networks is a problem of growing importance. On one hand, users wish to freely communicate and interact with their peers. On the other hand, the information they spread can bring to harmful consequences if it falls in the wrong hands. There is therefore a trade-off between utility, i.e., reaching as many intended nodes as possible, and privacy, i.e., avoiding the unintended ones. The problem has attracted the interest of the research community: some models have already been proposed to study how information propagate and to devise policies satisfying the intended privacy and utility requirements. In this paper we adapt the basic framework of Backes et al. to include more realistic features, that in practice influence the way in which information is passed around. More specifically, we consider: (a) the topic of the shared information, and (b) the time spent by users to forward information among them. For both features, we show a way to reduce our model to the basic one, thus allowing the methods provided in the original paper to cope with our enhanced scenarios. Furthermore, we propose an enhanced formulation of the utility/privacy policies, to maximize the expected number of reached users among the intended ones, while minimizing this number among the unintended ones, and we show how to adapt the basic techniques to these enhanced policies.

Keywords: Diffusion networks · Privacy/utility · Submodular functions

1 Introduction

In the last decade there has been a tremendous increase in the world-wide diffusion of social networks, leading to a situation in which a large part of the population is highly inter-connected. A consequence of such high connectivity is that, once a user shares a piece of information, it may spread very quickly. The implications of this phenomenon have attracted the attention of many researchers, interested in studying the potentials and the risks behind such implications.

R. M. Hierons and M. Mosbah (Eds.): ICTAC 2019, LNCS 11884, pp. 313–331, 2019.
https://doi.org/10.1007/978-3-030-32505-3_18

The involvement of the scientific community with this topic has already produced a large body of literature; see, for instance, [4,6,15,21,22], just to cite a few.

In general, *diffusion* [13] is a process by which information, viruses, gossips and any other behaviors spread over networks. Here, we follow a natural and common approach to modeling the net as a graph where nodes represent the users and edges are labeled by the likelihood of transmission between users.

One of the strengths, but also the main potential hazard, of social networks relies on the speed by which information can be diffused: once a piece of information becomes viral, there is no way to control it. This means that it can reach users that it was not meant to reach. If the information is a sensitive one, users naturally have an interest in controlling this phenomenon. In [1], this problem is addressed by defining two types of propagation policies that reconcile privacy (i.e., protecting the information from those who should not receive it) and utility (i.e., sharing the information with those who should receive it). Note that in the framework of [1], instead of considering privacy in terms of an adversary inferring sensitive information from the data published by the user, the authors consider privacy in terms of controlling the spreading of information within a network of users that share the information with each other. Thus the goal is to enable users to share information in social networks in such a way that, ideally, only the intended recipients receive the information.[1] *Utility-restricted privacy policies* minimize the risk, i.e., the expected number of malicious users that receive the information, while satisfying a constraint on the utility, i.e., a lower bound on the number of friends the user wants to reach. Dually, *privacy-restricted utility policies* maximize the number of friends with whom the information is shared, while respecting an upper bound on the number of malicious nodes reached by the information spread. The authors of [1] prove that *Maximum k-Privacy* - the minimization problem corresponding to the utility-restricted provacy policies - and *Maximum* τ*-Utility* - the maximization problem corresponding to the privacy-restricted utility policies - are NP-hard, and propose algorithms for approximating the solution.

Being one of the first framework to study the trade-off between privacy and utility, the model proposed in [1] is quite basic. One limitation is that the likelihood that governs the transmission along an edge is a constant, fixed in time and irrespective of any other features. We argue that this is not a realistic assumption, and we propose to enrich the framework for modeling the situations described in the following two scenarios.

First, imagine that you are a scientific researcher spending some time on a social network. Suddenly, you see a news about the proof of the century, stating that P = NP. Whom do you wish to share such an information with? Probably with a colleague or someone interested in the subject. To support this kind of scenario, following [7], we consider social networks in which a user may choose the peers to whom to send a piece of information based on the *topic* of that

---

[1] Even if this notion of privacy seems closer to the notion of *secrecy*, for the sake of continuity we adopt the terminology used in [1].

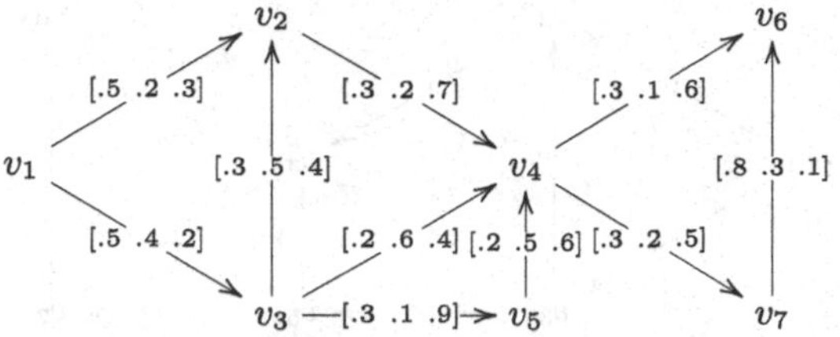

Fig. 1. A topic vector diffusion network, in which we use topic vectors with three components (*science, movies, society*)

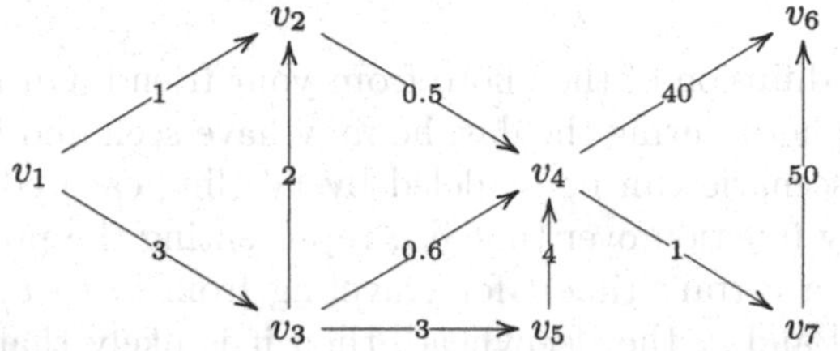

Fig. 2. A time diffusion network with sampled times for traversing the edge

information. To model such a situation, we label the edges of the net by *topic vectors*, defined as vectors in which each component represents the probability of a user to send an information of the corresponding topic (or tag) to the user at the other end of the edge. Furthermore, a piece of information is usually related to several topics, not just one. To model this latter aspect, we also tag a message with a probability distribution (topic distribution) over the topics, representing the weight of each topic in the message. To obtain the probability that a node v_i sends a message to another node v_j we then consider the scalar product of the topic vector of the edge (v_i, v_j) and the topic distribution of the message.

As an example, assume that there are three topics, *science*, *movies*, and *society*. Figure 1 represents a net whose edges are labeled with instances of these kinds of topic vectors. For example, if v_3 receives a message about a new movie of a director he likes, the probability that it will forward it to v_2 (rather than not) is 0.5, while the probability of forwarding it to v_4 is 0.6 and to v_5 is 0.1, representing the fact that v_2 and v_3 are much more interested than v_5 in the kind of movies that v_3 likes. Note that the sum of these probabilities is not 1, because these are independent events. Further, consider the P = NP message, and assume that its topic distribution is $(0.9, 0, 0.1)$. Since the edge (v_7, v_6) has topic vector $(0.8, 0.3, 0.1)$, the probability that v_7 sends the message to v_6 is $0.9 \times 0.8 + 0 \times 0.3 + 0.1 \times 0.1 = 0.73$. Note that, being the convex combination of probabilities, the result of such scalar product is always a probability.

Second, imagine that you are a night owl; at midnight, you see a funny photo and you want to share it with one of your friends. However, he is a sleepyhead and sleeps all night; thus, he will be able to forward such a photo only the next morning. If we are tracking the diffusion process until a few hours forward, there

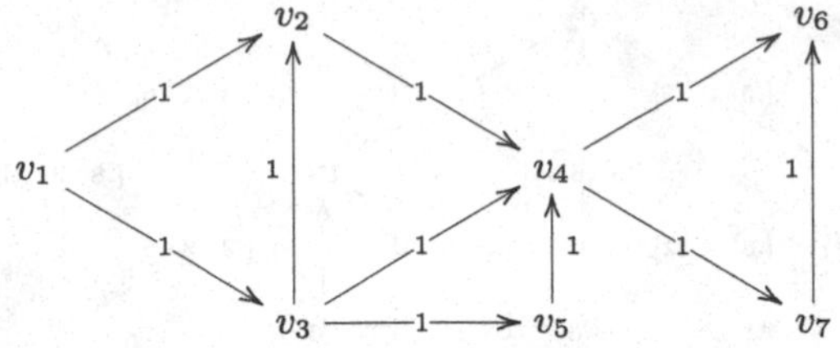

Fig. 3. A general diffusion network in which green nodes are friends and red nodes are malicious. (Color figure online)

will be no further diffusion of the photo from your friend. On the other hand, if you had sent the photo during the day, he may have seen and forwarded it soon afterwards. This scenario can be modeled by labeling each edge (v_i, v_j) with a probability density function over time δ_{ij}, representing the probability that the information takes a certain time t for traveling from v_i to v_j. For instance, if v_i is the night owl and v_j the sleepyhead, then it is likely that δ_{ij} will be a big amount of time, but there is still some probability that the information arrives at v_i when they are both awake, in which case the transmission time will be shorter. Each edge may have a different density function: for instance, if v_i has another friend v_z who is a night owl as well, then the moment in which v_z sees the information sent by v_i will be likely to be closer to the one in which v_i forwards the information; hence, the amount of time for the transmission from v_i to v_z will be small. By sampling the time for each edge, we obtain a snapshot of the net, which will have the same structure as a standard net. Figure 2 represents an instance of such a net.

Another limitation of the standard framework is in the way in which the trade-off problem is formulated in [1]: for maximizing privacy and utility, the corresponding problems try to minimize the number of malicious nodes infected *up to time* t (given a bound on the number of friends *initially* sharing the information), or to maximize the number of friends *initially* sharing the information (given a bound on the number of malicious nodes infected *up to time* t). By contrast, we argue that utility would be better expressed in terms of the friends reached by the information *up to time* t, instead of the initial friends only. Furthermore, privacy and utility would be more symmetric, in that both of them would be expressed in terms of nodes reached at time t.

As an example, consider Fig. 3 and suppose we want to monitor the diffusion up to time $t = 1$. Consider first the maximum utility problem under the constraint of reaching (at time $t = 1$) at most one malicious node. In the standard framework, there are two solutions for the set of initial nodes: either $\{v_1\}$ or $\{v_5\}$. They are considered equivalent because we only consider further infection of the malicious nodes (and in both cases, in 1 time unit just one malicious node gets infected). By contrast, we argue that $\{v_1\}$ is a better solution, because if we start with $\{v_1\}$ then in 1 time unit the information will reach also the friend node v_3, whereas no further friend will be reached if we start with $\{v_5\}$.

Consider now the maximum privacy problem. Assume that we want to minimize the number of malicious nodes infected up to time $t = 1$ under the constraint of having at least two friend sharing the information. The solution of the problem in [1] is any subset formed by two friend nodes. Any such subset, in fact, leads to infect two malicious nodes at time $t = 1$. By contrast, we argue that the optimal solution would be the (smaller) initial set $\{v_1\}$. In fact this solution would respect the constraint if, *as we propose*, we count also the friends infected at time $t = 1$, and would minimize the malicious nodes infected in the same time unit.

1.1 Related Work

There is a huge literature on information propagation in social networks, but most of the papers focus on maximizing the spread of information in the whole network. See for instance [5,9,11,14,18]. To make such works closer to real life situations, some papers revisit them on either the influence problem or the network model. For example, in [2,3,20], the problem is modified by considering the scenario where a company wants to use viral marketing to introduce a new product into a market when a competing product is simultaneously being introduced. Referring to A and B as the two technologies of interest, they denote with I_A (I_B) the initial set of users adopting technology A (B). Hence, they try to maximize the expected number of consumers that will adopt technology A, given I_A and I_B, under the assumption that consumers will use only one of the two products and will influence their friends on the product to use. In [2], the authors consider the problem of limiting the spread of misinformation in social networks. Considering the setting described before (with the two competitive companies), they refer to one of the two companies as the "bad" company and to the other one as the "good" company.

In the papers mentioned so far, authors always assume that all the selected top influential nodes propagate influence as expected. However, some of the selected nodes could not work well in practice, leading to influence loss. Thus, the objective of [24] is to find the set K of the most influential nodes with which initially the information should be shared, given a threshold on influence loss due to a failure of a subset of nodes $R \subseteq K$. This problem, as all the previous ones, are proven to be NP-hard; furthermore, all of [2,3,20,24] assume that the diffusion process is timeless.

A different research line consists in making the underlying network model closer to reality, instead of modifying the problem itself. For example, topic of information is handled in [7], where the authors infer what we call topic vector. Always considering the information item, the model in [23] endows each node with an influence vector (how authoritative they are on each topic) and a receptivity vector (how susceptible they are on each topic). While for diffusion networks there exists a good amount literature about the role of users' interests [7,23,25,26], the same is not true for the role of time with respect to user habits.

An orthogonal research line is represented by works like [7,10], aiming at inferring transmission likelihoods: given the observed infection times of nodes, they infer the edges of the global diffusion network and estimate the transmission rates of each edge that best explain the observed data. This leads to an interesting problem that can be solved with convex optimization techniques. Note that, as in [1], we are not dealing with this aspect, since we assume that the inference has already happened and we have an accurate estimate of the transmission likelihoods (whatever they are) for the whole network.

1.2 Contributions

The contributions of our paper are the following:

- We extend the basic graph diffusion model proposed in [1] by considering a more sophisticate labeling of the edges. This allows us to take into account, for the propagation of information, (a) the topics and (b) the probabilistic nature of the transmission rates.
- We reformulate the optimization goals of [1] by considering a notion of utility which takes into account the friend nodes reached up to a certain time t, rather than the initial set only. We argue that this notion is more natural, besides being more in line with that of privacy (the infected malicious nodes are counted up to time t as well).
- We prove that the resulting optimization problems are NP-hard and provide suitable approximation algorithms.

1.3 Paper Organization

This paper is organized as follows. In Sect. 2, we recall the basic notions and results from [1]. Then, in Sect. 3, we present the two enhanced models, one where information transmission is ruled by the topic of conversation, the other one based on the transmission time. In Sect. 4, we then modify the basic definitions of utility-restricted privacy policies and privacy-restricted utility policies, and show that all the theory developed by [1] with the original definitions can be smoothly adapted to these new (and more realistic) definitions. Finally, in Sect. 5, we conclude the paper, by also drawing lines for future research.

2 Background

In this section we recall the basic notions from [1], which will be used in the rest of the paper.

2.1 Submodular Functions

Definition 1 (Submodular function [8]). *A function* $f\colon 2^V \to \mathbb{R}$ *is* submodular *if, for all* $S, T \subseteq V$*, it holds that* $f(S) + f(T) \geq f(S \cup T) + f(S \cap T)$.

Defining $f(j|S) := f(S \cup \{j\}) - f(S)$ as the *profit* (or *cost*) of $j \in V$ in the context of $S \subseteq V$, then f is submodular iff $f(j|S) \geq f(j|T)$, for all $S \subseteq T$ and $j \notin T$. The function f is *monotone* iff $f(j|S) \geq 0$, for all $S \subseteq V$ and $j \notin S$. Moreover, f is *normalized* if $f(\emptyset) = 0$. An example of submodular function is given in [19]. Assume a factory able of making any finite set of products from a given set E and let $c : 2^E \rightarrow \mathbb{R}$ be the associated cost function. Let subset S be currently produced, and let us consider also producing product $e \notin S$. The marginal setup cost for adding e to S is $c(e|S)$. Suppose that $S \subset T$ and that $e \notin T$. Since T includes S, it is reasonable to guess that the marginal setup cost of adding e to T is not larger than the marginal setup cost of adding e to S; that is, for all $S \subset T \subset T \cup \{e\}$, it holds that $c(e|T) \leq c(e|S)$.

Given a submodular function f, the *curvature* κ_f of f is defined as

$$\kappa_f := min_{j \in V} \frac{f(j|V \setminus \{j\})}{f(\{j\})}$$

Intuitively, this factor measures how close a submodular function is to a modular function, where $f : 2^V \rightarrow \mathbb{R}$ is *modular* if, for all $S, T \subseteq V$, $f(S) + f(T) = f(S \cup T) + f(S \cap T)$. It has been noticed that the closer the function to being modular, the easier it is to optimize.

2.2 Diffusion Networks

Definition 2 (General Diffusion Network). *A general diffusion network is a tuple* $N = (V, \gamma)$, *where* $V = \{v_i\}_{i=1...n}$ *is the set of nodes and* $\gamma = (\gamma_{ij})_{i,j=1...n}$ *is the transmission matrix of the network (with* $\gamma_{ij} \geq 0$, *for all* i, j*).*

Thus, V and γ define a directed graph where each $\gamma_{ij} > 0$ represents an edge between nodes v_i and v_j along which the information can potentially flow, together with the flow likelihood. Let us now consider a general diffusion network N in which $F \subseteq V$ is the set of friendly nodes and $M \subseteq V$ is the set of malicious nodes, with $F \cap M = \emptyset$. The idea is to maximize the number of friends and minimizing the number of enemies reached by an information in a certain time window.

Definition 3 (Utility-restricted Privacy Policy). *A utility-restricted privacy policy* Π *is a 4-tuple* $\Pi = (F, M, k, t)$ *where* F *is the set of friend nodes,* M *is the set of malicious nodes,* k *is the number of nodes the information should be shared to, and* t *is the period of time in which the policy should be valid.*

Definition 4 (Privacy-restricted Utility Policy). *A privacy-restricted utility policy* Υ *is a 4-tuple* $\Upsilon = (F, M, \tau, t)$ *where* F *is the set of friend nodes,* M *is the set of malicious nodes,* τ *is the expected number of nodes in* M *receiving the information during the diffusion process, and* t *is the period of time in which the policy should be valid.*

Both the policies are focused on bounding the risk that a malicious node gets infected by time t, given that $F' \subseteq F$ is initially infected.

Definition 5 (Risk). *Let N be a diffusion network. The risk $\rho_N(F', M, t)$ caused by $F' \subseteq V$ with respect to $M \subseteq V$ within time t is given by*

$$\rho_N(F', M, t) = \sum_{m \in M} \Pr[t_m \leq t | F']$$

Here, $\Pr[t_m \leq t|F']$ is the likelihood that the infection time t_m of malicious node m is at most t, given that F' is infected at time $t = 0$.

Hence, the risk function gives us an upper bound on the number of malicious nodes receiving the information in a given time window, given that a subset of friendly nodes was initially infected. This definition of risk function recalls the one of *influence function* [1,9,11]. Here, instead of being interested in the expected number of infected nodes in a set of malicious nodes, we are interested in the infection in the whole network. Thus, given $A \subseteq V$, the influence in the network N within time t is denoted by $\sigma_N(A, t)$. Rodriguez et al. show in [11] that computing $\sigma_N(A, t)$ is #P-hard and they approach the problem of the influence estimation by using a randomized approximation algorithm. As already written in [1], since the risk function is just a generalization the regular influence function, computing $\rho_N(F', M, t)$ is also #P-hard; however, we can use the algorithm in [11] to approximate the risk function up to a constant factor: we simply ignore the infection times for nodes not in M.

To make notation lighter, we shall usually omit the subscript N from ρ_N, when clear from the context. To maximally satisfy a utility-restricted privacy policy and a privacy-restricted utility policy, the following two problems are defined.

Definition 6 (Maximum k-privacy – MP). *Given a utility-restricted privacy policy $\Pi = (F, M, k, t)$ and a general diffusion network N, the maximum k-privacy problem (MP, for short) is given by*

$$\begin{aligned} \underset{F' \subseteq F}{\textit{minimize}} \quad & \rho(F', M, t) \\ \textit{subject to} \quad & |F'| \geq k \end{aligned} \tag{1}$$

Definition 7 (Maximum τ-utility – MU). *Given a privacy-restricted utility policy $\Gamma = (F, M, \tau, t)$ and a general diffusion network N, the maximum τ-utility problem (MU, for short) is given by*

$$\begin{aligned} \underset{F' \subseteq F}{\textit{maximize}} \quad & |F'| \\ \textit{subject to} \quad & \rho(F', M, t) \leq \tau \end{aligned} \tag{2}$$

The idea behind MP is to look for a subset of at least k friendly nodes with which initially share the information, in order to minimize the diffusion between malicious nodes at time t. By contrast, MU looks for the maximum set of friendly nodes with which initially share the information, in order to infect at most τ malicious nodes at time t. Both problems are NP-hard. However, they can be approximated and the approximation algorithms rely on the submodularity of the risk function: since ρ is submodular, monotone and with a non-zero curvature, it is possible to derive an efficient constant factor approximation, where the approximation factor depends on the structure of the underlying network N.

Optimizing submodular functions is a difficult task, but we can get around the problem by choosing a proper surrogate function for the objective and optimize it; the surrogate functions usually are upper or lower bounds. For example, the *majorization-minimization* algorithms begin with an arbitrary solution Y to the optimization problem and then optimize a modular approximation formed via the current solution Y. Therefore, following the work in [1,12], we can solve MP (and MU) by choosing a surrogate function for ρ. In Algorithm 1, given a candidate solution $Y \subseteq F$, the modular approximation of the risk function ρ is given by

$$m_{g_Y}(X) = \rho(Y) + g_Y(X) - g_Y(Y)$$

where

$$g_Y(X) = \sum_{v \in X} g_Y(v) \qquad \text{and} \qquad g_Y(v) = \begin{cases} \rho(v|F \setminus \{v\}), & \text{if } v \in Y \\ \rho(v|Y), & \text{otherwise.} \end{cases}$$

Due to the submodularity of the risk function, we can use this submodular approximation as an upper bound for the risk, i.e. $m_{g_Y}(Y) \geq \rho(Y)$ [1,12].

Algorithm 1. Maximum k-Privacy

Require: Instance F, M, k of maximum k-privacy
Ensure: *satisfyingMP*(F, M, k)
$C \leftarrow \{X \subseteq F \,:\, |X| = k\}$
Select a random candidate solution $X^1 \in C$
$t \leftarrow 0$
repeat
  $t \leftarrow t + 1$
  $X^{t+1} \leftarrow argmin_{X \in C}\ m_{g_{X^t}}(X)$
until $X^{t+1} = X^t$
return X^t

At each iteration, Algorithm 1 finds the new set that minimizes the upper bound of the risk function. Clearly, since this set minimizes the upper bound of the risk function, it also minimizes the risk function.[2] Now, recall that the curvature $\kappa_{\rho(F,M,t)}$ of $\rho(F, M, t)$ is given by

$$\kappa_{\rho(F,M,t)} := min_{v \in F} \frac{\rho(v|F \setminus \{v\}, M, t)}{\rho(\{v\}, M, t)}$$

[2] This methodology can be seen as the gradient descent method for minimizing continuous differentiable functions: we start from a random point y and we iteratively move in the direction of the steepest descent, as defined by the negative of the gradient.

where $\rho(v|F \setminus \{v\}, M, t) := \rho(F, M, t) - \rho(F \setminus \{v\}, M, t)$. This quantity can be used to give the approximation factor.

Theorem 1. *Algorithm 1 approximates maximum k-privacy to a factor* $\frac{1}{\kappa_\rho}$. *That is, let* F' *be the output and* F^* *be the optimal solution; then,* $\rho(F', M, t) \leq \frac{1}{\kappa_\rho}\rho(F^*, M, t)$.

Starting from the approximation algorithm for maximum k-privacy, maximum τ-utility can be approximated through Algorithm 2.

Algorithm 2. Maximum τ-Utility

Require: Instance F, M, τ of maximum τ-utility
Ensure: $satisfyingMU(F, M, \tau)$

for $n \in [|F|, \ldots, 1]$ **do**
  $\tau' \leftarrow \min_{F' \subseteq F} \rho(F', M, t)$ s.t. $|F'| = n$
  if $\tau' \leq \tau$ **then**
    return n
return 0

Theorem 2. *Let* n^* *be the optimal solution to an instance of maximum* τ*-utility, and let* n *be the output of Algorithm 2 for the same instance, using a* $\frac{1}{\kappa_\rho}$*-approximation for maximum k-privacy. Then* $n \geq \kappa_\rho n^*$.

3 Enhanced Models

In this section, we provide two different models which modify the notion of general diffusion network by using different transmission matrices. In particular, in the first model, called *topic vector diffusion network*, we bind the likelihood of transmitting an information to the topic of that information; in the second one, called *time diffusion network*, we bind the likelihood to the amount of time an information takes for been transmitted. As in [1], we are not interested in the inference of transmission likelihoods, as the aim of the following two models is the reduction to the general model for which the two kinds of policies are defined.

3.1 Topic Vector Diffusion Network

We first consider a social network where edges are labeled by *topic vectors*, that are vectors in which each component represents the probability of a user to send an information of the corresponding topic (or tag) to another user.

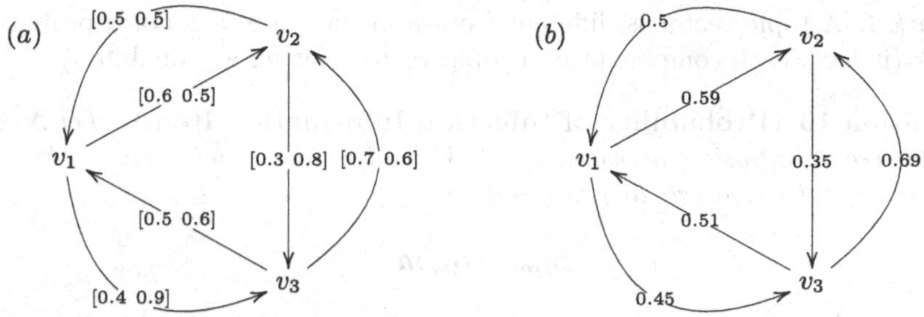

Fig. 4. From a topic vector diffusion network to the $\boldsymbol{m}$-diffusion network. (a) A topic vector diffusion network. (b) The associated (0.9 0.1)-Diffusion network

Definition 8 (Topic Vector Diffusion Network). *A topic vector diffusion network is a tuple* $N_{TV} = (V, \mathbf{A}, k)$, *where* $V = \{v_i\}_{i=1...n}$ *is the set of nodes in the network,* k *is the number of topics and* $\mathbf{A} = (\boldsymbol{\alpha_1}, \dots, \boldsymbol{\alpha_n})$ *is s.t.* $\boldsymbol{\alpha_i}$ *is the matrix of dimension* $n \times k$ *giving the topic vector that rules the transmission rates from node* v_i *to the other nodes in the network. That is,*

$$\boldsymbol{\alpha_i} := \begin{pmatrix} \alpha_{i1}^1 & \dots & \alpha_{i1}^k \\ \vdots & & \vdots \\ \alpha_{in}^1 & \dots & \alpha_{in}^k \end{pmatrix}.$$

where every $\boldsymbol{\alpha_{ij}} = (\alpha_{ij}^1 \dots \alpha_{ij}^k)$ *is called* topic vector *and each* α_{ij}^l *(for* $l = 1 \dots k$*) is the probability that user* i *sends an information of topic* l *to user* j.

Notice that a topic vector is not required to be a probability distribution and that, for every i, j and l, the probability of not sending an information of topic l from i to j is $1 - \alpha_{ij}^l$. Together, V and $\mathbf{A}$ define a weighted directed graph where each $\boldsymbol{\alpha_{ij}}$ (i.e., each row of $\boldsymbol{\alpha_i}$ having non zero components) represents an edge between v_i and v_j with weight $\boldsymbol{\alpha_{ij}}$. For example, consider the network N_{TV} in Fig. 4(a), with $V = \{v_1, v_2, v_3\}$, $k = 2$ and

$$\boldsymbol{\alpha_1} = \begin{pmatrix} 0 & 0 \\ 0.6 & 0.5 \\ 0.4 & 0.9 \end{pmatrix}, \boldsymbol{\alpha_2} = \begin{pmatrix} 0.5 & 0.5 \\ 0 & 0 \\ 0.3 & 0.8 \end{pmatrix}, \boldsymbol{\alpha_3} = \begin{pmatrix} 0.5 & 0.6 \\ 0.7 & 0.6 \\ 0 & 0 \end{pmatrix}$$

User v_1 will send to v_2 an information about topic 1 with probability 0.6 and an information about topic 2 with probability 0.5.

Definition 9 (Information Item). *An information item (or meme) is a* k*-dimensional probability vector, in which each component is the weight of a topic relating to the subject of the information. That is,* $\boldsymbol{m} := (m_1 \dots m_k)$ *such that* $m_1 + \dots + m_k = 1$.

For instance, consider vectors consisting of two components, *science* and *society*. The information item associated to a tweet on a scientific paper should be $\boldsymbol{m} = (0.9\ 0.1)$.

Remark 1. A topic vector is different from a meme since it is not a probability vector (indeed, each component of a topic vector is itself a probability).

Definition 10 (Probability of Infection Information Item). *Let N_{TV} be a topic vector diffusion network, $i, j \in V$ and $\boldsymbol{m}$ the input meme. Then, the probability that i sends $\boldsymbol{m}$ to j is given by:*

$$\beta_{ij\boldsymbol{m}} = \boldsymbol{\alpha_{ij}}\, \boldsymbol{m}^\top \tag{3}$$

Notice that, since each component of $\boldsymbol{\alpha_{ij}}$ is a probability and $\boldsymbol{m}$ is a probability vector, we obtain:

$$0 = \boldsymbol{0}\, \boldsymbol{m}^\top \leq \boldsymbol{\alpha_{ij}}\, \boldsymbol{m}^\top \leq \boldsymbol{1}\, \boldsymbol{m}^\top = 1$$

Definition 11 ($\boldsymbol{m}$-Diffusion Network). *An $\boldsymbol{m}$-diffusion network is a tuple $N_{\boldsymbol{m}} = (V, \beta_{\boldsymbol{m}})$, where $V = \{v_i\}_{i=1\ldots n}$ is the set of nodes and $\beta_{\boldsymbol{m}} = (\beta_{ij\boldsymbol{m}})_{i,j=1\ldots n}$ is the transmission matrix of the network that forwards $\boldsymbol{m}$ (with $\beta_{ij\boldsymbol{m}} \geq 0$).*

Given a topic vector diffusion network and an information item, we can derive the associated $\boldsymbol{m}$-diffusion network by determining the probability of infection between each node with respect to the information item (i.e., the transmission matrix $\beta_{\boldsymbol{m}}$). Resuming the example before, with $\boldsymbol{m} = (0.9\ 0.1)$ representing the information item of a scientific paper, consider the topic diffusion network in Fig. 4(b), in which we suppose the topic vectors have the same tag as $\boldsymbol{m}$ (science and society). By Definition 10, we have, e.g., that $\beta_{32\boldsymbol{m}} = (0.7\ 0.6)(0.9\ 0.1)^\top = 0.69$ and $\beta_{31\boldsymbol{m}} = (0.5\ 0.6)(0.9\ 0.1)^\top = 0.51$; hence, the probability that v_3 forwards $\boldsymbol{m}$ to v_2 is greater than the probability of forwarding it to v_1, since $\boldsymbol{m}$ is more focused on science than on society.

Even if $\boldsymbol{m}$-diffusion networks seem similar to general diffusion networks, they still have an important difference: in them, transmission depends on the information item. Thus, modeling real-life networks is more natural and more accurate with this enhanced framework. Nonetheless, the new framework can be reduced to the basic one via the transformation we are going to describe now. This of course has the advantage of reusing all the theory developed in [1] for free. To this aim, consider a sample of messages $T = \{\boldsymbol{m}_1, \ldots, \boldsymbol{m}_h\}$ and their associated $\boldsymbol{m}_l$-diffusion networks derived from the same topic vector diffusion network. Let us concentrate on two nodes i, j in V and define the independent events $E_{ijl} = \{i$ *sends* $\boldsymbol{m}_l$ *to* $j\}$; clearly, $Pr(E_{ijl}) = \beta_{ij\boldsymbol{m}_l}$. We can define a random variable X_{ij} counting the number of information items in T sent from i to j. Thus, we can compute the probability that i sends $0, 1, \ldots, h$ information items to j as follows:

$$Pr(X_{ij} = 0) = \prod_{l=1}^{h} (1 - \beta_{ij\boldsymbol{m}_l})$$

$$\vdots$$

$$Pr(X_{ij} = d) = \sum_{\{l_1,\ldots,l_d\} \subseteq \{1,\ldots,h\}} \beta_{ij\boldsymbol{m}_{l_1}} \cdots \beta_{ij\boldsymbol{m}_{l_d}} \left(\prod_{l' \in \{1,\ldots,h\} \setminus \{l_1,\ldots,l_d\}} (1 - \beta_{ij\boldsymbol{m}_{l'}}) \right)$$

$$\vdots$$

$$Pr(X_{ij} = h) = \prod_{l=1}^{h} \beta_{ij\boldsymbol{m}_l}$$

The derivation of the general diffusion network from a set of $\boldsymbol{m}_l$-diffusion networks (obtained from the same topic vector diffusion network) is given by first computing for each $i, j \in V$

$$E[X_{ij}] = \sum_{d=1}^{h} d Pr\,(X_{ij} = d)$$

Then, by starting from these expected values and by dividing by h (for the sake of normalization), we can recover a general diffusion network: the set of nodes remains V and $\gamma_{ij} := \frac{E[X_{ij}]}{h}$, for every i, j.

3.2 Time Diffusion Network

We now consider a diffusion network in which each edge (v_i, v_j) is equipped with a probability density function describing, for any given time interval (providing the time spent by the information in traveling along it), the probability of transmitting along that edge.

Definition 12 (Time transmission function). *A time transmission function $f(\delta)$ is a a probability density function over a time interval.*

Definition 13 (Time diffusion network). *A time diffusion network is a tuple $N_T = (V, \zeta)$, where $V = \{v_i\}_{i=1\ldots n}$ is the set of nodes in the network and $\zeta = (f_{ij}(\delta_{ij}))_{i,j=1\ldots n}$ is the transmission matrix of the network, with $f_{ij}(\cdot)$ a time transmission function and δ_{ij} a time interval (for every i and j).*

In contrast with the discrete-time model (which associates each edge with a fixed infection probability), this model associates each edge with a probability density function. Moreover, instead of considering parametric transmission functions such as exponential distribution, Pareto distribution or Rayleigh distribution, we consider the non-parametric ones because in real word scenarios the waiting times obey to different distributions. So, for example, if two nodes are usually logged-in simultaneously (hence, their respective delay in transmission is small), the time function will assign high probabilities to short intervals and negligible probabilities to long ones; the situation is dual for users that are usually logged-in during different moments of the day.

Now suppose that some external agent gives in input to some nodes of the network a certain information at time $t = 0$. Each of these nodes try to forward this information to their neighbors; clearly, this entails a certain amount of time.

Definition 14 (Transmission time). *Given two neighbor nodes i and j of a time diffusion network, the transmission time δ_{ij} is the amount of time the information requires for going from i to j during a diffusion process.*

Starting from a time diffusion network N_T, we can compute the random transmission times associated to each edge of the network by drawing them from the corresponding transmission functions. Consider now a diffusion process over a time diffusion network N_T with the sampled transmission times and suppose that the initial set of infected nodes is F'.

Definition 15 (Infection time of a node [11]). *The infection time of $v \in V$ is given by:*

$$t_v(\{\delta_{ij}\}_{(i,j)\in N_T}|F') := \min_{q\in Q_v(F')} \sum_{(i,j)\in q} \delta_{ij}$$

where F' is the set of nodes infected at time $t = 0$ and $Q_v(F')$ is the set of the directed paths from F' to v.

For preserving Theorems 1 and 2 also in this setting, we must first prove submodularity of the risk function on time diffusion networks. For this purpose, let us slightly modify Definition 5.

Definition 16 (Risk). *Let $N_T = (V, \zeta)$ be a time diffusion network. The risk $\rho_{N_T}(F', M, t)$ caused by $F' \subseteq V$ with respect to $M \subseteq V$ within time t is given by*

$$\rho(F', M, t) = \sum_{m\in M} Pr[t_m(\{\delta_{ij}\}_{(i,j)\in N_T}|F') \leq t]$$

Here, $Pr[t_m(\{\delta_{ij}\}_{(i,j)\in N_T}|F') \leq t]$ is the likelihood that the infection time t_m of malicious node m is at most t, given that F' is infected at time $t = 0$.

Theorem 3. *Given a time diffusion network $N_T = (V, \zeta)$, a set of friend nodes $F \subseteq V$, a set of malicious nodes $M \subseteq V$ and a time window t, the risk function $\rho_{N_T}(F, M, t)$ is monotonically nondecreasing and submodular in F.*

Proof. By definition, all nodes in F are infected at time $t = 0$. The infection time of a given node in the network only depends on the transmission times drawn from the transmission functions. Thus, given a sample $\{\delta_{ij}\}_{(i,j)\in N_T}$, we define $r_{\{\delta_{ij}\}}(F, M, t)$ as *the number* of nodes in M that can be reached from the nodes in F at time less than or equal to t for $\{\delta_{ij}\}$; and $R_{\{\delta_{ij}\}}(f, M, t)$ as *the set* of nodes in M that can be reached from the node f at time less than or equal to t for $\{\delta_{ij}\}$.

(i) $r_{\{\delta_{ij}\}}(F, M, t)$ is monotonically nondecreasing in F, for any sample $\{\delta_{ij}\}$. Indeed, $r_{\{\delta_{ij}\}}(F, M, t) = |\cup_{f \in F} R_{\{\delta_{ij}\}}(f, M, t)|$ and so, for any $n \notin V \setminus (F \cup M)$, $r_{\{\delta_{ij}\}}(F, M, t) \leq r_{\{\delta_{ij}\}}(F \cup \{n\}, M, t)$.

(ii) $r_{\{\delta_{ij}\}}(F, M, t)$ is submodular in F for a given sample $\{\delta_{ij}\}$. Let $R_{\{\delta_{ij}\}}(f|B, M, t)$ defined as the set of nodes in M that can be reached from node f in a time shorter than t, but cannot be reached from any node in the set of nodes $B \subseteq V$ for $\{\delta_{ij}\}$. For any $B \subseteq B'$ it holds that $|R_{\{\delta_{ij}\}}(f|B, M, t)| \geq |R_{\{\delta_{ij}\}}(f|B', M, t)|$. Consider now two sets of nodes $B \subseteq B'(\subseteq V)$ and a node $b \notin B'$:

$$\begin{aligned} & r_{\{\delta_{ij}\}}(B \cup \{b\}, M, t) - r_{\{\delta_{ij}\}}(B, M, t) \\ & = |R_{\{\delta_{ij}\}}(b|B, M, t)| \\ & \geq |R_{\{\delta_{ij}\}}(b|B', M, t)| \\ & = r_{\{\delta_{ij}\}}(B' \cup \{b\}, M, t) - r_{\{\delta_{ij}\}}(B', M, t) \end{aligned}$$

If we average over the probability space of possible transmission times,

$$\rho_{N_T}(F, M, t) = E_{\{\delta_{ij}\} \in N_T}[r_{\{\delta_{ij}\}}(F, M, t)]$$

is also monotonically nondecreasing and submodular. □

Given a time diffusion network, if the risk function has a nonzero curvature, then the results of [1] also hold for this model. Let $S_{ij}(\delta_{ij})$ be the *survival function*, expressing the probability of v_j not being infected by node v_i in less than δ_{ij} time units. Formally, $S_{ij}(\delta_{ij}) := 1 - \int_0^{\delta_{ij}} f_{ij}(\delta')d\delta'$.

Theorem 4. *Let $N_T = (V, \zeta)$ be a time diffusion network, for which $S_{ij}(\delta_{ij}) > 0$ until time t for all $v_i, v_j \in V$. Then $\kappa_{\rho(F,M,t)} > 0$.*

Proof. The infection time of a given node in the network only depends on the transmission times drawn from the transmission functions. Thus, given a sample $\{\delta_{ij}\}_{(i,j) \in N_T}$, we first remove all $v_i \in F$ s.t. $\rho_{N_T}(\{v_i\}, M, t) = 0$, since they can be safely infected at time $t = 0$. Now pick an arbitrary $v \in F$, thus there exists a dipath P from v to some $v_m \in M$. Since by hypothesis the survival function is nonzero until time t for all pairs of nodes on the path, then $\prod_{(i,j) \in P} S_{ij}(\delta_{ij}) > 0$. This fact, entails that the likelihood of infection of every node on this path is decreased if this path is removed. Moreover, this implies $\rho_{N_T}(F, M, t) - \rho_{N_T}(F \setminus \{v\}, M, t) > 0$. Thus, by definition of curvature, we obtain $\rho_{N_T}(v|F \setminus \{v\}, M, t) > 0$ and therefore $\kappa_{\rho_{N_T}(F,M,t)} > 0$. □

4 Policy Enhancements

Let us consider a general diffusion network $N = (V, \zeta)$, with fixed and disjoint sets of friend nodes F and of malicious nodes M. Starting from the propagation policies given for the basic framework in Sect. 2, we give a new definition for when an initial infection $F' \subseteq F$ within a network satisfies a utility-restricted privacy policy or a privacy-restricted utility policy. To this aim, we first introduce the notion of *gain.*

Definition 17 (Gain). *The gain $\pi(F', F, t)$ caused by $F' \subseteq F$ within time t is given by*

$$\pi(F', F, t) = \sum_{f_i \in F} \Pr[t_i \leq t | F']$$

Here, $\Pr[t_i \leq t|F']$ is the likelihood that the infection time t_i of a friend node f_i is at most t, given that F' is infected at time $t = 0$.

Hence, the gain function is similar to the risk function but, instead of determining the expected number of infected nodes in M, it gives us the expected number of infected nodes in F.

Remark 2. Clearly, since our definition of gain function is similar to the definition of risk function, we can state that:

1. the proof of submodularity for ρ in [1] can be easily adapted to show submodularity of π;
2. computing $\pi(F', M, t)$ is $\#P$-hard;
3. the gain function can be approximated up to a constant factor, by following the algorithm in [11].

Moreover, we follow the approach in [1] for the risk function and assume to have an oracle that exactly computes the gain function for a given initial infection F'.

Definition 18 (Satisfy a Utility-restricted Privacy Policy). *An initial infection F' satisfies a utility-restricted privacy policy $\Pi = (F, M, k, t)$ in a general diffusion network N if $F' \subseteq F$ and $\pi(F', M, t) \geq k$. A set F' maximally satisfies Π in N if there is no other set $F'' \subseteq F$ with $\pi(F'', F, t) \geq k$ and $\rho(F'', M, t) < \rho(F', M, t)$.*

Definition 19 (Satisfy a Privacy-restricted Utility Policy). *An initial infection F' satisfies an extended privacy-restricted utility policy $\Upsilon = (F, M, \tau, t)$ in a general diffusion network N if $F' \subseteq F$ and $\rho(F', M, t) \leq \tau$. A set F' maximally satisfies Υ in N if there is no other set $F'' \subseteq F$ with $\rho(F'', M, t) \leq \tau$ and $\pi(F'', F, t) > \pi(F', F, t)$.*

For finding an initial infection meeting Definitions 18 and 19, we define the following problems.

Definition 20 (Extended Maximum k-Privacy - EMP). *Given a utility-restricted privacy policy $\Pi = (F, M, k, t)$ and a general diffusion network N, the extended maximum k-privacy problem (EMP, for short) is given by*

$$\underset{F' \subseteq F}{\textit{minimize}} \quad \rho(F', M, t)$$
$$\textit{subject to} \quad \pi(F', F, t) \geq k$$

Definition 21 (Extended Maximum τ-Utility – EMU). *Given a privacy-restricted utility policy $\Upsilon = (F, M, \tau, t)$ and a general diffusion network N, the extended maximum τ-utility problem (EMU, for short) is given by*

$$\underset{F' \subseteq F}{\textit{maximize}} \quad \pi(F', F, t)$$
$$\textit{subject to} \quad \rho(F', M, t) \leq \tau$$

Clearly, if F' is an optimal solution to the EMU problem with respect to Υ, then F' maximally satisfies Υ; similarly, if F' is an optimal solution to the EMP problem with respect to Π, then F' maximally satisfies Π. Unfortunately, EMP and EMU problems are NP-hard; this can be proved by reducing MP and MU to them.

Theorem 5. *Extended maximum k-privacy and extended maximum τ-utility are NP-hard.*

Proof. We just show the reduction of MU to EMU since the other one is symmetric. Let ϕ be an instance of the MU problem, we can construct an instance of the EMU problem ω by setting the time parameter of the gain function to $t = 0$. Hence, F' is the seed set of ϕ, respecting the risk constraint, iff F' is the maximum set of initially infected nodes always respecting the risk constraint. As the MU problem is NP-hard [1], also EMU is NP-hard. □

However, like in the basic setting, both problems can be approximated, by slightly modifying Algorithms 1 and 2. For EMP, it suffices to replace line 1 in Algorithm 1 with

$$1a : C \leftarrow \{X \subseteq F \ : \ \pi(X, F, t) \geq k\}$$
$$1b : \textbf{if } C = \emptyset \textbf{ then return } \emptyset \quad //\textit{EMP cannot be satisfied.}$$

As already written in *Remark 2*, we can use the algorithms in [11] for the gain estimation: the randomized approximation algorithm for the influence estimation, when used as subroutines in the influence maximization algorithm, is guaranteed to find in polynomial time a set of X nodes with an influence of at least $(1 - \frac{1}{e})OPT - 2X\epsilon$, where ϵ is the accuracy parameter and OPT is the optimal value.

Similarly, Algorithm 2 can be easily adapted for handling EMU: it suffices to replace line 2 with

$$2 : \ \tau' \leftarrow min_{F' \subseteq F}\, \rho(F', M, t) \ \text{ s.t. } \pi(F', F, t) \geq n.$$

5 Conclusion

In this paper, we proposed some enhancements of the basic model in [1] for controlling utility and privacy in social networks. In particular, we added topics of conversation and time of the infection within the transmission likelihood. Furthermore, we modified the basic definitions of policy satisfaction, to make them closer to the intuitive meaning of such policies. Then, we extended the methods and results of [1] to our setting. We have demonstrated the applicability of our enhanced framework on various situations. Arguably these are toy examples, but nonetheless they reflect aspects of real-life social networks.

In the future, we are planning to extend this work and try to cope with the problems in Definitions 20 and 21, e.g. by finding a trade-off between the risk and the gain functions through *multiobjctive optimization* [16,17]. Clearly, one of the main problems could be the submodular nature of our objective functions. Furthermore, two other possible enhancements to be investigated are: (1) the sets of friend and malicious nodes (F and M) are topic dependent, and (2) let absolute time influence the likelihood of transmission (instead of letting time intervals influence this aspect). In the first setting, we think that the reduction to the basic framework becomes more involved, since a vector of memes cannot rely on a single topic vector diffusion network. In the second setting, a global time is needed for trying to reduce the new framework to the basic one. Finally, an orthogonally research line would be the setting up of a few experiments on real-life data, in order to empirically validate our results.

References

1. Backes, M., Gomez-Rodriguez, M., Manoharan, P., Surma, B.: Reconciling privacy and utility in continuous-time diffusion networks. In: 2017 IEEE 30th Computer Security Foundations Symposium (CSF), pp. 292–304 (2017)
2. Budak, C., Agrawal, D., El Abbadi, A.: Limiting the spread of misinformation in social networks. In: Proceedings of the 20th International Conference on World Wide Web, pp. 665–674. ACM (2011)
3. Carnes, T., Nagarajan, C., Wild, S.M., van Zuylen, A.: Maximizing influence in a competitive social network: a follower's perspective. In: Proceedings of the Ninth International Conference on Electronic Commerce, pp. 351–360 (2007)
4. Chen, W., Wang, C., Wang, Y.: Scalable influence maximization for prevalent viral marketing in large-scale social networks. In: Proceedings of the 16th International Conference on Knowledge Discovery and Data Mining, pp. 1029–1038. ACM (2010)
5. Chen, W., Wang, Y., Yang, S.: Efficient influence maximization in social networks. In: Proceedings of the 15th International Conference on Knowledge Discovery and Data Mining, pp. 199–208. ACM (2009)
6. De Choudhury, M., Mason, W., Hofman, J.M., Watts, D.: Inferring relevant social networks from interpersonal communication. In: Proceedings of the 19th International Conference on World Wide Web, WWW 2010, pp. 301–310 (2010)
7. Du, N., Song, L., Woo, H., Zha, H.: Uncover topic-sensitive information diffusion networks. In: AISTATS (2013)

8. Fujishige, S.: Submodular Functions and Optimization. Annals of Discrete Mathematics. Elsevier Science (2005)
9. Gomez Rodriguez, M., Schölkopf, B.: Influence maximization in continuous time diffusion networks. In: Proceedings of the 29th International Conference on Machine Learning, pp. 313–320. Omnipress (2012)
10. Gomez-Rodriguez, M., Balduzzi, D., Schölkopf, B.: Uncovering the temporal dynamics of diffusion networks. In: ICML (2011)
11. Gomez-Rodriguez, M., Song, L., Du, N., Zha, H., Schölkopf, B.: Influence estimation and maximization in continuous-time diffusion networks. ACM Trans. Inf. Syst. **34**(2), 9:1–9:33 (2016)
12. Iyer, R., Bilmes, J.: Submodular optimization with submodular cover and submodular knapsack constraints. In: Proceedings of the 26th International Conference on Neural Information Processing Systems, pp. 2436–2444. Curran Associates Inc. (2013)
13. Kasprzak, R.: Diffusion in networks. J. Telecommun. Inf. Technol. **2**, 99–106 (2012)
14. Kempe, D., Kleinberg, J., Tardos, E.: Maximizing the spread of influence through a social network. In: Proceedings of the Ninth ACM SIGKDD International Conference on Knowledge Discovery and Data Mining, pp. 137–146. ACM (2003)
15. Lappas, T., Terzi, E., Gunopulos, D., Mannila, H.: Finding effectors in social networks. In: Proceedings of the 16th ACM SIGKDD International Conference on Knowledge Discovery and Data Mining, pp. 1059–1068. ACM (2010)
16. Papadimitriou, C.H., Yannakakis, M.: On the approximability of trade-offs and optimal access of web sources. In: 41st Annual Symposium on Foundations of Computer Science, pp. 86–92. IEEE (2000)
17. Papadimitriou, C.H., Yannakakis, M.: Multiobjective query optimization. In: Proceedings of the Twentieth ACM SIGACT-SIGMOD-SIGART Symposium on Principles of Database Systems. ACM (2001)
18. Richardson, M., Domingos, P.: Mining knowledge-sharing sites for viral marketing. In: Proceedings of the Eighth ACM SIGKDD International Conference on Knowledge Discovery and Data Mining, pp. 61–70. ACM (2002)
19. Thomas McCormick, S., Iwata, S.: Introduction to submodular functions. In: Sauder School of Business, UBC Cargese Workshop on Combinatorial Optimization, September–October 2013 (2013)
20. Tzoumas, V., Amanatidis, C., Markakis, E.: A game-theoretic analysis of a competitive diffusion process over social networks. In: Goldberg, P.W. (ed.) WINE 2012. LNCS, vol. 7695, pp. 1–14. Springer, Heidelberg (2012). https://doi.org/10.1007/978-3-642-35311-6_1
21. Watts, D., Dodds, P.: Influentials, networks, and public opinion formation. J. Consum. Res. **34**, 441–458 (2007)
22. Watts, D.J., Strogatz, S.H.: Collective dynamics of 'small-world' networks. Nature **393**(6684), 440–442 (1998)
23. Yu, M., Gupta, V., Kolar, M.: An influence-receptivity model for topic based information cascades. In: International Conference on Data Mining (ICDM), pp. 1141–1146. IEEE (2017)
24. Zeng, Y., Chen, X., Cong, G., Qin, S., Tang, J., Xiang, Y.: Maximizing influence under influence loss constraint in social networks. Expert Syst. Appl. **55**(C), 255–267 (2016)
25. Zhou, D., Wenbao, H., Wang, Y.: Identifying topic-sensitive influential spreaders in social networks. Int. J. Hybrid Inf. Technol. **8**, 409–422 (2015)
26. Zhou, J., Zhang, Y., Cheng, J.: Preference-based mining of top k influential nodes in social networks. Future Gener. Comput. Syst. **31**, 40–47 (2014)

Equations, Types, and Programming Languages

Taylor Series Revisited

Xavier Thirioux(✉) and Alexis Maffart(✉)

IRIT, Toulouse, France
Xavier.Thirioux@enseeiht.fr, alexis.maffart@gmail.com

Abstract. We propose a renovated approach around the use of Taylor expansions to provide polynomial approximations. We introduce a coinductive type scheme and finely-tuned operations that altogether constitute an algebra, where our multivariate Taylor expansions are first-class objects. As for applications, beyond providing classical expansions of integro-differential and algebraic expressions mixed with elementary functions, we demonstrate that solving ODE and PDE in a direct way, without external solvers, is also possible. We also discuss the possibility of computing certified errors within our scheme.

Keywords: Taylor expansion · Certification · PDE

1 Motivations

1.1 Taylor Expansions

Our principal motivation is to provide an automatic way of approximating arbitrary multivariate numerical expressions, involving elementary functions, integrations, partial derivations and arithmetical operations. In terms of features, we propose an approach where Taylor expansions are first-class objects of our programming language, computed *lazily* on demand at any order. Finally, we also wish to obtain certified errors, which will by the end include errors of approximation and numerical errors, expressed in any suitable user-provided error domain, such as zero-centered intervals, intervals, zonotopes, etc. From a user's perspective, a typical workflow is first to compute a certified approximation at some order of some expression, second to evaluate the maximum error for the given domains of variables, and maybe third to compute a finer approximation at some higher order (without recomputing previous values) if the error is too coarse, and so on, until the approximation meets the user's expectations in terms of precision. We postulate that the expressions at hand are indeed analytical and possess a valid Taylor expansion around a given point and within variables' domains. If it is not the case, then the error computed at every increasing order won't show any sign of diminishing and could even diverge. Last but not least, our approach yields a direct means to express solutions to ODEs and PDEs and thus solve them, without complex numerical methods based on domains discretization.

Furthermore, we aim at bringing as much robustness and correction as possible to our library through a correct-by-construction approach. The type system

R. M. Hierons and M. Mosbah (Eds.): ICTAC 2019, LNCS 11884, pp. 335–352, 2019.
https://doi.org/10.1007/978-3-030-32505-3_19

is in charge of the correction as it ensures, at compile time, that dimensions of various tensors, functions, convolutions and power series conform to their specifications. This is of a particular importance in a complex and error-prone context involving a vast number of numerical computations such as ODEs and PDEs resolution. The type system which validates all dimension related issues greatly helps in reducing the focus on purely numerical concerns: correctness of approximation, precision, convergence. Moreover, correction could be proved more formally with a proof assistant such as COQ. This idea could be addressed in the future even if this work is likely to be laborious.

As a disclaimer, the current state of our contribution doesn't allow yet the computation of certified errors in the presence of differential equations, so we mainly focus here on infinite Taylor expansions without remainders. Still, as one of our prominent future goals, certified errors were taken into account in the design stage of our framework and we discuss them along this paper.

1.2 Applications

Among many possible applications, we more specifically aim at formally verifying systems dealing with complex numerical properties, such as controllers for embedded systems. Moreover, through certified integration of ODE, we may also consider hybrid systems, such as a continuous plant coupled to a discrete controller.

1.3 Outline

We start by recalling some related works around formalization and mechanization of Taylor expansions in Sect. 2. Then, we state a mathematical formulation of our on-demand multivariate Taylor expansions with errors in Sect. 3 before introducing our implementation of data structures and operations that form an algebra in Sect. 4. We separately discuss the more complex case of composition in Sect. 5. In Sect. 6, we present some experiments done on solving differential equations in a direct way. Finally, we open up some perspectives, notably about errors, then conclude, respectively in Sects. 7 and 8.

2 Related Works

2.1 Taylor Series

Although Taylor expansions are well known and form a very rich and interesting algebra, their realizations as software items are not widespread. From a mathematical perspective, some weaknesses may explain this lack of success: they only support analytical functions, a rather limited class of functions; they don't possess good convergence properties, uniform convergence is hardly guaranteed for instance; typical applications for polynomial approximations are usually not concerned with certified errors, mean error or integrated square error (through various norms) are more important and don't easily fit into Taylor expansion schemes. Finally, from a programming perspective, Taylor expansions are: hard to implement as they require many different operations to be implemented, from low-level pure numbers to high-level abstract Taylor expansions seen as first-class citizens; error-prone with lots of complex floating-point computations on non-trivial data structures; heavily resource demanding in our multi-dimensional setting because data structures rapidly grow as the precision order increases.

Here are a few works dealing with Taylor expansions. In [4], the author presents an early application of laziness to cleanly obtain Taylor polynomial approximations. Laziness allows to augment the degree of the resulting polynomial on demand. Yet, the setting is much simpler as it is strictly one-dimensional and certified errors are not in scope. With these restrictions, the author obtains nice formulations of automatic differentiation and polynomial approximations of classical phenomena in physics. Speaking about implementation, related works come in many flavors and date back to the now well established folklore of automatic differentiation (forward or backward modes). As for symmetric tensor algebra, which forms a well-suited representation basis for partial derivatives, a huge menagerie of (mostly C++) libraries exists, for tensors of arbitrary orders and dimensions (but some libraries put a very low upper-bound on these values). These implementations are clearly not oriented towards reliability and proof of correctness, but towards mere efficiency. This also comes at the expense of some user-friendliness, as memory management and user interface are more complex and error-prone than in our own library. Still, we may consider interfacing our code base with a trusted and stable tensor library, for much better performance.

One of the most prominent implementation of Taylor expansions is the COSY tool, *cf.* [5,8]. This tool has been used in industrial-scale engineering and scientific contexts, to modelize and predict the complex dynamics of particles in accelerators for instance. This tool supports $1D$ Taylor expansions with interval-based certified errors. Polynomial degree is not refinable on demand and Taylor expansions are not handled *per se* (*i.e.* not first-class citizens). The authors managed anyway to implement an error refinement scheme for solved form ordinary differential equations, that allows solving them with tight certified errors. Experiments show that this tool compares favorably to other traditional approximations and bounding techniques, such as branch-and-bound approaches and interval arithmetics, in terms of speed and precision. We also aim at implementing differential equation solving in our multi-dimensional setting.

At the other end of the spectrum, [7] proposes correct-by-construction univariate Taylor expansions with certified errors, which appears as a huge step. Integration of floating-point errors into this scheme is also a concern addressed in [6]. Still, apart from its limitation to the $1D$ case, this approach suffers from weaknesses: expansion degree is fixed and differential equations cannot be handled. The underlying algorithm won't be so easily turned into a co-inductive (lazy) equivalent version.

And in the middle of the spectrum comes [1], where the author defines a way to handle multivariate Taylor series and presents its implementation featuring on demand computation thanks to SCHEME laziness. The few points he did not implement and that we will try to cope with in our library are: errors certification which is not handled and efficiency which is not optimal. For instance, the author's method to multiply multivariate power series is to define a generic composition between a bivariate function and a power series and to instantiate it with the multiplication. This method is simply built upon the chain rule but has some drawbacks. First, the generic equation given can usually be drastically simplified for instance in the case of multiplication and second, such a generic scheme implies that some parts of the resulting coefficients will be computed several times differently. Conversely, in our solution, the pervasive multiplication operation is implemented with a strong concern on optimality.

Our work and specifically our data-structure is based on the dissertation [9, Part 2], with the nuance that a single unbounded tree will be used instead of an infinite sequence of finite trees, each such tree representing a symmetric tensor of a given order. This choice notably enables the resolution of partial differential equations, which was impossible in the setting of [9].

2.2 Differential Equations

Iterative methods are pervasive in integrating differential equations because they often provide an efficient way to find an approximation of an ODE solution. Some of them own validation aspects, such as [2] which relies on Runge-Kutta method to integrate ODE with a numerical validation. The main difference between these methods and our work as a direct method is that we don't need these next level iterations. We are able to yield a result in the equivalent of the first iteration.

3 Formalization

We recall the canonical presentation of a multivariate Taylor expansion at order R in dimension N. This expansion converges to $f(\mathbf{x})$ when $R \rightarrow +\infty$ for an **analytical** function f only in a chosen neighbourhood of point $\mathbf{0}$.

$$f(\mathbf{x}) = \sum_{|\alpha|<R} \mathbf{D}_f^{\alpha}(\mathbf{0}) \cdot \frac{\mathbf{x}^{\alpha}}{\alpha!} + \sum_{|\alpha|=R} \mathbf{D}_f^{\alpha}(\lambda * \mathbf{x}) \cdot \frac{\mathbf{x}^{\alpha}}{\alpha!}$$

In the above formulation, $\mathbf{x} = (\mathbf{x}_0, \ldots, \mathbf{x}_{N-1}) \in \mathbb{R}^N$, $\alpha = (\alpha_0, \ldots, \alpha_{N-1}) \in \mathbb{N}^N$ indexes the derivation order of f in the symmetric tensor of partial derivatives $\mathbf{D}_f^{\alpha}$ and $\lambda \in [0, 1]$ is an unknown coefficient that characterizes the exact Taylor remainder. We have to compute derivatives both at point $\mathbf{0}$ for the polynomial part and at point $\lambda * \mathbf{x}$ for the error part. We choose to use a single co-inductive data-structure that encodes all possible derivatives, indexed by some α. As for the elements of this structure, we handle $\langle value, error \rangle$ pairs. Our framework is error-agnostic as the value-error domain is user-defined and only requires arithmetical operations. Several solutions are available in the literature: zero-centered intervals, intervals, zonotopes, etc. In the remainder, we only assume that elements of our structures form an algebra (including addition, multiplication and some elementary functions), disregarding whether they are pure values or values with errors.

This co-inductive structure, that we coin a "cotensor", enables to compute finer approximations on demand and also to lazily represent expansions of solutions to ODEs and PDEs, when they are expressed in solved form, i.e. not implicit (as it would for instance be the case if the solution were specified as a zero of a polynomial form in a functional space).

4 An Algebra of Taylor Series

4.1 Data Structure

Coefficients are present in each node of a unique tree structure and are written as $s_{o_0,\ldots,o_{N-1}}$ where every o_i is the number of occurrences of the variable x_i in the path that leads to the considered coefficient $s_{o_0,\ldots,o_{N-1}}$.

The principle is quite simple: at each node, we choose either to keep the same variable accounting for the final Taylor series, or we drop it and repeat the same process for lower dimension variables. This is pictured in tree branches of the following example as x_i for the first case and $\overline{x_i}$ for the second case. The variable at the root of the tree is X_n if the dimension is $n + 1$. This tree is developed below and represents a symmetric cotensor s of dimension 4:

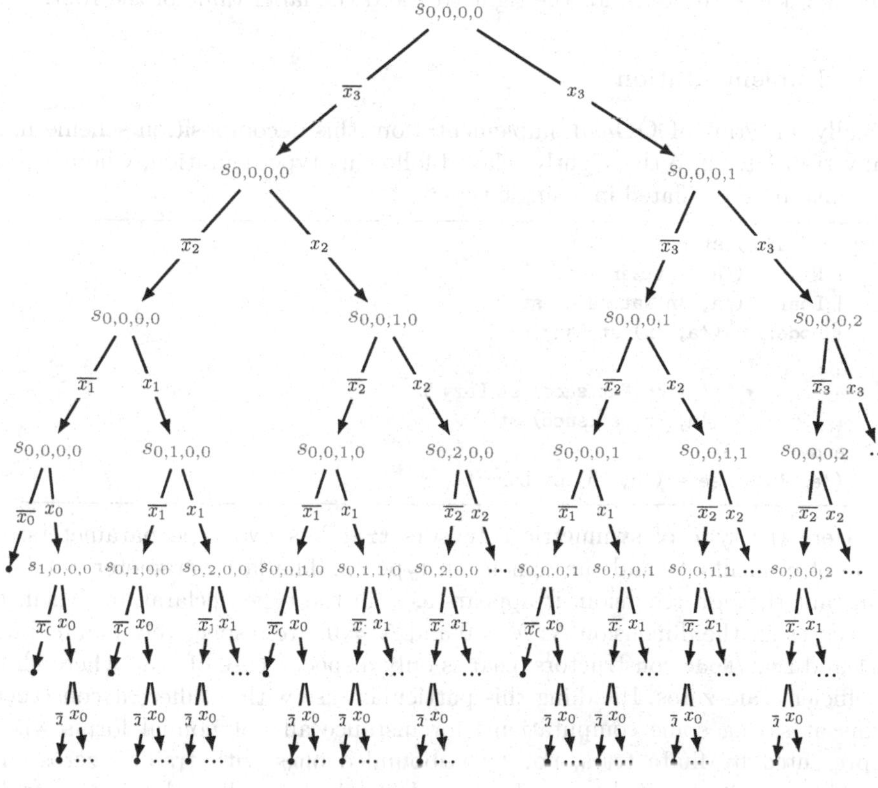

4.2 Structural Decomposition

We will introduce for this co-inductive structure a few notations inspired from the computation of the quotient and the remainder with respect to variable X_n. We will call a left cotensor a cotensor which is the left branch of another cotensor and we will denote L_{n+1} the set of left symmetric cotensors and R_{n+1} the set of right symmetric cotensors in dimension $n+1$. If V is the set of labels at the root of the tree, we have the following definitions:

$$\begin{aligned} L_{n+1} &\triangleq L_n + X_n.R_{n+1} \\ R_{n+1} &\triangleq L_n + X_n.R_{n+1} + V \\ \text{Hence :} \quad R_{n+1} &= L_{n+1} + V \end{aligned}$$

We note that the only difference between left and right cotensors is the constant part $v \in V$ and from now, we are going to consider that right case is the general one and that left case is the specification of the right case with constant part equal to 0. This will prevent us from writing similar redundant equations for all algebraic operations we will describe later. A cotensor is, then, considered a right cotensor by default, even if it has no parent because it contains a significant value $v \in V$ which is the constant part of the Taylor series. It comes then that a tree is interpreted as a Taylor series by adding together the term for the left tree, X_n times the term for the right tree and the label value of the root.

4.3 Implementation

Finally, in terms of OCAML implementation, this decomposition scheme naturally translates into the slightly relaxed following type definition, where L_n and R_n have been conflated in a single type:

```
type ('a, _) st =
  | Nil:   ('a, Nat.zero) st
  | Leaf:  ('a, 'n Nat.succ) st
  | Node:    ('a, 'n) st Lazy.t
           * 'a
           * ('a, 'n Nat.succ) st Lazy.t
           -> ('a, 'n Nat.succ) st
and
  ('a, 'n) tree = ('a, 'n) st Lazy.t
```

Here the type of symmetric cotensors `tree` has two type parameters: the type of elements `'a` and the dimension type `'n`. The last parameter not being constant through recursion, it appears as `_` in the type declaration. Then, the two cases for the dimension N: $N = 0$ and $N \neq 0$, are respectively handled with `Nil` and `Leaf`/`Node` constructors. `Leaf` is only a special case of `Node` where all the coefficients are zeros. Handling this particular case with a different constructor aims at saving some computations, for instance all polynomial forms will be represented by finite trees, not by unbounded ones with trailing zeros. And `Leaf` constructor is used to mark the end of a branch when the dimension has decreased to 0, namely all the variables has been consumed. Type parameters of constructors' arguments behave accordingly to the decomposition of R_{n+1}.

The `Nat.zero` and `Nat.succ` type constructors encode the dimensions of manipulated cotensors, as we use GADT[1] allowed by OCaml. We use a standard type-level encoding of Peano numbers and operations that we don't detail here. We hereby enforce a correct-by-construction use of our data-structures.

4.4 Component-Wise Operations

From this section onward, we assume cotensor elements form a field, with arithmetical operations on it. It may be in practice a field of coefficients or/and errors. These elements are denoted by V_A and V_B. Functions "$\lambda.\bullet$" and "$\bullet + \bullet$" straightforwardly witness the vector space structure of cotensors. The Hadamard product "$\bullet \odot \bullet$" is the component-wise product of two cotensors of same dimension. Hence with the notation $A_{n+1} \triangleq A_n^L + X_n.A_{n+1}^R + V_A$:

$$\begin{aligned} A_{n+1} + B_{n+1} &= (A_n^L + B_n^L) + X_n.(A_{n+1}^R + B_{n+1}^R) + (V_A + V_B) \\ \lambda.A_{n+1} &= \lambda.A_n^L + \lambda.X_n.A_{n+1}^R + \lambda.V_A \\ A_{n+1} \odot B_{n+1} &= (A_n^L \odot B_n^L) + X_n.(A_{n+1}^R \odot B_{n+1}^R) + (V_A * V_B) \end{aligned}$$

4.5 Multiplication

Let us define a new notation for cotensors in order to specify the multiplication. We are now going to consider that the error term is no longer separated in a precise term of the equation but is distributed in all the terms of the equation. Which gives:

$$\begin{aligned} S(X_0, ..., X_N) &= (S_0 + S_1 \odot X + S_2 \odot X^2 + ... + S_m \odot X^m + ...) \quad \text{shortened in} \\ &= (S_0 + S_1 X + S_2 X^2 + ... + S_m X^m + ...) \\ \text{where} \quad X &= (X_0, ..., X_N) \end{aligned}$$

This notation is inspired by derivation order; even if we do not consider order of cotensors; because it will be of a great help when defining the multiplication and introducing the convolution product. Product of Taylor expansions is really pervasive and appears in many operations (derivation formulas, composition of Taylor series, etc.). It is naturally defined with an explicit convolution. Concretely:

$$\begin{aligned} S(X_0, ..., X_N) \times T(X_0, ..., X_N) &= (S_0 + S_1 X + ... + S_p X^p + ...) \\ &\quad \times (T_0 + T_1 X + ... + T_q X^q + ...) \\ &= R_0 + R_1 X + R_2 X^2 + ... + R_k X^k + ... \\ \text{where} \quad \forall k \in \mathbb{N}, \quad R_k &= \sum_{i=0}^{k} S_i T_{k-i} \end{aligned}$$

[1] Generalized Algebraic Data Types.

To compute the coefficients at order k, we need to consider every product that will produce an order k, *i.e.* every coefficient of order i by every coefficient of order $k-i$, i ranging from 0 to k.

In our setting, we maintain a typed convolution structure to express computation of the term $\sum_{i=0}^{k} S_i T_{k-i}$. This structure, while geared towards static guarantees and proof of correctness, still allows for some efficient implementation. Informally, we may specify our structure as an array containing couples of cotensors of a specific dimension and that will represent absolute paths. The same structure is used to represent relative paths. We introduce a path notation, illustrated by the following examples in dimension n:

- () is the considered tree
- (n) is the tree we get when we take the n-th variable (X_n) once in the considered tree
- $(n.n.n{-}1)$ when we take the X_n variable twice and then X_{n-1} once

The so called "considered tree" is the original tree given in parameter if considering absolute paths or a specific tree (descendant of the original one) if considering relative paths. Through the relative paths (left part of the semi-colon), we will store the number of times we went down a right branch since the last left branch, namely relative paths are about the current variable and absolute paths are about all previous variables with respect to the order. Initially, the structure contains a couple of the two original trees given in parameter for both absolute paths (and the part for relative paths is empty):

$$\begin{vmatrix} () \; ; \; () \\ () \quad () \end{vmatrix}$$

Then, at each step of the algorithm:

- If the current node is a right branch, we will update the relative paths by adding the current node ($k.k$ here) and shifting the lines as follows:

$$\begin{vmatrix} () & (k) & ; \dots \\ (k) & () & \dots \end{vmatrix} \quad \text{becomes} \quad \begin{vmatrix} () & (k) & (k.k) & ; \dots \\ (k.k) & (k) & () & \dots \end{vmatrix}$$

- if the current node is a left branch, we will combine the relative paths with the absolute ones, store the result as the new absolute paths and empty the new relative paths:

$$\begin{vmatrix} () & (n{-}1) & ; & () & (n) \\ (n{-}1) & () & & (n) & () \end{vmatrix} \quad \text{becomes} \quad \begin{vmatrix} () & ; & () & (n) & (n{-}1) & (n{-}1.n) \\ () & & (n{-}1.n) & (n{-}1) & (n) & () \end{vmatrix}$$

Folding this structure to compute a term of a product simply consists in combining relative paths with absolute paths, multiplying cotensors roots column-wise and then summing these intermediate results altogether. Associating a relative path to an absolute one means concatenating them. Speaking in terms of trees, it means that the relative path begins where the absolute one ends in the tree.

4.6 Differential Operations

Cotensors of dimension N may not only be structurally decomposed on X_{N-1} but also on any other X_k, which we would call a non-structural decomposition. For that purpose, the "$\bullet[\bullet]$" function specializes a cotensor, *i.e.* drops some index by specializing it to a specific dimension k, and therefore represents the division by a monomial X_k. Conversely, the "$\bullet \uparrow \bullet$" function represents the multiplication by a monomial X_k. For a cotensor of dimension N, they are defined in terms of polynomials as:

$$\begin{aligned}
(\mathbf{S}[k])(X_0,\ldots,X_{N-1}) &\triangleq \frac{\mathbf{S}(X_0,\ldots,X_{N-1})-\mathbf{S}(X_0,\ldots,X_{k-1},0,X_{k+1},\ldots,X_{N-1})}{X_k}\\
(\mathbf{S}\uparrow k)(X_0,\ldots,X_{N-1}) &\triangleq X_k.\mathbf{S}(X_0,\ldots,X_{N-1})
\end{aligned}$$

Using the same notations as for component-wise operations, we show how these operators simply fit the structural decomposition:

$$\begin{aligned}
\mathbf{S}[k] &= (\mathbf{S}^L + X_{N-1}.\mathbf{S}^R + V_S)[k]\\
&= \begin{cases} \mathbf{S}^R, \text{ for } k = N-1 \\ \dfrac{\mathbf{S}^L + X_{N-1}.\mathbf{S}^R - \mathbf{S}^L{}_{|X_k \leftarrow 0} - X_{N-1}.\mathbf{S}^R{}_{|X_k \leftarrow 0} + V_S - V_S}{X_k} = \mathbf{S}^L[k] + X_{N-1}.\mathbf{S}^R[k], \\ \qquad\qquad\qquad\qquad\qquad\qquad \text{for } k < N \end{cases}\\
\mathbf{S}\uparrow k &= \begin{cases} \mathbf{0} + X_{N-1}.\mathbf{S}, \text{ for } k = N-1 \\ (\mathbf{S}^L + X_{N-1}.\mathbf{S}^R + V_S).X_k = \mathbf{S}^L \uparrow k + X_{N-1}.(\mathbf{S}^R \uparrow k) + V_S.X_k, \\ \qquad\qquad\qquad\qquad \text{for } k < N \end{cases}
\end{aligned}$$

Differential operations introduce partial differentiation and integration in the cotensor algebra. These differentiation and integration operators respectively refer to $\mathbf{S}[\bullet]$ and $\mathbf{S} \uparrow \bullet$. They also use the cotensor of integration/derivation factors "Δ_k", where the o_i are the variable occurrence number, such that:

$$\begin{aligned}
(\Delta_k)_{(o_0,\ldots,o_{N-1})} &\triangleq 1 + o_k, \text{ for } \textstyle\sum_i o_i = R\\
\frac{d\mathbf{S}(X_0,\ldots,X_{N-1})}{dX_k} &\triangleq \mathbf{S}[k] \odot \Delta_k\\
\int_0^{X_k} \mathbf{S}(X_0,\ldots,x_k,\ldots,X_{N-1})dx_k &\triangleq (\mathbf{S} \odot \Delta_k^{-1}) \uparrow k
\end{aligned}$$

5 The Composition Operator

5.1 Differential Method

Principle. The Taylor series algebra with the previous operations still remains basic, and that is why we are now interested in composing Taylor series with elementary functions. To do so, we only need to apply elementary functions to arbitrary arguments, *i.e.* to compose univariate Taylor series with multivariate ones. A general composition scheme of Taylor series is also possible in our setting but out of the scope of our current concerns. This method lies on a differential decomposition, namely a function is the sum of the integrals of its derivatives with respect to all its variables, plus a constant term:

$$H : \mathbb{R}^N \to \mathbb{R}, \qquad H = H(0) + \sum_{i<N} \int^{X_i} \frac{\partial H}{\partial X_i}\Big|_{\substack{X_k=0 \\ k>i}} dX_i$$

Example. We need to partially evaluate the derivatives at **0** to avoid counting several times the parts shared by different variables, as illustrates the following concrete example:

let $F : \mathbb{R}^3 \to \mathbb{R}, \qquad F(x, y, z) = x^3 + 2x^2y + xz + 5y^2 + 3yz^2$

$$\begin{cases} \frac{\partial f}{\partial x} = 3x^2 + 4xy + z & \int_0^x \frac{\partial f}{\partial x} dx = x^3 + 2x^2y + xz \\ \frac{\partial f}{\partial y} = 2x^2 + 10y + 3z^2 & \int_0^y \frac{\partial f}{\partial y} dy = 2x^2y + 5y^2 + 3yz^2 \\ \frac{\partial f}{\partial z} = x + 6yz & \int_0^z \frac{\partial f}{\partial z} dz = xz + 3yz^2 \end{cases}$$

The blue terms are redundant and that is why we have:

$$F(x, y, z) = F(0, 0, 0) + \int_0^x \frac{\partial f}{\partial x} dx + \int_0^y \frac{\partial f}{\partial y}\Big|_{x=0} dy + \int_0^z \frac{\partial f}{\partial z}\Big|_{\substack{x=0 \\ y=0}} dz$$

Composition. As we are in the specific case of composition, we will use the classic chain rule:

$$\frac{\partial (f \circ g)}{\partial X_i}_{i<N} = (\frac{\partial g}{\partial X_i})_{i<N} \times (f' \circ g)$$

Hence :

$$f \circ g = f \circ g(0) + \sum_{i<N} \int^{X_i} (\frac{\partial g}{\partial X_i} \times f' \circ g)\Big|_{\substack{X_k=0 \\ k>i}} dX_i$$

The computation of the partial derivatives $\frac{\partial (f \circ g)}{\partial X_i}_{i<N}$ is done case by case with respect to the elementary function f at use, each such function having a well-known derivative f'. The cases where $f = \exp, \sin, \cos, \log, \text{atan}, x^a, \ldots$ are easily handled. So, according to the above equation, we only need to partially evaluate these derivatives, to integrate them then and to finally sum the results.

This method will bring us satisfying results as detailed below, but one must bear in mind that despite the method is very short in terms of code and then easily implemented, it is not optimal in terms of computation. This differential method for the composition is not canonical in that it does not compute the minimum number of operations to produce the coefficients of the result. As a witness of non canonicity in the definition of composition, the Δ_k coefficients will be used for multiplication and division consecutively, which could be avoided. Besides, as long as we do not handle certified errors, the method does not need an additive decomposition of f but it will be the case as soon as we handle the errors and we will have to deal with this constraint.

5.2 Elementary Functions

Elementary functions, limited to one argument functions, are specified as univariate Taylor series. Therefore, as only one branch of the cotensor will be meaningful, such series are treated separately. This is only a matter of efficiency and obviously not mandatory. To obtain a Taylor expansion of an elementary function, we need to be able to compute any n-th derivative. Taylor series for elementary functions are well known, so the first way to produce such a series is to compute the coefficients iteratively and lazily with respect to the known formulas, such as the following ones:

$$\begin{aligned}
\exp(x) &= \sum_{i\in\mathbb{N}} \frac{x^i}{i!}\\
\log(1+x) &= \sum_{i\in\mathbb{N}} \frac{-(-x)^i}{i}\\
(1+x)^p &= \sum_{i\in\mathbb{N}} \binom{p}{i} x^i\\
\sin(x) &= \sum_{i\in\mathbb{N}} \frac{(-1)^i}{(2i+1)!} x^{2i+1}\\
\cos(x) &= \sum_{i\in\mathbb{N}} \frac{(-1)^i}{(2i)!} x^{2i}
\end{aligned}$$

Similar formulations are available for elementary functions not presented here.

6 Experimentation

Now that the main operations are available in our algebra, we can start using it. Differential equations are pervasive in dynamical systems and our point is to propose a direct (*i.e.* non-iterative) way to solve them. By direct method, we mean that coefficients are computed once and for all and therefore there is no need to iterate over their values until a specific precision is reached. Precision in our case is seen differently: coefficients are computed only once and if the user wants a finer precision, the user will increase the order of derivation which means that new and deeper coefficients will be computed.

6.1 Airy Equation

To illustrate this direct approach for solving ODEs and PDEs, we will use the first dimension Airy equation which stands as follows:

$$f'' - xf = 0$$

As the equation contains a second derivative, we split it for convenience in two first order equations introducing *f_dot* as f derivative:

$$\begin{cases} f\_dot = f\_dot_0 + \int^x xf \\ f = f_0 + \int^x f\_dot \end{cases}$$

Then, thanks to OCaml laziness, we express and solve this mutually recursive system directly, with the following principle:

- According to the second equation, computing the first coefficient of f, the constant part, means summing the constant part of f_0 with the constant part of $\int_x f\_dot$. We know that the constant part of an integral will be 0, whatever the integrand is.
- the first coefficient of f_dot, or equivalently the second coefficient of f, is computed the same way (no need to evaluate the argument of the integral).
- then the mutual recursion works and the third coefficient of f, or the second one of f_dot, is simply the result of integrating the constant part of xf, actually 0. The other coefficients are also computed in finite time.

So the trick is to stay a step ahead by computing a first coefficient of a recursive Taylor series without having to evaluate itself, thanks to the integral operator, and then to keep this advance all along the computation so that the recursion will always end. Indeed, if the computation scheme respects the causality, for example in one dimension: computing a coefficient requires only strictly lower order coefficients, then we can ensure the recursion will end.

Once we get the solution up to a specific order, we evaluate it as a polynomial function so that we can draw its graph (Figs. 1 and 2):

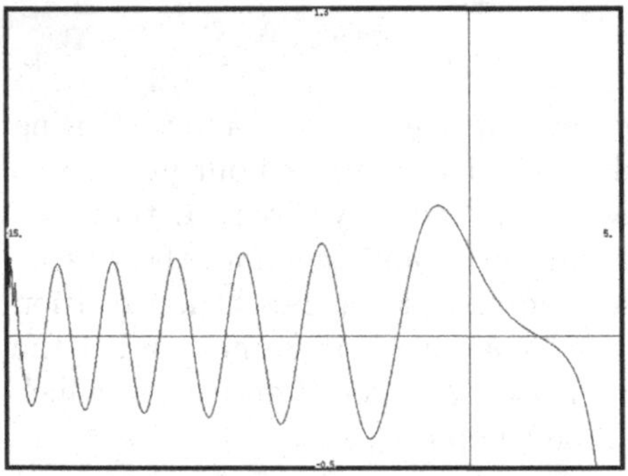

Fig. 1. Our function (at order 150)

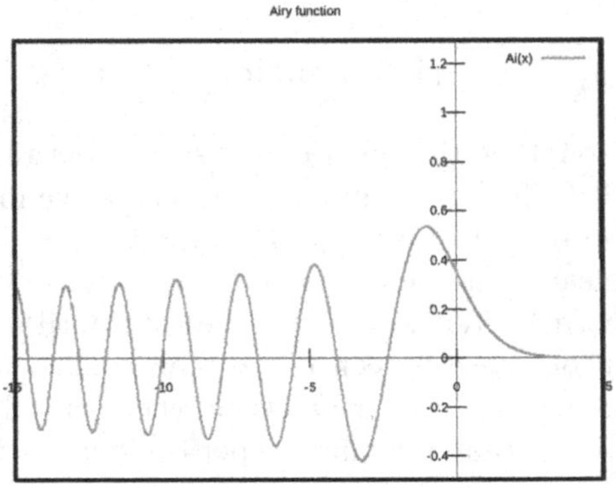

Fig. 2. Theoretical result

We can observe that the approximation is reliable on a specific interval and diverge outside of it. We can have this conclusion because we know the theoretical result in this case, but we won't know it in most cases. This is what will motivate the necessary handling of certified errors. Intervals of errors, which are only an example of error representation, will give the user information about how far the theoretical function could be from the returned approximation.

6.2 Heat Equation

In order to explain the principle of causality more precisely and to show a more general case, we are going to present the 2-dimensional heat equation example:

$$\frac{\partial u}{\partial t} = \alpha \frac{\partial^2 u}{\partial x^2}$$

There are 2 different ways of integrating this equation and we chose to integrate it with respect to variable t so that initial conditions are a function of variable x at initial time $t = 0$. Here is the new form of the equation:

$$u(x,t) = u_0(x) + \alpha \times \int^t \frac{\partial^2 u(x,t)}{\partial x^2}$$

where $u_0(x)$ will be a data we have. The causality is respected if computing any derivative $\frac{\partial^{i+j} u}{\partial x^i \partial t^j}$ boils down to compute elements of initial condition $u_0(x)$. And in the case of the heat equation, we can ensure it will be possible thanks to *Schwarz*'s theorem about switching partial derivatives:

$$\frac{\partial^{i+j} u}{\partial x^i \partial t^j} = \frac{\partial^{i+j-1} u}{\partial x^i \partial t^{j-1}} \left(\frac{\partial u}{\partial t}\right) = \frac{\partial^{i+j-1} u}{\partial x^i \partial t^{j-1}} \left(\frac{\partial^2 u}{\partial x^2}\right) = \frac{\partial^{i+j+1} u}{\partial x^{i+2} \partial t^{j-1}} = ... = \frac{\partial^{i+2j} u}{\partial x^{i+2j}}$$

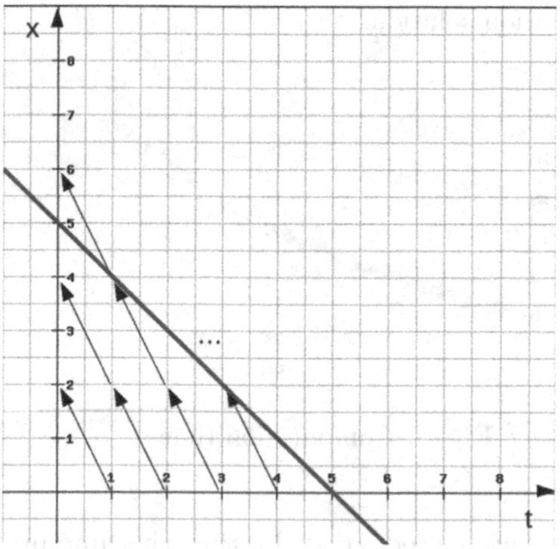

This graph illustrates the dependencies between the partial derivatives and we see that all arrows will end up on the vertical axis which represents the derivatives with respect to x only, namely the different parts of $u_0(x)$. The causality being respected ensures that the recursion will end. This example in 2 dimensions shows how the principle of causality is more flexible than it was presented with the Airy equation. Indeed, we said that coefficients of specific order should require strictly lower order coefficients, which is graphically represented by arrows crossing the blue line from the top right-hand corner down to bottom left-hand corner. But we state now that it is not a necessary condition as we can see with the heat equation where higher order coefficients are required but with respect to other variables. So arrows are allowed to cross the blue line in the opposite direction as long as they end on the vertical axis.

Figure 3 shows our heat equation solution developed at order 25. The vertical axis is the temperature. We set the initial conditions to a sinus, which concretely means we impose the temperature on one axis to be an alternation of warm and cold at initial time. The graph converges to a uniform average value along the time which is consistent with the physical interpretation.

What we call order here and denote by R is only the unrolling depth of the infinite tree we build. The graph in Fig. 4 shows the computation times (in seconds, on a common laptop computer) of the heat equation solution according to order and the graph in Fig. 5 shows this computation time divided by the number of coefficients of the solution, which lies in $\theta(R^N)$ with N the dimension, according to [9]. By dividing the computation time by the number of computed

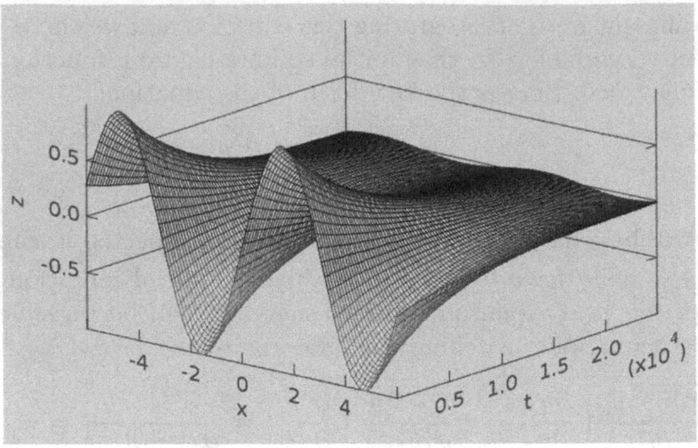

Fig. 3. Heat equation solution

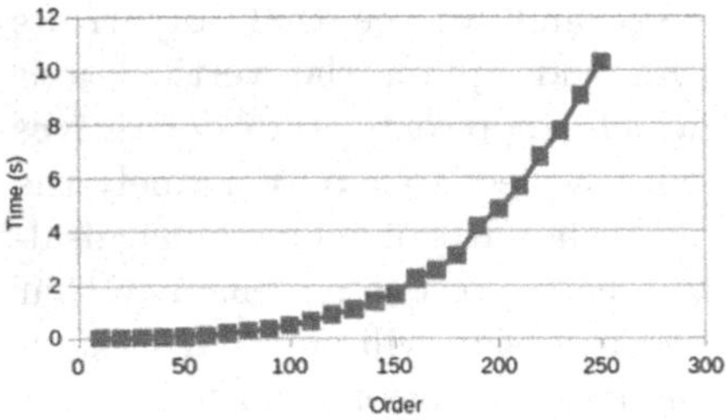

Fig. 4. Computation time

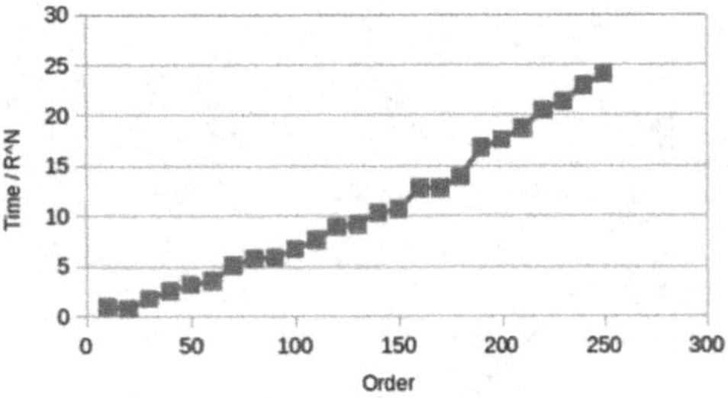

Fig. 5. Computation time/R^2

coefficients (normalized to 1 for $R = 0$), we aimed at evaluating the amount of additional computation done per useful coefficient, *i.e.* the "administrative" overhead induced by the resolution of the equation, due to auxiliary data structures, memory allocations, etc. We observe only a linear overhead and despite the relative simplicity of the heat equation, it comforts us in the decisions taken so far for implementing our framework.

7 Perspectives

7.1 Canonical Method for Composition

As defined in Sect. 5, composition involves the resolution of a partial differential equation. This hinders the computation of error bounds. Indeed, as far as we know, there is no established general method to solve such equations with certified errors, beyond ad-hoc situations such as elliptic, parabolic, etc., equations with specific initial conditions.

In order to devise a direct more tractable and non recursive way to compose Taylor series, following schemes such as Faà di Bruno's formula, we first need to handle errors. As in formal power series, composition $(f \circ g)$ may be achieved only when g has no constant part. To factorize out the constant part of g (so that we fall back to evaluation at point $\mathbf{0}$), we depend on an additive decomposition of f, when available.

Again, we sum up some decompositions of standard elementary functions. For every $A_{n+1} \in R_{n+1}$, we have the following equations, where we remark that their right-hand sides are built from a constant part (V_A) and another term without a constant part:

$$\begin{aligned}
\exp\big(A_n^L + X_n.A_{n+1}^R + V_A\big) &= \exp(V_A)\exp\big(A_n^L + X_n.A_{n+1}^R\big)\\
\log\big(A_n^L + X_n.A_{n+1}^R + V_A\big) &= \log(V_A) + \log\Big(1 + \tfrac{A_n^L + X_n.A_{n+1}^R}{V_A}\Big)\\
\sin\big(A_n^L + X_n.A_{n+1}^R + V_A\big) &= \sin(V_A)\cos\big(A_n^L + X_n.A_{n+1}^R\big)\\
&\quad + \cos(V_A)\sin\big(A_n^L + X_n.A_{n+1}^R\big)\\
\arctan\big(A_n^L + X_n.A_{n+1}^R + V_A\big) &= \arctan(V_A) + \arctan\Big(\tfrac{A_n^L + X_n.A_{n+1}^R}{1 + V_A.(A_n^L + X_n.A_{n+1}^R)}\Big)
\end{aligned}$$

We are currently developing a canonical composition operator $f \circ g$ following decomposition schemes that are all well known to strongly involve combinatorial reasoning. Our preliminary results already show that the administrative content of such heavy combinatorial computations, such as iterating over partitions, combinations, permutations and so on, have a great cost and are not yet on a par with the differential approach in terms of efficiency, at least for the tested instances. More investigation is required in that respect. We still expect to obtain an efficient canonical solution, with a simpler error propagation scheme and furthermore less computations to reduce such propagation.

7.2 Certified Errors

Taylor Models. Differential equations put aside, we are already able to compute certified errors in our framework. It merely requires the introduction of an arithmetical domain for errors. We introduce below a very simple error domain based upon symmetric zero-centered monotonic error functions.

Let us assume $\mathbb{K}$ stands for the value domain. Error functions are then elements of the following domain $\mathbb{E}$, assuming we work in dimension N:

$$\mathbb{E} \triangleq \{f \in (\mathbb{K}^+)^N \to \mathbb{K}^+ \mid f(\mathbf{0}) = 0,\ f \text{ monotonous}\}$$

The error model is then the product $\mathbb{K}\times\mathbb{E}$. The semantics $[\![\,.\,]\!]$ of an element of this model represents a function from variable bounds to sets of possible values:

$$[\![\langle v, \epsilon\rangle]\!] \triangleq \mathbf{X} \in (\mathbb{K}^+)^N \mapsto \{k \in \mathbb{K} \mid |k - v| \leq \epsilon(\mathbf{X})\}$$

The error model has $N+1$ constructors: $(k, \mathbf{0})$ for $k \in \mathbb{K}$, denoted "k" and the $i \in [0, N-1]$ indexed family $(0, \mathbf{X} \mapsto \mathbf{X}_i)$, denoted "$\mathbf{X}_i$". It is endowed with a $\mathbb{K}$-algebra structure and is further turned into an full-fledged domain using suitable definitions of elementary functions on $\mathbb{K} \times \mathbb{E}$, as illustrated below. Similar definitions may be devised for other elementary functions:

$$\begin{array}{ll}\langle v_1, \epsilon_1\rangle + \langle v_2, \epsilon_2\rangle & \triangleq \langle v_1 + v_2, \epsilon_1 + \epsilon_2\rangle \\ \alpha \times \langle v, \epsilon\rangle & \triangleq \langle \alpha \times v, |\alpha| \times \epsilon\rangle \\ \langle v_1, \epsilon_1\rangle \times \langle v_2, \epsilon_2\rangle & \triangleq \langle v_1 \times v_2, |v_1| \times \epsilon_2 + |v_2| \times \epsilon_1 + \epsilon_1 \times \epsilon_2\rangle \\ e^{\langle v, \epsilon\rangle} & \triangleq \langle e^v, e^v \times (e^\epsilon - 1)\rangle \\ \log\langle v, \epsilon\rangle & \triangleq \langle \log v, \log(1 + \frac{\epsilon}{v})\rangle \quad (v \neq 0)\end{array}$$

Taylor models are then built from cotensors of $\langle value, error\rangle$ terms. We consider a function $f \in \mathbb{R}^N \to \mathbb{R}$, assumed analytical at point $\mathbf{0}$ and note respectively f_α and ϵ_α as the value and error at derivation multi-index α.

A Taylor model predicate $\mathcal{TM}(f, R, \delta)$ at order R in a δ-neighbourhood of point $\mathbf{0}$ (where $\delta \in {\mathbb{R}^+}^N$) is defined as the following:

$$\mathcal{TM}(f, R, \delta) \triangleq \forall \mathbf{x} \in \mathbb{R}^N . |\mathbf{x}| \leq \delta \implies |f(\mathbf{x}) - \sum_{\alpha=0}^{|\alpha| \leq R} f_\alpha \mathbf{x}^\alpha| \leq \sum_{|\alpha|=R} \epsilon_\alpha(\delta)|\mathbf{x}|^\alpha$$

A Taylor model for parameters R and δ is then the set of functions f such that $\mathcal{TM}(f, R, \delta)$ holds true.

Issues with Recursive Definitions. We recall that the above definitions must be amended in order to account for errors in (recursive) differential equations. Indeed, in that case, dependencies between errors at different derivation orders do not respect the causality relation fulfilled by pure values. So we need to compute another fixed point, different from the one for pure values. We illustrate this discrepancy between values and errors, considering the following partial development of a Taylor series with errors for a bivariate function f:

$$f(X, Y) \triangleq \langle f_0, \epsilon_0\rangle + X.\langle f_X, \epsilon_X\rangle + Y.\langle f_Y, \epsilon_Y\rangle + \ldots$$

Then, integrating f along X, accounting for errors, yields the following series:

$$\int^X f = \langle 0, |X|.(|f_0| + \epsilon_0)\rangle + X.\langle f_0, \epsilon_0\rangle + Y.\langle 0, |X|.(|f_Y| + \epsilon_Y)\rangle + \ldots$$

Unfortunately, we remark that the error term $|X|.(|f_0| + \epsilon_0)$ at order 0, while still a zero-centered monotonic error function, directly depends on ϵ_0, the error function of f at order 0. The same problem occurs at order Y. On the contrary, the value part of the integrand is always 0, so is independent of f. As we wish to define f recursively through such an integrand, setting for instance $f = \int^X f$, we face the necessity to find a different computation scheme for errors than for values. This is left for future work, but we feel that it would probably imply to transpose in our multivariate setting the kind of argumentation found in the Picard-Lindelôf theorem (that determines existence and unicity of solutions to ODEs in solved forms).

Going Further. Many other sensible choices for computing errors are also possible such as arbitrary intervals, zonotopes, etc., but we haven't experimented with these solutions yet. We chose to stick to the lightweight zero-centered error domain, giving up some precision to save computation time, mostly because it is much simpler to implement and also because we rely on on-demand cotensor exploration to increase precision, by computing deeper coefficients of Taylor expansions. We nevertheless plan to address the problem of finding a well-suited error domain, in terms of precision with respect to computation time.

Accounting for numerical errors is also on our roadmap. As a first approach, we postulate that we would only have to represent every real number with an interval of lower and upper approximations given as two floating-point numbers, lifting every computation from an algebra of real numbers to an algebra of floating-point intervals. The main question will be to test whether accumulating numerical errors along a huge number of computations could significantly degrade precision, as the derivation order increases, jeopardizing the core feature of our framework.

Another method, closely related to our own functional language framework exploiting laziness, would be to consider using a setup for exact real number algebra, as illustrated for instance in [3]. Besides its lack of efficiency wrt. floating-point numbers, it would not suffer from a potential untamable accumulation of errors and would also open the way for a complete formal verification (including tensorial structure and numerical aspects). This is left for future work.

8 Conclusion

With a renovated view on Taylor series, we provide an implementation of a genuine full-fledged algebra of such series, in the multivariate case. Even if the work is far from being completed, it has been proven useful already as we are able to deal smoothly with partial differential equations in solved form, without any input from domain expert. To the best of our knowledge, implementing such an algebra of Taylor series with a concern on efficiency through carefully crafted algorithmics but also on correctness through strong typing has not been tried before. Indeed, although not presented here, our implementation puts an emphasis on strong typing, through extensive use of advanced OCaml GADT features. This proved really helpful in designing correct-by-construction code, at least with respect to dimensions and derivation orders, while implementing complex and error-prone numerical computations.

The next big challenges to take up are: first, the introduction of a better composition scheme; second, error domains and computation schemes compatible with every construction of our algebra. This would pave the way for applying our library in the paradigm of guaranteed integration for instance, notwithstanding other pervasive usages of Taylor series in various scientific fields.

References

1. Pearlmutter, B.A., Siskind, J.: Lazy multivariate higher-order forward-mode AD. In: POPL 2007, January 2007
2. Sandretto, J.A.D., Chapoutot, A.: Validated explicit and implicit Runge-Kutta methods. Reliable Comput. **22**(1), 79–103 (2016)
3. Geuvers, H., Niqui, M., Spitters, B., Wiedijk, F.: Constructive analysis, types and exact real numbers. Math. Struct. Comput. Sci. **17**(1), 3–36 (2007). https://doi.org/10.1017/S0960129506005834
4. Karczmarczuk, J.: Functional differentiation of computer programs. High. Order Symbolic Comput. **14**(1), 35–57 (2001). https://doi.org/10.1023/A:1011501232197
5. Makino, K., Berz, M.: Rigorous integration of flows and ODEs using Tylor models. In: Symbolic Numeric Computation, SNC 2009, Kyoto, 03–05 August 2009, pp. 79–84 (2009). https://doi.org/10.1145/1577190.1577206
6. Martin-Dorel, É., Hanrot, G., Mayero, M., Théry, L.: Formally verified certificate checkers for hardest-to-round computation. J. Autom. Reasoning **54**(1), 1–29 (2015). https://doi.org/10.1007/s10817-014-9312-2
7. Martin-Dorel, É., Rideau, L., Théry, L., Mayero, M., Pasca, I.: Certified, efficient and sharp univariate Taylor models in COQ. In: 15th International Symposium on Symbolic and Numeric Algorithms for Scientific Computing, SYNASC 2013, Timisoara, 23–26 September 2013, pp. 193–200 (2013). https://doi.org/10.1109/SYNASC.2013.33
8. Revol, N., Makino, K., Berz, M.: Taylor models and floating-point arithmetic: proof that arithmetic operations are validated in COSY. J. Log. Algebraic Program. **64**(1), 135–154 (2005). https://doi.org/10.1016/j.jlap.2004.07.008
9. Thirioux, X.: Verifying embedded systems. Habilitation thesis, Institut National Polytechnique de Toulouse, France, September 2016

Solving the Expression Problem in C++, á la LMS

Seyed Hossein Haeri[1(✉)] and Paul Keir[2]

[1] Université catholique de Louvain, Louvain-la-Neuve, Belgium
hossein.haeri@uclouvain.be
[2] University of the West of Scotland, Glasgow, UK
paul.keir@uws.ac.uk

Abstract. We give a C++ solution to the Expression Problem that takes a components-for-cases approach. Our solution is a C++ transliteration of how Lightweight Modular Staging solves the Expression Problem. It, furthermore, gives a C++ encoding to object algebras and object algebra interfaces. We use our latter encoding by tying its recursive knot as in Datatypes à la Carte.

1 Introduction

The Expression Problem (EP) [6,31,37] is a recurrent problem in Programming Languages, for which a wide range of solutions have been proposed. Consider those of Torgersen [35], Odersky and Zenger [20], Swierstra [34], Oliveira and Cook [23], Bahr and Hvitved [2], Wang and Oliveira [38], Haeri and Schupp [16], and Haeri and Keir [12], to name a few. EP is recurrent because it is repeatedly faced over embedding DSLs – a task commonly taken in the PL community. Embedding a DSL is often practised in phases, each having its own Algebraic Datatype (ADT) and functions defined on it. For example, take the base and extension to be the type checking and the type erasure phases, respectively. One wants to avoid recompiling, manipulating, and duplicating one's type checker if type erasure adds more ADT cases or defines new functions on them.

Haeri [11] phrases EP as the challenge of implementing an ADT – defined by its cases and the functions on it – that:

E1. is *extensible in both dimensions*: Both new cases and functions can be added.
E2. provides *weak static type safety*: Applying a function f on a statically[1] constructed ADT term t should fail to compile when f does not cover all the cases in t.

[1] If the guarantee was for dynamically constructed terms too, we would have called it strong static type safety.

This work is partially funded by the LightKone European H2020 Project under Grant Agreement No. 732505 and partially by the Belgian National Fund for Scientific Research (F.R.S.-FNRS).

R. M. Hierons and M. Mosbah (Eds.): ICTAC 2019, LNCS 11884, pp. 353–371, 2019.
https://doi.org/10.1007/978-3-030-32505-3_20

E3. upon extension, forces *no manipulation or duplication* to the existing code.
E4. accommodates the extension with *separate compilation*: Compiling the extension imposes no requirement for repeating compilation or type checking of existing ADTs and functions on them. Compilation and type checking of the extension should not be deferred to the link or run time.

On the other hand, Rompf and Odersky [32] coin Lightweight Modular Staging (LMS) for Polymorphic Embedding [17] of DSLs in Scala. They employ a fruitful combination of the Scala features detailed in [21] that, as a side-product, offers a very simple yet effective solution to the EP. We call that side-product the "Scala LMS-EPS." In this paper, we offer a new C++ solution that is greatly inspired by the Scala LMS-EPS. We call our own solution the "C++ LMS-EPS."

Amongst the EP solutions, LMS is distinctive for its ease of extension: both in adding new ADT cases and functions defined on them. We chose to implement LMS in C++ to show the independence of LMS from Scala's combination of following features: `traits`, abstract type members, and `super` calls. Instead, the C++ LMS-EPS makes use of the following C++ features: curiously recurring template pattern (Sect. 3.2), abbreviated function templates of C++20[2] (Sect. 3.3), user-defined deduction guides and variadic `templates` (Sect. 5.3), and, most notably, `std::variant` (Sect. 3.1). (For the unfamiliar reader, an introduction to those C++ features comes in Appendix A.) Unlike Scala, C++ is a mainstream language which is well-known for its efficiency. Similar to Scala, C++ is a multi-paradigm language with a high level of abstraction (from C++17 onward).

Given its presentation in C++, the C++ LMS-EPS machinery may look as an EP solution that is too specific to C++. In order to correct that impression, we recall that it is typical for EP solutions to be presented with tactful uses of a single language. Take Datatypes à la Carte [34], CDTs [1], PCDTs [2], and MRM [25] in Haskell, Polymorphic Variants [9] in OCaml, and LMS [32] and MVCs [22] in Scala. The C++ LMS-EPS is amongst the few EP solutions which are presented in a mainstream programming language.

Here is a list of our contributions:

The C++ LMS-EPS takes a *components-for-cases* (C4C) [11] approach (Sects. 2.1 and 3.1). It implements ADTs (Sect. 3.2) using an encoding of object algebra interfaces [26] that is akin to Swierstra's sum of functors [34]. We tie the recursive knot using F-Bounding [4]. To implement functions on ADTs (Sect. 3.3), the C++ LMS-EPS gets a simulation of Haskell's Combinator Pattern [28, Sect. 16] (Sect. 5.3) to first acquire an encoding of object algebras. Our latter encoding, however, does not use *self*-references [27]. The C++ LMS-EPS outperforms its Scala predecessor by ensuring strong static type safety (Sect. 4.2). The way to distinguish between the C++ LMS-EPS and EP solutions that use Generalised Algebraic Datatypes (GADTs) is in Sect. 6. Detailed discussion on the related work comes in Sect. 7.

---

[2] Although our codebase remains fully functional without that (using ordinary type parametrisation), we retain its usage here for enhanced readability.

2 Background

2.1 Formal Notation

In this paper, we use parts of the $\gamma\Phi C_0$ calculus developed for solving the Expression Compatibility Problem [15]. $\gamma\Phi C_0$ was developed after observing that sharing ADT cases amongst ADTs is not limited to ADTs only extending one another. For example, consider the ADTs α_1, α_2, and α_3 defined as: $\alpha_1 ::= \mathit{Num}(\mathbb{Z}) \mid \mathit{Add}(\alpha_1)$, $\alpha_2 ::= \mathit{Num}(\mathbb{Z}) \mid \mathit{Add}(\alpha_2) \mid \mathit{Mul}(\alpha_2)$, and $\alpha_3 ::= \mathit{Num}(\mathbb{Z}) \mid \mathit{Add}(\alpha_3) \mid \mathit{Sub}(\alpha_3)$. Both α_2 and α_3 extend α_1. But, neither of them is an extension to the other. In order to share the implementation effort required for encoding α_2 and α_3, then $\gamma\Phi C_0$ promotes the ADT cases to *components* (in their Component-Based Software Engineering [33, Sect. 17], [29, Sect. 10] sense).

In $\gamma\Phi C_0$, ADT cases are independent of ADTs but still parameterised by them. In the $\gamma\Phi C_0$ notation, one would write $\alpha_1 = \mathit{Num} \oplus \mathit{Add}$, $\alpha_2 = \mathit{Num} \oplus \mathit{Add} \oplus \mathit{Mul}$, and $\alpha_3 = \mathit{Num} \oplus \mathit{Add} \oplus \mathit{Sub}$. In the $\gamma\Phi C_0$ ADT definitions, what comes to the r.h.s. of the "=" is called the *case list* of the ADT on the l.h.s. of the "=". The connection between $\gamma\Phi C_0$ and C4C becomes more clear in Sect. 3.1. Hereafter, we refer to α_1 as *NA* (for Numbers and Addition) and to α_2 as *NAM* (for Numbers, Addition, and Multiplication).

2.2 The Scala LMS-EPS

Suppose one is interested in encoding *NA* and in evaluating its expressions. One possible Scala implementation is:

```
1 trait NA {
2   trait Exp                                          //Exp ::=
3   case class Num(n: Int)            extends Exp      //     Num(n) |
4   case class Add(l: Exp, r: Exp) extends Exp         //     Add(Exp, Exp)
5   def eval: Exp => Int = {
6     case Num(n)     => n
7     case Add(l, r) => eval(l) + eval(r) } }
```

Scala uses inheritance for definition of ADT cases. In lines 3 and 4 above, for example, `Num` and `Add` inherit from their ADT type, i.e., `Exp`. Implementing *NAM* without manipulation or duplication of `NA` can now be done as:

```
1 trait NAM extends NA {                             //Exp ::= ... |
2   case class Mul(l: Exp, r: Exp) extends Exp       //     Mul(Exp, Exp)
3   override def eval: Exp => Int = {
4     case Mul(l, r) => eval(l) * eval(r)
5     case e          => super.eval(e)} }
```

Line 2 above adds the new case (`Mul`). Line 4 above handles its evaluation. And, line 5 above makes a **`super`** call to employ the evaluation already defined at `NA`. Note that `NAM` inherits `Num` and `Add` because it **`extends`** `NA`.

Addition of a function on *NA* whilst addressing E3 and E4 is similar. For example, here is how to provide pretty printing:

```
trait NAPr extends NA {
  def to_str: Exp => String = {
    case Num(n)     => n.toString
    case Add(l, r) => to_str(l) + " + " + to_str(r)} }
```

3 The C++ Version

C++ offers no built-in support for ADTs. Neither does it support mixin-composition for a **super** call to be possible. The C++ LMS-EPS mitigates those by exercising a coding discipline that is explained in Sects. 3.1 to 3.3. Term creation and application of functions on that comes in Sect. 3.4.

3.1 Cases

An EP solution takes a C4C approach when each ADT case is implemented using a standalone component that is ADT-parameterised. In the C++ LMS-EPS, the ADT-parametrisation translates into type-parametrisation by ADT. For example, here are the C++ counterparts of `Num` and `Add` in Sect. 2.2:

```
1 template<typename ADT> struct Num {//Num α : ℤ → α
2   Num(int n): n_(n) {}
3   int n_;
4 };
```

Above comes a C4C equivalent of `Num` in Sect. 2.2. Verbosity aside, an important difference to notice is that `Num` in Sect. 2.2 is a case for the ADT `Exp` of Sect. 2.2, **exclusively**. On the contrary, the above `Num` is a case for the encoding of **every** ADT α such that $Num \in cases(\alpha)$. The `Add` below is similar.

```
1 template<typename ADT> struct Add {//Add α : α × α → α
2   using CsVar = typename ADT::cases;
3   Add(const CsVar& l, const CsVar& r):
4     l_(std::make_shared<CsVar>(l)), r_(std::make_shared<CsVar>(r)) {}
5   const std::shared_ptr<CsVar> l_, r_;
6 };
```

Terms created using `Add`, however, are recursive w.r.t. their ADT. That is reflected in line 5 with the `l_` and `r_` data members of `Add` being shared pointers to the case list of `ADT`, albeit packed in a `std::variant`. (See line 2 in `NATemp` below.) Line 2 is a type alias that will become more clear in Sect. 3.2. The need for storing `l_` and `r_` in `std::shared_ptrs` is discussed in Sect. 5.1.

We follow the terminology of C++ IDPAM[3] [12] in calling `Num` and `Add` of this section and similar C4C encodings of ADT cases the *case components*.

3.2 ADTs

Defining ADTs in the C++ LMS-EPS is less straightforward:

```
1 template<typename ADT> struct NATemp
2 {using cases = std::variant<Num<ADT>, Add<ADT>>;};
3 struct NA: NATemp<NA> {};
```

In Swierstra's terminology [34], lines 1 and 2 define a recursive knot that line 3 ties. In the terminology of Oliveira et al. [26], `NATemp` is an object algebra interface. That is because `NATemp` declares a set of algebraic signatures (namely, that of `Num` and `Add`) but does not define (implement) them. In other words, those signatures do not pertain to a fixed ADT.

[3] Integration of a Decentralised Pattern Matching.

What matters to the C++ LMS-EPS is that `NATemp` underpins every `ADT`, for which instances of `Num<ADT>` or `Add<ADT>` are valid terms. (Using $\gamma\Phi C_0$, one denotes that by $\forall\alpha.\ \alpha \lhd \mathit{Num} \oplus \mathit{App}$.) Given `NATemp`, in line 3, we introduce `NA` as an instance of such `ADTs`. That introduction is done in a specific way for F-Bounding [4] commonly referred to in C++ as the Curiously Recurring Template Pattern (CRTP) [36, Sect. 21.2]. See Sect. 5.2 for why employing CRTP is required here.

The nested type name `cases` at line 2 above is what we used in the definition of `CsVar` at line 2 of `Add` in Sect. 3.1.

3.3 Functions

Just like that for ADTs, defining functions on ADTs takes two steps in the C++ LMS-EPS:

First, for a function `f` on an ADT `A`, one implements an auxiliary function that takes a continuation as an argument. Suppose that one chooses the name `a_plus_f_matches` for the auxiliary function. (See below for the intention behind the naming of the auxiliary function.) Using the continuation, `a_plus_f_matches` implements the *raw* pattern matching for `f` on every extension to `A`. Once called with `f` substituted for the continuation, `a_plus_f_matches` returns the pattern matching of `f`, now exclusively *materialised* for `A`.

Second, one implements `f` itself, which, by passing itself to `a_plus_f_matches`, acquires the right pattern matching; and, then, visits `f`'s parameter using the acquired pattern matching.

As an example for the above two steps, we implement below an evaluator for `NA` expressions:

```
1 template<typename ADT> auto na_plus_ev_matches(auto eval)
2 {//na_plus_ev_matches<α ◁ Num ⊕ Add>
3   return match {
4     []     (const Num<ADT>& n) {return n.n_;},            //λNum(n). n
5                                             //λAdd(l,r). eval(l) + eval(r)
6     [eval](const Add<ADT>& a) {return eval(*a.l_) + eval(*a.r_);}
7   };
8 }
```

Above is the first step: `na_plus_ev_matches` is the auxiliary function for evaluation. `eval` in line 1 is the continuation. `na_plus_ev_matches` produces the raw pattern matching for every `ADT` that extends `NA`. It does so by passing match statements for `Num<ADT>` and `Add<ADT>` to the `match` combinator. In line 4, a λ-abstraction is used for matching `Num<ADT>` instances. Line 6, on the other hand, use a λ-abstraction to match `Add<ADT>` instances. The difference is that the latter λ-abstraction is recursive and captures the variable `eval` (by mentioning it between square brackets in line 6). Furthermore, rather than using `na_plus_ev_matches`, it uses the continuation `eval` for recursion.

In short, the `match` combinator bundles a set of match statements together. Such a match statement can be any callable C++ object. In this paper, we only use λ-abstractions for our match statements. More on `match` in Sect. 5.3.

```
1 int na_eval(const NA::cases& expr) {
2   auto pm = na_plus_ev_matches<NA>(na_eval);
3   return std::visit(pm, expr);
4 }
```

Above is the second step for provision of evaluation for `NA` expressions. In line 2, it acquires the right pattern matching for `NA` by passing itself as the continuation to `na_plus_ev_matches`. Then, in line 3, it visits the expression to be evaluated using the acquired pattern matching.

We would like to end this subsection by emphasising on the following: In the terminology of Oliveira et al. [26], `na_plus_ev_matches` is an object algebra. In the latter work, compositionality of object algebras comes at the price of a generalisation of *self*-references [27]. (In short, inside the body of an instance of a given class, a *self*-references is a pointer/reference to the very instance itself. Such a pointer/reference needs to also deal with virtual construction [8].) Notably, however, we achieve that (Sect. 4.1) without resorting to *self*-references.

3.4 Tests

Using the following two pieces of syntactic sugar for literals and addition

```
auto operator"" _n (ulonglong n) {return Num<NA>(n);}
auto operator + (const NA::cases& l, const NA::cases& r) {return Add<NA>(l, r);}
```

`na_eval(5_n + 5_n + 4_n)` returns 14, as expected.

4 Addressing the EP Concerns

We now show how our technology is an EP solution.

4.1 E1 (Bidimensional Extensibility)

Extensibility in the dimension of ADTs is simple. Provided the `Mul` case component below

```
template<typename ADT> struct Mul{ //Mul α :: α × α → α
  using CsVar = typename ADT::cases;
  Mul(const CsVar& l, const CsVar& r):
    l_(std::make_shared<CsVar>(l)), r_(std::make_shared<CsVar>(r)) {}
  const std::shared_ptr<CsVar> l_, r_;
};
```

encoding *NAM* using the C++ LMS-EPS can be done just like that for *NA*:

```
template<typename ADT> struct NAMTemp
{using cases = std::variant<Num<ADT>, Add<ADT>, Mul<ADT>>;};
struct NAM: NAMTemp<NAM> {};
```

But, one can also extend `NA` to get `NAM`:

```
template<typename ADT> struct NAMTemp
{using cases = ext_variant_by_t<NATemp<ADT>, Mul<ADT>>;};
```

In the absence of a built-in **extends** for **traits**, that is the C++ LMS-EPS counterpart for extending an ADT to another. See Sect. 5.4 for the definition of `ext_variant_by_t`.

Extensibility in the dimension of functions is not particularly difficult. For example, here is how one does pretty printing for `NA`:

```
template<typename ADT> auto na_plus_to_str_matches(auto to_string) {
  return match {
    []          (const Num<ADT>& n) {return std::to_string(n.n_);},
    [to_string](const Add<ADT>& a) {return to_string(*a.l_) + " + " +
                                                    to_string(*a.r_);}
  };
}
std::string na_to_string(const NA::cases& expr) {
  auto pm = na_plus_to_str_matches<NA>(na_to_string);
  return std::visit(pm, expr);
}
```

`na_plus_to_str_matches` is the auxiliary function with `to_string` being the continuation. `na_to_string` is the pretty printing for `NA`.

```
1 template<typename ADT> auto nam_plus_to_str_matches(auto to_string) {
2   return match {
3     na_plus_to_str_matches<ADT>(to_string),
4     [to_string](const Mul<ADT>& m) {return to_string(*m.l_) + " * " +
5                                                     to_string(*m.r_);}
6   };
7 }
```

On the other hand, the above auxiliary function called `nam_plus_to_str_matches` reuses the match statements already developed by `na_plus_to_str_matches` (line 3). It does so by including the latter function in the list of match statements it includes in its `match` combinator. Note that the former function, moreover, passes its own continuation (i.e., `to_string`) as an argument to the latter function. Such a reuse is the C++ LMS-EPS counterpart of the **super** call in line 5 of `NAM` in Sect. 2.2.

The similarity becomes more clear when one observes that both the Scala LMS-EPS and the C++ LMS-EPS scope the match statements and have mechanisms for reusing the existing ones. In the Scala LMS-EPS, the match statements are scoped in a method of the base **trait**. That method, then, can be **override**n at the extension and reused via a **super** call. On the other hand, in the C++ LMS-EPS, the match statements are scoped in the auxiliary functions. That auxiliary function, then, can be mentioned in the `match` of the extension's auxiliary function (just like the new match statements), enabling its reuse.

4.2 E2 (Static Type Safety)

Suppose that in the pretty printing for `NAM`, one mistakenly employs `na_plus_to_str_matches` instead of `nam_plus_to_str_matches`. (Note that the latter name starts with `nam` whilst the former only starts with `na`.) That situation is like when the programmer attempts pretty printing for a `NAM` expression without having provided the pertaining match statement of `Mul`. Here is the erroneous code:

```
std::string nam_to_string(const NAM::cases& expr) {//WRONG!
  auto pm = na_plus_to_str_matches<NAM>(nam_to_string);
  return std::visit(pm, expr);
}
```

As expected, the above code fails to compile. As an example, GCC 7.1 produces three error messages. In summary, those error messages state that `na_plus_to_str_matches` only has match statements for `Num` and `Add` (but not `Mul`). Note that the code fails to compile even without passing a concrete argument into `nam_to_string`. That demonstrates our **strong** static type safety. The C++ LMS-EPS can guarantee that because the compiler chooses the right match statement using overload resolution, i.e., at compile-time. C.f. Sect. 5.3 for more.

4.3 E3 (No Manipulation/Duplication)

Notice how nothing in the evidence for our support for E1 and E2 requires manipulation, duplication, or recompilation of the existing codebase. Our support for E3 follows.

4.4 E4 (Separate Compilation)

Our support for E4, in fact, follows just like E3. It turns out, however, that C++ **`template`**s enjoy two-phase translation [36, Sect. 14.3.1]: Their parts that depend on the type parameters are type checked (and compiled) only when they are instantiated, i.e., when concrete types are substituted for all their type parameters. As a result, type checking (and compilation) will be redone for every instantiation. That type-checking peculiarity might cause confusion w.r.t. our support for E4.

In order to dispel that confusion, we need to recall that `Add`, for instance, is a class **`template`** rather than a class. In other words, `Add` is not a type (because it is of kind $* \rightarrow *$) but `Add<NA>` is. The interesting implication here is that `Add<NA>` and `Add<NAM>` are in no way associated to one another. Consequently, introduction of `NAM` in presence of `NA`, causes no repetition in type checking (or compilation) of `Add<NA>`. (`Add<NAM>`, nonetheless, needs to be compiled in presence of `Add<NA>`.) The same argument holds for every other case component already instantiated with the existing ADTs.

More generally, consider a base ADT $\Phi_b = \oplus\overline{\gamma}$ and its extension $\Phi_e = (\oplus\overline{\gamma}) \oplus (\oplus\overline{\gamma}')$. Let $\#(\overline{\gamma}) = n$ and $\#(\overline{\gamma}') = n'$, where $\#(.)$ is the number of components in the component combination. Suppose a C++ LMS-EPS codebase that contains case components for $\gamma_1, \ldots, \gamma_n$ and $\gamma'_1, \ldots, \gamma'_{n'}$. Defining Φ_b in such a codebase incurs compilation of n case components. Defining Φ_e on top incurs compilation of $n+n'$ case components. Nevertheless, that does not disqualify our EP solution because defining the latter component combination does not incur recompilation of the former component **combination**. Note that individual components differ from their combination. And, E4 requires the combinations not to be recompiled.

Here is an example in terms of DSL embedding. Suppose availability of a type checking phase in a codebase built using the C++ LMS-EPS. Adding a type erasure phase to that codebase, does not incur recompilation of the type checking phase. Such an addition will, however, incur recompilation of the case components common between the two phases. Albeit, those case components will be recompiled for the type erasure phase. That addition leaves the compilation of the same case components for the type checking phase intact. Hence, our support for E4.

A different understanding from separate compilation is also possible, in which: an EP solution is expected to, upon extension, already be done with the type checking and compilation of the "core part" of the new ADT. Consider extending *NA* to *NAM*, for instance. With that understanding, *Num* and *Add* are considered the "core part" of *NAM*. As such, the argument is that the type checking and compilation of that "core part" should not be repeated upon the extension.

However, before instantiating `Num` and `Add` for `NAM`, both `Num<NAM>` and `Add<NAM>` are neither type checked nor compiled. That understanding, hence, refuses to take our work for an EP solution. We find that understanding wrong because the core of *NAM* is *NA*, i.e., the $Num \oplus Add$ **combination**, as opposed to both *Num* and *Add* but individually. Two quotations back our mindset up:

The definition Zenger and Odersky [20] give for separate compilation is as follows: "Compiling datatype extensions or adding new processors should not encompass re-type-checking the original **datatype** or existing processors [functions]." The datatypes here are `NA` and `NAM`. Observe how compiling `NAM` does not encompass repetition in the type checking and compilation of `NA`.

Wang and Oliveira [38] say an EP solution should support: "software evolution in both dimensions in a modular way, without modifying the code that has been written previously." Then, they add: "Safety checks or compilation steps must not be deferred until link or runtime." Notice how neither definition of new case components or ADTs, nor addition of case components to existing ADTs to obtain new ADTs, implies modification of the previously written code. Compilation or type checking of the extension is not deferred to link or runtime either.

For more elaboration on the take of Wang and Oliveira on (bidimensional) modularity, one may ask: If *NA*'s client becomes a client of *NAM*, will the client's code remain intact under E3 and E4? Let us first disregard code that is exclusively written for *NA* for it is not meant for reuse by *NAM*:

```
void na_client_f(const NA&) {...}
```

If on the contrary, the code only counts on the availability of *Num* and *Add*:

```
template <
 typename ADT, typename = std::enable_if_t<adt_contains_v<ADT, Num, Add>>
> void na_plus_client_f(const ADT& x) {...}
```

Then, it can expectedly be reused upon transition from *NA* to *NAM*. (We drop the definition of `adt_contains_v` due to space restrictions.)

5 Technicality

5.1 Why std::shared_ptr?

Although not precisely what the C++ specification states, it is not uncommon for the current C++ compilers to require the types participating in the formation of a `std::variant` to be default-constructable. That requirement is, however, not fulfilled by our case components. As a matter of fact, ADT cases, in general, are unlikely to fulfil that requirement.

But, as shown in line 2 of `NATemp`, the C++ LMS-EPS needs the case components to participate in a `std::variant`. Wrapping the case components in a default-constructable type seems inevitable. We choose to wrap them inside a `std::shared_ptr` because, then, we win sub-expression sharing as well.

5.2 Why CRTP?

The reader might have noticed that, in the C++ LMS-EPS, defining ADTs is also possible without CRTP. For example, one might try the following for *NA*:

```
struct OtherNA { using cases = std::variant<Num<OtherNA>, Add<OtherNA>>; };
```

Then, extending `OtherNA` to an encoding for *NAM* will, however, not be possible as we extended `NATemp` to `NAMTemp` in Sect. 4.1. In addition to employing a different extension metafunction than `ext_variant_by_t` in Sect. 5.4, we would need some extra work in the case components. For example, here is how to enrich `Add`:

```
template<typename ADT> struct Add
{/*... like before ... */ template<typename A> using case_component = Add<A>;};
```

Then, we can still extend `NATemp` to get `NAM`:

```
struct NAM { using cases = ext_variant_by_t<NATemp<NAM>, Mul<NAM>>; };
```

If one wishes to, it is even possible to completely abolish `NATemp` – and, in fact, all the CRTP:

```
struct NAM { using cases = ext_to_by_t<NA, NAM, Mul<NAM>>; };
```

where `ext_to_by_t` is defined in Sect. 5.4.

5.3 The match Combinator

The definition of our `match` combinator is as follows[4]:

```
template<typename... Ts> struct match: Ts...
{using Ts::operator()...;};
template<typename... Ts> match(Ts...) -> match<Ts...>;
```

[4] This is a paraphrase of the `overloaded` combinator taken from the `std::visit`'s online specification at the C++ Reference: https://en.cppreference.com/w/cpp/utility/variant/visit.

As one can see above, `match` is, in fact, a type parameterised **`struct`**. In lines 1 and 2 above, `match` derives from all its type arguments. At line 2, it also makes all the **`operator`**`()`s of its type arguments accessible via itself. Accordingly, `match` is *callable* in all ways its type arguments are.

Line 3 uses a C++ feature called *user-defined deduction guides.* Recall that C++ only offers automatic type deduction for **`template`** functions. Without line 3, thus, `match` is only a **`struct`**, missing the automatic type deduction. The programmer would have then needed to list all the type arguments explicitly to instantiate `match`. That would have been cumbersome and error-prone – especially, because those types can rapidly become human-unreadable. Line 3 helps the compiler to deduce type arguments for the **`struct`** (i.e., the `match` to the right of "->") in the same way it would have done that for the function (i.e., the `match` to the left of "->").

One may wonder why we need all those `Ts::`**`operator`** `()`s. The reason is that, according to the C++ specification, the first argument of `std::visit` needs to be a *callable.* The compiler tries the second `std::visit` argument against all the call pieces of syntax that the first argument provides. The mechanism is that of C++'s overload resolution. In this paper, we use `match` only for combining λ-abstractions. But, all other sorts of callable are equally acceptable to `match`.

Finally, we choose to call `match` a combinator because, to us, its usage is akin to Haskell's Combinator Pattern [28, Sect. 16].

5.4 Definitions of ext_variant_by_t and ext_to_by_t

Implementation of `ext_variant_by_t` is done using routine **`template`** metaprogramming:

```
1 template<typename, typename...> struct evb_helper;
2 template<typename... OCs, typename... NCs>
3 struct evb_helper<std::variant<OCs...>, NCs...>
4 {using type = std::variant<OCs..., NCs...>;};
5 template<typename ADT, typename... Cs> struct ext_variant_by
6 {using type = typename evb_helper<typename ADT::cases, Cs...>::type;};
7 template<typename ADT, typename... Cs>
8 using ext_variant_by_t = typename ext_variant_by<ADT, Cs...>::type;
```

`ext_variant_by_t` (line 8) extends an `ADT` by the cases `Cs...`. To that end, `ext_variant_by_t` is a syntactic shorthand for the `type` nested type of `ext_variant_by`. `ext_variant_by` (line 5) works by delegating its duty to `evb_helper` after acquiring the case list of `ADT` (line 6). Given a `std::variant` of old cases (`OCs...`) and a series of new cases (`NCs...`), the metafunction `evb_helper` type-evaluates to a `std::variant` of old and new cases (line 4).

Implementing `ext_to_by_t` is not particularly more complicated. So, we drop explanation and only provide the code:

```
template<typename, typename> struct materialise_for_helper;
template<typename ADT, typename... Cs>
struct materialise_for_helper<ADT, std::variant<Cs...>>
{using type = std::variant<typename Cs::template case_component<ADT>...>;};

template<typename ADT1, typename ADT2> struct materialise_for {
```

```
  using type = typename materialise_for_helper<ADT2, typename ADT1::cases>::type;
};

template<typename ADT1, typename ADT2, typename... Cs> struct ext_to_by {
  using type = typename evb_helper<typename materialise_for<ADT1, ADT2>::type,
                                   Cs...>::type;
};

template<typename ADT1, typename ADT2, typename... Cs>
using ext_to_by_t = typename ext_to_by<ADT1, ADT2, Cs...>::type;
```

6 C4C versus GADTs

When embedding DSLs, it is often convenient to piggyback on the host language's type system. In such a practice, GADTs are a powerful means to guarantee the absence of certain type errors. For example, here is a Scala transliteration[5] of the running example Kennedy and Russo [18] give for GADTs in object-oriented languages:

```
sealed abstract class Exp[T]
case class Lit(i: Int)                                        extends Exp[Int]
case class Plus(e1: Exp[Int], e2: Exp[Int])                   extends Exp[Int]
case class Equals(e1: Exp[Int], e2: Exp[Int])                 extends Exp[Boolean]
case class Cond(e1: Exp[Boolean], e2: Exp[Int], e3: Exp[Int]) extends Exp[Int]
/* ... more case classes ... */
def eval[T](exp: Exp[T]): T = exp match {...}
```

Notice first that `Exp` is type parameterised, where `T` is an arbitrary Scala type. That is how `Lit` can derive from `Exp[Int]` whilst `Equals` derives from `Exp [Boolean]`. Second, note that `Plus` takes two instances of `Exp[Int]`. Contrast that with the familiar encodings of $\alpha = \mathit{Plus}(\alpha, \alpha) \mid \ldots$, for some ADT α. Unlike the GADT one, the latter encoding cannot outlaw nonsensical expressions such as `Plus(Lit(5), Lit(true))`. Third, note that `eval` is polymorphic in the carrier type of `Exp`, i.e., `T`.

The similarity between the above case definitions and our case components is that they are both type parameterised. Nevertheless, the former are parameterised by the type of the Scala expression they carry. Whereas, our case components are parameterised by their ADT types. The impact is significant. Suppose, for example, availability of a case component `Bool` with the corresponding `operator""_b` syntactic sugar. In their current presentation, our case components cannot statically outlaw type-erroneous expressions like `12_n + "true" _b`. On the other hand, the GADT `Cond` is exclusively available as an ADT case of `Exp` and cannot be used for other ADTs.

Note that, so long as statically outlawing `12_n + "true"_b` is the concern, one can always add another layer in the `Exp` grammar so that the integral cases and Boolean cases are no longer at the exact same ADT. That workaround, however, will soon become unwieldy. That is because, it involves systematically separating syntactic categories for every carrier type – resulting in the craft of a new type system. GADTs employ the host language's type system instead.

[5] Posted online by James Iry on Wed, 22/10/2008 at http://lambda-the-ultimate.org/node/1134.

The bottom line is that GADTs and C4C encodings of ADTs are orthogonal. One can always generalise our case components so they too are parameterised by their carrier types and so they can guarantee similar type safety.

7 Related Work

The support of the Scala LMS-EPS for E2 can be easily broken using an incomplete pattern matching. Yet, given that Scala pattern matching is dynamic, whether LMS really relaxes E2 is debatable. Note that the problem in the Scala LMS-EPS is not an "Inheritance is not Subtyping" one [7]: The polymorphic function of a deriving **trait** does specialise that of the base.

In comparison to the Scala LMS-EPS, we require one more step for defining ADTs: the CRTP. Nevertheless, given that the C++ LMS-EPS is C4C, specifying the cases of an ADT is by only listing the right case components in a `std::variant`. Defining functions on ADTs also requires one more step in the C++ LMS-EPS: using the continuation. When extending a function for new ADT cases, the C++ LMS-EPS, however, needs no explicit **super** call, as required by the Scala LMS-EPS.

A note on the Expression Compatibility Problem is appropriate here. As detailed earlier [11, Sect. 4.2], the Scala LMS-EPS cannot outlaw incompatible extensions. Neither can the current presentation of the C++ LMS-EPS. Nonetheless, due to its C4C nature, that failure is not inherent in the C++ LMS-EPS. One can easily constrain the ADT type parameter of the case components in a similar fashion to the Scala IDPAM [16] to enforce compatibility upon extension.

The first C4C solution to the EP is the Scala IDPAM [16]. ADT creation in the Scala IDPAM too requires F-Bounding. But, the type annotation required when defining an ADT using their case components is heavier.

In the Scala IDPAM, the number of type annotations required for a function taking an argument of an ADT with n cases is $\mathcal{O}(n)$. That is $\mathcal{O}(1)$ in the C++ LMS-EPS. The reason is that, in C++, with programming purely at the type level, types can be computed from one another. In particular, an ADT's case list can be computed programmatically from the ADT itself. That is not possible in Scala without **implicits**, which are not always an option. In the Scala IDPAM too, implementation of functions on ADTs is *nominal*: For every function on a given ADT α, all the corresponding match components—i.e., match statements also delivered as components—of α's cases need to be manually mixed in to form the full function implementation. The situation is similar for the C++ LMS-EPS in that all the match statements are required to be manually listed in the `match` combinator. However, instead of using a continuation, in the Scala IDPAM, one mixes in a base case as the last match component. Other than F-Bounding, the major language feature required for the Scala IDPAM is stackability of **traits**. In the C++ LMS-EPS, that is variadic **templates**. The distinctive difference between the C++ LMS-EPS and the Scala IDPAM is that the latter work relaxes E2 in the absence of a default [30]. On the contrary, the C++ LMS-EPS guarantees strong static type safety.

The second C4C solution to the EP is the C++ IDPaM [12]. There are two reasons to prefer the C++ IDPaM over the C++ LMS-EPS: Firstly, in the C++ IDPaM, definition of a function f on ADTs amounts to provision of simple (C++) function overloads, albeit plus a one-off macro instantiation for f. (Those function overloads are called match components of the C++ IDPaM.) Secondly, in the C++ IDPaM, function definition is *structural*: Suppose the availability of all the corresponding match components of α's case list and the macro instantiation for f. Then, unlike the C++ LMS-EPS, to define f on α, the programmer need not specify which match statements to include in the pattern matching. The compiler deductively obtains the right pattern matching using α's structure, i.e. α's case list.

There are two reasons to prefer the C++ LMS-EPS over the C++ IDPaM. Firstly, implementing ADTs and functions on them is only possible in the C++ IDPaM using a metaprogramming facility shipped as a library. That library was so rich in its concepts that it was natural to extend [13] for multiple dispatch. Behind the scenes, the library performs iterative pointer introspection to choose the right match statements. In the C++ LMS-EPS, that pointer introspection is done using the compiler's built-in support for `std::variant`. That saves the user from having to navigate the metaprogramming library upon mistakes (or bugs). Furthermore, when it comes to orchestrating the pattern matching, the compiler is likely to have more optimisation opportunities than the library. Secondly, unlike their C++ IDPaM equivalents, case components of the C++ LMS-EPS do not inherit from their ADT. This entails weaker coupling between case components and ADT definitions.

Instead of `std::variant`, one can use `boost::variant`[6] to craft a similar solution to the C++ LMS-EPS. Yet, the solution would have not been as clean with its auxiliary functions as here. In essence, for a function f, one would have needed to manually implement each match statement as a properly-typed overload of `F::operator ()`. Extending f to handle new ADT cases, nevertheless, would have been more akin to the Scala LMS-EPS. That is because, then, providing the new match statements would have amounted to implementing the corresponding `FExtended::operator ()` overloads, for some `FExtended` that derives from `F`. (Compare with Sect. 2.2.) Moreover, `boost::variant` requires special settings for working with recursive types (such as ADTs) that damage readability.

Using object algebras [10] to solve EP has become popular over recent years. Oliveira and Cook [23] pioneered that. Oliveira et al. [26] address some awkwardness issues faced upon composition of object algebras. Rendel, Brachthäuser and Ostermann [30] add ideas from attribute grammars to get reusable tree traversals. As also pointed out by Black [3], an often neglected factor about solutions to EP is the complexity of term creation. That complexity increases from one work to the next in the above literature. The symptom develops to the extent that it takes Rendel, Brachthäuser and Ostermann 12 non-trivial lines of code to create a term representing "$3 + 5$". Of course, those 12 lines are not for the latter task

6 https://www.boost.org/doc/libs/1_67_0/doc/html/variant.html.

exclusively and enable far more reuse. Yet, those 12 lines are inevitable for term creation for "3+5", making that so heavyweight. The latter work uses automatic code generation for term creation. So, the ADT user has a considerably more involved job using the previous object algebras technologies for EP than that of ours. Additionally, our object algebras themselves suffer from much less syntactic noise. Defining functions on ADTs is slightly more involved in the C++ LMS-EPS than object algebras for the EP. For example, pretty-printing for *NA* takes 12 (concise) Scala lines in the latter work, whereas that is 14 (syntactically noisy) C++ lines in ours.

Garrigue [9] solves EP using global case definitions that, at their point of definition, become available to every ADT defined afterwards. Per se, a function that pattern matches on a group of these global cases can serve any ADT containing the selected group. OCaml's built-in support for Polymorphic Variants [9] makes definition of both ADTs and functions on them easier. However, we minimise the drawbacks [3] of ADT cases being global by promoting them to components.

Swierstra's Datatypes à la Carte [34] uses Haskell's type classes to solve EP. In his solution too, ADT cases are ADT-independent but ADT-parameterised. He uses Haskell Functors to that end. Defining functions on ADTs amounts to defining a type class, instances of which materialising match statements for their corresponding ADT cases. Without syntactic sugaring, term creation can become much more involved than that for ordinary ADTs of Haskell. Defining the syntactic sugar takes many more steps than us, but, makes term creation straightforward. Interestingly enough, using the Scala type classes [24] can lead to simpler syntactic sugar definition but needs extra work for the lack of direct support in Scala for type classes. In his machinery, Swierstra offers a `match` that is used for monadically inspecting term structures.

Bahr and Hvitved extend Swierstra's work by offering Compositional Datatypes (CDTs) [1]. They aim at higher modularity and reusability. CDTs support more recursion schemes, and, extend to mutually recursive data types and GADTs. Besides, syntactic sugaring is much easier using CDTs because smart constructors can be automatically deduced for terms.

Later on, they offer Parametric CDTs (PCDTs) [2] for automatic α-equivalence and capture-avoiding variable bindings. PCDTs achieve that using Difuntors [19] (instead of functors) and a CDT encoding of Parametric Higher-Order Abstract Syntax [5]. Case definitions take two phases: First an equivalent of our case components need to be defined. Then, their case components need to be materialised for each ADT, similar to but different from that of Haeri and Schupp [11,14].

The distinctive difference between C4C and the works of Swierstra, Bahr, and Hvitved is the former's inspiration by CBSE. Components, in their CBSE sense, ship with their 'requires' and 'provides' interfaces. Whereas, even though the latter works too parametrise cases by ADTs, the interface that CDTs, for instance, define do not go beyond algebraic signatures. Although we do not present those for C++ LMS EPS here, C4C goes well beyond that, enabling easy solutions to the Expression Families Problem [23] and Expression Compatibility Problem [16] as well as GADTs. The respective article is in submission.

8 Conclusion

In this paper we show how a new C4C encoding of ADTs in C++ can solve EP in a way that is reminiscent to the Scala LMS-EPS. On its way, our solution gives rise to simple encodings for object algebras and object algebra interfaces and relates to Datatypes à la Carte ADT encodings.

Given the simplicity of our encoding for object algebras and object algebra interfaces in the absence of heavy notation for term creation, an interesting future work is mimicking the earlier research on object algebra encodings for EP. We need to investigate whether our technology still remains simple when we take all the challenges those works take. Another possible future work is extension of our (single dispatch) mechanism for implementing functions on ADTs to multiple dispatch. Of course, C++ LMS-EPS needs far more experimentation with real-size test cases to study its scalability. Finally, we are working on a C++ LMS-EPS variation that, unlike our current presentation, structurally implements functions on ADTs. The latter variation has thus far presented itself as a promising vehicle for also delivering multiple dispatch.

A C++ Features Used

A C++ **`struct`** (or **`class`**) can be type parameterised. The **`struct`** `S` below, for example, takes two type parameters `T1` and `T2`:

```
template<typename T1, typename T2> struct S {...};
```

Likewise, C++ functions can take type parameters:

```
template<typename T1, typename T2> void f(T1 t1, T2 t2) {...}
```

From C++20 onward, certain type parameters need not to be mentioned explicitly. For example, the above function `f` can be *abbreviated* as:

```
void f(auto t1, auto t2) {...}
```

A (**`template`** or non-**`template`**) **`struct`** can define nested type members. For example, the **`struct`** `T` below defines `T::type` to be **`int`**:

```
struct T {using type = int;};
```

Nested types can themselves be type parameterised, like `Y::`**`template`** `type`:

```
struct Y {template<typename> using type = int;};
```

C++17 added `std::variant` as a type-safe representation for unions. An instance of `std::variant`, at any given time, holds a value of one of its alternative types. That is, the static type of such an instance is that of the `std::variant` it is defined with; whilst, the dynamic type is one and only one of those alternative types. As such, a function that is to be applied on a `std::variant` needs to be applicable to its alternative types. Technically, a *visitor* is required for the alternative types. The function `std::visit`, takes a visitor in addition to a pack of arguments to be visited.

```
auto twice = [](int n){return n * 2;}
```

The variable `twice` above is bound to a λ-abstraction that, given an **int**, returns its value times two. λ-abstractions can also capture unbound names. In such a case, the captured name needs to be mentioned in the opening square brackets before the list of parameters. For example, the λ-abstraction `times` below captures the name m:

```
auto times = [m](int n){return n * m;}
```

References

1. Bahr, P., Hvitved, T.: Compositional data types. In: Järvi, J., Mu, S.-C. (eds.) 7th WGP, Tokyo, Japan, pp. 83–94. ACM, September 2011
2. Bahr, P., Hvitved, T.: Parametric compositional data types. In: Chapman, J., Levy, P.B. (eds.) 4th MSFP. ENTCS, vol. 76, pp. 3–24, February 2012
3. Black, A.P.: The expression problem, gracefully. In: Sakkinen, M. (ed.) MASPEGHI@ECOOP 2015, pp. 1–7. ACM, July 2015
4. Canning, P., Cook, W.R., Hill, W., Olthoff, W., Mitchell, J.C.: F-bounded polymorphism for object-oriented programming. In: 4th FPCA, pp. 273–280, September 1989
5. Chlipala, A.: Parametric higher-order abstract syntax for mechanized semantics. In: Hook, J., Thiemann, P. (eds.) 13th ICFP, Victoria, BC, Canada, pp. 143–156, September 2008
6. Cook, W.R.: Object-oriented programming versus abstract data types. In: de Bakker, J.W., de Roever, W.P., Rozenberg, G. (eds.) REX 1990. LNCS, vol. 489, pp. 151–178. Springer, Heidelberg (1991). https://doi.org/10.1007/BFb0019443
7. Cook, W.R., Hill, W.L., Canning, P.S.: Inheritance is not Subtyping. In: 17th POPL, San Francisco, CA, USA, pp. 125–135. ACM (1990)
8. Ernst, E., Ostermann, K., Cook, W.R.: A virtual class calculus. In: Morrisett, J.G., Peyton Jones, S.L. (eds.) 33rd POPL, pp. 270–282. ACM, January 2006
9. Garrigue, J.: Code reuse through polymorphic variants. In: FSE, vol. 25, pp. 93–100 (2000)
10. Guttag, J.V., Horning, J.J.: The algebraic specification of abstract data types. Acta Inform. **10**, 27–52 (1978)
11. Haeri, S.H.: Component-based mechanisation of programming languages in embedded settings. Ph.D. thesis, STS, TUHH, Germany, December 2014
12. Haeri, S.H., Keir, P.W.: Metaprogramming as a Solution to the Expression Problem, November 2019
13. Haeri, S.H., Keir, P.W.: Multiple dispatch using compile-time metaprogramming. Submitted to 16th ICTAC, November 2019
14. Haeri, S.H., Schupp, S.: Reusable components for lightweight mechanisation of programming languages. In: Binder, W., Bodden, E., Löwe, W. (eds.) SC 2013. LNCS, vol. 8088, pp. 1–16. Springer, Heidelberg (2013). https://doi.org/10.1007/978-3-642-39614-4_1
15. Haeri, S.H., Schupp, S.: Expression compatibility problem. In: Davenport, J.H., Ghourabi, F. (eds.) 7th SCSS. EPiC Comp., vol. 39, pp. 55–67. EasyChair, March 2016

16. Haeri, S.H., Schupp, S.: Integration of a decentralised pattern matching: venue for a new paradigm intermarriage. In: Mosbah, M., Rusinowitch, M. (eds.) 8th SCSS. EPiC Comp., vol. 45, pp. 16–28. EasyChair, April 2017
17. Hofer, C., Ostermann, K., Rendel, T., Moors, A.: Polymorphic embedding of DSLs. In: Smaragdakis, Y., Siek, J.G. (eds.) 7th GPCE, Nashville, TN, USA, pp. 137–148. ACM, October 2008
18. Kennedy, A., Russo, C.V.: Generalized algebraic data types and object-oriented programming. In: Johnson, R.E., Gabriel, R.P. (eds.) 20th OOPSLA, San Diego, CA, USA, pp. 21–40. ACM, October 2005
19. Meijer, E., Hutton, G.: Bananas in space: extending fold and unfold to exponential types. In: Williams, J. (ed.) 7th FPCA, La Jolla, California, USA, pp. 324–333. ACM, June 1995
20. Odersky, M., Zenger, M.: Independently extensible solutions to the expression problem. In: FOOL, January 2005
21. Odersky, M., Zenger, M.: Scalable component abstractions. In: 20th OOPSLA, San Diego, CA, USA, pp. 41–57. ACM (2005)
22. Oliveira, B.C.d.S.: Modular visitor components. In: Drossopoulou, S. (ed.) ECOOP 2009. LNCS, vol. 5653, pp. 269–293. Springer, Heidelberg (2009). https://doi.org/10.1007/978-3-642-03013-0_13
23. Oliveira, B.C.d.S., Cook, W.R.: Extensibility for the masses – practical extensibility with object algebras. In: Noble, J. (ed.) ECOOP 2012. LNCS, vol. 7313, pp. 2–27. Springer, Heidelberg (2012). https://doi.org/10.1007/978-3-642-31057-7_2
24. Oliveira, B.C.d.S., Moors, A., Odersky, M.: Type classes as objects and implicits. In: Cook, W.R., Clarke, S., Rinard, M.C. (eds.) 25th OOPSLA, pp. 341–360. ACM, October 2010
25. Oliveira, B.C.d.S., Mu, S.-C., You, S.-H.: Modular reifiable matching: a list-of-functors approach to two-level types. In: Lippmeier, B. (ed.) 8th Haskell, pp. 82–93. ACM, September 2015
26. Oliveira, B.C.d.S., van der Storm, T., Loh, A., Cook, W.R.: Feature-oriented programming with object algebras. In: Castagna, G. (ed.) ECOOP 2013. LNCS, vol. 7920, pp. 27–51. Springer, Heidelberg (2013). https://doi.org/10.1007/978-3-642-39038-8_2
27. Ostermann, K.: Dynamically composable collaborations with delegation layers. In: Magnusson, B. (ed.) ECOOP 2002. LNCS, vol. 2374, pp. 89–110. Springer, Heidelberg (2002). https://doi.org/10.1007/3-540-47993-7_4
28. O'Sullivan, B., Goerzen, J., Stewart, D.: Real World Haskell: Code You Can Believe in. O'Reilly, Sebastopol (2008)
29. Pressman, R.S.: Software Engineering: A Practitioner's Approach, 7th edn. McGraw-Hill, New York (2009)
30. Rendel, T., Brachthäuser, J.I., Ostermann, K.: From object algebras to attribute grammars. In: Black, A.P., Millstein, T.D. (eds.) 28th OOPSLA, pp. 377–395. ACM, October 2014
31. Reynolds, J.C.: User-defined types and procedural data structures as complementary approaches to type abstraction. In: Schuman, S.A. (ed.) New Direc. Algo. Lang., pp. 157–168. INRIA (1975)
32. Rompf, T., Odersky, M.: Lightweight modular staging: a pragmatic approach to runtime code generation and compiled DSLs. In: 9th GPCE, Eindhoven, Holland, pp. 127–136. ACM (2010)
33. Sommerville, I.: Software Engineering, 9th edn. Addison-Wesley, Reading (2011)
34. Swierstra, W.: Data types à la Carte. JFP **18**(4), 423–436 (2008)

35. Torgersen, M.: The expression problem revisited. In: Odersky, M. (ed.) ECOOP 2004. LNCS, vol. 3086, pp. 123–146. Springer, Heidelberg (2004). https://doi.org/10.1007/978-3-540-24851-4_6
36. Vandevoorde, D., Josuttis, N.M., Gregor, D.: C++ Templates: The Complete Guide, 2nd edn. Addison Wesley, Reading (2017)
37. Wadler, P.: The Expression Problem. Java Genericity Mailing List, November 1998
38. Wang, Y., Oliveira, B.C.d.S.: The expression problem, Trivially! In: 15th MODULARITY, pp. 37–41. ACM, New York (2016)
39. Zenger, M., Odersky, M.: Extensible algebraic datatypes with defaults. In: 6th ICFP, pp. 241–252, Florence, Italy. ACM (2001)

Laws of Monadic Error Handling

Härmel Nestra(✉)

Institute of Computer Science, University of Tartu,
J. Liivi 2, 50409 Tartu, Estonia
harmel.nestra@ut.ee

Abstract. One of the numerous applications of monads in functional programming is error handling. This paper proposes several new axiomatics for equational reasoning about monadic computations with error handling and studies their relationships with each other and with axiomatics considered earlier. The primary intended application area is deterministic top-down parsing.

Keywords: Monads · Error handling · Parsing · Equational reasoning

1 Introduction

Monads provide a uniform language for elegant specification of computations in many different application areas, including top-down parsing and error handling [10,16,17]. The pure functional programming language Haskell [7] has the following general interface for monadic computations (some minor details omitted):

$$\begin{array}{l}\textbf{class}\ \mathit{Applicative}\ m \Rightarrow \mathit{Monad}\ m\ \textbf{where}\\ \quad \mathit{return} :: a \to m\ a\\ \quad (\gg\!=)\ \ :: m\ a \to (a \to m\ b) \to m\ b\end{array}$$

Here, m denotes the monad instance, meaning a specific kind of computational structures. A type of the form $m\ a$ consists of computations that return values of type a. As intended, the method *return* creates a trivial computation that returns its argument value, whereas the operator $\gg\!=$ sequentially invokes a computation and its continuation that depends on the result value of the computation.

The module *Control.Monad.Except* of the library of GHC [8] defines the following uniform extended interface for monads with error handling capabilities:

$$\begin{array}{l}\textbf{class}\ \mathit{Monad}\ m \Rightarrow \mathit{MonadError}\ e\ m \mid m \to e\ \textbf{where}\\ \quad \mathit{throwError} :: e \to m\ a\\ \quad \mathit{catchError} :: m\ a \to (e \to m\ a) \to m\ a\end{array}$$

Here, e is the error type associated to the monad. By intention, *throwError* creates a computation that throws the specified error, whereas *catchError* combines a computation and an error handler to a new computation that tries the given computation and, if it halts with an error, passes this error to the handler.

Functional programming is often advocated because of its close relationship with equational reasoning. Popular Haskell type classes such as *Monad*,

R. M. Hierons and M. Mosbah (Eds.): ICTAC 2019, LNCS 11884, pp. 372–391, 2019.
https://doi.org/10.1007/978-3-030-32505-3_21

MonadPlus, *Applicative* and *Alternative* have standard axiomatics that all instances are expected (though not forced) to meet. Surprisingly, the error monad class has no standard axiomatics. This paper aims at filling this gap. It develops several new axiomatics for monads with error handling and studies the relationships of the axiomatics with each other and with those proposed earlier. The design of the axiomatics is primarily guided by semantic patterns arising in deterministic top-down parsing which seem to be universal enough.

A key observation (which is not new) is that the intentional meaning of the error monad operations mimics that of *return* and $\gg\!=$ with errors taking over the role of result values. Hence an axiomatics we are looking for would naturally incorporate two copies of monad axioms, one dealing with normal values and the other with errors. However, the types of the methods of *MonadError* are less general than needed for a perfect correspondence. Indeed, while $\gg\!=$ can switch from one result type to another, *catch* cannot change the error type because the monad is parametric in the result type but not in the error type.

Error handling in its more general form has attracted some interest recently. For example, Malakhovski [13] proposes *conjoined monads* that can be specified by the following Haskell type class (ignoring the fact that in this form the code would be rejected as it reuses existing method names):

$$
\begin{array}{l}
\textbf{class}\ \mathit{ConjoinedMonads}\ m\ \textbf{where} \\
\quad \mathit{return} :: a \to m\ e\ a \\
\quad (\gg\!=)\ :: m\ e\ a \to (a \to m\ e\ a') \to m\ e\ a' \\
\quad \mathit{throw}\ :: e \to m\ e\ a \\
\quad \mathit{catch}\ :: m\ e\ a \to (e \to m\ e'\ a) \to m\ e'\ a
\end{array}
$$

Here, m is a type constructor with two arguments, out of which the second denotes the type of expected normal results and the first denotes the possible error type. As *catch* consumes an error handler that can change the error type, the expected generality and symmetry between the two type arguments is achieved. In [15], we studied bifunctors equipped with operations that obey monad laws in both arguments of the bifunctor and some additional axioms. Handlers that can change error types were introduced for controlling the numerous error types involved when smart error handling is used within the expressive power similar to that of the Applicative-Alternative interface.

Following the spirit of [15], we assume error handlers being able to change the error type. While achieving more generality along with beauty of symmetry, the approach still applies to Haskell error monads as a special case.

Despite the full symmetry in the types, the axioms need not be symmetric. Parsing normally keeps track of the part of input already processed; semantic asymmetry between normal values and errors may arise if failures disrupt the coherent processing of the input even if the errors are caught. The standard monad transformers of Haskell enable one to construct both parsers where successes and failures behave symmetrically and parsers for which this is not the case. Unlike in [15], here we consider both symmetric and asymmetric versions, whereby in the asymmetric case, we obtain axiomatics that capture the common semantic properties of error handling more precisely than that in [15].

The structure of the paper is as follows: Sect. 2 introduces the mathematical machinery used. Section 3 introduces, in mathematical terms, the operations on computations considered in this paper. Section 4 briefly studies the case where failures and successes behave symmetrically. Sections 5 and 6 study axiomatics for the asymmetric case, Sect. 7 discusses models of the axiomatics. Section 8 addresses the difference between the symmetric and asymmetric axiomatics. Section 9 compares the results of this paper to previous work, in particular summarizing the contributions that are new w.r.t. [15].

2 Preliminaries

Throughout the paper, we use notation from category theory for brevity. If E and A are arbitrary data types then $E + A$ denotes the corresponding sum type (obtained by type constructor *Either* in Haskell). Functions inl and inr are the canonical injections of types E and A, respectively, into $E + A$. Whenever $h : E \to X$ and $f : A \to X$, the "case study" $h \triangledown f$ is the unique function of type $E+A \to X$ that satisfies both $(h \triangledown f) \circ \mathsf{inl} = h$ and $(h \triangledown f) \circ \mathsf{inr} = f$; in particular, $\mathsf{inl} \triangledown \mathsf{inr} = \mathsf{id}$. (Equivalently, $h \triangledown f$ maps inputs of the form $\mathsf{inl}\, e$ to $h\, e$ and inputs of the form $\mathsf{inr}\, a$ to $f\, a$. In Haskell, the meaning of operation $\triangledown$ is implemented by function *either*.) Moreover, $+$ is defined on functions $h : E \to E'$ and $f : A \to A'$ by $h+f = \mathsf{inl} \circ h \triangledown \mathsf{inr} \circ f : E+A \to E'+A'$. (We treat the composition operator $\circ$ as having higher priority in expressions than the operators $\triangledown$ and $+$.)

Using concepts of category theory does not mean an ambition for being applicable in many categories, neither is our aim to reckon with partiality. Everything is to be interpreted in category ***Set***. A "moral" justification for reasoning about Haskell programs without considering partiality is given by Danielsson et al. [4].

We will use the following notations for some special functions:

$$
\begin{array}{lcll}
\mathsf{assocr} & = & (\mathsf{id} + \mathsf{inl}) \triangledown \mathsf{inr} \circ \mathsf{inr} & : (E + E') + A \to E + (E' + A) \\
\mathsf{assocl} & = & \mathsf{inl} \circ \mathsf{inl} \triangledown (\mathsf{inr} + \mathsf{id}) & : E + (E' + A) \to (E + E') + A \\
\mathsf{swap} & = & \mathsf{inr} \triangledown \mathsf{inl} & : E + A \to A + E \\
\mathsf{swasr} & = & \mathsf{assocl} \circ (\mathsf{id} + \mathsf{swap}) \circ \mathsf{assocr} & : (E + A) + A' \to (E + A') + A \\
\mathsf{swasl} & = & \mathsf{assocr} \circ (\mathsf{swap} + \mathsf{id}) \circ \mathsf{assocl} & : E + (E' + A) \to E' + (E + A)
\end{array}
$$

In addition, $\mathsf{absurd} : \varnothing \to X$ is the unique function from the empty set to set X.

Whereas data types are objects and functions are morphisms, type constructors play the role of functors. In particular, binary type constructors give rise to bifunctors (which technically are just functors whose domain is the Cartesian product of two categories). The corresponding morphism mappings are implemented by specific polymorphic functions that satisfy the functor laws. For example, the operation $+$ as defined above on both sets and functions is a bifunctor. Indeed, $+$ on functions preserves types, as well as identities (i.e., $\mathsf{id} + \mathsf{id} = \mathsf{id}$) and composition (i.e., $(h' + f') \circ (h + f) = h' \circ h + f' \circ f$).

If F is a bifunctor then fixing one of its arguments to a given object A (on objects) and its identity (on morphisms) while allowing the other argument vary results in a unary functor; we call such functors *sections* of F.

The error transformer:

$$
\begin{aligned}
\mathcal{E}_E\, M\, A &= M\,(E + A)\\
\mathcal{E}_E\, M\, f &= M\,(\mathsf{id} + f)\\
\mathsf{return}\, a &= \mathsf{return}(\mathsf{inr}\, a)\\
k^*\, t &= (\mathsf{return} \circ \mathsf{inl} \triangledown k)^*\, t
\end{aligned}
$$

The writer transformer:

$$
\begin{aligned}
\mathcal{W}_W\, M\, A &= M\,(A \times W)\\
\mathcal{W}_W\, M\, f &= M\,(f \times \mathsf{id})\\
\mathsf{return}\, a &= \mathsf{return}(a, \mathit{1})\\
k^*\, t &= (\lambda(a, w) \,.\, M\,(\mathsf{id} \times (w\cdot))\,(k\, a))^*\, t
\end{aligned}
$$

The reader transformer:

$$
\begin{aligned}
\mathcal{R}_R\, M\, A &= R \to M\, A\\
\mathcal{R}_R\, M\, f &= \lambda g \,.\, M\, f \circ g\\
\mathsf{return}\, a &= \lambda r \,.\, \mathsf{return}\, a\\
k^*\, t &= \lambda r \,.\, (\lambda a \,.\, k\, a\, r)^*\, (t\, r)
\end{aligned}
$$

The state transformer:

$$
\begin{aligned}
\mathcal{S}_S\, M\, A &= S \to M\,(A \times S)\\
\mathcal{S}_S\, M\, f &= \lambda g \,.\, M\,(f \times \mathsf{id}) \circ g\\
\mathsf{return}\, a &= \lambda s \,.\, \mathsf{return}(a, s)\\
k^*\, t &= \lambda s \,.\, (\lambda(a, s') \,.\, k\, a\, s')^*\, (t\, s)
\end{aligned}
$$

The update transformer:

$$
\begin{aligned}
\mathcal{U}_{S,W}\, M\, A &= S \to M\,(A \times W)\\
\mathcal{U}_{S,W}\, M\, f &= \lambda g \,.\, M\,(f \times \mathsf{id}) \circ g\\
\mathsf{return}\, a &= \lambda s \,.\, \mathsf{return}(a, \mathit{1})\\
k^*\, t &= \lambda s \,.\, (\lambda(a, w) \,.\, M\,(\mathsf{id} \times (w\cdot))\,(k\, a\, (s \bullet w)))^*\, (t\, s)
\end{aligned}
$$

Fig. 1. Definitions of the error, reader, writer, state and update monad transformers

If F is an endofunctor then a family of morphisms $\mathsf{return} : A \to F\,A$, one for each object A, satisfying the *naturality* law $F\,f \circ \mathsf{return} = \mathsf{return} \circ f$, is called *unit* of F. A *monad* is an endofunctor M equipped with unit $\mathsf{return} : A \to M\,A$ and a mapping $(\_)^*$ of morphisms of type $A \to M\,B$ to morphisms of type $M\,A \to M\,B$ for every pair of objects (A, B), such that the *monad laws* $k^* \circ \mathsf{return} = k$, $\mathsf{return}^* = \mathsf{id}$ and $l^* \circ k^* = (l^* \circ k)^*$ are fulfilled and M on morphisms satisfies $M\,f = (\mathsf{return} \circ f)^*$. The mapping $(\_)^*$ is called *bind* of the monad. In the Haskell monad class referred to in Sect. 1, bind is denoted by $\gg\!=$. Note that here and in the rest of the paper, we use the order of arguments of monad bind common in category theory, which is the opposite to their order in Haskell. We preferred this choice because of the concise form of laws it results in.

Recall from the folk knowledge of the community of functional programming the concepts of error, reader, writer and state monad transformer; we include also the *update* monad transformer. They map monads to new monads. In practice, they provide a means for building up complex monads with many-sided functionality from simple monads each dealing with a specific kind of effects [9,11,14]. In particular, successful parsing and parse error handling, as well as other smaller tasks typically involved in parsing, can be combined in a modular way.

Figure 1 presents the precise definitions. For each transformer, the functor application on objects and morphisms, as well as the monad operations, are specified. Here E, R, W and S stand for arbitrary fixed sets, whereby W is equipped with monoid unit $\mathit{1}$ and multiplication $\_\cdot\_$. In the definition of update transformers, we additionally assume an action $\_\bullet\_$ of the monoid $(W; \cdot, \mathit{1})$ on S, i.e., a function $S \times W \to S$ satisfying the laws $s \bullet \mathit{1} = s$ and $(s \bullet w) \bullet w' = s \bullet (w \cdot w')$. For simplicity, monad unit and bind are used without explicit reference to the particular monad they represent; in the l.h.s. of the equality sign they are from the monad being defined and in the r.h.s. they are from the underlying monad (i.e., M). The Haskell-like section notation $(w\cdot)$ means $\lambda w' \mathbin{.} w \cdot w'$.

Update transformers are defined in the lines of update monads introduced by Ahman and Uustalu [1]. Like update monads capturing the functionality of reader, writer and state monads as shown in [1], update transformers incorporate all reader, writer and state transformers in similar sense: The result of applying any reader, writer or state transformer to some monad M can be obtained as a homomorphic image of the result of applying an update transformer to M. This observation helps us to reuse proofs of common properties of these transformers.

The order of transformer applications matters in general. For example, applying an error transformer after a state transformer to monad M gives the monad $\mathcal{E}_E\,(\mathcal{S}_S\,M)\,A = S \to M\,((E + A) \times S)$ that associates states to both normal values and errors. Applying the same transformers in the opposite order leads to the monad $\mathcal{S}_S\,(\mathcal{E}_E\,M)\,A = S \to M\,(E + A \times S)$ where states are associated to normal values only; the state is forgotten upon failure. Hence an error catching operation fully symmetric to monad bind is impossible in the second case.

3 Setting the Scene

In the rest of this paper, types of the form $F\,(E, A)$ denote computations that either succeed with a result of type A or fail with an error of type E. Technically, F is a suitably fixed bifunctor, meaning that for any two functions $h : E \to E'$ and $f : A \to A'$, there also exists a function $F\,(h, f) : F\,(E, A) \to F\,(E', A')$, and the family of these functions meets functor laws. In all axiomatics we develop, functor laws (as well as sum laws) are assumed implicitly. Functions of the form $F\,(h, f)$ will be called *functor maps* (after the Haskell method *fmap*). We will see examples of F in Sects. 4 and 7.

Assume that there are two (polymorphic) operations $\mathsf{return} : A \to F\,(E, A)$ and $\mathsf{throw} : E \to F\,(E, A)$ for creating trivial computations that succeed and fail, respectively, with the value consumed as parameter. Both these operations are units of functors: return for any left section and throw for any right section of F. Like in our prior work [15], we put these operations together to obtain a joint unit

$$\eta \,:\, E + A \to F\,(E, A).$$

Then $\mathsf{return} = \eta \circ \mathsf{inr}$ and $\mathsf{throw} = \eta \circ \mathsf{inl}$. On the other hand, $\eta = \mathsf{throw} \triangledown \mathsf{return}$.

Assume F also being equipped with bind and catch operations, both of the type of monad bind but with different sections of F in the role of the monad. Like in [15], we denote the bind operation by $(\_)^{\curlyvee}$ and the catch operation by $(\_)^{\curlywedge}$:

$$\begin{aligned} (\_)^{\curlyvee} &: (A \to F\,(E, A')) \to (F\,(E, A) \to F\,(E, A')) \\ (\_)^{\curlywedge} &: (E \to F\,(E', A)) \to (F\,(E, A) \to F\,(E', A)) \end{aligned}$$

The argument function of type $A \to F\,(E, A')$ of bind specifies how to proceed if the computation of type $F\,(E, A)$ succeeds with a result value of type A; similarly the argument of type $E \to F\,(E', A)$ of catch specifies how to proceed if the computation of type $F\,(E, A)$ fails with an error of type E. As in the unit case, functionalities of bind and catch can be packed together into a joint handle

$$(\_)^* \;:\; (E + A \to F\,(E', A')) \to (F\,(E, A) \to F\,(E', A')).$$

Here, the argument of type $E + A \to F\,(E', A')$ specifies the continuation of the computation for all cases. The initial operations can be expressed as follows:

$$k^{\curlyvee} = (\eta \circ \mathsf{inl} \triangledown k)^* \qquad \text{(BND-HDL)}$$

$$x^{\curlywedge} = (x \triangledown \eta \circ \mathsf{inr})^* \qquad \text{(CCH-HDL)}$$

Indeed, if $k^{\curlyvee}$ for some k is applied to a failing computation then, by intention, the same error is rethrown. In the r.h.s. of the Equ. BND-HDL, this is achieved by giving $\eta \circ \mathsf{inl}$ as the first argument of $\triangledown$. The second argument of $\triangledown$ specifies the main case, i.e., that normal values are handled by k. Similarly, if $x^{\curlywedge}$ for some x is applied to a computation that succeeds then no catching takes place, i.e., the same normal value is returned. This justifies the Equ. CCH-HDL.

Pure function applications that only modify the result of a given computation constitute an important subset of all monadic actions; we specify

$$\phi \;:\; (E + A \to E' + A') \to (F\,(E, A) \to F\,(E', A')).$$

Intuitively, $\phi(g)$ reinterprets the result of a computation according to the given pure function g, whereby both the previous and the modified result can be either a normal value or an error. Such functions $\phi(g)$ were called *mixmaps* in [15]. We continue using this term. Here we can define ϕ via the joint unit and handle:

$$\phi(g) = (\eta \circ g)^* \qquad \text{(MIXMAP-HDL)}$$

As errors may be mapped to normal values and vice versa, failures may become successes and successes may become failures. For this reason, asymmetry of semantics (if defined so) appears even in the case of pure function applications.

We will pay attention to the following special cases of mixmap later:

$$\mathsf{turnr} = \phi(\mathsf{assocr}) \;:\; F\,(E + E', A) \to F\,(E, E' + A) \qquad \text{(TURNR-MIXMAP)}$$

$$\mathsf{turnl} = \phi(\mathsf{assocl}) \;:\; F\,(E, E' + A) \to F\,(E + E', A) \qquad \text{(TURNL-MIXMAP)}$$

$$\mathsf{fuser} = \phi(\mathsf{id} \triangledown \mathsf{inr}) \;:\; F\,(E + A, A) \to F\,(E, A) \qquad \text{(FUSER-MIXMAP)}$$

$$\mathsf{fusel} = \phi(\mathsf{inl} \triangledown \mathsf{id}) \;:\; F\,(E, E + A) \to F\,(E, A) \qquad \text{(FUSEL-MIXMAP)}$$

$$! = \phi(\mathsf{swap}) \;:\; F\,(E, A) \to F\,(A, E) \qquad \text{(NEG-MIXMAP)}$$

The operation turnr enables partial requalification of errors as normal values; similarly turnl requalifies some normal values as errors. These operations are important because all such requalifications can be expressed in terms of turnr, turnl and functor maps. The operations fuser and fusel are similar but forget the origin of values on the side where they are moved to. They deserve special attention apart from turnr, turnl because they in some sense represent bind and catch in the world of pure function applications (more details in Sect. 6). The operation ! is called *negation*, after the negation operator of parsing expression grammars (PEG) [5] that also turns around the result status of (parsing) computation.

4 An Abstract View of the Symmetric Case

Our previous work [15] requires throw and catch to satisfy monad laws. The following are analogous equations about the joint unit and joint handle:

$$k^* \circ \eta = k \qquad \text{(HDL-LUNIT)}$$
$$(\eta \circ (h + f))^* = F\,(h, f) \qquad \text{(HDL-RUNIT)}$$
$$l^* \circ k^* = (l^* \circ k)^* \qquad \text{(HDL-ASSOC)}$$

The identity law

$$\eta^* = \mathsf{id} \qquad \text{(HDL-ID)}$$

can be obtained as a special case of HDL-RUNIT. Monad laws of all section functors of F, with return, $(\_)^{\curlyvee}$, throw and $(\_)^{\curlywedge}$ defined in Sect. 3 as monad operations, easily follow from these equations. Although HDL-LUNIT, HDL-ID and HDL-ASSOC look like monad laws, they do not make F a monad because F is not an endofunctor. Instead, they make F a relative monad on the sum functor $+$. *Relative monads* introduced by Altenkirch et al. [2,3] generalize the notion of monad to non-endofunctors. The generalization adapts monad laws without change but requires the unit to have type $J\,A \to F\,A$ for a fixed underlying functor J and the bind operation to map morphisms of type $J\,A \to F\,B$ to morphisms of type $F\,A \to F\,B$. Our laws match the definition with $J = +$.

The set of laws HDL-LUNIT, HDL-RUNIT, HDL-ASSOC is symmetric w.r.t. the first and second argument of F. Hence it is a suitable candidate for axiomatization of monadic computations with error handling where successes and failures behave symmetrically. Applications of the error monad transformer form a family of bifunctors matching the axiomatics. Indeed for any monad M, define $F_M\,(E, A) = \mathcal{E}_E\,M\,A = M\,(E + A)$ along with $\eta = \mathsf{return}$ and $k^* = k^*$, where return and $(\_)^*$ are those of M. One can recognize F_M equipped with η and $(\_)^*$ being the *restriction* of the monad M to a relative monad on $+$ (in the sense of [2,3]). Hence all functors of the form F_M satisfy the symmetric axiomatics.

Let now F be a functor satisfying the symmetric axiomatics.[1] Define monad $M\,X = F\,(\varnothing, X)$ and consider $F_M(E, A) = M\,(E + A) = F\,(\varnothing, E + A)$. Then $\phi(\mathsf{inr}) : F\,(E, A) \to F_M\,(E, A)$ and $\mathsf{fusel} \circ F\,(\mathsf{absurd}, \mathsf{id}) : F_M\,(E, A) \to F\,(E, A)$ are natural transformations that are inverses of each other and preserve the operations η and $(\_)^*$. Hence every functor satisfying the symmetric axiomatics is of the form F_M for some monad M (up to isomorphism).

For mixmaps, this axiomatics enables us to derive commutation with unit, definition of functor maps, and preservation of composition and identity:

$$\phi(g) \circ \eta = \eta \circ g \qquad \text{(MIXMAP-UNITNAT)}$$
$$\phi(h + f) = F\,(h, f) \qquad \text{(MIXMAP-FUN)}$$
$$\phi(g') \circ \phi(g) = \phi(g' \circ g) \qquad \text{(MIXMAP-COMP)}$$
$$\phi(\mathsf{id}) = \mathsf{id} \qquad \text{(MIXMAP-ID)}$$

---

[1] The observations in this paragraph were suggested by an anonymous reviewer of an earlier version of this paper.

These laws can be proven using the axioms and language of relative monads at the most abstract level. Indeed, assume the definition Mixmap-Hdl and substitute a general base functor J for $+$ in the laws. Then Mixmap-UnitNat is proven by $\phi(g) \circ \eta = (\eta \circ g)^{*} \circ \eta = \eta \circ g$, Mixmap-Fun in its general form is proven by $\phi(J\,f) = (\eta \circ J\,f)^{*} = F\,f$, whereas Mixmap-Comp is proven by $\phi(g') \circ \phi(g) = (\eta \circ g')^{*} \circ (\eta \circ g)^{*} = ((\eta \circ g')^{*} \circ \eta \circ g)^{*} = (\eta \circ g' \circ g)^{*} = \phi(g' \circ g)$. The identity law now follows from $\phi(\mathsf{id}) = \phi(J\,\mathsf{id}) = F\,\mathsf{id} = \mathsf{id}$.

The laws Mixmap-Comp and Mixmap-Id look exactly like functor laws, but ϕ is not a functor (or the morphism mapping of a functor) since its type is $(J\,A \to J\,B) \to (F\,A \to F\,B)$ rather than $(A \to B) \to (F\,A \to F\,B)$. Following the notion of relative monad, ϕ might be called *relative functor*. Continuing the analogy with functor, the law Mixmap-UnitNat states the corresponding naturality property of the unit. Naturality of η in the usual sense is

$$F\,(h, f) \circ \eta \;=\; \eta \circ (h + f) \qquad \text{(Unit-Nat)}$$

In the abstract language, it follows from the above by $F\,f \circ \eta = \phi(J\,f) \circ \eta = \eta \circ J\,f$.

5 Axiomatics for Asymmetric Mixmap

If handling of successes and failures is asymmetric then the axiomatics developed in Sect. 4 does not apply since Hdl-Assoc need not hold; in fact, even Mixmap-Comp may be invalid. For example, if $g' = g = \mathsf{swap}$ then $\phi(g' \circ g) = \phi(\mathsf{id}) = \mathsf{id}$, whereas $\phi(g') \circ \phi(g) = {!} \circ {!} \neq \mathsf{id}$ since negation forgets the consumed input.

Next we seek for weaker replacements of Hdl-Assoc and Mixmap-Comp that would be suitable for reasoning in the asymmetric case. We also find equivalent axiomatics expressed in terms of other operations. We start with mixmap in this section and proceed with the general handle operation in the next section.

We use the term *state* for anything that, besides the result values, can affect the later computation (in parsing, state is or contains the consumed input). The intuition of pure asymmetric mixmaps is expressed by the following conventions:

- A mixmap application resulting in a success keeps the state unchanged;
- A mixmap application resulting in a failure loses the state;
- The state has no impact on the result value of a mixmap application;
- All violations of symmetry are caused by treating the state differently.

The conventions allow us to abstract from the state and observe just its binary status (kept vs lost), as the input of a mixmap application where the state is replaced with its status determines the output whose state is similarly abstracted. We use functor $\ddagger$ defined by $E \ddagger A = (E + A) + A$ for representing values with abstracted state: Normal values whose computation history has kept the state are encoded in the form $\mathsf{inr}\,a$ and normal values whose computation history has lost the state in the past are of the form $\mathsf{inl}(\mathsf{inr}\,a)$.

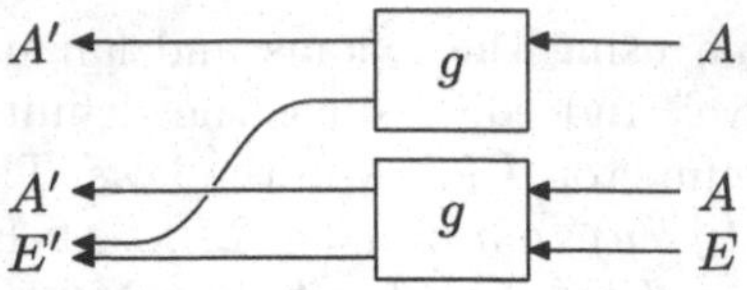

Fig. 2. The schema of evaluation of $\mathsf{sep}\, g$

The dependency of the abstract output of a mixmap application on its abstract input is captured by function $\mathsf{sep} : (E{+}A \to E'{+}A') \to (E{\ddagger}A \to E'{\ddagger}A')$ defined by

$$\mathsf{sep}\, g \;=\; \mathsf{inl} \circ g \triangledown (\mathsf{inl} + \mathsf{id}) \circ g \circ \mathsf{inr} \qquad (\textsc{Sep})$$

The function is illustrated in Fig. 2. (The diagrams show computation tracks as right-to-left paths, following the way function compositions are written. Data with tag inl occur below data with tag inr of the same level.)

Hence the following implication could be a suitable replacement to Mixmap-Comp for the asymmetric case: $\phi(g_m) \circ \ldots \circ \phi(g_1) = \phi(g'_n) \circ \ldots \circ \phi(g'_1)$ whenever $\mathsf{sep}\, g_m \circ \ldots \circ \mathsf{sep}\, g_1 = \mathsf{sep}\, g'_n \circ \ldots \circ \mathsf{sep}\, g'_1$. However, we can find equivalent finite sets of laws in the equational form. Note that, to violate preservation of a composition $g' \circ g$ by sep, an input value $\mathsf{inr}\, a$ must be mapped by g to $\mathsf{inl}\, e$ which in turn must be mapped by g' to $\mathsf{inr}\, a'$, because then $\mathsf{sep}(g' \circ g)$ and $\mathsf{sep}\, g' \circ \mathsf{sep}\, g$ map the input to different copies of the type of normal values. Relying on this observation, we can establish three important cases of sep actually preserving composition:

$$\mathsf{sep}\, g' \circ \mathsf{sep}(h \triangledown \mathsf{inr} \circ f) \;=\; \mathsf{sep}(g' \circ (h \triangledown \mathsf{inr} \circ f)) \qquad (\textsc{Sep-Comp-RPres})$$

$$\mathsf{sep}(\mathsf{inl} \circ h \triangledown f) \circ \mathsf{sep}\, g \;=\; \mathsf{sep}((\mathsf{inl} \circ h \triangledown f) \circ g) \qquad (\textsc{Sep-Comp-LPres})$$

$$\mathsf{sep}(\mathsf{id} \triangledown \mathsf{inr}) \circ \mathsf{sep}(h \triangledown (\mathsf{inl} + \mathsf{id}) \circ f) \;=\; \mathsf{sep}((\mathsf{id} \triangledown \mathsf{inr}) \circ (h \triangledown (\mathsf{inl} + \mathsf{id}) \circ f)) \qquad (\textsc{Sep-Comp-Cross})$$

Consequently, the following special cases of Mixmap-Comp must hold:

$$\phi(g') \circ \phi(h \triangledown \mathsf{inr} \circ f) \;=\; \phi(g' \circ (h \triangledown \mathsf{inr} \circ f)) \qquad (\textsc{Mixmap-Comp-RPres})$$

$$\phi(\mathsf{inl} \circ h \triangledown f) \circ \phi(g) \;=\; \phi((\mathsf{inl} \circ h \triangledown f) \circ g) \qquad (\textsc{Mixmap-Comp-LPres})$$

$$\phi(\mathsf{id} \triangledown \mathsf{inr}) \circ \phi(h \triangledown (\mathsf{inl} + \mathsf{id}) \circ f) \;=\; \phi((\mathsf{id} \triangledown \mathsf{inr}) \circ (h \triangledown (\mathsf{inl} + \mathsf{id}) \circ f)) \qquad (\textsc{Mixmap-Comp-Cross})$$

The schemata in Fig. 3 depict the data flow in these three cases. Indeed in the l.h.s. of Sep-Comp-RPres, the rightmost sep is applied to a function that preserves inr, so inr cannot switch to inl at the first step. Dually in Sep-Comp-LPres, sep in the left is applied to a function that preserves inl, meaning that inl cannot switch to inr at the second step. In Sep-Comp-Cross, both steps can do the risky switch; nevertheless, inl that is obtained from inr at the first step is marked with an extra tag inl causing the second step to preserve inl. Hence the opposite switches apply to different data, the computation tracks cross safely.

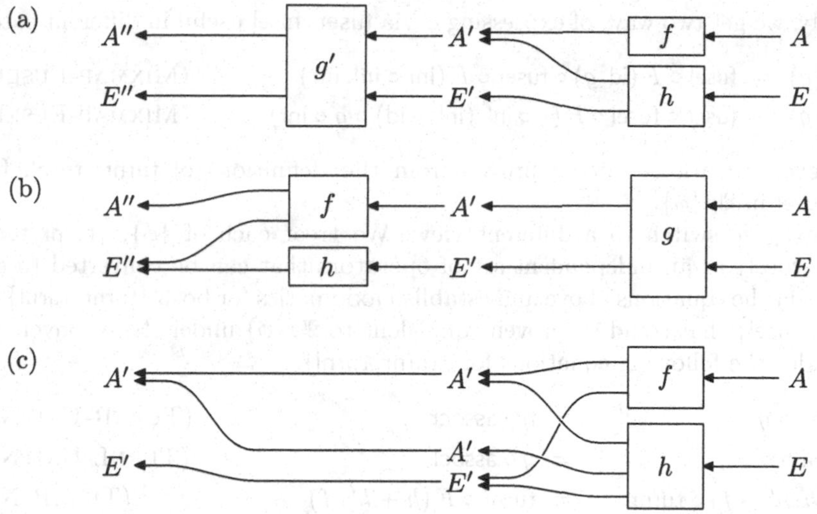

Fig. 3. Data flow of composition of two underlying functions in the case of: (a) Sep-Comp-RPres; (b) Sep-Comp-LPres; (c) Sep-Comp-Cross

In our work, we proved that, if Mixmap-Fun is assumed, the conjunction of Mixmap-Comp-RPres, Mixmap-Comp-LPres and Mixmap-Comp-Cross is equivalent to the general implication of $\phi(g_m) \circ \ldots \circ \phi(g_1) = \phi(g'_n) \circ \ldots \circ \phi(g'_1)$ by $\mathsf{sep}\, g_m \circ \ldots \circ \mathsf{sep}\, g_1 = \mathsf{sep}\, g'_n \circ \ldots \circ \mathsf{sep}\, g'_1$. Hence the set of equations consisting of Mixmap-UnitNat, Mixmap-Fun, Mixmap-Comp-RPres, Mixmap-Comp-LPres and Mixmap-Comp-Cross is a suitable axiomatics of mixmap for asymmetric semantics. We denote this set of axioms as $\mathfrak{Ax}\,\{\phi\}$ for brevity.

In Sect. 3, operations turnr, turnl, fuser, fusel were defined as special cases of mixmap. We can express fuser, fusel in terms of turnr, turnl and vice versa:

$$\begin{aligned} \mathsf{fuser} &= F\,(\mathsf{id}, \mathsf{id} \triangledown \mathsf{id}) \circ \mathsf{turnr} && \text{(FuseR-TurnR)} \\ \mathsf{fusel} &= F\,(\mathsf{id} \triangledown \mathsf{id}, \mathsf{id}) \circ \mathsf{turnl} && \text{(FuseL-TurnL)} \\ \mathsf{turnr} &= \mathsf{fuser} \circ F\,(\mathsf{id} + \mathsf{inl}, \mathsf{inr}) && \text{(TurnR-FuseR)} \\ \mathsf{turnl} &= \mathsf{fusel} \circ F\,(\mathsf{inl}, \mathsf{inr} + \mathsf{id}) && \text{(TurnL-FuseL)} \end{aligned}$$

On the other hand, ϕ can be defined via turnr and turnl by

$$\phi(g) = F\,(\mathsf{id} \triangledown \mathsf{id}, \mathsf{id}) \circ \mathsf{turnl} \circ F\,(\mathsf{id}, g) \circ \mathsf{turnr} \circ F\,(\mathsf{inr}, \mathsf{id}) \quad \text{(Mixmap-TurnLR)}$$

In Mixmap-TurnLR, the application of g is performed via functor map after shifting everything to the right by $\mathsf{turnr} \circ F\,(\mathsf{inr}, \mathsf{id})$. After applying g, the true roles of data are restored by $F\,(\mathsf{id} \triangledown \mathsf{id}, \mathsf{id}) \circ \mathsf{turnl}$. This way of computing $\phi(g)$ is correct as no data flow from the right to the left and back occurs. A dual computation shifting everything to the left rather than to the right is not equivalent.

Finally, we get two ways of expressing ϕ via fuser, fusel useful in different cases:

$$\phi(g) = \mathsf{fusel} \circ F(\mathsf{id}, g) \circ \mathsf{fuser} \circ F(\mathsf{inr} \circ \mathsf{inl}, \mathsf{inr}) \qquad (\text{MIXMAP-FUSELR})$$
$$\phi(g) = \mathsf{fuser} \circ \mathsf{fusel} \circ F(g \circ \mathsf{inl}, (\mathsf{inl} + \mathsf{id}) \circ g \circ \mathsf{inr}) \qquad (\text{MIXMAP-FUSERL})$$

All seven equations can be proven from the definitions of turnr, turnl, fuser and fusel in $\mathfrak{Ax}\,\{\phi\}$.

Next, we switch to a different view: We treat each of $\{\phi\}$, {turnr, turnl}, {fuser, fusel} as an independent set of operators that can be converted to each other via the equations above and establish axiomatics for both {turnr, turnl} and {fuser, fusel} that could be proven equivalent to $\mathfrak{Ax}\,\{\phi\}$ under these conversions. Consider the following equations for {turnr, turnl}:

$$\mathsf{turnr} \circ \eta = \eta \circ \mathsf{assocr} \qquad (\text{TURNR-UNITNAT})$$
$$\mathsf{turnl} \circ \eta = \eta \circ \mathsf{assocl} \qquad (\text{TURNL-UNITNAT})$$
$$F(h, h' + f) \circ \mathsf{turnr} = \mathsf{turnr} \circ F(h + h', f) \qquad (\text{TURNR-NAT})$$
$$F(h + h', f) \circ \mathsf{turnl} = \mathsf{turnl} \circ F(h, h' + f) \qquad (\text{TURNL-NAT})$$
$$\mathsf{turnr} \circ \mathsf{turnr} = F(\mathsf{id}, \mathsf{assocr}) \circ \mathsf{turnr} \circ F(\mathsf{assocr}, \mathsf{id}) \qquad (\text{TURNR-TURNR})$$
$$\mathsf{turnl} \circ \mathsf{turnl} = F(\mathsf{assocl}, \mathsf{id}) \circ \mathsf{turnl} \circ F(\mathsf{id}, \mathsf{assocl}) \qquad (\text{TURNL-TURNL})$$
$$\mathsf{turnr} \circ F(\mathsf{swasr}, \mathsf{id}) \circ \mathsf{turnl} = \mathsf{turnl} \circ F(\mathsf{id}, \mathsf{swasl}) \circ \mathsf{turnr} \qquad (\text{TURNRL-TURNLR})$$
$$\mathsf{turnl} \circ \mathsf{turnr} = \mathsf{id} \qquad (\text{TURNLR-ID})$$

For {fuser, fusel}, we can use the following set of laws:

$$\mathsf{fuser} \circ \eta = \eta \circ (\mathsf{id} \triangledown \mathsf{inr}) \qquad (\text{FUSER-UNITNAT})$$
$$\mathsf{fusel} \circ \eta = \eta \circ (\mathsf{inl} \triangledown \mathsf{id}) \qquad (\text{FUSEL-UNITNAT})$$
$$F(h, f) \circ \mathsf{fuser} = \mathsf{fuser} \circ F(h + f, f) \qquad (\text{FUSER-NAT})$$
$$F(h, f) \circ \mathsf{fusel} = \mathsf{fusel} \circ F(h, h + f) \qquad (\text{FUSEL-NAT})$$
$$\mathsf{fuser} \circ \mathsf{fuser} = \mathsf{fuser} \circ F(\mathsf{id} \triangledown \mathsf{inr}, \mathsf{id}) \qquad (\text{FUSER-FUSER})$$
$$\mathsf{fusel} \circ \mathsf{fusel} = \mathsf{fusel} \circ F(\mathsf{id}, \mathsf{inl} \triangledown \mathsf{id}) \qquad (\text{FUSEL-FUSEL})$$
$$\mathsf{fusel} \circ \mathsf{fuser} = \mathsf{fuser} \circ \mathsf{fusel} \circ F(\mathsf{inl} \triangledown \mathsf{id}, \mathsf{inl} + \mathsf{id}) \qquad (\text{FUSEL-FUSER})$$
$$\mathsf{fuser} \circ F(\mathsf{inl}, \mathsf{id}) = \mathsf{id} \qquad (\text{FUSER-ID})$$
$$\mathsf{fusel} \circ F(\mathsf{id}, \mathsf{inr}) = \mathsf{id} \qquad (\text{FUSEL-ID})$$

Denote these sets of axioms by $\mathfrak{Ax}$ {turnr, turnl} and $\mathfrak{Ax}$ {fuser, fusel}, respectively. Denote by $\mathfrak{Ax}\,\{\eta\}$ the set consisting of one law UNIT-NAT. In our work, we have proven the following result:

Theorem 1. *Let F be a bifunctor equipped with operations η, ϕ,* turnr, turnl, fuser, fusel. *Under the mutually defining equations given above, $\mathfrak{Ax}\,\{\phi\}$ is equivalent to $\mathfrak{Ax}$* {turnr, turnl} $\cup\, \mathfrak{Ax}\,\{\eta\}$*, as well as to $\mathfrak{Ax}$* {fuser, fusel} $\cup\, \mathfrak{Ax}\,\{\eta\}$*. Moreover, defining two of the operation sets $\{\phi\}$,* {turnr, turnl}, {fuser, fusel} *in terms of the third one makes all mutually defining equations valid; in particular,* MIXMAP-FUSELR *and* MIXMAP-FUSERL *are equivalent.* □

One can encounter an evident correspondence between single axioms or segments of consecutive axioms of $\{\phi\}$, $\{\mathsf{turnr}, \mathsf{turnl}\}$ and $\{\mathsf{fuser}, \mathsf{fusel}\}$ based on means they involve and roles they play. Still the correspondence is not perfect. Indeed, asymmetry is introduced into $\mathfrak{Ax}\,\{\mathsf{turnr}, \mathsf{turnl}\}$ by the identity law TURNLR-ID (the dual equation cannot be proven), whereas the interchange law FUSEL-FUSER introduces asymmetry into $\mathfrak{Ax}\,\{\mathsf{fuser}, \mathsf{fusel}\}$ (the rest of both $\mathfrak{Ax}\,\{\mathsf{turnr}, \mathsf{turnl}\}$ and $\mathfrak{Ax}\,\{\mathsf{fuser}, \mathsf{fusel}\}$ is symmetric). Also, $\mathfrak{Ax}\,\{\phi\}$ contains no identity axiom since MIXMAP-ID is implied by MIXMAP-FUN, whereas the identity laws of both $\{\mathsf{turnr}, \mathsf{turnl}\}$ and $\{\mathsf{fuser}, \mathsf{fusel}\}$ have independent significance.

Other equivalent axiomatics exist of course. The principles according to which the choice in favour of these sets of laws was made are simplicity of the equations, systematic build-up of the sets and symmetry as much as possible. In addition, $\mathfrak{Ax}\,\{\mathsf{fuser}, \mathsf{fusel}\}$ constitutes rewrite rules for transforming every expression in the form of the composition of a finite number of fuser, fusel and functor maps to the canonical form $\mathsf{fuser} \circ \mathsf{fusel} \circ F\,(h, f)$. The canonical form can be proven unique: If an expression is provably equal to both $\mathsf{fuser} \circ \mathsf{fusel} \circ F\,(h, f)$ and $\mathsf{fuser} \circ \mathsf{fusel} \circ F\,(h', f')$ then $\mathsf{fuser} \circ \mathsf{fusel} \circ F\,(h, f) = \mathsf{fuser} \circ \mathsf{fusel} \circ F\,(h', f')$ is derivable from the axioms and must therefore be valid for all models of the axioms. As $\mathfrak{Ax}\,\{\phi\}$ is satisfied for $F = \ddagger$ by defining $\eta = \mathsf{inl}$, $\phi = \mathsf{sep}$, the equality $\mathsf{sep}(\mathsf{id} \triangledown \mathsf{inr}) \circ \mathsf{sep}(\mathsf{inl} \triangledown \mathsf{id}) \circ \mathsf{sep}(h + f) = \mathsf{sep}(\mathsf{id} \triangledown \mathsf{inr}) \circ \mathsf{sep}(\mathsf{inl} \triangledown \mathsf{id}) \circ \mathsf{sep}(h' + f')$ must be valid. Calculation shows that this implies both $f = f'$ and $h = h'$.

Note that $\mathfrak{Ax}\,\{\phi\}$ is complete w.r.t. the model used in the previous paragraph due to the injectivity of inl and the fact that $\phi(g_m) \circ \ldots \circ \phi(g_1) = \phi(g'_n) \circ \ldots \circ \phi(g'_1)$ is provable in $\mathfrak{Ax}\,\{\phi\}$ whenever $\mathsf{sep}\, g_m \circ \ldots \circ \mathsf{sep}\, g_1 = \mathsf{sep}\, g'_n \circ \ldots \circ \mathsf{sep}\, g'_1$.

6 Axiomatics for Asymmetric Handle, Bind and Catch

In Sect. 5, we saw that for mixmap it is enough to assume preservation of composition in three cases captured by axioms MIXMAP-COMP-RPRES, MIXMAP-COMP-LPRES and MIXMAP-COMP-CROSS. The cases correspond to specific data flow patterns depicted in Fig. 3. Guided by this, we replace HDL-ASSOC with its three special cases that represent similar data flow patterns:

$$l^* \circ (x \triangledown \eta \circ \mathsf{inr} \circ f)^* = (l^* \circ (x \triangledown \eta \circ \mathsf{inr} \circ f))^* \qquad (\text{HDL-ASSOC-RPRES})$$

$$(\eta \circ \mathsf{inl} \circ h \triangledown l)^* \circ k^* = ((\eta \circ \mathsf{inl} \circ h \triangledown l)^* \circ k)^* \qquad (\text{HDL-ASSOC-LPRES})$$

$$(\eta \circ (\mathsf{id} \triangledown \mathsf{inr}))^* \circ (x \triangledown F\,(\mathsf{inl}, \mathsf{id}) \circ k)^* = ((\eta \circ (\mathsf{id} \triangledown \mathsf{inr}))^* \circ (x \triangledown F\,(\mathsf{inl}, \mathsf{id}) \circ k))^* \qquad (\text{HDL-ASSOC-CROSS})$$

As a result, we obtain an axiomatics $\mathfrak{Ax}\,\{(\_)^*\}$ that consists of five laws HDL-LUNIT, HDL-RUNIT, HDL-ASSOC-RPRES, HDL-ASSOC-LPRES and HDL-ASSOC-CROSS (HDL-ID is a theorem). Assuming that ϕ is given by MIXMAP-HDL, deriving the axioms of $\mathfrak{Ax}\,\{\phi\}$ from $\mathfrak{Ax}\,\{(\_)^*\}$ is straightforward.

The bind operation $(\_)^{\curlyvee}$ and the catch operation $(\_)^{\curlywedge}$ were specified via $(\_)^*$ by BND-HDL and CCH-HDL in Sect. 3. Interestingly, analogies exist between

$(\_)^{\curlyvee}$ and fusel, similarly between $(\_)^{\curlywedge}$ and fuser. Firstly, both $(\_)^{\curlyvee}$ and fusel keep errors unchanged, whereas both $(\_)^{\curlywedge}$ and fuser keep normal values unchanged. But there is more behind: Assuming the definitions MIXMAP-HDL, BND-HDL and CCH-HDL, it is easy to see that the defining equations FUSER-MIXMAP and FUSEL-MIXMAP are equivalent to $\mathsf{fuser} = \eta^{\curlywedge}$ and $\mathsf{fusel} = \eta^{\curlyvee}$, respectively. The axiom set we next propose for $\{(\_)^{\curlyvee}, (\_)^{\curlywedge}\}$ also has characteristics in common with $\mathfrak{Ax}\,\{\mathsf{fuser}, \mathsf{fusel}\}$, most notably providing means for rewriting all successive compositions of binds and catches except those of the form $x^{\curlywedge} \circ k^{\curlyvee}$.

We require bind to satisfy the monad laws with $\eta \circ \mathsf{inr}$ playing the role of unit. In addition, $\eta \circ \mathsf{inl}$ followed by a bind should act as zero and functor maps on error values should act as homomorphisms, i.e., preserve bind:

$$k^{\curlyvee} \circ \eta \circ \mathsf{inr} = k \qquad \text{(BND-LUNIT)}$$
$$(\eta \circ \mathsf{inr} \circ f)^{\curlyvee} = F\,(\mathsf{id}, f) \qquad \text{(BND-RUNIT)}$$
$$l^{\curlyvee} \circ k^{\curlyvee} = (l^{\curlyvee} \circ k)^{\curlyvee} \qquad \text{(BND-ASSOC)}$$
$$k^{\curlyvee} \circ \eta \circ \mathsf{inl} = \eta \circ \mathsf{inl} \qquad \text{(BND-LZERO)}$$
$$F\,(h, \mathsf{id}) \circ k^{\curlyvee} = (F\,(h, \mathsf{id}) \circ k)^{\curlyvee} \circ F\,(h, \mathsf{id}) \qquad \text{(BND-FUNHOM)}$$

Catch is expected to meet the corresponding dual laws:

$$x^{\curlywedge} \circ \eta \circ \mathsf{inl} = x \qquad \text{(CCH-LUNIT)}$$
$$(\eta \circ \mathsf{inl} \circ h)^{\curlywedge} = F\,(h, \mathsf{id}) \qquad \text{(CCH-RUNIT)}$$
$$y^{\curlywedge} \circ x^{\curlywedge} = (y^{\curlywedge} \circ x)^{\curlywedge} \qquad \text{(CCH-ASSOC)}$$
$$x^{\curlywedge} \circ \eta \circ \mathsf{inr} = \eta \circ \mathsf{inr} \qquad \text{(CCH-LZERO)}$$
$$F\,(\mathsf{id}, f) \circ x^{\curlywedge} = (F\,(\mathsf{id}, f) \circ x)^{\curlywedge} \circ F\,(\mathsf{id}, f) \qquad \text{(CCH-FUNHOM)}$$

Asymmetry comes in via an additional law allowing us to change the order between bind and catch. Two axiom candidates are both deserving this role:

$$k^{\curlyvee} \circ \eta^{\curlywedge} = (\eta \circ \mathsf{inl} \bigtriangledown k)^{\curlywedge} \circ (F\,(\mathsf{inl}, \mathsf{id}) \circ k)^{\curlyvee} \qquad \text{(BND-UNITCCH)}$$
$$k^{\curlyvee} \circ x^{\curlywedge} = (\eta \circ \mathsf{inl} \bigtriangledown k^{\curlyvee} \circ x)^{\curlywedge} \circ (F\,(\mathsf{inl}, \mathsf{id}) \circ k)^{\curlyvee} \circ F\,(\mathsf{inr}, \mathsf{id}) \qquad \text{(BND-CCH)}$$

In BND-CCH, we have $x : E \to F\,(E', A)$ and $k : A \to F\,(E', A')$, whence $x^{\curlywedge} : F\,(E, A) \to F\,(E', A)$ and $k^{\curlyvee} : F\,(E', A) \to F\,(E', A')$. Figure 4 depicts the data flows of both sides of the equation. Intuitively, applying $k^{\curlyvee}$ after $x^{\curlywedge}$ in the l.h.s. means first catching errors by x and then applying k to normal values (which can be either having existed in the beginning or introduced by x). To obtain the same result after applying k to normal values first, errors that exist in the beginning must be distinguished from errors introduced by k in order to catch by x the errors of the first kind only. Distinguishing of errors of two different kinds can be achieved by tagging them with inl and inr, respectively, which is done by $F\,(\mathsf{inl}, \mathsf{id})$ and $F\,(\mathsf{inr}, \mathsf{id})$ in the r.h.s. of BND-CCH.

One can interpret BND-UNITCCH in a similar manner (here $E = E' + A$).

Although BND-CCH applies more generally, it is actually implied by BND-UNITCCH and the other axioms. Hence let $\mathfrak{Ax}\{(\_)^{\curlyvee}, (\_)^{\curlywedge}\}$ denote the set of all equations of $(\_)^{\curlyvee}$ and $(\_)^{\curlywedge}$ introduced above except BND-CCH.

As the only role of $F\,(\mathsf{inl}, \mathsf{id})$ and $F\,(\mathsf{inr}, \mathsf{id})$ in BND-CCH is to separate errors of different origin, one can easily find equivalent corollaries of $\mathfrak{Ax}\{(\_)^{\curlyvee}, (\_)^{\curlywedge}\}$ that achieve the same aim. For example, one can interchange $F\,(\mathsf{inl}, \mathsf{id})$ and $F\,(\mathsf{inr}, \mathsf{id})$ and also the sides of $\triangledown$. This can be proven by inserting $F\,(\mathsf{swap}, \mathsf{id}) \circ F\,(\mathsf{swap}, \mathsf{id})$ (which equals identity) between the two leftmost computations in the r.h.s. of BND-CCH, applying CCH-RUNIT and CCH-ASSOC to join the two leftmost terms and rewriting the rest by BND-FUNHOM.

It turns out that $\mathfrak{Ax}\{(\_)^{*}\}$ and $\mathfrak{Ax}\{(\_)^{\curlyvee}, (\_)^{\curlywedge}\}$ restrict the class of bifunctors with joint unit equivalently. To establish this, we need a definition of handle in terms of bind and catch. Both following equations are good:

$$k^{*} = k^{\curlyvee} \circ \eta^{\curlywedge} \circ F\,(\mathsf{inr} \circ \mathsf{inl}, \mathsf{inr}) \qquad \text{(HDL-BNDCCH)}$$

$$k^{*} = (\eta \circ \mathsf{inl} \triangledown k \circ \mathsf{inl})^{\curlywedge} \circ (F\,(\mathsf{inl}, \mathsf{id}) \circ k \circ \mathsf{inr})^{\curlyvee} \circ F\,(\mathsf{inr}, \mathsf{id}) \qquad \text{(HDL-CCHBND)}$$

The idea behind HDL-BNDCCH is easy: Note that $\eta^{\curlywedge} \circ F\,(\mathsf{inr} \circ \mathsf{inl}, \mathsf{inr})$ is equal to $\phi(\mathsf{inr})$ (provable after rewriting $\eta^{\curlywedge} = \mathsf{fuser} = \phi(\mathsf{id} \triangledown \mathsf{inr})$) which means shifting everything to the right. Hence the r.h.s. of HDL-BNDCCH mimics the functionality of k^{*} by applying k to the input after requalifying it as a surely normal value. The equation HDL-CCHBND can be understood similarly to BND-CCH.

We have proven the following theorem:

Theorem 2. *Let F be a bifunctor equipped with operations η, $(\_)^{*}$, $(\_)^{\curlyvee}$, $(\_)^{\curlywedge}$. Under the mutually defining equations given above, $\mathfrak{Ax}\{(\_)^{*}\}$ is equivalent to $\mathfrak{Ax}\{(\_)^{\curlyvee}, (\_)^{\curlywedge}\} \cup \mathfrak{Ax}\{\eta\}$; the same holds even if* CCH-FUNHOM *and either* BND-LZERO *or* CCH-LZERO *are excluded. Moreover, defining one of the operation sets $\{(\_)^{*}\}$, $\{(\_)^{\curlyvee}, (\_)^{\curlywedge}\}$ in terms of the other makes all mutually defining equations valid; in particular,* HDL-BNDCCH *and* HDL-CCHBND *are equivalent.* □

Equivalence of the axiomatics of $\{(\_)^{*}\}$ and $\{(\_)^{\curlyvee}, (\_)^{\curlywedge}\}$ implies that any composition of a finite number of functions of the form k^{*} is equal to the composition

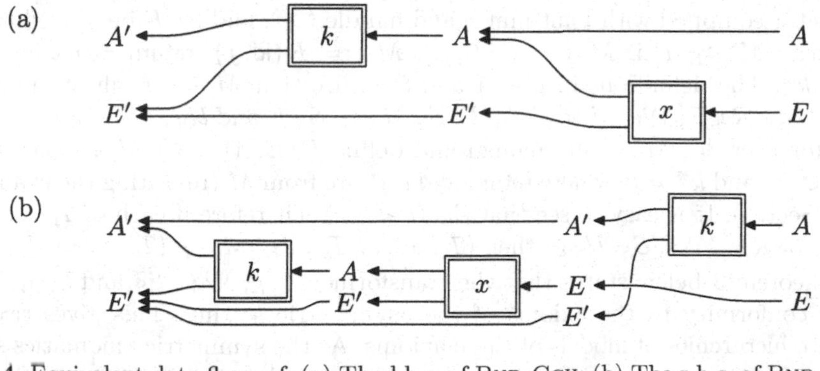

Fig. 4. Equivalent data flows of: (a) The l.h.s. of BND-CCH; (b) The r.h.s. of BND-CCH

of two functions of this form. Indeed, each function of this form can be rewritten in the form $x^{\nwarrow} \circ t^{\veebar} \circ F\,(h, \mathsf{id})$ by HDL-CCHBND. Marking catches, binds and functor maps of errors by letters C, B, F, respectively, we obtain a word consisting of CBF repeated finitely many times. We can rewrite $BB \rightarrow B$ (BND-ASSOC), $CC \rightarrow C$ (CCH-ASSOC), $FF \rightarrow F$ (functor), $BC \rightarrow CBF$ (BND-CCH), $FB \rightarrow BF$ (BND-FUNHOM) and $F \rightarrow C$ (CCH-RUNIT). Thus the chain $CBFCBF \rightarrow CBCCBF \rightarrow CBCBF \rightarrow CCBFBF \rightarrow CCBBFF \rightarrow^3 CBF$ of rewrites eliminates one repetition, implying that the whole composition equals an expression of the form $x^{\nwarrow} \circ t^{\veebar} \circ F\,(h, \mathsf{id})$. Here, all three functions are expressible in the form k^*, among which the two rightmost functions can be joined.

The axiomatics of $\{(\_)^{\veebar}, (\_)^{\nwarrow}\}$ considered in our earlier work [15] does not contain interchange laws like BND-UNITCCH or BND-CCH. Instead, it assumes an extra homomorphism law and two De Morgan laws:

$$\mathsf{turnr} \circ \mathsf{turnl} \circ x^{\nwarrow} = (\mathsf{turnr} \circ \mathsf{turnl} \circ x)^{\nwarrow} \circ \mathsf{turnr} \circ \mathsf{turnl} \qquad \text{(CCH-TURNRLHOM)}$$

$$! \circ k^{\veebar} \circ ! = (! \circ k)^{\nwarrow} \circ ! \circ ! \qquad \text{(BND-CCH-DEM)}$$

$$! \circ h^{\nwarrow} = (! \circ h)^{\veebar} \circ ! \qquad \text{(CCH-BND-DEM)}$$

All three are provable in $\mathfrak{Ax}\,\{(\_)^{\veebar}, (\_)^{\nwarrow}\} \cup \mathfrak{Ax}\,\{\eta\}$. The fact that they only impose properties of a single bind or catch in connection with special cases of mixmap suggests that replacing the cited laws of [15] with the interchange law makes the axiomatics strictly stronger.

7 Models of the Axiomatics

Our prior work [15] constructs a hierarchy of bifunctors F that fulfill all laws considered there. Figure 5 revisits the construction using notational conventions of Sect. 2. Unlike in [15], we define $(\_)^*$ here instead of $(\_)^{\veebar}$ and $(\_)^{\nwarrow}$. The construction generalizes the reader, writer, state and update monad transformers to bifunctors equipped with joint unit and handle. The accent $\hat{}$ is used in denotations of bifunctor transformations for clarity. The well-known monad transformers can be obtained from them via fixing the error type E.

We can formalize the relationship between monad and bifunctor transformers more precisely. Let M be a monad with operations return and $(\_)^*$, let F be a bifunctor equipped with joint unit η and handle $(\_)^*$, and let E be a fixed type; we write $M \prec_E F$ if $M\,A = F\,(E, A)$, $M\,f = F\,(\mathsf{id}, f)$, $\mathsf{return} = \eta \circ \mathsf{inr}$ and $k^* = k^{\veebar}$. The definitions in Figs. 1 and 5 ensure that $M \prec_E F$ always implies $\mathcal{R}_R\,M \prec_E \hat{\mathcal{R}}_R\,F$, $\mathcal{W}_W\,M \prec_E \hat{\mathcal{W}}_W\,F$, $\mathcal{S}_S\,M \prec_E \hat{\mathcal{S}}_S\,F$ and $\mathcal{U}_{S,W}\,M \prec_E \hat{\mathcal{U}}_{S,W}\,F$.

Moreover, let M be any monad and define $F\,(E, A) = \mathcal{E}_E\,M\,A$ along with $\eta = \mathsf{return}$ and $k^* = k^*$ where return and $(\_)^*$ are from M (revisiting the example from Sect. 4). It is easy to see that $\mathcal{E}_E\,M \prec_E F$. Therefore if each of $\mathcal{T}_1, \ldots, \mathcal{T}_m$ is one of $\mathcal{R}_R$, $\mathcal{W}_W$, $\mathcal{S}_S$, $\mathcal{U}_{S,W}$ then $(\mathcal{T}_m \circ \ldots \circ \mathcal{T}_1 \circ \mathcal{E}_E)\,M \prec_E (\hat{\mathcal{T}}_m \circ \ldots \circ \hat{\mathcal{T}}_1)\,F$.

Theorem 3 below states that the transformers $\hat{\mathcal{R}}_R$, $\hat{\mathcal{W}}_W$, $\hat{\mathcal{S}}_S$ and $\hat{\mathcal{U}}_{S,W}$ preserve conformity to the axioms of the asymmetric setting. This gives rise to infinite hierarchies of models of these axioms. As the symmetric axiomatics subsumes the asymmetric axiomatics, every bifunctor of the family shown in Sect. 4

The reader transformation:

$$\hat{\mathcal{R}}_R\, F\,(E,A) = R \to F\,(E,A) \qquad \hat{\mathcal{R}}_R\, F\,(h,f) = \lambda g \,.\, F\,(h,f) \circ g$$
$$\eta(x) = \lambda r \,.\, \eta(x)$$
$$k^*\, t = \lambda r \,.\, (\lambda a \,.\, k\, a\, r)^*\,(t\, r)$$

The writer transformation:

$$\hat{\mathcal{W}}_W\, F\,(E,A) = F\,(E, A \times W) \qquad \hat{\mathcal{W}}_W\, F\,(h,f) = F\,(h, f \times \mathsf{id})$$
$$\eta(x) = \eta((\mathsf{id} + (,1))\, x)$$
$$k^*\, t = (k \circ \mathsf{inl} \triangledown (\lambda(a,w) \,.\, F\,(\mathsf{id}, \mathsf{id} \times (w\cdot))\,(k\,(\mathsf{inr}\, a))))^*\, t$$

The state transformation:

$$\hat{\mathcal{S}}_S\, F\,(E,A) = S \to F\,(E, A \times S) \qquad \hat{\mathcal{S}}_S\, F\,(h,f) = \lambda g \,.\, F\,(h, f \times \mathsf{id}) \circ g$$
$$\eta(x) = \lambda s \,.\, \eta((\mathsf{id} + (,s))\, x)$$
$$k^*\, t = \lambda s \,.\, ((\lambda e \,.\, k\,(\mathsf{inl}\, e)\, s) \triangledown (\lambda(a,s') \,.\, k\,(\mathsf{inr}\, a)\, s'))^*\,(t\, s)$$

The update transformation:

$$\hat{\mathcal{U}}_{S,W}\, F\,(E,A) = S \to F\,(E, A \times W) \qquad \hat{\mathcal{U}}_{S,W}\, F\,(h,f) = \lambda g \,.\, F\,(h, f \times \mathsf{id}) \circ g$$
$$\eta(x) = \lambda s \,.\, \eta((\mathsf{id} + (,1))\, x)$$
$$k^*\, t = \lambda s \,.\, ((\lambda e \,.\, k\,(\mathsf{inl}\, e)\, s) \triangledown (\lambda(a,w) \,.\, F\,(\mathsf{id}, \mathsf{id} \times (w\cdot))\,(k\,(\mathsf{inr}\, a)\,(s \bullet w))))^*\,(t\, s)$$

Fig. 5. Definitions of the reader, writer, state and update bifunctor transformers

to fulfill the symmetric axiomatics is a suitable base case. Hence the hierarchy constructed in [15] also satisfies the asymmetric axiomatics of this paper.

Theorem 3. *Let F be a bifunctor equipped with operations η and $(\_)^*$. If F fulfills $\mathfrak{Ar}\,\{(\_)^*\}$ then each of $\hat{\mathcal{R}}_R\, F$, $\hat{\mathcal{W}}_W\, F$, $\hat{\mathcal{S}}_S\, F$, $\hat{\mathcal{U}}_{S,W}\, F$ fulfills $\mathfrak{Ar}\,\{(\_)^*\}$.* □

Section 5 was concluded by showing that taking $F = \ddagger$, $\phi = \mathsf{sep}$, $\eta = \mathsf{inl}$ provides a model of $\mathfrak{Ar}\,\{\phi\}$ w.r.t. which $\mathfrak{Ar}\,\{\phi\}$ is complete. One can extend this example to models of $\mathfrak{Ar}\,\{(\_)^*\}$ by defining either $k^* = k \triangledown (\mathsf{inl}{+}\mathsf{id}) \circ (\mathsf{id} \triangledown \mathsf{inr}) \circ k \circ \mathsf{inr}$ or $k^* = k \triangledown \mathsf{swasr} \circ k \circ \mathsf{inr}$. However, $\mathfrak{Ar}\,\{(\_)^*\}$ is not complete w.r.t. these models. For example, the equation

$$((\eta \circ \mathsf{inl} \circ f)^{\curlyvee} \circ x \triangledown l)^* \circ k^* \;=\; (((\eta \circ \mathsf{inl} \circ f)^{\curlyvee} \circ x \triangledown l)^* \circ k)^*$$

is valid in both obtained models but invalid in the model $\hat{\mathcal{S}}_S\,(+)$ of the axioms (where S is an arbitrary set containing at least two elements).

8 Laws that Distinguish the Two Settings

Above, we have defined axiomatics of mixmap and handle for both symmetric and asymmetric behaviour of successes and failures. In the asymmetric setting, we have found equivalent axiomatics for other operation sets. We now do the same in the symmetric setting. We show that, for each of the other operation sets ($\{\mathsf{turnr}, \mathsf{turnl}\}$ and $\{\mathsf{fuser}, \mathsf{fusel}\}$ in the case of mixmap and $\{(\_)^{\curlyvee}, (\_)^{\curlywedge}\}$ in the case of handle), adding the law dual to the only axiom that introduces asymmetry makes the reasoning power equal to the axiomatics for symmetric setting.

We start from mixmap as above. For {turnr, turnl} and {fuser, fusel}, the dual identity and dual interchange laws, respectively, are to be added:

$$\mathsf{turnr} \circ \mathsf{turnl} \;=\; \mathsf{id} \qquad (\textsc{TurnRL-Id})$$

$$\mathsf{fuser} \circ \mathsf{fusel} \;=\; \mathsf{fusel} \circ \mathsf{fuser} \circ F\,(\mathsf{id} + \mathsf{inr}, \mathsf{id} \triangledown \mathsf{inr}) \qquad (\textsc{FuseR-FuseL})$$

Obviously TurnRL-Id is implied by Mixmap-Fun and Mixmap-Comp. As adding TurnRL-Id makes the axiomatics of {turnr, turnl} symmetric and the defining equations FuseR-TurnR and FuseL-TurnL are symmetric, too, dualizing everything in a proof of FuseL-FuseR establishes FuseR-FuseL. The proof of Mixmap-Comp relying on the extended axiomatics of {fuser, fusel} is straightforward by transforming both sides of Mixmap-Comp to the canonical form, applying FuseR-FuseL on both sides and calculating.

Altogether, we have established the following:

Theorem 4. *Let F be a bifunctor equipped with operations η, ϕ,* turnr, turnl, fuser, fusel *meeting their asymmetric axiomatics and mutually defining equations. Then* Mixmap-Comp, TurnRL-Id *and* FuseR-FuseL *are equivalent.* □

In the case of bind and catch, the dual interchange law to be added is:

$$x^{\barwedge} \circ \eta^{\veebar} \;=\; (x \triangledown \eta \circ \mathsf{inr})^{\veebar} \circ (F\,(\mathsf{id}, \mathsf{inr}) \circ x)^{\barwedge} \qquad (\textsc{Cch-UnitBnd})$$

Proving that Cch-UnitBnd holds if Hdl-Assoc is available is straightforward. For the other direction, one can first establish the dual of Bnd-Cch by dualizing everything in a derivation of Bnd-Cch in $\mathfrak{Ax}\,\{(\_)^{\veebar}, (\_)^{\barwedge}\} \cup \mathfrak{Ax}\,\{\eta\}$. Now transform both sides of Hdl-Assoc to the form $x^{\barwedge} \circ k^{\veebar} \circ F\,(\mathsf{inr}, \mathsf{id})$ and apply the dual of Bnd-Cch on each side. The results convert to equal expressions.

Hence the following holds:

Theorem 5. *Let F be a bifunctor equipped with operators η, $(\_)^{*}$, $(\_)^{\veebar}$, $(\_)^{\barwedge}$ fulfilling their asymmetric axiomatics and mutually defining equations. Then the laws* Hdl-Assoc *and* Cch-UnitBnd *are equivalent.* □

9 Related Work

The class *MonadPlus* extends the Haskell monad interface with additional methods, one of which is $mzero :: m\ a$ where m is the monad. Several different classic axiomatics for *MonadPlus* occur in the literature that all require *mzero* to be the left zero of $\gg\!=$. Our laws Bnd-LZero and Cch-LZero are analogous requirements lifted to the function level.

Equational reasoning about monadic computations with additional axioms specific to different effects has been discussed by Gibbons and Hinze [6]. For error handling, that paper uses operations of monad with zero and addition, the latter being less powerful than our operation $(\_)^{\barwedge}$.

Our previous work [15] considered bifunctors equipped with monadic bind and catch with symmetric types and made the first attempt to axiomatize their asymmetric semantics in parsing. It also introduced ϕ and its special cases (turnr, turnl, fuser, fusel, negation). Studying monad-level operations was not the primary goal. They were introduced as an intermediate step towards smart error handling operations with usage and expressive power resembling those of the *Applicative* and *Alternative* class methods. A few laws (like distributivity) considered there are omitted in this paper as not universally valid in parsing.

Naming of the laws here mostly follows that of [15], except for MIXMAP-COMP that denotes an asymmetric law in that paper. The main achievements of this work compared to [15] are the following:

- Axiomatics equivalent to the axiomatics of asymmetric mixmap for both operation sets $\{\mathsf{turnr}, \mathsf{turnl}\}$ and $\{\mathsf{fuser}, \mathsf{fusel}\}$ (in particular, the law FUSEL-FUSER is absent in [15] and axioms for $\{\mathsf{turnr}, \mathsf{turnl}\}$ are not considered);
- Noting $\mathsf{fuser} \circ \mathsf{fusel} \circ F\,(h, f)$ as a unique canonical form that any sequence of compositions of asymmetric mixmap applications can be converted to;
- Introduction of the joint handle along with its axiomatics in both symmetric and asymmetric setting;
- Introduction of BND-UNITCCH and giving better grounds for this choice of axioms by proving the axiomatics equivalent to that of handle;
- Contrasting the symmetric and asymmetric case with each other (the earlier paper only considers asymmetric semantics).

Bind and catch operations with generalized type of catch were advocated by Malakhovski [13]. That work also proposes an axiomatics that is a strict subset of ours (it consists of the standard laws of monad and the left zero law for both bind and catch). Recently, yet another proposal to axiomatize Haskell *MonadError* class (and also other classes in the monad transformer library) was discussed in the Haskell Libraries mailing list [12].

10 Conclusion

In this work, we proposed some axiomatics, expressed in terms of different operations, for equational reasoning about monadic computations with error handling and studied relationships between them. We considered both the case where successes and failures behave symmetrically and the case where failures erase the program state. We found that the laws of relative monad on the sum functor are appropriate for describing the symmetric setting. Axiomatics of the asymmetric setting was obtained by suitably weakening that of the symmetric setting.

The axiomatics for the asymmetric setting was chosen in two stages. Axiomatization of pure function application (mixmap) was based on truth in a simple abstract model ($F = \ddagger$, $\phi = \mathsf{sep}$, $\eta = \mathsf{inl}$). The obtained axiomatics was then generalized to impure computations by analogy. For the general axiomatics, we have not managed to prove completeness w.r.t. any particular model.

In well-known libraries of functional languages, error handler functions do not allow changing the error type. We ignored this restriction as an unnecessary incumbrance in theory. Despite this, our asymmetric axiomatics of $\{(\_)^{\curlyvee}, (\_)^{\curlywedge}\}$ can be applied to Haskell *MonadError* type class (although proofs may have to use more general types than those in Haskell, just like solving cubic equations in real numbers sometimes can be done only with help of complex numbers).

Acknowledgement. This work was partially supported by the Estonian Research Council grant PSG61. The author thanks all anonymous reviewers for their valuable feedback.

References

1. Ahman, D., Uustalu, T.: Update monads: cointerpreting directed containers. In: Matthes, R., Schubert, A. (eds.) 19th International Conference on Types for Proofs and Programs, TYPES 2013, Toulouse, France, 22–26 April 2013. LIPIcs, vol. 26, pp. 1–23. Schloss Dagstuhl - Leibniz-Zentrum fuer Informatik (2013). https://doi.org/10.4230/LIPIcs.TYPES.2013.1
2. Altenkirch, T., Chapman, J., Uustalu, T.: Monads need not be endofunctors. In: Ong, L. (ed.) FoSSaCS 2010. LNCS, vol. 6014, pp. 297–311. Springer, Heidelberg (2010). https://doi.org/10.1007/978-3-642-12032-9_21
3. Altenkirch, T., Chapman, J., Uustalu, T.: Monads need not be endofunctors. Log. Methods Comput. Sci. **11**(1) (2015). https://doi.org/10.2168/LMCS-11(1:3)2015
4. Danielsson, N.A., Hughes, J., Jansson, P., Gibbons, J.: Fast and loose reasoning is morally correct. In: Morrisett, J.G., Jones, S.L.P. (eds.) Proceedings of the 33rd ACM SIGPLAN-SIGACT Symposium on Principles of Programming Languages, POPL 2006, Charleston, South Carolina, USA, 11–13 January 2006, pp. 206–217. ACM (2006). https://doi.org/10.1145/1111037.1111056
5. Ford, B.: Parsing expression grammars: a recognition-based syntactic foundation. In: Jones, N.D., Leroy, X. (eds.) Proceedings of the 31st ACM SIGPLAN-SIGACT Symposium on Principles of Programming Languages, POPL 2004, Venice, Italy, 14–16 January 2004, pp. 111–122. ACM (2004). https://doi.org/10.1145/964001.964011
6. Gibbons, J., Hinze, R.: Just do it: simple monadic equational reasoning. In: Chakravarty, M.M.T., Hu, Z., Danvy, O. (eds.) Proceeding of the 16th ACM SIGPLAN International Conference on Functional Programming, ICFP 2011, Tokyo, Japan, 19–21 September 2011, pp. 2–14. ACM (2011). https://doi.org/10.1145/2034773.2034777
7. Haskell. https://www.haskell.org
8. Haskell hierarchical libraries. https://downloads.haskell.org/~ghc/latest/docs/html/libraries/index.html
9. Hutton, G., Meijer, E.: Monadic parser combinators. Technical report NOTTCS-TR-96-4, University of Nottingham (1996)
10. Hutton, G., Meijer, E.: Monadic parsing in Haskell. J. Funct. Program. **8**(4), 437–444 (1998). https://doi.org/10.1017/S0956796898003050
11. Liang, S., Hudak, P., Jones, M.P.: Monad transformers and modular interpreters. In: Cytron, R.K., Lee, P. (eds.) Conference Record of POPL 1995: 22nd ACM SIGPLAN-SIGACT Symposium on Principles of Programming Languages, San Francisco, California, USA, 23–25 January 1995, pp. 333–343. ACM Press (1995). https://doi.org/10.1145/199448.199528

12. The Libraries archives. Proposal: Laws for MTL classes. https://mail.haskell.org/pipermail/libraries/2019-April/029549.html
13. Malakhovski, J.: Exceptionally monadic error handling. CoRR abs/1810.13430 (2018). http://arxiv.org/abs/1810.13430
14. Moggi, E.: An abstract view of programming languages. Technical report ECS-LFCS-90-113, University of Edinburgh (1990)
15. Nestra, H.: Double applicative functors. In: Fischer, B., Uustalu, T. (eds.) ICTAC 2018. LNCS, vol. 11187, pp. 333–353. Springer, Cham (2018). https://doi.org/10.1007/978-3-030-02508-3_18
16. Wadler, P.: Comprehending monads. In: LISP and Functional Programming, pp. 61–78 (1990). https://doi.org/10.1145/91556.91592
17. Wadler, P.: The essence of functional programming. In: Sethi, R. (ed.) Conference Record of the Nineteenth Annual ACM SIGPLAN-SIGACT Symposium on Principles of Programming Languages, Albuquerque, New Mexico, USA, 19–22 January 1992, pp. 1–14. ACM Press (1992). https://doi.org/10.1145/143165.143169

Solving of Regular Equations Revisited

Martin Sulzmann[1(✉)] and Kenny Zhuo Ming Lu[2]

[1] Karlsruhe University of Applied Sciences, Karlsruhe, Germany
martin.sulzmann@hs-karlsruhe.de
[2] Nanyang Polytechnic, Singapore, Singapore
luzhuomi@gmail.com

Abstract. Solving of regular equations via Arden's Lemma is folklore knowledge. We first give a concise algorithmic specification of all elementary solving steps. We then discuss a computational interpretation of solving in terms of coercions that transform parse trees of regular equations into parse trees of solutions. Thus, we can identify some conditions on the shape of regular equations under which resulting solutions are unambiguous. We apply our result to convert a DFA to an unambiguous regular expression. In addition, we show that operations such as subtraction and shuffling can be expressed via some appropriate set of regular equations. Thus, we obtain direct (algebraic) methods without having to convert to and from finite automaton.

Keywords: Regular equations and expressions · Parse trees · Ambiguity · Subtraction · Shuffling

1 Introduction

The conversion of a regular expression (RE) into a deterministic finite automaton (DFA) is a well-studied topic. Various methods and optimized implementations exist. The opposite direction has received less attention. In the literature, there are two well-known methods to translate DFAs to REs, namely, state elimination [5] and solving of equations via Arden's Lemma [2].

The solving method works by algebraic manipulation of equations. Identity laws are applied to change the syntactic form of an equation's right-hand side such that Arden's Lemma is applicable. Thus, the set of equations is reduced and in a finite number of steps a solution can be obtained. State elimination has a more operational flavor and reduces states by introducing transitions labeled with regular expressions. The state elimination method appears to be better studied in the literature. For example, see the works [1,10,13] that discuss heuristics to obtain short regular expressions.

In this paper, we revisit solving of regular equations via Arden's Lemma. Specifically, we make the following contributions:

- We give a concise algorithmic description of solving of regular equations where we give a precise specification of all algebraic laws applied (Sect. 3).

R. M. Hierons and M. Mosbah (Eds.): ICTAC 2019, LNCS 11884, pp. 392–409, 2019.
https://doi.org/10.1007/978-3-030-32505-3_22

- We give a computational interpretation of solving by means of coercions that transform parses tree of regular equations into parse trees of solutions. We can identify simple criteria on the shape of regular equations under which resulting solutions are unambiguous (Sect. 4).
- We apply our results to the following scenarios:
  - We show that regular expressions obtained from DFAs via Brzozowski's algebraic method are always unambiguous (Sect. 5).
  - We provide direct, algebraic methods to obtain the subtraction and shuffle among two regular expressions (Sect. 6). Correctness follows via some simple coalgebraic reasoning.

We conclude in Sect. 7 where we also discuss related works.

The online version of this paper contains an appendix with further details such as proofs and a parser for regular equations. We also report on an implementation for solving of regular equations in Haskell [12] including benchmark results.[1]

2 Preliminaries

Let Σ be a finite set of symbols (literals) with x, y, and z ranging over Σ. We write Σ^* for the set of finite words over Σ, ε for the empty word, and $v \cdot w$ for the concatenation of words v and w. A language is a subset of Σ^*.

Definition 1 (Regular Languages). *The set $\mathcal{R}$ of regular languages is defined inductively over some alphabet Σ by*

$$R, S ::= \emptyset \mid \{\varepsilon\} \mid \{x\} \mid (R + S) \mid (R \cdot S) \mid (R^*) \qquad \textit{where } x \in \Sigma.$$

Each regular language is a subset of Σ^* where we assume that $R \cdot S$ denotes $\{v \cdot w \mid v \in R \wedge w \in S\}$, $R + S$ denotes $R \cup S$ and R^* denotes $\{w_1 \cdot \dots \cdot w_n \mid n \geq 0 \wedge \forall i \in \{1, \dots, n\}.\ w_i \in R\}$.

We write $R \equiv S$ if R and S denote the same set of words.

We often omit parentheses by assuming that * binds tighter than $\cdot$ and $\cdot$ binds tighter than $+$. As it is common, we assume that $+$ and $\cdot$ are right-associative. That is, $R + S + T$ stands for $(R + (S + T))$ and $R + S + R \cdot S \cdot T$ stands for $R + (S + (R \cdot (S \cdot T)))$.

Definition 2 (Regular Expressions). *The set RE of* regular expressions *is defined inductively over some alphabet Σ by*

$$r, s ::= \phi \mid \varepsilon \mid x \mid (r + s) \mid (r \cdot s) \mid (r^*) \qquad \textit{where } x \in \Sigma.$$

Definition 3 (From Regular Expressions to Languages). *The meaning function $\mathcal{L}$ maps a regular expression to a language. It is defined inductively as follows:*
$\mathcal{L}(\phi) = \{\}$. $\mathcal{L}(\varepsilon) = \{\varepsilon\}$. $\mathcal{L}(x) = \{x\}$. $\mathcal{L}(r + s) = (\mathcal{L}(r) + \mathcal{L}(s))$. $\mathcal{L}(r \cdot s) = (\mathcal{L}(r) \cdot \mathcal{L}(s))$. $\mathcal{L}(r^*) = (\mathcal{L}(r)^*)$.

[1] http://arxiv.org/abs/1908.03710.

We say that regular expressions r and s are equivalent, $r \equiv s$, if $\mathcal{L}(r) = \mathcal{L}(s)$.

Definition 4 (Nullability). *A regular expression r is* nullable *if $\varepsilon \in \mathcal{L}(r)$.*

Lemma 5 (Arden's Lemma [2]). *Let R, S, T be regular languages where $\varepsilon \notin S$. Then, we have that $R \equiv S \cdot R + T$ iff $R \equiv S^* \cdot T$.*

The direction from right to left holds in general. For the direction from left to right, pre-condition $\varepsilon \notin S$ is required. For our purposes, we only require the direction from right to left.

3 Solving Regular Equations

Definition 6 (Regular Equations). *We write E to denote a* regular equation *of the form $R \approx \alpha$ where the form of the right-hand side α is as follows.*

$$\alpha \ ::= r \cdot R \mid r \mid \alpha + \alpha$$

In addition to α, we will sometimes use β to denote right-hand sides.

We will treat regular language symbols R like variables. We write r, s, t to denote expressions that do not refer to symbols R.

We write $R \in \alpha$ to denote that R appears in α. Otherwise, we write $R \notin \alpha$.

We write $\mathcal{E}$ to denote a set $\{R_1 \approx \alpha_1, \ldots, R_n \approx \alpha_n\}$ of regular equations. We assume that (1) left-hand sides are distinct by requiring that $R_i \neq R_j$ for $i \neq j$, and (2) regular language symbols on right-hand sides appear on some left-hand side by requiring that for any $R \in \alpha_j$ for some j there exists i such that $R = R_i$. We define $dom(\mathcal{E}) = \{R_1, \ldots, R_n\}$.

Regular languages are closed under union and concatenation, hence, we can guarantee the existence of solutions of these variables in terms of regular expressions.

Definition 7 (Solutions). *We write $\{R_1 \mapsto \gamma_1, \ldots, R_n \mapsto \gamma_n\}$ to denote an idempotent substitution mapping R_i to γ_i where γ_i denote expressions that may consist of a mix of regular expressions and regular language symbols R.*

Let $\psi = \{R_1 \mapsto \gamma_1, \ldots, R_n \mapsto \gamma_n\}$ be a substitution and γ some expression. Then, $\psi(\gamma)$ is derived from γ by replacing each occurrence of R_i by γ_i.

Let $\mathcal{E} = \{R_1 \approx \alpha_1, \ldots, R_n \approx \alpha_n\}$. Then, we say that ψ is a solution *for $\mathcal{E}$ if $\psi(R_i)$, $\psi(\alpha_i)$ are regular expressions where $\psi(R_i) \equiv \psi(\alpha_i)$ for $i = 1, \ldots, n$.*

We solve equations as follows. We apply Arden's Lemma on equations that are of a certain (normal) form $R \approx s \cdot R + \alpha$ where $R \notin \alpha$. Thus, we can eliminate this equation by substituting R with $s^* \cdot \alpha$ on all right-hand sides. In case of $R \approx \alpha$ where $R \notin \alpha$ we can substitute directly. We repeat this process until all equations are solved. Below, we formalize the technical details.

Definition 8 (Normal Form). *We say that $R \approx \alpha$ is in* normal form *iff either (1) $R \notin \alpha$, or (2) $\alpha = s_1 \cdot R_1 + \cdots + s_n \cdot R_n + t$ such that $R = R_1$ and $R_i \neq R_j$ for $i \neq j$.*

Recall that t does not refer to symbols R. Every equation can be brought into normal form by applying the following algebraic equivalence laws.

Definition 9 (Equivalence). *We say two expressions* γ_1 *and* γ_2 *are* equivalent, *written* $\gamma_1 \simeq \gamma_2$, *if one can be transformed into the other by application of the following rules.*

$$\text{(E1)}\ \gamma_1 \cdot (\gamma_2 + \gamma_3) \simeq \gamma_1 \cdot \gamma_2 + \gamma_1 \cdot \gamma_3 \qquad \text{(E2)}\ \gamma_1 \cdot (\gamma_2 \cdot \gamma_3) \simeq (\gamma_1 \cdot \gamma_2) \cdot \gamma_3$$

$$\text{(E3)}\ \gamma_1 + (\gamma_2 + \gamma_3) \simeq (\gamma_1 + \gamma_2) + \gamma_3 \qquad \text{(E4)}\ \gamma_2 \cdot \gamma_1 + \gamma_3 \cdot \gamma_1 \simeq (\gamma_2 + \gamma_3) \cdot \gamma_1$$

$$\text{(E5)}\ \gamma_1 + \gamma_2 \simeq \gamma_2 + \gamma_1 \qquad \text{(E6)}\ \frac{\gamma_1 \simeq \gamma_2}{\beta[\gamma_1] \simeq \beta[\gamma_2]} \qquad \text{(E7)}\ \frac{\gamma_1 \simeq \gamma_2 \quad \gamma_2 \simeq \gamma_3}{\gamma_1 \simeq \gamma_3}$$

Rule (E6) assumes expressions with a hole.

$$\beta[] ::= [] \mid \beta[] + \beta \mid \beta + \beta[]$$

We write $\beta[\gamma]$ *to denote the expression where the hole* $[]$ *is replaced by* γ.

We formulate solving of equations in terms of a rewrite system among a configuration $\langle \psi, \mathcal{E} \rangle$ where substitution ψ represents the so far accumulated solution and $\mathcal{E}$ the yet to be solved set of equations.

Definition 10 (Solving). *Let* $\mathcal{E} = \{R_1 \approx \alpha_1, \ldots, R_n \approx \alpha_n\}$. *Then, we write* $R \approx \alpha \uplus \mathcal{E}'$ *to denote the set that equals to* $\mathcal{E}$ *where* $R \approx \alpha$ *refers to some equation in* $\mathcal{E}$ *and* $\mathcal{E}'$ *refers to the set of remaining equations.*

$$\text{(Arden)}\ \frac{R \notin \alpha}{\langle \psi, R \approx s \cdot R + \alpha \uplus \mathcal{E} \rangle \Rightarrow \langle \psi, R \approx s^* \cdot \alpha \uplus \mathcal{E} \rangle}$$

$$\text{(Subst)}\ \frac{\begin{array}{c} R \notin \alpha \\ \psi' = \{R \mapsto \alpha\} \cup \{S \mapsto \{R \mapsto \alpha\}(\gamma) \mid S \mapsto \gamma \in \psi\} \\ \mathcal{E}' = \{R' \approx \alpha'' \mid R' \approx \alpha' \in \mathcal{E} \wedge \{R \mapsto \alpha\}(\alpha') \simeq \alpha''\} \end{array}}{\langle \psi, R \approx \alpha \uplus \mathcal{E} \rangle \Rightarrow \langle \psi', \mathcal{E}' \rangle}$$

We write $\Rightarrow^*$ *to denote the transitive and reflexive closure of solving steps* $\Rightarrow$.

Initially, all equations are in normal form. Rule (Arden) applies Arden's Lemma on some equation in normal form. Rule (Subst) removes an equation $R \approx \alpha$ where $R \notin \alpha$. The substitution $\{R \mapsto \alpha\}$ implied by the equation is applied on all remaining right-hand sides. To retain the normal form property of equations, we normalize right-hand sides by applying rules (E1-7). The details of normalization are described in the proof of the upcoming statement. We then extend the solution accumulated so far by adding $\{R \mapsto \alpha\}$. As we assume substitutions are idempotent, $\{R \mapsto \alpha\}$ is applied on all expressions in the codomain of ψ.

Theorem 11 (Regular Equation Solutions). *Let $\mathcal{E}$ be a set of regular equations in normal form. Then, $\langle\{\},\mathcal{E}\rangle \Rightarrow^* \langle\psi,\{\}\rangle$ for some substitution ψ where ψ is a solution for $\mathcal{E}$.*

Proof. We first observe that rule (Arden) and (Subst) maintain the normal form property for equations. This immediately applies to rule (Arden).

Consider rule (Subst). Consider $R' \approx \alpha'$. We need to show that $\{R \mapsto \alpha\}(\alpha')$ can be transformed to some form α'' such that $R' \approx \alpha''$ is in normal form.

If $R \notin \alpha'$ nothing needs to be done as we assume that equations are initially in normal form.

Otherwise, we consider the possible shapes of α and α'. W.l.o.g. α' is of the form $t_1 \cdot R_1 + \cdots + r \cdot R + \cdots + t_n \cdot R_n + t'$ and α is of the form $s_1 \cdot T_1 + \cdots + s_k \cdot T_k + t''$. We rely here on rule (E3) that allows us to drop parentheses among summands.

R is replaced by α in α'. This generates the subterm $r \cdot (s_1 \cdot T_1 + \cdots + s_k \cdot T_k + t'')$. On this subterm, we exhaustively apply rules (E1-2). This yields the subterm $(r \cdot s_1) \cdot T_1 + \cdots + (r \cdot s_k) \cdot T_k + t''$.

This subterm is one of the sums in the term obtained from $\{R \mapsto \alpha\}(\alpha')$. Via rules (E6-7) the above transformation steps can be applied on the entire term $\{R \mapsto \alpha\}(\alpha')$. Hence, this term can be brought into the form $r_1 \cdot S_1 + \cdots + r_m \cdot S_m + t$. Subterm t equals $t' + t''$ and subterms $r_i \cdot S_i$ refer to one of the subterms $t_j \cdot R_j$ or $(r \cdot s_l) \cdot T_l$.

We are not done yet because subterms $r_i \cdot S_i$ may contain duplicate symbols. That is, $S_i = S_j$ for $i \neq j$. We apply rule (E4) in combination with rule (E3) and (E4) to combine subterms with the same symbol. Thus, we reach the form $r'_1 \cdot R'_1 + \cdots + r'_o \cdot R'_o + t$ such that $R_i \neq R_j$ for $i \neq j$.

If $R' \neq R'_i$ for $i = 1, \ldots, o$ we are done. Otherwise, $R = R'_i$ for some i. We apply again (E3) and (E5) to ensure that the component $s_i \cdot R_i$ appears first in the sum.

Next, we show that within a finite number of (Arden) and (Subst) rule applications we reach the configuration $\langle\psi,\{\}\rangle$. For this purpose, we define an ordering relation among configurations $\langle\psi,\mathcal{E}\rangle$.

For $\mathcal{E} = \{R_1 \approx \alpha_1, \ldots, R_n \approx \alpha_n\}$ we define

$$vars(\mathcal{E}) = (\{R_1, \ldots, R_n\}, \{\{S_1, \ldots, S_m\}\})$$

where $\{\{\ldots\}\}$ denotes a multi-set and S_j are the distinct occurrences of symbols appearing on some right-hand side α_i. Recall that by construction $\{S_1, \ldots, S_m\} \subseteq \{R_1, \ldots, R_n\}$. See (2) in Definition 6. We define $\langle\psi,\mathcal{E}\rangle < \langle\psi',\mathcal{E}'\rangle$ iff either (a) $M \subsetneq M'$ or (b) $M = M'$ and the number of symbols in N is strictly smaller than the number of symbols in N' where $vars(\mathcal{E}) = (M, N)$ and $vars(\mathcal{E}') = (M', N')$.

For sets $\mathcal{E}$ of regular equations as defined in Definition 6 this is a well-founded order. Each of the rules (Subst) and (Arden) yield a smaller configuration w.r.t this order. For rule (Subst) case (a) applies whereas for rule (Arden) case (b) applies. Configuration $\langle\psi,\{\}\rangle$ for some ψ is the minimal element. Hence, in a finite number of rule applications we reach $\langle\psi,\{\}\rangle$.

Substitution ψ must be a solution because (1) normalization steps are equivalence preserving and (2) based on Arden's Lemma we have that every solution for $R \approx s^* \cdot \alpha$ is also a solution for $R \approx s \cdot R + \alpha$. □

Example 1. Consider $\mathcal{E} = \{R_1 \approx x \cdot R_1 + y \cdot R_2 + \varepsilon, R_2 \approx y \cdot R_1 + x \cdot R_2 + \varepsilon\}$. For convenience, we additionally make use of associativity of concatenation $(\cdot)$.

$$
\begin{array}{ll}
 & \langle\{\}, \{R_1 \approx x \cdot R_1 + y \cdot R_2 + \varepsilon, R_2 \approx y \cdot R_1 + x \cdot R_2 + \varepsilon\}\rangle \\
\text{(Arden)} & \\
\Rightarrow & \langle\{\}, \{R_1 \approx x^* \cdot (y \cdot R_2 + \varepsilon), R_2 \approx y \cdot R_1 + x \cdot R_2 + \varepsilon\}\rangle \\
\text{(Subst)} & \\
\Rightarrow & (y \cdot (x^* \cdot (y \cdot R_2 + \varepsilon)) + x \cdot R_2 + \varepsilon \simeq (y \cdot x^* \cdot y + x) \cdot R_2 + y \cdot x^* \cdot \varepsilon + \varepsilon) \\
 & \langle\{R_1 \mapsto x^* \cdot (y \cdot R_2 + \varepsilon)\}, \{R_2 \approx (y \cdot x^* \cdot y + x) \cdot R_2 + y \cdot x^* \cdot \varepsilon + \varepsilon\}\rangle \\
\text{(Arden)} & \\
\Rightarrow & \langle\{R_1 \mapsto x^* \cdot (y \cdot R_2 + \varepsilon)\}, \{R_2 \approx (y \cdot x^* \cdot y + x)^* \cdot (y \cdot x^* \cdot \varepsilon + \varepsilon)\}\rangle \\
\text{(Subst)} & \\
\Rightarrow & \langle\{R_1 \mapsto x^* \cdot (y \cdot (y \cdot x^* \cdot y + x)^* \cdot (y \cdot x^* \cdot \varepsilon + \varepsilon) + \varepsilon), \\
 & R_2 \mapsto (y \cdot x^* \cdot y + x)^* \cdot (y \cdot x^* \cdot \varepsilon + \varepsilon)\}, \{\}\rangle
\end{array}
$$

The formulation in Definition 10 leaves the exact order in which equations are solved unspecified. Semantically, this form of non-determinism has no impact on the solution obtained. However, the syntactic shape of solutions is sensitive to the order in which equations are solved.

Suppose we favor the second equation which then yields the following.

$$
\begin{array}{ll}
 & \langle\{\}, \{R_1 \approx x \cdot R_1 + y \cdot R_2 + \varepsilon, R_2 \approx y \cdot R_1 + x \cdot R_2 + \varepsilon\}\rangle \\
\Rightarrow^* & \langle\{R_1 \mapsto (x + y \cdot x^* \cdot y)^* + y \cdot x^* + \varepsilon, \\
 & R_2 \mapsto x^* \cdot (y \cdot ((x + y \cdot x^* \cdot y)^* + y \cdot x^* + \varepsilon) + \varepsilon)\}, \{\}\rangle
\end{array}
$$

where for convenience, we exploit the law $r \cdot \varepsilon \equiv r$.

4 Computational Interpretation

We characterize under which conditions solutions to regular equations are unambiguous. By unambiguous solutions we mean that the resulting expressions are unambiguous. An expression is ambiguous if there exists a word which can be matched in more than one way. That is, there must be two distinct parse trees which share the same underlying word [3].

We proceed by establishing the notion of a parse tree. Parse trees capture the word that has been matched and also record which parts of the regular expression have been matched. We follow [7] and view expressions as types and parse trees as values.

Definition 12 (Parse Trees)

$$u, v ::= \textsc{Eps} \mid \textsc{Sym}\ x \mid \textsc{Seq}\ v\ v \mid \textsc{Inl}\ v \mid \textsc{Inr}\ v \mid vs \mid \textsc{Fold}\ v \qquad vs ::= [] \mid v : vs$$

The valid relations among parse trees and regular expressions are defined via a natural deduction style proof system.

$$\mathcal{E} \vdash [] : r^* \qquad \mathcal{E} \vdash \textsc{Eps} : \epsilon \qquad \frac{x \in \Sigma}{\mathcal{E} \vdash \textsc{Sym}\ x : x}$$

$$\frac{\mathcal{E} \vdash v : r \quad \mathcal{E} \vdash vs : r^*}{\mathcal{E} \vdash (v : vs) : r^*} \qquad \frac{\mathcal{E} \vdash v_1 : r_1 \quad \mathcal{E} \vdash v_2 : r_2}{\mathcal{E} \vdash \textsc{Seq}\ v_1\ v_2 : r_1 \cdot r_2}$$

$$\frac{\mathcal{E} \vdash v_1 : r_1}{\mathcal{E} \vdash \textsc{Inl}\ v_1 : r_1 + r_2} \qquad \frac{\mathcal{E} \vdash v_2 : r_2}{\mathcal{E} \vdash \textsc{Inr}\ v_2 : r_1 + r_2} \qquad \frac{\mathcal{E} \vdash v : \alpha \quad R \approx \alpha \in \mathcal{E}}{\mathcal{E} \vdash \textsc{Fold}\ v : R}$$

For expressions not referring to variables we write $\vdash v : r$ *as a shorthand for* $\{\} \vdash v : r$.

Parse tree values are built using data constructors. The constant constructor Eps represents the value belonging to the empty word regular expression. For letters, we use the unary constructor Sym to record the symbol. In case of choice $(+)$, we use unary constructors Inl and Inr to indicate if either the left or right expression is part of the match. For repetition (Kleene star) we use Haskell style lists where we write $[v_1, ..., v_n]$ as a short-hand for the list $v_1 : ... : v_n : []$. In addition to the earlier work [7], we introduce a Fold constructor and a proof rule to (un)fold a regular equation.

Example 2. Consider $\mathcal{E} = \{R \approx x \cdot R + y\}$. Then, we find that

$$\mathcal{E} \vdash \textsc{Fold}\ (\textsc{Inl}\ (\textsc{Seq}\ \ (\textsc{Sym}\ x)\ (\textsc{Fold}\ (\textsc{Inr}\ (\textsc{Sym}\ y))))) : R$$

The equation is unfolded twice where we first match against the left part $x \cdot R$ and then against the right part y.

The relation established in Definition 12 among parse trees, expressions and equations is correct in the sense that (1) flattening of the parse tree yields a word in the language and (2) for each word there exists a parse tree.

Definition 13 (Flattening). *We can flatten a parse tree to a word as follows:*

$$\begin{array}{llll} |\textsc{Eps}| = \epsilon & |\textsc{Sym}\ x| \quad = x & |\textsc{Inl}\ v| = |v| & |v : vs| \quad = |v| \cdot |vs| \\ |[]| \quad = \epsilon & |\textsc{Seq}\ v_1\ v_2| = |v_1| \cdot |v_2| & |\textsc{Inr}\ v| = |v| & |\textsc{Fold}\ v| = |v| \end{array}$$

Proposition 14. *Let* $\mathcal{E}$ *be a set of regular equations and* ψ *a solution. Let* $R \in dom(\mathcal{E})$. *(1) If* $w \in \mathcal{L}(\psi(R))$ *then* $\mathcal{E} \vdash v : R$ *for some parse tree* v *such that* $|v| = w$. *(2) If* $\mathcal{E} \vdash v : R$ *then* $|v| \in \mathcal{L}(\psi(R))$.

The above result follows by providing a parser for regular equations. For (1) it suffices to compute a parse tree if one exists. For (2) we need to enumerate all possible parse trees. This is possible by extending our prior work [18,19] to the regular equation setting. Details are given in the online version of this paper.

Parse trees may not be unique because some equations/expressions may be ambiguous in the sense that a word can be matched in more than one way. This means that there are two distinct parse trees representing the same word. We extend the notion of ambiguous expressions [3] to the setting of regular equations.

Definition 15 (Ambiguity). *Let $\mathcal{E}$ be a set of regular equations and r be an expression. We say r is* ambiguous *w.r.t. $\mathcal{E}$ iff there exist two distinct parse trees v_1 and v_2 such that $\mathcal{E} \vdash v_1 : r$ and $\mathcal{E} \vdash v_2 : r$ where $|v_1| = |v_2|$.*

Example 3. [Inl (Seq (Sym x) (Sym y))] and [Inr (Inl (Sym x)), Inr (Inr (Sym y))] are two distinct parse trees for expression $(x \cdot y + x + y)^*$ (where $\mathcal{E} = \{\}$) and word $x \cdot y$.

On the other hand, the equation from Example 2 is unambiguous due to the following result.

Definition 16 (Non-Overlapping Equations). *We say an equation E is* non-overlapping *if E is of the following form $R \approx x_1 \cdot R_1 + \cdots + x_n \cdot R_n + t$ where $x_i \neq x_j$ for $i \neq j$ and either $t = \varepsilon$ or $t = \phi$.*

Equation $R \approx x \cdot R + y$ does not exactly match the above definition. However, we can transform $\mathcal{E} = \{R \approx x \cdot R + y\}$ into the equivalent set $\mathcal{E}' = \{R \approx x \cdot R + y \cdot S, S \approx \varepsilon\}$ that satisfies the non-overlapping condition.

Proposition 17 (Unambiguous Regular Equations). *Let $\mathcal{E}$ be a set of non-overlapping equations where $R \in dom(\mathcal{E})$. Then, we have that R is unambiguous.*

Ultimately, we are interested in obtaining a parse tree for the resulting solutions rather than the original set of equations. For instance, the solution for Example 2 is $x^* \cdot y$. Hence, we wish to transform the parse tree

Fold (Inl (Seq (Sym x) (Fold (Inr (Sym y)))))

into a parse tree for $x^* \cdot y$. Furthermore, we wish to guarantee that if equations are unambiguous so are solutions. We achieve both results by explaining each solving step among regular equations in terms of a (bijective) transformation among the associated parse trees.

We refer to these transformations as coercions as they operate on parse trees. We assume the following term language to represent coercions.

Definition 18 (Coercion Terms). *Coercion terms c and patterns pat are inductively defined by*

$$\begin{array}{ll} c & ::= v \mid k \mid \lambda v.c \mid c\ c \mid \mathsf{rec}\ x.c \mid \mathsf{case}\ c\ \mathsf{of}\ [pat_1 \Rightarrow c_1, \ldots, pat_n \Rightarrow c_n] \\ pat & ::= y \mid k\ pat_1\ ...pat_{arity(k)} \end{array}$$

*where pattern variables y range overs a denumerable set of variables disjoint from Σ and constructors k are taken from the set $\mathcal{K} = \{$*Eps, Seq, Inl, Inr, Fold*$\}$. The function $arity(k)$ defines the arity of constructor k. Patterns are linear (i.e., all pattern variables are distinct) and we write $\lambda pat.c$ as a shorthand for $\lambda v.$*case v of $[pat \Rightarrow c]$.

We give meaning to coercions in terms of a standard big-step operational semantics. Given a coercion (function) f and some (parse tree) value u, we write $f\ u \Downarrow v$ to denote the evaluation of f for input u with resulting (parse tree) value v. We often write $f(u)$ as a shorthand for v. We say a coercion f is *bijective* if there exists a coercion g such that for every u, v where $f\ u \Downarrow v$ we have that $g\ v \Downarrow u$. We refer to g as the *inverse* of f.

We examine the three elementary solving steps, Arden, normalization and substitution. For each solving step we introduce an appropriate (bijective) coercion to carry out the transformation among parse trees.

Lemma 19 (Arden Coercion). *Let $\mathcal{E}$ be a set of regular equations where $R \approx s \cdot R + \alpha \in \mathcal{E}$ such that $R \notin \alpha$ and $\mathcal{E} \vdash \textsc{Fold}\ v : R$ for some parse tree v. Then, there exists a bijective coercion f_A such that $\mathcal{E} \vdash f_A(v) : s^* \cdot \alpha$ where $|v| = |f_A(v)|$.*

Proof. By assumption $\mathcal{E} \vdash v : s \cdot R + \alpha$. The following function f_A satisfies $\mathcal{E} \vdash f_A(v) : s^* \cdot \alpha$ where $|v| = |f_A(v)|$. For convenience we use symbols v and u as pattern variables.

$$
\begin{aligned}
f_A = \mathsf{rec}\ f.\lambda x.\ \mathsf{case}\ x\ \mathsf{of}\ & \\
& [\textsc{Inr}\ u \Rightarrow \textsc{Seq}\ []\ u, \\
& \textsc{Inl}\ (\textsc{Seq}\ u\ (\textsc{Fold}\ v) \Rightarrow \mathsf{case}\ f(v)\ \mathsf{of} \\
& \qquad\qquad\qquad\qquad\qquad [\textsc{Seq}\ us\ u_2 \Rightarrow \textsc{Seq}\ (u : us)\ u_2]]
\end{aligned}
$$

Function f_A is bijective. Here is the inverse function.

$$
\begin{aligned}
f_A^{-1} = \mathsf{rec}\ g.\lambda x.\ \mathsf{case}\ x\ \mathsf{of}\ & \\
& [\textsc{Seq}\ []\ u \Rightarrow \textsc{Fold}\ (\textsc{Inr}\ u), \\
& \textsc{Seq}\ (v : vs)\ u \Rightarrow \textsc{Fold}\ (\textsc{Inl}\ (\textsc{Seq}\ v\ (g\ (\textsc{Seq}\ vs\ u))))]
\end{aligned}
$$

□

Lemma 20 (Normalization Coercion). *Let γ_1, γ_2 be two expressions such that $\gamma_1 \simeq \gamma_2$ and $\mathcal{E} \vdash v : \gamma_1$ for some set $\mathcal{E}$ and parse tree v. Then, there exists a bijective coercion f such that $\mathcal{E} \vdash f(v) : \gamma_2$ where $|v| = |f(v)|$.*

Proof. For each of the equivalence proof rules, we introduce an appropriate (bijective) coercion. For rule (E1) we employ

$$
\begin{aligned}
f_{E_1} = \lambda v.\ \mathsf{case}\ v\ \mathsf{of}\ & \\
& [\textsc{Seq}\ u\ (\textsc{Inl}\ v) \Rightarrow \textsc{Inl}\ (\textsc{Seq}\ u\ v), \\
& \textsc{Seq}\ u\ (\textsc{Inr}\ v) \Rightarrow \textsc{Inr}\ (\textsc{Seq}\ u\ v)]
\end{aligned}
$$

where the inverse function is as follows.

$$
\begin{aligned}
f_{E_1}^{-1} = \lambda v.\ \mathsf{case}\ v\ \mathsf{of}\ & \\
& [\textsc{Inl}\ (\textsc{Seq}\ u\ v) \Rightarrow \textsc{Seq}\ (\textsc{Inl}\ u)\ v, \\
& \textsc{Inr}\ (\textsc{Seq}\ u\ v) \Rightarrow \textsc{Seq}\ (\textsc{Inr}\ u)\ v]
\end{aligned}
$$

Coercions for rules (E2-5) can be defined similarly. Rule (E7) corresponds to function composition and rule (E6) requires to navigate to the respective hole position. Details are omitted for brevity. □

We will write $\gamma_1 \overset{f}{\simeq} \gamma_2$ to denote the coercion f to carry out the transformation of γ_1's parse tree into γ_2's parse tree.

What remains is to define coercions to carry out substitution where we replace subterms.

Definition 21 (Substitution Context). *We define expressions with multiple holes to characterize substitution of a subterm by another.*

$$\delta\langle\rangle ::= r \cdot \langle\rangle \mid \delta\langle\rangle + \delta\langle\rangle \mid \delta\langle\rangle + \alpha \mid \alpha + \delta\langle\rangle$$

We refer to $\delta\langle\rangle$ *as a* substitution context.

We define a set of functions indexed by the shape of a substitution context. For $\delta\langle\rangle$ *we transform* $\alpha\langle R\rangle$*'s parse tree into* $\alpha\langle\alpha\rangle$*'s parse tree assuming the equation* $R \approx \alpha$.

$$f_{r\cdot\langle\rangle} = \begin{array}{l}\lambda u.\ \mathsf{case}\ u\ \mathsf{of} \\ [\textsc{Seq}\ u\ (\textsc{Fold}\ v) \Rightarrow \textsc{Seq}\ u\ v]\end{array}$$

$$f_{\delta\langle\rangle+\delta\langle\rangle} = \begin{array}{l}\lambda u.\ \mathsf{case}\ u\ \mathsf{of} \\ [\textsc{Inl}\ v \Rightarrow \textsc{Inl}\ (f_{\delta\langle\rangle}(v)), \\ \textsc{Inr}\ v \Rightarrow \textsc{Inr}\ (f_{\delta\langle\rangle}(v))]\end{array}$$

$$f_{\delta\langle\rangle+\alpha} = \begin{array}{l}\lambda u.\ \mathsf{case}\ u\ \mathsf{of} \\ [\textsc{Inl}\ v \Rightarrow \textsc{Inl}\ (f_{\delta\langle\rangle}(v)), \\ \textsc{Inr}\ v \Rightarrow \textsc{Inr}\ v]\end{array} \qquad f_{\alpha+\delta\langle\rangle} = \begin{array}{l}\lambda u.\ \mathsf{case}\ u\ \mathsf{of} \\ [\textsc{Inl}\ v \Rightarrow \textsc{Inl}\ v, \\ \textsc{Inr}\ v \Rightarrow \textsc{Inr}\ (f_{\delta\langle\rangle}(v))]\end{array}$$

Functions $f_{\delta\langle\rangle}$ navigate to the to-be-replaced subterm and drop the FOLD constructor if necessary. There are inverse functions which we omit for brevity.

Lemma 22 (Substitution Coercion). *Let* $\mathcal{E}$ *be a set of equations,* $R \approx \alpha \in \mathcal{E}$ *such that* $\mathcal{E} \vdash v : \delta\langle R\rangle$ *for some parse tree* v *and substitution context* $\delta\langle\rangle$*. Then, we find that* $\mathcal{E} \vdash f_{\delta\langle\rangle}(v) : \delta\langle\alpha\rangle$ *where* $|v| = |f_{\delta\langle\rangle}(v)|$.

Proof. Follows by induction over the structure of $\delta\langle\rangle$. □

We integrate the elementary coercions into the solving process. For this purpose, we assume that regular equations and substitutions are annotated with parse trees. For example, we write $\{v_1 : R_1 \approx \alpha_1, \ldots, v_n : R_n \approx \alpha_n\}$ to denote a set of regular equations $\mathcal{E}$ where for each i we have that $\mathcal{E} \vdash v_i : R_i$. Similarly, we write $\{v_1 : R_1 \mapsto \gamma_1, \ldots, v_n : R_n \mapsto \gamma_n\}$ for substitutions.

Definition 23 (Coercive Solver)

(C-Arden)
$$\frac{R \notin \alpha}{\langle \psi, \textsc{Fold}\ v : R \approx s \cdot R + \alpha \uplus \mathcal{E} \rangle \Rightarrow \langle \psi, \textsc{Fold}\ f_A(v) : R \approx s^* \cdot \alpha \uplus \mathcal{E} \rangle}$$

(C-Subst)
$$\frac{\begin{array}{l} R \notin \alpha \\ \psi' = \{v : R \mapsto \alpha\} \\ \quad \cup \{v : R' \mapsto \alpha' \mid v : R' \mapsto \alpha' \in \psi \wedge R \notin \alpha'\} \\ \quad \cup \{\textsc{Fold}\ f(f_{\delta\langle\rangle}(v)) : R' \mapsto \alpha'' \mid \textsc{Fold}\ v : R' \mapsto \delta'\langle R \rangle \in \psi \wedge \\ \qquad\qquad R \notin \delta'\langle \alpha \rangle \wedge \\ \qquad\qquad \delta'\langle \alpha \rangle \overset{f}{\simeq} \alpha''\} \\ \mathcal{E}' = \{v : R' \approx \alpha' \mid v : R' \approx \alpha' \in \mathcal{E} \wedge R \notin \alpha'\} \\ \quad \cup \{\textsc{Fold}\ f(f_{\delta\langle\rangle}(v)) : R' \approx \alpha'' \mid \textsc{Fold}\ v : R' \approx \delta'\langle R \rangle \in \mathcal{E} \wedge \\ \qquad\qquad R \notin \delta'\langle \alpha \rangle \wedge \\ \qquad\qquad \delta'\langle \alpha \rangle \overset{f}{\simeq} \alpha''\} \end{array}}{\langle \psi, \textsc{Fold}\ v : R \approx \alpha \uplus \mathcal{E} \rangle \Rightarrow \langle \psi', \mathcal{E}' \rangle}$$

In the coercive Arden rule, we apply the Arden coercion introduced in Lemma 19. During substitution we uniformly normalize right-hand sides of equations *and* the codomains of substitutions. Side condition $R \notin \delta'\langle \alpha \rangle$ guarantees that all occurrences of R are replaced. Parse trees are transformed by first applying the substitution coercion followed by the normalization coercion. Thus, we can transform parse trees of regular equations into parse trees of solutions.

Proposition 24 (Coercive Solving). *Let $\mathcal{E} = \{v_1 : R_1 \approx \alpha_1, \ldots, v_n : R_n \approx \alpha_n\}$ be a parse tree annotated set of regular equations in normal form where $\mathcal{E} \vdash v_i : R_i$ for $i = 1, \ldots, n$. Then, $\langle \{\}, \mathcal{E} \rangle \Rightarrow^* \langle \psi, \{\} \rangle$ for some substitution ψ where $\psi = \{u_1 : R_1 \mapsto s_1, \ldots, u_n : R_n \mapsto s_n\}$ such that $\vdash u_i : s_i$ and $|u_i| = |v_i|$ for $i = 1, \ldots, n$.*

Proof. Follows immediately from Lemmas 19, 20 and 22. □

Theorem 25 (Unambiguous Solutions). *Let $\mathcal{E}$ be a set of non-overlapping equations where $\langle \{\}, \mathcal{E} \rangle \Rightarrow^* \langle \psi, \{\} \rangle$ for some substitution ψ. Then, for each $R \in dom(\mathcal{E})$ we find that $\psi(R)$ is unambiguous.*

Proof. Follows from Propositions 17 and 24 and the fact that coercions are bijective. □

5 Brzozowski's Algebraic Method

We revisit Brzozowski's algebraic method [4] to transform an automaton into a regular expression. Based on our results we can show that resulting regular expressions are always unambiguous.

Definition 26 (Deterministic Finite Automata (DFA)). *A deterministic finite automaton (DFA) is a 5-tuple* $M = (Q, \Sigma, \delta, q_0, F)$ *consisting of a finite set* Q *of states, a finite set* Σ *of symbols, a transition function* $\delta : Q \times \Sigma \rightarrow Q$, *an initial state* $q_0 \in Q$, *and a set* F *of accepting states. We say* M *accepts word* $w = x_1 \ldots x_n$ *if there exists a sequence of states* $p_1, \ldots, p_{n+1}$ *such that* $p_{i+1} = \delta(p_i, x_n)$ *for* $i = 1, \ldots, n$, $p_1 = q_0$ *and* $p_{n+1} \in F$.

Brzozowski turns a DFA into an equivalent set of (characteristic) regular equations.

Definition 27 (Characteristic Equations). *Let* $M = (Q, \Sigma, \delta, q_0, F)$ *be a DFA. We define* $\mathcal{E}_M = \{R_q \approx \sum_{x \in \Sigma} x \cdot R_{\delta(q,x)} + f(q) \mid q \in Q\}$ *where* $f(q) = \varepsilon$ *if* $q \in F$. *Otherwise,* $f(q) = \phi$. *We refer to* $\mathcal{E}_M$ *as the* characteristic equations *obtained from* M.

He suggests solving these equations via Arden's Lemma but the exact details (e.g. normalization) are not specified. Assuming we use the solving method specified in Definition 10 we can conclude the following. By construction, characteristic equations are non-overlapping. From Theorem 25 we can derive the following result.

Corollary 1. *Solutions obtained from characteristic equations are unambiguous.*

Instead of a DFA we can also turn a non-deterministic automaton (NFA) into an equivalent regular expression. Each ε transitions is represented by the component $\varepsilon \cdot R$. For two non-deterministic transitions via symbol x to follow states R_1 and R_2, we generate the component $x \cdot R_1 + x \cdot R_2$. Resulting characteristic equations will be overlapping in general. Hence, we can no longer guarantee unambiguity.

6 Subtraction and Shuffle

We introduce direct methods to subtract and shuffle regular expressions. Instead of turning the regular expressions into a DFA and carrying out the operation at the level of DFAs, we generate an appropriate set of equations by employing Brzozowski derivatives. Solving the equations yields then the desired result. In essence, our method based on solutions resulting from derivative-based equations is isomorphic to building a derivative-based DFA from expressions, applying the product automaton construction among DFAs and then turn the resulting DFA into an expression via Brzozowski's algebraic method.

For subtraction, equations generated are non-overlapping. Hence, resulting expressions are also unambiguous. First, we recall the essential of derivatives before discussing each operation including some optimizations.

6.1 Brzozowski's Derivatives

The *derivative* of a regular expression r with respect to some symbol x, written $d_x(r)$, is a regular expression for the left quotient of $\mathcal{L}(r)$ with respect to x. That is, $\mathcal{L}(d_x(r)) = \{w \in \Sigma^* \mid x \cdot w \in \mathcal{L}(r)\}$. A derivative $d_x(r)$ can be computed by recursion over the structure of the regular expression r.

Definition 28 (Brzozowski Derivatives [4]**)**

$$d_x(\phi) = \phi \qquad\qquad d_x(\varepsilon) = \phi$$

$$d_x(y) = \begin{cases} \varepsilon \ \mathit{if}\ x = y \\ \phi \ \mathit{otherwise} \end{cases} \qquad\qquad d_x(r+s) = d_x(r) + d_x(s)$$

$$d_x(r \cdot s) = \begin{cases} d_x(r) \cdot s & \mathit{if}\ \varepsilon \notin \mathcal{L}(r) \\ d_x(r) \cdot s + d_x(s) & \mathit{otherwise} \end{cases} \qquad d_x(r^*) = d_x(r) \cdot r^*$$

Example 4. The derivative of $(x+y)^*$ with respect to symbol x is $(\varepsilon+\phi)\cdot(x+y)^*$. The calculation steps are as follows:

$$d_x((x+y)^*) = d_x(x+y)\cdot(x+y)^* = (d_x(x)+d_x(y))\cdot(x+y)^* = (\varepsilon+\phi)\cdot(x+y)^*$$

Theorem 29 (Expansion [4]**).** *Every regular expression r can be represented as the sum of its derivatives with respect to all symbols. If $\Sigma = \{x_1, \ldots, x_n\}$, then*

$$r \equiv x_1 \cdot d_{x_1}(r) + \cdots + x_n \cdot d_{x_n}(r) \ (+\varepsilon \ \mathit{if}\ r\ \mathit{nullable})$$

Definition 30 (Descendants and Similarity). *A* descendant *of r is either r itself or the derivative of a descendant. We say r and s are* similar, *written $r \sim s$, if one can be transformed into the other by finitely many applications of the rewrite rules (Idempotency) $r + r \sim r$, (Commutativity) $r + s \sim s + r$, (Associativity) $r + (s + t) \sim (r + s) + t$, (Elim1) $\varepsilon \cdot r \sim r$, (Elim2) $\phi \cdot r \sim \phi$, (Elim3) $\phi + r \sim r$, and (Elim4) $r + \phi \sim r$.*

Lemma 31. *Similarity is an equivalence relation that respects regular expression equivalence: $r \sim s$ implies $r \equiv s$.*

Theorem 32 (Finiteness [4]**).** *The elements of the set of descendants of a regular expression belong to finitely many similarity equivalence classes.*

Similarity rules (Idempotency), (Commutativity), and (Associativity) suffice to achieve finiteness. Elimination rules are added to obtain a compact *canonical representative* for equivalence class of similar regular expressions. The canonical form is obtained by systematic application of the similarity rules in Definition 30. We enforce right-associativity of concatenated expressions, sort alternative expressions according to their size and their first symbol, and concatenations lexicographically, assuming an arbitrary total order on Σ. We further remove duplicates and apply elimination rules exhaustively (the details are standard [8]).

Definition 33 (Canonical Representatives). *For a regular expression* r*, we write* $cnf(r)$ *to denote the canonical representative among all expressions similar to* r*. We write* $\mathcal{D}(r)$ *for the set of canonical representatives of the finitely many dissimilar descendants of* r*.*

Example 5. We find that $cnf((\varepsilon + \phi) \cdot (x + y)^*) = (x + y)^*$ where $x < y$.

6.2 Subtraction

Definition 34 (Equations for Subtraction). *Let* r, s *be two regular expressions. For each pair* $(r', s') \in \mathcal{D}(r) \times \mathcal{D}(s)$ *we introduce a variable* $R_{r',s'}$*. For each such* $R_{r',s'}$ *we define an equation of the following form. If* $\mathcal{L}(r') = \emptyset$*, we set* $R_{r',s'} \approx \phi$*. Otherwise,* $R_{r',s'} \approx \sum_{x\in\Sigma} x \cdot R_{cnf(d_x(r')),cnf(d_x(s'))} + t$ *where* $t = \varepsilon$ *if* $\varepsilon \in \mathcal{L}(r'), \varepsilon \notin \mathcal{L}(s')$*, otherwise* $t = \phi$*. All equations are collected in a set* $\mathcal{S}_{r,s}$*.*

Let $\psi = solve(\mathcal{S}_{r,s})$*. Then, we define* $r - s = \psi(R_{r,s})$*.*

As the set of canonical derivatives is finite, the set $solve(\mathcal{S}_{r,s})$ is finite as well. Hence, a solution must exist. Hence, $r - s$ is well-defined.

Lemma 35. *Let* r, s *be two regular expressions. Then, we find that*

$$\mathcal{L}(r) - \mathcal{L}(s) \equiv \textstyle\sum_{x\in\Sigma} x \cdot (\mathcal{L}(cnf(d_x(r))) - \mathcal{L}(cnf(d_x(s)))) + T$$

where $T = \{\varepsilon\}$ *if* $\varepsilon \in \mathcal{L}(r), \varepsilon \notin \mathcal{L}(s)$*, otherwise* $T = \emptyset$*.*

Proof. By the Expansion Theorem 29 and Lemma 31, we find that $r \equiv \sum_{x\in\Sigma} x \cdot cnf(d_x(r)) + t$ and $s \equiv \sum_{x\in\Sigma} x \cdot cnf(d_x(s)) + t'$ where $t = \varepsilon$ if r is nullable. Otherwise, $t = \phi$. For t' we find $t' = \varepsilon$ if s is nullable. Otherwise, $t' = \phi$.

By associativity, commutativity of $+$ and some standard algebraic laws

$$\begin{aligned} (x \cdot R) - (x \cdot S) &\equiv x \cdot (R - S) \\ (x \cdot R) - (y \cdot S) &\equiv x \cdot R \qquad \text{where } x \neq y \\ R - \phi &\equiv R \\ (R + S) - T &\equiv (R - T) + (S - T) \\ R - (S + T) &\equiv (R - S) - T \end{aligned}$$

the result follows immediately. □

Theorem 36 (Subtraction). *Let* r, s *be two regular expressions. Then, we find that* $r - s$ *is unambiguous and* $\mathcal{L}(r - s) \equiv \mathcal{L}(r) - \mathcal{L}(s)$*.*

Proof. By construction, equations are non-overlapping. Unambiguity follows from Theorem 25.

We prove the equivalence claim via a coalgebraic proof method [15]. We show that the relation $\{(\mathcal{L}(\psi(R_{r',s'})), \mathcal{L}(r') - \mathcal{L}(s')) \mid (r', s') \in \mathcal{D}(r) \times \mathcal{D}(s)\}$ is a bisimulation where $\psi = solve(\mathcal{S}_{r,s})$. For that to hold two elements are in relation if either (1) they are both nullable, or (2) their derivatives, i.e. taking away the same leading literal, are again in relation.

Consider a pair $(\mathcal{L}(\psi(R_{r',s'})), \mathcal{L}(r') - \mathcal{L}(s'))$. For $\mathcal{L}(r') = \emptyset$ we have that $R_{r',s'} \approx \phi$. The conditions imposed on a bisimulation follow immediately.

Otherwise, $R_{r',s'}$ is defined by the equation

$$R_{r',s'} \approx \sum_{x \in \Sigma} x \cdot R_{cnf(d_x(r')),cnf(d_x(s'))} + t \qquad (E1)$$

where $t = \varepsilon$ if $\varepsilon \in \mathcal{L}(r'), \varepsilon \notin \mathcal{L}(s')$, otherwise $t = \phi$. From Lemma 35 we can conclude that

$$\mathcal{L}(r) - \mathcal{L}(s) \equiv \sum_{x \in \Sigma} x \cdot (\mathcal{L}(cnf(d_x(r))) - \mathcal{L}(cnf(d_x(s)))) + T \qquad (E2)$$

where $T = \{\varepsilon\}$ if $\varepsilon \in \mathcal{L}(r), \varepsilon \notin \mathcal{L}(s)$, otherwise $T = \emptyset$. Immediately, we find that if one component of the pair is nullable, the other one must be nullable as well.

We build the derivative for each component w.r.t. some literal x. Given that ψ is a solution and via (E1) and (E2) the resulting derivatives are equal to $\mathcal{L}(\psi(R_{cnf(d_x(r')),cnf(d_x(s'))}))$ and $\mathcal{L}(cnf(d_x(r))) - \mathcal{L}(cnf(d_x(s)))$. Hence, derivatives are again in relation. This concludes the proof. □

Example 6. We consider $r_1 = (x + y)^*$ and $r_2 = (x \cdot x)^*$. Let us consider first the canonical descendants of both expressions.

$$\begin{array}{lll}
\mathcal{C}(d_x((x+y)^*)) = (x+y)^* & & \\
\mathcal{C}(d_y((x+y)^*)) = (x+y)^* & & \\
\mathcal{C}(d_x((x \cdot x)^*)) & = x \cdot (x \cdot x)^* & = r_3 \\
\mathcal{C}(d_y((x \cdot x)^*)) & = \phi & = r_4 \\
d_x(x \cdot (x \cdot x)^*) & = (x \cdot x)^* & \\
d_y(x \cdot (x \cdot x)^*) & = \phi &
\end{array}$$

The resulting equations are as follows.

$$\begin{array}{l}
R_{1,2} = x \cdot R_{1,3} + y \cdot R_{1,4} + \phi \\
R_{1,3} = x \cdot R_{1,2} + y \cdot R_{1,4} + \varepsilon \\
R_{1,4} = r_1
\end{array}$$

Solving of the above proceeds as follows. We first apply $R_{1,4} = r_1$.

$$\begin{array}{l}
R_{1,2} = x \cdot R_{1,3} + y \cdot r_1 + \phi \\
R_{1,3} = x \cdot R_{1,2} + y \cdot r_1 + \varepsilon
\end{array}$$

Next, we remove the equation for $R_{1,3}$ and apply some simplifications.

$$R_{1,2} = x \cdot x \cdot R_{1,2} + x \cdot y \cdot r_1 + x + y \cdot r_1$$

Via Arden's Lemma we find that $R_{1,2} = (x \cdot x)^* \cdot (x \cdot y \cdot r_1 + x + y \cdot r_1)$ and we are done.

6.3 Shuffle

Definition 37 (Shuffle). *The* shuffle operator $\| :: \Sigma^* \times \Sigma^* \to \wp(\Sigma^*)$ *is defined inductively as follows:*

$$\begin{array}{ll} \epsilon \| w & = \{w\} \\ w \| \epsilon & = \{w\} \\ x \cdot v \| y \cdot w & = \{x \cdot u \mid u \in v \| y \cdot w\} \cup \{y \cdot u \mid u \in x \cdot v \| w\} \end{array}$$

We lift shuffling to languages by

$$L_1 \| L_2 = \{u \mid u \in v \| w \wedge v \in L_1 \wedge w \in L_2\}$$

For example, we find that $x \cdot y \| z = \{x \cdot y \cdot z, x \cdot z \cdot y, z \cdot x \cdot y\}$.

Definition 38 (Equations for Shuffling). *Let r, s be two regular expressions. For each pair $(r', s') \in \mathcal{D}(r) \times \mathcal{D}(s)$ we introduce a variable $R_{r',s'}$. For each such $R_{r',s'}$ we define an equation of the following form. If $\mathcal{L}(r') = \emptyset$, we set $R_{r',s'} \approx \phi$. Otherwise, $R_{r',s'} \approx \sum_{x \in \Sigma}(x \cdot R_{cnf(d_x(r')),s'} + x \cdot R_{r',cnf(d_x(s'))}) + t$ where $t = t_1 + t_2$. Expression $t_1 = s'$ if $\varepsilon \in \mathcal{L}(r')$, otherwise $t_1 = \phi$. Expression $t_2 = r'$ if $\varepsilon \in \mathcal{L}(s')$, otherwise $t_2 = \phi$. All equations are collected in a set $\mathcal{H}_{r,s}$.*

Let $\psi = solve(\mathcal{H}_{r,s})$. Then, we define $r \| s = \psi(R_{r,s})$.

Lemma 39. *Let r, s be two regular expressions. Then, we find that*

$$\mathcal{L}(r) \| \mathcal{L}(s) \equiv \textstyle\sum_{x \in \Sigma}(x \cdot (\mathcal{L}(cnf(d_x(r))) \| \mathcal{L}(s)) + x \cdot (\mathcal{L}(r) \| \mathcal{L}(cnf(d_x(s))))) + T$$

where $T = T_1 + T_2$. $T_1 = s$ if $\varepsilon \in \mathcal{L}(r)$, otherwise $T_1 = \phi$. $T_2 = r$ if $\varepsilon \in \mathcal{L}(s)$, otherwise $T_2 = \phi$.

Theorem 40 (Shuffling). *Let r, s be two regular expressions. Then, we find that $\mathcal{L}(r \| s) \equiv \mathcal{L}(r) \| \mathcal{L}(s)$.*

7 Related Works and Conclusion

Our work gives a precise description of solving of regular equations including a computational interpretation by means of parse tree transformations. Thus, we can characterize conditions under which regular equations and resulting regular expressions are unambiguous.

Earlier work by Gruber and Holzer [9] gives a comprehensive overview on the conversion of finite automaton to regular expressions and vice versa. Like many other works [4, 14], the algorithmic details of solving regular equations based on Arden's Lemma are not specified in detail.

Brzozowski's and McCluskey's [5] state elimination method appears to be the more popular and more widespread method. For example, consider work by Han [10] and in collaboration with Ahn [1], as well as work by Moreira, Nabais and Reis [13] that discuss state elimination heuristics to achieve short regular expressions.

Sakarovitch [16,17] shows that the state elimination and solving via regular equation methods are isomorphic and produce effectively the same result. Hence, our (unambiguity) results are transferable to the state elimination setting. The other way around, state elimination heuristics are applicable as demonstrated by our implementation.

It is well understood how to build the subtraction and intersection among DFAs via the product automaton construction [11]. If we wish to apply these operations among regular expressions we need to convert expressions back and forth to DFAs. For example, we can convert a regular expression into a DFA using Brzozowski's derivatives [4] and then use Brzozowski's algebraic method to convert back the product automaton to a regular expression.

To build the shuffle among two regular expressions, the standard method is to (1) build the shuffle derivative-based DFA, (2) turn this DFA into some regular equations and then (3) solve these regular equations. Step (1) relies on the property that the canonical derivatives for shuffle expressions are finite.

In our own work [20], we establish finiteness for several variations of the shuffle operator. Caron, Champarnaud and Mignot [6] and Thiemann [21] establish finiteness of derivatives for an even larger class of regular expression operators.

We propose direct methods to build the intersection and the shuffle among two regular expressions. For each operation we generate an appropriate set of equations by employing Brzozowski derivatives. We only rely on finiteness of canonical derivatives for standard regular expressions. Solving of these equations then yields the desired expression. Correctness follows via some simple (co)algebraic reasoning and we can guarantee that resulting expressions are unambiguous.

Acknowledgments. We thank referees for CIAA'18, ICTAC'18 and ICTAC'19 for their helpful comments on previous versions of this paper.

References

1. Ahn, J.-H., Han, Y.-S.: Implementation of state elimination using heuristics. In: Maneth, S. (ed.) CIAA 2009. LNCS, vol. 5642, pp. 178–187. Springer, Heidelberg (2009). https://doi.org/10.1007/978-3-642-02979-0_21
2. Arden, D.N.: Delayed-logic and finite-state machines. In: 2nd Annual Symposium on Switching Circuit Theory and Logical Design, Detroit, Michigan, USA, 17–20 October 1961, pp. 133–151 (1961)
3. Brabrand, C., Thomsen, J.G.: Typed and unambiguous pattern matching on strings using regular expressions. In: Proceedings of PPDP 2010, pp. 243–254. ACM (2010)
4. Brzozowski, J.A.: Derivatives of regular expressions. J. ACM **11**(4), 481–494 (1964)
5. Brzozowski, J.A., McCluskey, E.J.: Signal flow graph techniques for sequential circuit state diagrams. IEEE Trans. Electronic Computers **12**(2), 67–76 (1963)
6. Caron, P., Champarnaud, J.-M., Mignot, L.: A general framework for the derivation of regular expressions. RAIRO Theor. Inf. Appli. **48**(3), 281–305 (2014)

7. Frisch, A., Cardelli, L.: Greedy regular expression matching. In: Díaz, J., Karhumäki, J., Lepistö, A., Sannella, D. (eds.) ICALP 2004. LNCS, vol. 3142, pp. 618–629. Springer, Heidelberg (2004). https://doi.org/10.1007/978-3-540-27836-8_53
8. Grabmayer, C.: Using proofs by coinduction to find "traditional" proofs. In: Fiadeiro, J.L., Harman, N., Roggenbach, M., Rutten, J. (eds.) CALCO 2005. LNCS, vol. 3629, pp. 175–193. Springer, Heidelberg (2005). https://doi.org/10.1007/11548133_12
9. Gruber, H., Holzer, M.: From finite automata to regular expressions and back - a summary on descriptional complexity. Int. J. Found. Comput. Sci. **26**(8), 1009–1040 (2015)
10. Han, Y.-S.: State elimination heuristics for short regular expressions. Fundam. Inf. **128**(4), 445–462 (2013)
11. Hopcroft, J.E., Motwani, R., Ullman, J.D.: Introduction to Automata Theory, Languages, and Computation, 3rd edn. Addison-Wesley Longman Publishing Co. Inc., Boston (2006)
12. Lu, K.Z.M., Sulzmann, M.: Solving regular expression equations. http://github.com/luzhuomi/regex-symb
13. Moreira, N., Nabais, D., Reis, R.: State elimination ordering strategies: some experimental results. In: Proceedings of DCFS 2010. EPTCS, vol. 31, pp. 139–148 (2010)
14. Neumann, C.: Converting deterministic finite automata to regular expressions, March 2005. http://citeseerx.ist.psu.edu/viewdoc/summary?doi=10.1.1.85.2597
15. Rot, J., Bonsangue, M., Rutten, J.: Coinductive proof techniques for language equivalence. In: Dediu, A.-H., Martín-Vide, C., Truthe, B. (eds.) LATA 2013. LNCS, vol. 7810, pp. 480–492. Springer, Heidelberg (2013). https://doi.org/10.1007/978-3-642-37064-9_42
16. Sakarovitch, J.: Elements of Automata Theory. Cambridge University Press, Cambridge (2009)
17. Sakarovitch, J.: Automata and rational expressions (2015). https://arxiv.org/abs/1502.03573
18. Sulzmann, M., Lu, K.Z.M.: POSIX regular expression parsing with derivatives. In: Codish, M., Sumii, E. (eds.) FLOPS 2014. LNCS, vol. 8475, pp. 203–220. Springer, Cham (2014). https://doi.org/10.1007/978-3-319-07151-0_13
19. Sulzmann, M., Lu, K.Z.M.: Derivative-based diagnosis of regular expression ambiguity. Int. J. Found. Comput. Sci. **28**(5), 543–562 (2017)
20. Sulzmann, M., Thiemann, P.: Derivatives and partial derivatives for regular shuffle expressions. J. Comput. Syst. Sci. **104**, 323–341 (2019)
21. Thiemann, P.: Derivatives for enhanced regular expressions. In: Han, Y.-S., Salomaa, K. (eds.) CIAA 2016. LNCS, vol. 9705, pp. 285–297. Springer, Cham (2016). https://doi.org/10.1007/978-3-319-40946-7_24

Author Index